Charge Transport in Disordered Solids with Applications in Electronics

Wiley Series in Materials for Electronic and Optoelectronic Applications

Series Editors

Dr Peter Capper, *SELEX Sensors and Airborne Systems Infrared Ltd, Southampton, UK*
Professor Safa Kasap, *University of Saskatchewan, Canada*
Professor Arthur Willoughby, *University of Southampton, Southampton, UK*

Published Titles

Bulk Crystal Growth of Electronic, Optical and Optoelectronic Materials, Edited by P. Capper

Properties of Group-IV, III–V and II–VI Semiconductors, S. Adachi

Optical Properties of Condensed Matter and Applications, Edited by J. Singh

Forthcoming Titles

Thin Film Solar Cells: Fabrication, Characterization and Applications, Edited by J. Poortmans and V. Arkhipov

Liquid Phase Epitaxy of Electronic, Optical and Optoelectronic Materials, Edited by P. Capper and M. Mauk

Dielectric Films for Advanced Microelectronics, Edited by K. Maex, M. R. Baklanov and M. Green.

Charge Transport in Disordered Solids with Applications in Electronics

Edited by

Sergei Baranovski

Philipps University Marburg, Germany

John Wiley & Sons, Ltd

Telephone (+44) 1243 779777

Email (for orders and customer service enquiries): cs-books@wiley.co.uk
Visit our Home Page on www.wileyeurope.com or www.wiley.com

Other Wiley Editorial Offices

John Wiley & Sons Inc., 111 River Street, Hoboken, NJ 07030, USA

Jossey-Bass, 989 Market Street, San Francisco, CA 94103-1741, USA

Wiley-VCH Verlag GmbH, Boschstr. 12, D-69469 Weinheim, Germany

John Wiley & Sons Australia Ltd, 42 McDougall Street, Milton, Queensland 4064, Australia

John Wiley & Sons (Asia) Pte Ltd, 2 Clementi Loop #02-01, Jin Xing Distripark, Singapore 129809

John Wiley & Sons Canada Ltd, 6045 Freemont Blvd, Mississauga, Ontario, Canada L5R 4J3

Wiley also publishes its books in a variety of electronic formats. Some content that appears in print may not be available in electronic books.

Library of Congress Cataloging-in-Publication Data

Charge transport in disordered solids with applications in electronics / edited by Sergei Baranovski.
p. cm. – (Wiley series in materials for electronic and optoelectronic applications)
Includes bibliographical references and index.
ISBN-13: 978-0-470-09504-1 (cloth : alk. paper)
ISBN-10: 0-470-09504-0 (cloth : alk. paper)
1. Amorphous semiconductors–Electric properties. 2. Solids–Electric properties.
3. Semiconductros–Materials. I. Baranovski, Sergei. II. Series.
TK7871.99.A45C53 2006
621.3815′2–dc22

2006014686

British Library Cataloguing in Publication Data

A catalogue record for this book is available from the British Library

ISBN-13 978-0-470-09504-1 (HB)
ISBN-10 0-470-09504-0 (HB)

Typeset in 10/12 pt Times by SNP Best-set Typesetter Ltd., Hong Kong
Printed and bound in Great Britain by Antony Rowe, Chippenham, Wiltshire
This book is printed on acid-free paper responsibly manufactured from sustainable forestry in which at least two trees are planted for each one used for paper production.

Contents

Series Preface

WILEY SERIES IN MATERIALS FOR ELECTRONIC AND OPTOELECTRONIC APPLICATIONS

This book series is devoted to the rapidly developing class of materials used for electronic and optoelectronic applications. It is designed to provide much-needed information on the fundamental scientific principles of these materials, together with how these are employed in technological applications. The books are aimed at postgraduate students, researchers and technologists, engaged in research, development and the study of materials in electronics and photonics, and industrial scientists developing new materials, devices and circuits for the electronic, optoelectronic and communications industries.

The development of new electronic and optoelectronic materials depends not only on materials engineering at a practical level, but also on a clear understanding of the properties of materials, and the fundamental science behind these properties. It is the properties of a material that eventually determine its usefulness in an application. The series therefore also includes such topics as electrical conduction in solids, optical properties, thermal properties, etc., all with applications and examples of materials in electronics and optoelectronics. The characterization of materials is also covered within the series in as much as it is impossible to develop new materials without the proper characterization of their structure and properties. Structure–property relationships have always been fundamentally and intrinsically important to materials science and engineering.

Materials science is well known for being one of the most interdisciplinary sciences. It is the interdisciplinary aspect of materials science that has led to many exciting discoveries, new materials and new applications. It is not unusual to find scientists with a chemical engineering background working on materials projects with applications in electronics. In selecting titles for the series, we have tried to maintain the interdisciplinary aspect of the field, and hence its excitement to researchers in this field.

PETER CAPPER
SAFA KASAP
ARTHUR WILLOUGHBY

Preface

This book is written to meet the growing interest of researchers (physicists, chemists, and engineers) in charge transport properties of disordered materials. This interest is mostly caused by successful current applications of such materials in various devices and by their promise for future applications. The term 'disordered materials' usually describes solids without perfect crystalline atomic structure. Study of charge transport and optical properties of such systems has dominated material sciences in recent years and it is still a rapidly growing and developing research area. A brief historical overview of research in this field can be found in the introductory parts to several chapters in our book, particularly in Chapter 2.

Manufacturability and low production costs of disordered materials, along with their specific charge transport properties, make such systems extremely favorable and in some cases unique for various applications, particularly for large-area devices, where demands on the mobilities of charge carriers are not very high. Disordered materials are commercially used in electrophotography, solar cells, field transistors for flat-panel displays, optical memories, light-emitting diodes, solid batteries, etc. Current device applications of disordered solids and the potential of such materials for future applications are described in several chapters of our book, particularly in Chapters 3, 4, 7, and 11. These chapters, written by top-level researchers in the corresponding fields, should be of interest for the broad community of scientists, particularly for engineers working on the development of new devices.

Intensive study of disordered materials is, however, driven not only by their usefulness for valuable devices, but also by the exclusive interest of researchers in fundamental mechanisms determining the physical properties of such materials. The theory of how solid materials perform charge transport has been for many years mostly confined to crystalline systems with the constituent atoms in regular arrays. The discovery of how to make charge-conducting glasses, amorphous materials and alloys led to an explosion of measurements of electrical conductivity in these new materials. However, the well-known concepts for description of charge transport in crystalline solids, which are usually offered in university courses and which can be found in numerous textbooks, are not applicable to disordered materials, since these concepts are essentially based on the assumption of long-range atomic order. It has been, and still is, a challenging task to develop a consistent theory for charge transport in disordered systems without translation symmetry in the atomic structure.

In our book, we mostly focus on disordered semiconductor materials, leaving out the broad class of disordered metals. The latter systems are more traditional with respect to the description of charge transport properties. One can usually perform such a description on the basis of extended and modified concepts developed for electrical conduction in ordered crystalline systems, for example, using the Boltzmann kinetic equation.

The common feature of the disordered semiconductors considered in our book is the strong effect of the spatial localization of charge carriers on the electrical conduction. We consider materials which differ drastically from each other by their chemical content. Some of them are organic systems, such as molecularly doped and conjugated polymers; some other materials are inorganic solids, such as glasses, amorphous and microcrystalline semiconductors. Furthermore, biological systems, such as DNA molecules, are also considered in our book with respect to their charge transport properties. In some of these disordered materials, electrical conduction is due to electrons and, in other materials, the conduction is due to ions. Nevertheless, in spite of the different chemistry and even the different nature of the charge carriers, all these materials have rather similar charge conducting properties. Electrical conduction in all these materials can often be described in the framework of rather universal theoretical concepts. This universality in description of charge transport in chemically different solids is one of the main messages of our book. Scientific communities are working sometimes separated and isolated from each other, differentiating between 'inorganic', 'organic', and 'biological' materials with respect to the charge transport properties. Furthermore, a theoretical description of ion transport is often opposed to that of electron transport. Our book shows, however, that charge transport properties of chemically different materials are not necessarily so different from each other.

The leading concept of our book is to present complementary chapters describing various features of analogous materials from different points of view. For instance, Chapters 1 and 2 present theoretical concepts for the description of electrical conduction in inorganic (mostly amorphous) disordered semiconductors. Chapters 3 and 5 present the techniques and results of the experimental investigation of such materials, along with some device applications, while Chapter 4 is exclusively devoted to the device applications of disordered inorganic semiconductors in modern electronics. Chapters 6 and 7 are complementary to each other with respect to the description of charge transport mechanisms in organic disordered solids and the device applications of such solids, respectively. Chapters 10 and 11 are complementary to each other, describing transport mechanisms of ion-conducting glasses and various device applications of such glasses.

The book starts with a theoretical description of weak localization effects caused by disorder potential in semiconductors. This topic is currently intensively studied both theoretically and experimentally. Therefore Chapter 1 will be of interest to the broad community of physicists, including students, who would like to learn more about the modern concepts in solid state physics and perhaps make an attempt at developing new theoretical models for charge transport. The chapters following are mostly devoted to the 'hopping mode' of electrical conduction. This transport regime is inherent to all disordered materials with localized charged carriers. In hopping transport, electrons move by tunneling and ions move by overcoming the potential barriers between spatially localized states. Local transition probabilities between such states have a very broad distribution of magnitudes, being exponentially dependent on the site energies, as well as on the intersite distances (for electrons) or on the intersite potential barriers (for ions). For the description of electrical conduction in such cases, one cannot use traditional averaging procedures for local scattering rates, known from theories developed for charge transport in ordered crystalline materials, where scattering rates usually have rather narrow distributions of magnitudes. In disordered materials, it is not the average rates that determine electrical conductivity, but rather the rates of the most difficult transitions that are still necessary in order to provide charge transport to a given distance. This new philosophy, in comparison to the traditional theoretical approaches developed for crystalline materials, was first used in the works of N.F. Mott and his co-workers. The corresponding concepts are introduced in the monograph by N.F. Mott and E.A. Davis, *Electronic Processes in Non-crystalline Materials*, Clarendon,

Oxford, 1979, which we strongly recommend to the reader. While these concepts are given in a schematic form, the comprehensive theoretical treatment of charge transport in disordered systems was described in detail in the monograph by B.I. Shklovskii and A.L. Efros, *Electronic Properties of Doped Semiconductors*, Springer, Berlin, 1984, which we also strongly recommend as a perfect textbook on the theory of hopping transport. A detailed description of charge transport processes specific for amorphous semiconductors is given in the monograph by H. Overhof and P. Thomas, *Electronic Transports in Hydrogenated Semiconductors*, Springer, Berlin, 1989.

Although not only the quantitative theory, but also the analysis of possible erroneous treatments of hopping transport can be found in these monographs, some researchers, dealing with new disordered materials such as organic disordered solids, sometimes tend to use inappropriate approaches because they are probably not aware of the approved theoretical methods presented in textbooks. One possible reason for this might be the above-mentioned belief of many researchers that theoretical methods developed for the description of charge transport in some materials are not of use for the description of electrical conduction in chemically different systems. Our book demonstrates, on the contrary, that similar theoretical concepts can be applied to the description of physically similar transport processes in chemically different solids. For instance, methods in Chapters 2 and 3 for inorganic amorphous semiconductors are analogous to those for organic disordered solids described in Chapters 6 and 7. In many cases, it is not the local chemistry, but rather the long-range geometrical structure of the transport path for charge carriers that is significant, particularly for direct current, for which the long travel distances of charge carriers are decisive. The corresponding theoretical description is usually based on the percolation theory, which plays as important role for charge transport in disordered materials as the kinetic equation does for the description of charge transport in ordered crystalline solids.

In the case of direct current, the decisive spatial scale for charge transport corresponds to the correlation length of the infinite percolation cluster, which is usually very large (see the monograph by Shklovskii and Efros). In the case of alternating current, the scale of distances, through which charge carriers have to travel in order to respond to the alternating electric field, decreases with increasing frequency. For example, in the limit of the very high field frequencies, only transitions of charge carriers within pairs of localized states with very high rates can contribute to the conductivity. The intermediate regime between low frequencies (direct current) and very high frequencies (pair model) is still *terra incognita* for many researchers. The corresponding comprehensive theory of hopping transport in the broad frequency range is therefore presented in our book in Chapter 9. The theory in this chapter clearly shows that the general description of charge transport via localized states, based particularly on the percolation theory, is valid not only for chemically different systems, but also for systems in which electrical current is provided by different charged species—electrons or ions. The complementary information on mechanisms of ion conduction can be found in Chapter 10.

For analysis of electrical conduction as well as for device applications of any system, one of the crucial questions is how the charge carriers recombine within the material. Two chapters of our book are devoted solely to the problems of carrier recombination. In Chapter 5, a very advanced method for studying carrier recombination in disordered materials is introduced, based on the measurements of the pulsed electrically detected magnetic resonance. This investigation technique is rapidly developing and promises to become one of the most powerful tools to study carrier recombination in disordered systems. In Chapter 8, transport of charge carriers and their recombination is studied in polymer–semiconductor nanocomposites. This chapter shows how useful a good theory can be for designing new devices based particularly on nanocomposite materials.

Currently we are witnessing a real burst in the multidisciplinary applications of physical investigation techniques. In recent years, particular attention of researchers has been devoted to studying the charge transport processes in biological systems. Such processes as transport of ions through cell membranes or transport of electrons via long biological molecules, play a vital role for living organisms. Understanding the charge transport processes in cells and biological molecules is of great importance for biology and medicine. Furthermore, many attempts are currently being made to apply biological objects, for instance DNA molecules, in electronic devices. Biological systems can be considered as noncrystalline objects, in which charge carriers are spatially localized, i.e., the wavefunctions describing the charge carriers are not spread over the whole volume of the system, as would be the case in crystalline solids, but they are rather restricted to single molecules or molecular complexes. Therefore, the physical concepts and investigation methods already developed for the description of charge transport in inorganic and organic disordered solids with spatially localized states described in this book can be of great interest for researchers dealing with charge transport in biological systems. Chapter 12 is devoted to charge transport in DNA molecules, biological systems with rather peculiar electrical conducting properties.

Our book is addressed to various groups of readers. Postgraduate students and young scientists can use this book as an introduction to the field of charge transport in a very broad class of disordered solids, while experienced scientists will find rather complete descriptions of many advanced research techniques along with experimental results and theoretical concepts, which might be of use for their scientific work. Information on current device applications of disordered materials and their potential for future applications will be of special interest for engineers.

The editor is deeply grateful to all researchers who collaborated in this project. The idea for such a book—to bring under one roof similar scientific results obtained in parallel, though sometimes isolated research fields—looks very attractive and rather simple. However, it could not have been realized without the strong effort of the international team of leading researchers who contributed to this book. Particularly, the role of Professor Safa Kasap (University of Saskatchewan) who launched and encouraged the work on this ambitious project should be emphasized. The editor is deeply grateful to his colleagues Professor Peter Thomas and Professor Florian Gebhard at the Philipps University Marburg, who provided excellent conditions for the work on this project and who were always open to discussion of various scientific issues related to this book. The concept for the book has been developed in numerous discussions with my close colleague Dr Oleg Rubel at the Philipps University Marburg, who not only co-authored two chapters, but also contributed to editing several other chapters. The editor is indebted to numerous colleagues from all over the world for valuable discussions and for providing the latest information on the subjects related to this book. Valuable help of the editors at Wiley, Ms Jenny Cossham, Ms Lynette James, Ms Alexandra Carrick and Ms Gill Whitley for realization of this project cannot be overestimated. I would also like to thank my wife Svetlana and daughter Julia for their patience and support during my work on this book.

SERGEI BARANOVSKI
Marburg, Germany

1 Charge Transport via Delocalized States in Disordered Materials

Igor Zvyagin

Faculty of Physics, Moscow State University, 119899 Moscow, Russia

Charge Transport in Disordered Solids with Applications in Electronics Edited by S. Baranovski.

1.1 INTRODUCTION[1]

The standard theory of electron transport phenomena in ideal crystalline semiconductors is based on the band theory that determines single-electron eigenstates (Bloch waves) and energies forming energy bands (with a quasi-continuous level distribution) separated by bandgaps. Static disorder modifies the electron wavefunctions and the energy spectrum of the system. A useful characteristic of the electron states is the density of states that can be applied to the description of both ordered solids and solids with arbitrarily strong disorder. In the presence of weak disorder, the eigenstates of a disordered material differ only slightly from the eigenstates of an ideal crystal and the density of states appears to be only slightly different from that of an ideal crystal. Therefore, the description of the electronic processes in weakly disordered crystalline solids may use the concept of almost free quasi-particles (Bloch waves) modified by the interaction with the random field and by the electron–electron interaction. Interacting electrons in extended states may be often described using the *Fermi-liquid approach*, which assumes that a system of electrons is described by using a self-consistent field that determines the properties of almost free quasi-particles, whose energy spectrum is a function of the interaction, and the distribution function. The wavefunctions remain extended over the entire sample and have a random component that may be described as a random phase variation of the electronic wave. Accordingly, transport in a system with weak disorder may be described using the basis of the unperturbed Bloch eigenstates and assuming that a random scattering potential (related to random static deviations of the potential in the crystal from periodicity, to lattice vibrations, etc.) induces transitions between these eigenstates. The corresponding theory is based on the Boltzmann transport equation, whose condition of applicability is

$$\hbar/\tau << \bar{\varepsilon}, \tag{1.1.1}$$

where τ is the quasi-momentum relaxation time and $\bar{\varepsilon}$ is the characteristic quasi-particle energy (equal to kT for nondegenerate electron gas or to the Fermi energy ε_F for the degenerate case). Condition (1.1.1) allows one to combine the classical description of statistical properties of the gas of quasi-particles with quantum-mechanical treatment of individual scattering events. Condition (1.1.1) may be written in the alternative form

$$\lambda << l, \tag{1.1.2}$$

where $\lambda = \hbar / \bar{p} = \bar{k}^{-1}$ is the de Broglie wavelength of quasi-particles, $\bar{p}$ and $\bar{k}$ are their characteristic quasi-momentum and quasi-wave vector, $l = \bar{v}\tau$ is their mean free path, and $\bar{v}$ is the characteristic velocity of the quasi-particles.

In noncrystalline materials (amorphous solids and liquids), short-range order exists, even for strong structural disorder, and the electron energy spectrum is known to retain the characteristic features of the band spectrum: the regions of high electron density of states (corresponding to the allowed bands of the crystal) exist, separated by the regions where the density of states is lower (often by several orders of magnitude) corresponding to the

[1]Some of the problems discussed in this chapter are also treated in books [1–7] and reviews [8–11] and references therein.

energy gaps of the band spectrum of the crystal. With increasing disorder, however, extended electron states inside the energy bands of the crystal may become localized (Anderson localization [12]); eventually, for sufficiently strong disorder, all electron states become localized. Localization implies that the envelope of the wave function decays exponentially away from some localization point $\boldsymbol{R}$, $\psi(r) \sim \exp(-|\mathbf{r} - \mathbf{R}|/\alpha)$, where α is the localization length. Mott has pointed out that the role of disorder can be different for states of different energies; for intermediate disorder, localized and extended states may coexist in the same system at different energies. Typically, the states in the middle of the band may remain extended (extended states are 'current-carrying', i.e., the average velocity in these states is nonzero); on the other hand, near the band edges, the states are localized (Figure 1.1). Following Mott, one may introduce the energy ε_c, corresponding to the localization threshold that separates extended and localized states; this localization threshold is called *the mobility edge* [13]. Extended states above the mobility edge contribute to the DC current even at $T = 0\,\text{K}$ and if $\varepsilon_F > \varepsilon_c$, the conductivity $\sigma|_{T\to 0K}$ (and the resistivity $\rho|_{T\to 0K}$) remains finite. A finite conductivity at $T = 0\,\text{K}$ is a signature of the metallic state, whereas a vanishing conductivity (or a divergent resistivity $\rho|_{T\to 0K}$) of an infinite system as $T \to 0\,\text{K}$ indicates that the system is an insulator. Therefore, a *metal–insulator transition* occurs as the Fermi level crosses the mobility edge ε_c passing from extended to localized states.

Mott has argued that the concept of the conduction band (the region of extended states above the mobility edge) may be used, even for materials with strong disorder (say, for amorphous semiconductors), the mobility edge ε_c playing the role of the conduction band edge. It should be noted that there is no singularity in the density of states at the mobility edge ε_c. Likewise, one may introduce the concept of the mobility edge for the valence band ε_v. The energy interval between ε_c and ε_v is called *the mobility gap*, and the states in the mobility gap are localized.

For extended states, once the electron scattering is weak, the Boltzmann equation makes it possible to calculate the transport coefficients. At finite temperatures, the usual expression

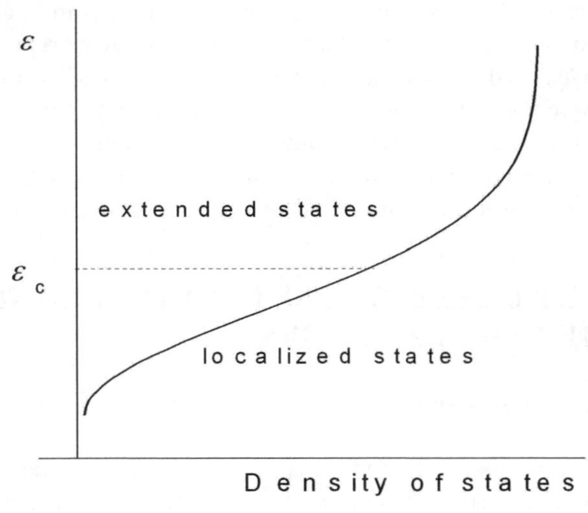

Figure 1.1 Density of states in a disordered solid (schematic); ε_c denotes the mobility edge that separates extended and localized states

for the conductivity in the Boltzmann transport theory (linear with respect to the electric field) may be written [1] as

$$\sigma = \int \sigma(\varepsilon)\left(-\frac{\partial f(\varepsilon)}{\partial \varepsilon}\right) d\varepsilon, \tag{1.1.3}$$

where $f(\varepsilon)$ is the electron distribution function, $\sigma(\varepsilon)$ is the energy-dependent conductivity, and the integration in Equation (1.1.3) is performed over the extended states in the conduction band. (To be specific, in what follows, we usually speak about electrons and the conduction band; the modification for the case of holes and the valence band is straightforward.) Equation (1.1.3) immediately follows from the Boltzmann equation for elastic electron scattering; for inelastic scattering, $\sigma(\varepsilon)$ is only formally introduced by Equation (1.1.3) and must be determined from the corresponding transport theory. It follows from Equation (1.1.3) that in the region of extended states, $\sigma|_{T\to 0K} = \sigma(\varepsilon_F)$, where ε_F is the Fermi level.

In the absence of long-range crystalline order, electronic states are different from Bloch states, quasi-momentum is no longer a good quantum number, the concept of the Brillouin zone cannot be used anymore, etc. Strictly speaking, for disordered (amorphous or liquid) metals and disordered degenerate semiconductors, where the Fermi level lies deeply inside the conduction band in the region of extended states, transport cannot be generally described by the Boltzmann transport theory. In disordered semiconductors, where the Fermi level lies in the region of localized states, a new transport mechanism is possible, related to inelastic tunneling between localized states (*hopping*). Since localized eigenstates have different energies, transitions between them must be inelastic, and the required energy is usually provided by the interaction with vibrations of the atomic matrix (phonons). Therefore, in contrast to conduction by electrons in extended states, where the interaction with phonons reduces the conductivity, the interaction with phonons stimulates hopping conductivity that vanishes as $T \to 0\,\mathrm{K}$.

In this chapter we consider some of the existing concepts and approaches to the description of conduction in the region of extended states; hopping processes will be considered in subsequent chapters. The basis of our discussion is the Fermi-liquid approximation, which appeared to be successful in understanding numerous transport properties of strongly disordered solids. We also discuss some important aspects of the effect of electron–electron correlations in the presence of a random potential and their effect on transport properties of an electron gas. The electron–electron interaction effects can appear to be important, especially at low electron concentrations and in this chapter we discuss some of the features of the interplay between disorder and interactions.

1.2 TRANSPORT BY ELECTRONS IN EXTENDED STATES FAR FROM THE MOBILITY EDGES

1.2.1 Weak-scattering theories

A situation typical of metals corresponds to conduction by a degenerate gas of charge carriers whose energies are close to the Fermi level. Once the disorder is weak, the transport can be described by using the standard Fermi-liquid approach and the Boltzmann theory. In this theory, the quasi-particles (electrons) treated in the nearly free electron approximation are assumed to be *weakly* scattered by disorder (static random potential of structural defects, impurities, etc.). Once the interaction with the random potential is weak and Condition (1.1.1) or (1.1.2) is satisfied, it suffices to keep only the lowest-order terms and to use the Born approximation in the collision integral. The standard transport theory leads to the following expression for the conductivity:

$$\sigma = \frac{ne^2}{m}\tau. \tag{1.2.1}$$

Here n is the concentration of the free charge carriers (quasi-particles), m is the carrier effective mass, and τ is the transport momentum relaxation time expressed in terms of the scattering cross-section. This expression has a form appearing in the simple Drude theory for the conductivity of a gas of free particles. The theory also makes it possible to calculate other transport coefficients (diffusion coefficient, thermopower, etc.).

For strong structural disorder, transport is not expected to be described by the conventional Boltzmann theory; nevertheless, surprisingly, sometimes the weak scattering approach works quite well for this case. An example is provided by *Ziman's theory of liquid metals* [14, 15], which appeared to be quite successful in describing the properties of many liquid metals. In these systems, the electrons are scattered by the ions of the metal, and one might expect that the electron mean free path is of the order of an ineratomic distance. However, in Ziman's theory, the electrons treated in the nearly free electron approximation are assumed to *weakly* interact with ions of the disordered matrix of the liquid and their wave functions are taken to be plane waves. Using the expression for the conductivity in terms of the mean free path l from the Boltzmann theory and the measured conductivities, one can estimate l for different liquid metals; it appears that l can be one to two orders of magnitude greater than the interatomic spacing. This indicates that the scattering is indeed weak in spite of substantial disorder. Therefore, one is justified in using the standard transport theory for a weak scattering potential; the scattering is essentially similar to the scattering of X-rays or neutrons by liquids. In this case, the calculation of the conductivity presents no difficulties.

In Ziman's theory, the Born approximation is used; in addition, correlations in the positions of the scattering ions described by the structure factor are taken into account. The conductivity is related to the probability of scattering between the plane wave eigenstates $|\mathbf{k}\rangle$ and $|\mathbf{k}'\rangle$, which is proportional to the square of the matrix element $|\langle\mathbf{k}|V|\mathbf{k}'\rangle|^2$, where $V(\mathbf{x}) = \sum_i v_a(\mathbf{x} - \mathbf{R}_i)$ is the scattering potential equal to the sum of atomic potentials $v_a(\mathbf{x} - \mathbf{R}_i)$ of individual centers randomly distributed in space. As for scattering of X-rays or neutrons by liquids, we may write

$$|\langle\mathbf{k}|V|\mathbf{k}'\rangle|^2 = |V(\mathbf{q})|^2 = N\Omega|v_{\mathrm{a}}(\mathbf{q})|^2 S(\mathbf{q}), \tag{1.2.2}$$

where N is the concentration of scattering centers; Ω is the volume of the system; $V(\mathbf{q})$ and $v_{\mathrm{a}}(\mathbf{q})$ are the Fourier transforms of the total and atomic potentials, respectively; $\mathbf{q} = \mathbf{k}' - \mathbf{k}$; $|v_a(\mathbf{q})|^2$ is the atomic form factor; $S(\mathbf{q})$ is the structure factor

$$S(\mathbf{q}) = 1 + \langle\exp(i\mathbf{qR})\rangle = 1 + N^{-1}\int d\mathbf{R}\exp(i\mathbf{qR})h(R), \tag{1.2.3}$$

where $\langle\ldots\rangle$ denotes the averaging over impurity configurations and $h(R)$ is the pair correlation function related to the radial distribution function $g(R)$ by the expression $h(R) = g(R) - 1$ (here $Ng(R)d\mathbf{R}$ is the probability of finding an atom in the volume element $d\mathbf{R}$ at a distance R from the origin $\mathbf{R} = 0$, provided that there is an atom at the origin). For an isotropic case, we easily obtain the standard expression

$$S(q) = 1 + 4\pi N \int \frac{\sin qR}{qR} h(R) R^2 \, dR. \tag{1.2.4}$$

The standard transport theory leads to Equation (1.2.1) for the conductivity, where τ is the transport relaxation time, which, for elastic carrier scattering, is given by the expression

$$\begin{aligned} \tau_p^{-1} &= \frac{2\pi}{\hbar} \sum_{\mathbf{q}} \left\langle |V(\mathbf{q})|^2 \right\rangle (1 - \cos\theta) \delta(\varepsilon_{\mathrm{p+q}} - \varepsilon_{\mathrm{p}}) \\ &= 2\pi \left(\frac{m}{2\pi\hbar^2} \right)^2 N\Omega v \int_0^{\pi} d\theta \sin\theta (1 - \cos\theta) |v_a(q)|^2 S(q), \end{aligned} \tag{1.2.5}$$

where $\theta = 2 \arcsin(q/2k)$ is the scattering angle, and v_{F} is the electron velocity. For a degenerate electron gas, we have

$$\tau^{-1} = \frac{8}{\pi} \left(\frac{m}{\hbar^2} \right)^2 N v_{\mathrm{F}} \int_0^1 S(2k_F x) (\Omega v_a(2k_F x))^2 x^3 dx, \tag{1.2.6}$$

where v_{F} is the Fermi velocity and k_{F} is the Fermi momentum.

One of the spectacular applications of Ziman's theory of liquid metals is the interpretation of the temperature dependence of the resistivity of liquid metals. It appears that the resistivity of monovalent metals (Li, Na, K, Rb, Cs) increases with temperature, whereas the resistivity of bivalent metals decreases. The reason for this is clear if one considers the form of the structure factor $S(q)$ that was measured in numerous studies; $S(q)$ is schematically shown in Figure 1.2. The dashed vertical lines correspond to the values $q_{1,2} = 2k_{\mathrm{F}}$ for monovalent and bivalent metals. Since the integrand in Equation (1.2.6) is a rapidly increasing function of x, the main contribution to the integral comes from the region near the upper limit, i.e., near the point $q = 2k_{\mathrm{F}}$. With increasing temperature, the correlation in ion positions becomes weaker and $S(q)$ approaches the limiting value $S(q) = 1$, corresponding to absolutely random positions of the ions. Thus the value $S(q_1)$ increases with temperature and the value $S(q_2)$ decreases.

The above simplified theory using the Born approximation works quite well for many liquid metals, indicating that the scattering may indeed be considered as weak. There are several reasons for this. First, it is well known in the electronic theory of metals that the nearly free electron model may be used if the actual ion potential is replaced by a smooth pseudopotential incorporating the effect of the core states on the states of free electrons [16, 17]. Second, the pseudopotential is screened due to electron spatial redistribution. Screening substantially suppresses the pseudopotential compared with the pseudopotential of 'bare' ions [15]. Third, the theory of liquid metals takes into account the correlation in spatial positions of neighboring atoms by introducing the structure factor $S(q)$ (Equation

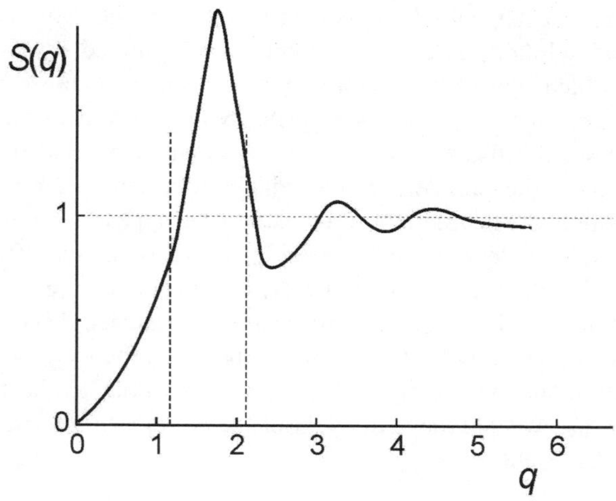

Figure 1.2 Sructure factor $S(q)$ for liquid metals (schematic)

1.2.4). It appears that for many simple metals, the structure factor $S(q)$ is small at small q just in the region where the screened pseudopotential has its maximum, whereas for the values of q, where the structure factor is large, the pseudopotential is small.

A weak-scattering theory may be also applied to degenerate heavily doped semiconductors, where the Fermi level lies in the region of extended states not very close to the mobility edge. The condition for applicability of this theory is $na_B^3 >> 1$, where n is the electron concentration, assumed to be equal to the concentration of the doping impurity, and a_B is the Bohr radius; it appeared to be possible to construct an asymptotic theory for the electron energy spectrum and electron transport [18]. Once the impurity concentration is not too high (the average interimpurity distance is much greater than the lattice period), one may consider Bloch electrons scattered by screened atomic impurity potentials; moreover, the impurity atoms may be considered as randomly distributed in space. Due to screening, the scattering appears to be weak in the asymptotic limit $na_B^3 >> 1$ and the Born approximation may be used giving Equation (1.2.1). A straightforward calculation of the higher-order terms describing both the terms corresponding to higher Born approximations for scattering by an impurity atom and to correlated many-impurity scattering were analyzed, thus establishing the applicability conditions for the transport theory in question [19, 20]. The result of the perturbation theory for transport coefficients of a degenerate semiconductor with screened ionized impurities is that the perturbation series is expected to be slowly converging and lowest-order corrections to the conductivity are of the form [19]

$$0.3\ln^{-1}\left(na_B^3\right)+0.3\left(na_B^3\right)^{-1/6}. \qquad (1.2.7)$$

These corrections obtained for $na_B^3 >> 1$ decrease very slowly with increasing na_B^3. Thus using a weak scattering theory, one can expect to obtain only qualitative agreement with experiment [9, 21]. In addition, for germanium and silicon, one has to take into account the multi-valley structure of the conduction band [22].

Another example of a disordered system, for which a reliable theory exists for charge carrier transport via extended states above the mobility edge, is provided by semiconductor solid solutions with short-range random disorder potential of the white-noise type [23–26]. Semiconductor solid solutions A_xB_{1-x} or, in other words, mixed crystals, are crystalline semiconductors in which sites of the crystalline lattice can be occupied by atoms of different types A and B; here $x(0 \le x \le 1)$ is the probability that a site is occupied by an A atom. Due to the random spatial distribution of A and B atoms, local statistical fluctuations of the composition x along the sample are unavoidable. Since the position of the band edge depends on composition, the fluctuations of the band edge appear, similar to those produced by electrostatic disorder potential. To be specific, we consider the effect of the random potential on conduction band electrons. Let $\varepsilon_c(x)$ be the conduction band minimum for a crystal with composition x. In Figure 1.3, a possible dependence $\varepsilon_c(x)$ is shown schematically. If the average composition for the whole sample is x_0, the local positions of the band edge $\varepsilon_c(x)$ fluctuate about the average value $\varepsilon_c(x_0)$, according to the fluctuations of the composition x about x_0. At small deviations Δx of the composition from the average value, one can use the linear relation

$$\varepsilon_c(x_0 + \Delta x) = \varepsilon_c(x_0) + \alpha \Delta x , \tag{1.2.8}$$

where

$$\alpha = \left. \frac{d\varepsilon_c(x)}{dx} \right|_{x=x_0} . \tag{1.2.9}$$

If the deviation of the concentration of A atoms from its mean value in some region of a sample is $\xi(\mathbf{r})$ and the total concentration of lattice sites is N, the deviation of the composition in this region is $\Delta x = \xi(\mathbf{r})/N$ and the potential energy of an electron at the bottom of the conduction band is

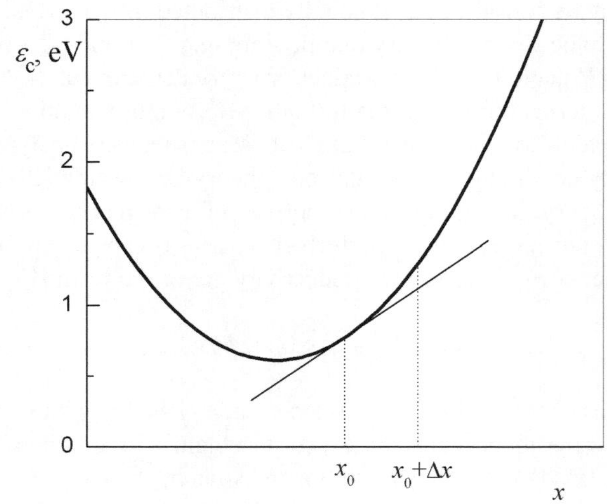

Figure 1.3 Composition dependence of the conduction band edge $\varepsilon_c(x)$ in a mixed crystal (schematic)

$$V(r) = \alpha \frac{\xi(\mathbf{r})}{N}. \tag{1.2.10}$$

Although the disorder in such systems is called a 'short-range' disorder, it should be realized that this description is valid only for size fluctuations much larger than the lattice constant of the material. The term 'short-range' is due to the assumption on the absolutely uncorrelated statistical properties of the disorder. This means that potential amplitudes at adjacent spatial points are completely uncorrelated. Indeed, it is usually assumed that the correlation function of disorder in mixed crystals can be approximated by a white-noise correlation function of the form

$$\langle \xi(\mathbf{r})\xi(\mathbf{r}') \rangle = x(1-x)N\delta(\mathbf{r}-\mathbf{r}'). \tag{1.2.11}$$

Then the random potential created by such composition fluctuations is described by the correlation function [26]

$$\langle V(\mathbf{r})V(\mathbf{r}') \rangle = \gamma\delta(\mathbf{r}-\mathbf{r}'), \tag{1.2.12}$$

where

$$\gamma = \frac{\alpha^2}{N} x(1-x). \tag{1.2.13}$$

Charge carriers in mixed crystals are scattered by composition fluctuations. As usual in the kinetic description of free electrons, the fluctuations with a space scale of the order of the electron wavelength are most efficient in scattering. We have

$$\left\langle |V_q|^2 \right\rangle = \frac{1}{\Omega} \int \mathrm{dr}\, \exp(i\mathbf{qr}) \langle V(\mathbf{r})V(0) \rangle \tag{1.2.14}$$

and using the correlation function (Equation 1.2.12), we obtain the relation

$$\left\langle |V_q|^2 \right\rangle = \frac{\alpha^2 x(1-x)}{\Omega N}, \tag{1.2.15}$$

which shows that the scattering by composition fluctuations is equivalent to that by a short-range potential [25]. Substituting Equation (1.2.15) into Equation (1.2.5), we find [26]

$$\tau_p^{-1} = \frac{\alpha^2 x(1-x)mk}{\pi\hbar^3 N}. \tag{1.2.16}$$

This formula leads to the electron mobility of the form [25, 26]

$$\mu_C = \frac{\pi^{3/2}}{2\sqrt{2}} \frac{e\hbar^4 N}{\alpha^2 x(1-x)m^{5/2}(kT)^{1/2}}. \tag{1.2.17}$$

Very similar formulas can be found in many recent publications (see, for example [27]). It has also been modified for applications to two-dimensional systems [28] and to disordered dilute magnetic semiconductors [29].

This theoretical description can be applied to other disordered systems, provided that the correlation function of the disorder potential has the form of Equation (1.2.12) with a known amplitude γ. However, it should be noted that this short-range disorder of a white-noise type is only a simple model, which can hardly be justified for most disordered materials.

The Boltzmann theory predicts that, in the low-temperature region, charge carrier scattering is dominated by random static potential (to be specific, that of impurities). This scattering is elastic, its probability remains finite at $T = 0$, and the conductivity is almost independent of temperature. Some temperature dependence of the conductivity may exist due to the temperature dependence of the scattering processes. For most scattering mechanisms, it follows from the classical Bolzmann equation that scattering probability and, hence, the resistivity increase with temperature, i.e., the temperature coefficient of the resistivity is positive: $\mathrm{d}\rho/\mathrm{d}T > 0$. This is usually called a *normal metallic behavior* of the resistivity.

The temperature dependence of the conductivity may be due to phonon scattering whose probability increases with increasing temperature, giving rise to a positive temperature coefficient, predominantly due the increase in the number of phonons. At high temperatures, it is proportional to $kT/(\hbar\omega)$ (where ω is the phonon frequency); accordingly, we have $\sigma \sim T^{-1}$. At low temperatures, the temperature dependence of the conductivity is stronger (the Bloch–Grüneisen law, $\sigma \sim T^{-5}$).

Scattering by static disorder (impurities and defects) may also give rise to some temperature dependence of the conductivity. This is related mainly to screening effects. Screening becomes weaker with increasing temperature so that, for this scattering mechanism, one can also expect a positive temperature coefficient of the resistivity, i.e., normal metallic behavior.

Yet another scattering mechanism is electron–electron interaction. Since for a degenerate electron gas, only the electrons in the layer of width of the order of kT near the Fermi level can be scattered, the scattering probability is proportional to T^2. The electron–electron collisions are substantially inelastic, the energy variation being of the order of kT. Thus classical electron–electron scattering is again expected to result in normal metallic behavior of a degenerate electron gas at low temperatures.

However, in the low-temperature range, the temperature dependence of the classical conductivity described by the Boltzmann theory becomes weaker and the effects related to quantum interference of scattered electron waves can become important. These effects that lie beyond the classical approach are discussed in Section 1.2.2.

1.2.2 Weak localization

An important feature of scattering by static disorder is that the elastic scattering is coherent and, therefore, when considering higher-order terms in the calculation of the conductivity, one must take into account the interference of the electron waves scattered by different impurities. This interference gives rise to quantum interference corrections to the Drude conductivity producing a *weak localization* of the electronic states [30]. Inelastic effects such as scattering by phonons and by other electrons destroy the phase coherence, and the weak localization, being an interference phenomenon, can be suppressed by inelastic collisions. An explicit form of the interference corrections to the conductivity obtained by summing the singular backscattering terms of the perturbation expansion in powers of the impurity concentration are [30]

$$\Delta\sigma_{3D} = -\left(\frac{e^2}{\pi^3\hbar}\right)\left(\frac{1}{l} - \frac{1}{L_\varphi}\right), \tag{1.2.18}$$

$$\Delta\sigma_{2D} = -\left(\frac{e^2}{\pi^2\hbar}\right)\ln\left(\frac{L_\varphi}{l}\right) \tag{1.2.19}$$

for three- and two-dimensional systems, respectively (we do not discuss one-dimensional systems here). Here $L_\varphi = \sqrt{D\tau_\varphi}$, D is the diffusion coefficient determined by the impurity elastic scattering, τ_φ is the phase-breaking time, and l is the elastic mean free path. In Equations (1.2.18, 1.2.19), L_φ and l appear as natural cutoffs for the size of closed electron paths when summing over all closed paths corresponding to coherent electron propagation. It should be noted that the phase-breaking time τ_φ does not always coincide with the inelastic collision time τ_{in} [31, 32]. Indeed, if the energy change $\Delta\varepsilon$ in an individual collision is small compared with $\hbar/\tau_{in}$, the phase variation in a single scattering event is about $\Delta\varepsilon\tau_{in}/\hbar$ and may be small compared with 2π. This may be the case for electron–electron scattering. For phonon scattering, the phase variation is usually not small, and $L_\varphi \simeq l_{in}$, where l_{in} is the inelastic mean free path.

As discussed in Section 1.2.1, the inelastic scattering time depends on temperature, increasing as temperature decreases. Let $\tau_{in} \sim T^{-p}$ where p is an index depending on the scattering mechanism, dimensionality, etc. Using $l_{in} = aT^{-p/2}$ in place of the cutoff length L_φ in corrections (Equations 1.2.18, 1.2.19) to the conductivity, we obtain

$$\Delta\sigma_{3D}(T) = -\left(\frac{e^2}{\pi^3\hbar}\right)\left(\frac{1}{l} - \frac{1}{a}T^{p/2}\right), \tag{1.2.20}$$

$$\Delta\sigma_{2D} = \frac{pe^2}{2\pi^2\hbar}\ln\left(\frac{T}{T_0}\right), \tag{1.2.21}$$

where $T_0 = (a/l)^{2/p}$. In the low-temperature range where the Drude conductivity is independent of temperature, the temperature dependence of the quantum interference corrections (Equations 1.2.20, 1.2.21) is important, although corrections must be small in the region of applicability of the perturbation theory. This temperature dependence corresponds to the increase in conductivity with T, similar to that typical of localized state conduction.

Other phase-breaking mechanisms can also affect the conductivity, in particular, related to the dephasing effect of a magnetic field, magnetic impurity, and spin–orbit scattering. A magnetic field B destroys time-reversal symmetry and provides a magnetic length cutoff $L_H = (eB/\hbar)^{1/2}$ [33]. Indeed, the magnetic field introduces phase shifts for electrons traveling along a closed path in opposite directions. Since one must average over all possible paths, the average interference correction vanishes for paths of size exceeding L_H, i.e., magnetic field suppresses the localization effect. It turns out that the magnetoresistance due to this effect is always *negative*. Furthermore, since L_φ can be quite large, the characteristic magnetic field can be very small, of the order of several mT. According to Equation (1.2.19), in two dimensions, the magnetoresistance varies logarithmically with B. Lee [34] has shown that magnetic impurities destroy the coherence, so that on a length scale greater than $L_s = \sqrt{D\tau_s}$, where τ_s is the spin-flip time, the conductivity is no longer dependent on length. In the case of spin–orbit coupling, the quantum correction to the classical (Drude) conductivity is positive and the effect of the spin–orbit interaction is sometimes called *weak anti-localization* [35].

It should be noted that for low-mobility samples and at low temperatures, which are needed for the observation of weak-localization corrections so that condition $L_\varphi >> l$ can be satisfied, the cutoff length can be greater than the sample size L. In this case, the quantum correction to the conductivity given by (Equations 1.2.18, 1.2.19), where L_φ is replaced by L, does not depend on temperature.

1.2.3 Interaction effects

The Fermi-liquid theory is currently used to describe the effect of the electron–electron interaction on the electronic properties of pure metals. For nonideal 'dirty' metals, the role of interaction effects can be substantially different from the case of almost free quasi-particles. Apart from the weak-localization effects, elastic scattering by impurities in metals can substantially modify the electron–electron interaction. Let the elastic momentum relaxation time τ be small compared with the characteristic time $\hbar/(kT)$ required for two interacting quasi-particles to change their energy by a value of about kT, i.e., let

$$kT\tau/\hbar << 1. \tag{1.2.22}$$

This means that two interacting particles diffuse coherently (are scattered many times by impurities) before they exchange an energy of about kT; this electron–electron interaction regime is '*diffusive*'. In the diffusive regime, the electrons spend a longer time in a given region of space compared with the plane-wave states, and their interaction is enhanced.

Under Condition (1.2.22), the motion of electrons during the characteristic time of electron–electron collisions is characterized by the diffusion coefficient $D = v_F^2\tau/3$, where v_F is the Fermi velocity. The electron–electron interaction produces a cusp in the electronic density of states at the Fermi level [36, 37]. The physical reason for such a variation in the density of extended states near the Fermi level is related to the shift of the energy of a particle added into the system due to the Coulomb interaction with electrons of the occupied states. Thus, the nature of the cusp is similar to that of the Coulomb gap appearing at the Fermi level in the region of localized states [5]. This feature produces the anomaly in the tunneling current at zero bias which is often observed when studying the current–voltage characteristics of tunneling contacts (see, e.g., [38–40]).

It has long been recognized that, in addition to the weak-localization correction, the interaction between the electrons that coherently diffuse due to impurity scattering also gives rise to the quantum correction to the conductivity, which has the form [8, 33, 36]

$$\Delta\sigma_{3D}^{(\mathrm{int})}(T) = A\frac{e^2}{\hbar}\left(1 - \frac{9}{8}F_0^\alpha\right)\left(\frac{k_B T}{\hbar D}\right)^{1/2}, \tag{1.2.23}$$

$$\Delta\sigma_{2D}^{(\text{int})} = \frac{e^2}{2\pi^2\hbar}\left\{1+3\left(1-\frac{\ln\left(1+F_0^\alpha\right)}{F_0^\alpha}\right)\right\}\ln\left(kT\tau/\hbar\right) \quad (1.2.24)$$

in three-dimensional and two-dimensional cases, respectively. Here A is a constant ($A = 1.3/3\sqrt{2}\pi^2$, [8]) and the constants F_0^α are the Fermi-liquid interaction parameters (in the triplet channel) that depend on the screening radius. Thus the interaction term has a $T^{1/2}$ dependence, but its sign depends on the relative size of the exchange and Hartree terms, which is a function of the screening length. In doped multi-valley semiconductors, such effects as the presence of several conduction band minima, scattering, and mass anisotropy must be considered if a detailed quantitative comparison is desired.

We see that the temperature dependence of the conductivity is similar for localization and interaction corrections, both for the three-dimensional and two-dimensional cases. However, it is possible to distinguish between the corresponding contributions experimentally, using the measurements in a magnetic field. Localization corrections are suppressed by a magnetic field, producing a negative magnetoresistance, whereas the interaction magnetoresistance is positive (isotropic for spin splitting and transverse for the orbital part). The Hall coefficient R_H is another quantity that behaves differently for localization and interaction effects. In particular, there is no weak-localization correction to the Hall coefficient [41], whereas in the interaction theory $\delta R_H/R_H = 2\delta\rho/\rho$, where $\delta\rho$ is the logarithmic correction to the resistivity ρ.

The regime corresponding to the condition

$$kT\tau/\hbar >> 1 \quad (1.2.25)$$

is called '*ballistic*'; in this regime, the time of the energy exchange is much shorter than τ and the electron–electron interaction is mediated by a single impurity. The Friedel electron density oscillations appear around an impurity with a short-range scattering potential so that an electron is backscattered from the impurity as well as from the Friedel oscillations. Constructive interference of the two scattered waves, which gives rise to a linear correction to the Drude conductivity, dependent on the Fermi-liquid interaction constant in the triplet channel F_0^α, in two-dimensions is [42]

$$\delta\sigma(T) = \sigma(0)\left(1+\frac{3F_0^\alpha}{1+F_0^\alpha}\right)\frac{kT}{\varepsilon_F}. \quad (1.2.26)$$

The expression in the brackets in Equation (1.2.26) consists of two contributions: one of the exchange processes and the second of the Hartree interaction. The sign of the temperature dependence of the conductance is determined by the sign and magnitude of the interaction parameter F_0^α. If F_0^α is sufficiently large and negative, we have $d\delta\sigma/dT < 0$ and this corresponds to metallic-type conduction.

Experimental studies of the corrections to the Drude conductivity has shown that in most cases, interaction corrections cannot be disregarded in both three-dimensional and two-dimensional systems [8]. This is not unexpected at low concentrations, where the role of Coulomb interaction is known to become important. The role of interaction may be characterized by the dimensionless Wigner–Seitz parameter r_s, defined as the average dimensionless distance between charge carriers measured in units of the effective Bohr radii $a_B = 4\pi\kappa_0\kappa\hbar^2/(me^2)$, where κ is the permittivity and m is the effective mass. For a three-dimensional degenerate electron gas, we have $r_s^{(3D)} = (3/4\pi)^{1/3}(n^{(3D)}a_B^3)^{-1/3}$ and for two-dimensional systems, $r_s^{(2D)} = \pi^{-1/2}(n^{(2D)}a_B^2)^{-1/2}$. The ratio U_C/ε_F of the Coulomb energy of electrons at an average distance to the Fermi energy is expressed in terms of the parameter r_s. Thus, for a degenerate three-dimensional electron gas, we have $U_C \simeq \varepsilon_B/r_s^{(3D)}$, $E_F = (9\pi/4)^{2/3}\varepsilon_B(r_s^{(3D)})^{-2/3}$, and $U_C/\varepsilon_F \simeq (4/9\pi)^{2/3}(r_s^{(3D)})^{-1/3}$; here, $\varepsilon_B = e^2/(2\kappa a_B)$ is the effective Bohr energy. For some semiconductor structures (e.g., silicon-based), one must additionally take into account the valley degeneracy. For the two-dimensional case, we have $U_C \simeq \varepsilon_B/r_s^{(2D)}$, $\varepsilon_F = \varepsilon_B(r_s^{(2D)})^{-2}$, and $U_C/\varepsilon_F \simeq r_s^{(3D)}$. In both three-dimensional and two-dimensional systems, the interaction effects are seen to be important at large r_s, i.e., at low electron concentrations, in particular, near the metal–insulator transition. For semiconductor structures typically studied, the values of r_s are often quite large (thus, for GaAs/AlGaAs heterostructures, the values of r_s are in the range 10–17 for a two-dimensional hole gas and 1–2.5 for a two-dimensional electron gas [42]).

Since the experiments clearly indicate the presence of weak-localization corrections, electron–electron interaction effects cannot be generally disregarded, in particular, in the energy region near the mobility edge. Nevertheless, the concept of noninteracting charge carries has appeared to be very fruitful for understanding the main features of extended-state conduction, including the vicinity of the metal–insulator transition. A basis for the description of transport by noninteracting charge carriers is provided by the one-parameter scaling theory of localization considered in Section 1.3.

1.3 SCALING THEORY OF LOCALIZATION

1.3.1 Main ideas of the scaling theory of localization

Transition from the region $\varepsilon_F > \varepsilon_c$ to the region $\varepsilon_F < \varepsilon_c$ is in a sense similar to a phase transition, and the difficulties arising in the description of the properties of the system in the vicinity of the localization threshold are similar to those in the theory of thermodynamic phase transitions. On the other hand, the analogy with phase transitions suggests the possibility of applying some methods of the modern theory of critical phenomena to the problem of Anderson localization. One of them is the method of the renormalization group, including scaling transformations. The scaling theory of the localization suggested by Abrahams *et al.* [43] is based on certain assumptions on the behavior of the conductance G of the system under scaling transformations. The conductance G is expressed in terms of the conductivity by

$$G = \sigma L^{d-2}, \tag{1.3.1}$$

where L is the linear size of the system chosen in the form of a hypercube and d is its dimensionality.

The conductance of the system G is directly related to the localization of electronic wavefunctions. According to Thouless [44], the degree of localization of wavefunctions in the bulk of the system is related to the sensitivity of energy levels to the variation in boundary conditions, and the shift of the levels with varying boundary conditions is expressed in terms of the conductance. Let ζ be the shift of an energy level if the boundary conditions are changed, e.g., from periodic to antiperiodic. Then for the states localized in the bulk, the quantity ζ is exponentially small, provided that the localization radius α is smaller than the system size L, $\zeta \sim \exp(-L/\alpha)$. The effect of boundary conditions on extended states can be estimated using the following argument [45]. For $L >> l$, where l is the mean free path with respect to elastic collisions, the electron has time to be multiply scattered before it reaches the boundary of the system. The motion of an electron undergoing frequent elastic coherent collisions (without breaking the phase of the wave function) has a diffusive character. Let D be the corresponding diffusion coefficient; then the effect of the boundaries on the electron motion becomes important in time $t \sim L^2/D$ when the electron reaches the boundary. On the other hand, one can estimate the time t after which the effect of boundary conditions becomes important in a different way. This is the time t after which the variation $\zeta t/\hbar$ in the phase of the wave electron function produced by a change in the boundary conditions becomes of the order of unity, i.e., $t \sim \hbar/\zeta$. Comparing the two expressions for the time t, we obtain

$$\zeta L^2/D\hbar \sim 1. \tag{1.3.2}$$

It turns out that the quantity ζ is related to the diffusion coefficient, and by Einstein's relation, to the conductivity. For degenerate d-dimensional electron gas, we may write Einstein's relation as

$$\sigma = e^2 D \rho_F = e^2 D L^{-d} w^{-1},$$

where n is the concentration of charge carriers, ρ_F is the density of states at the Fermi level, and $w = (\rho_F L^d)^{-1}$ is the characteristic interlevel separation. Then we obtain from Equation (1.3.2)

$$G = \sigma L^{d-2} = (e^2/\hbar)\zeta/w. \tag{1.3.3}$$

Accordingly, the dimensionless conductance

$$g = G\hbar/e^2 = \zeta/w \tag{1.3.4}$$

is directly proportional to the shift of the energy level at the change in boundary conditions and characterizes the degree of localization of electronic wave functions.

From Equation (1.3.3), we see that $d = 2$ is the critical dimensionality for the problem in question, and the localization properties of electronic states in systems with $d < 2$ and $d > 2$ are different. Indeed, for a system of size $L >> l$ and extended electron states, one can define conductivity, independent of L. According to Equation (1.3.3), for $d < 2$, the relative level shift decreases with increasing L, i.e., the tendency to localization is realized, whereas for $d > 2$, the quantity ζ/w increases with the system size.

1.3.2 The main equations of one-parameter scaling

A qualitative analysis of the behavior of the function $g(L)$, i.e., of the localization properties, can be performed using the assumption of scaling invariance. To this aim, one can perform a scaling transformation similar to Kadanoff's transformation in the theory of phase transitions [46]. Consider a large hypercube (with an edge $L >> l$) of volume L^d in the d-dimensional space, where the function $g(L)$ is defined by Equations (1.3.3) and (1.3.4), and then pass to a large system of linear size bL constructed from b^d such hypercubes. Assuming that the shift of energy levels when constructing the large system of size bL from the original hypercubes is determined only by the quantity $g(L)$, one can write

$$g(bL) = f[b, g(L)]. \tag{1.3.5}$$

This is a basic assumption of the scaling theory of localization [43], related to the assumption that there is a unique scale (of length or energy) that determines the behavior of physical parameters in the critical region (*one-parameter scaling*).

Equation (1.3.5) can be written in a differential form similar to the Gell-Mann–Low equation in quantum electrodynamics. To this aim, we assume that the parameter b in Equation (1.3.5) varies continuously and differentiate with respect to b, setting then $b = 1$. Thus we find

$$\frac{d \ln g(L)}{d \ln L} = \beta[g(L)], \tag{1.3.6}$$

where

$$\beta[g(L)] = g^{-1} [\partial f(b, g)/\partial b]_{b=1}$$

is the scaling function similar to the Gell-Mann–Low function. The differential form (Equation 1.3.6) often appears to be more convenient for the analysis of the conductance $g(L)$ than the scaling relation (Equation 1.3.5).

The function $\beta(g)$, which appears in Equation (1.3.6), depends only on the dimensionality of the system. Its asymptotic form at small and large g can be found by the following arguments. If g is small, i.e., ζ/w is small, then the states are localized, and in a space of any dimensionality we have $g(L) \sim g_1 \exp(-L/\alpha)$ at large L. From this, we obtain

$$\beta(g) \to \ln(g/g_1) \qquad \text{as } g \to 0. \tag{1.3.7}$$

On the contrary, for large $\beta(g)$, macroscopic theory of transport phenomena may be used. The asymptotic form of the function $\beta(g)$ at large g can be found taking account of the quantum interference corrections to the result of the transport theory based on the Boltzmann equation (see Section 1.2). For the three-dimensional case, Equation (1.2.18) gives, for the size-dependent correction to the conductivity, $g(L) = L(g_0 + A/L)$, where g_0 is the conductance of the infinitely large system and $A = \pi^{-3}$. From this, we easily obtain that at large g

$$\beta(g) \simeq 1 - A/g. \tag{1.3.8}$$

For a two-dimensional metal, the asymptotic form of the function $\beta(g)$ at large g is determined by the correction (Equation 1.2.19) to the transport equation, related to the backscattering, whose sum logarithmically diverges at small wave vectors k. Using Equation (1.2.19), we obtain

$$\beta(g) \simeq -A/g, \tag{1.3.9}$$

at large g. Combining the results for systems of different dimensionalities, at large g we obtain

$$\beta(g) \to d - 2 - A/g, \tag{1.3.10}$$

where d is the system dimensionality and A is, generally, different for $d = 2$ and $d = 3$. Using asymptotic expressions (1.3.7) and (1.3.10) and assuming that the function $\beta(g)$ is continuous and monotonic, we can qualitatively represent its shape in the entire range of variation of g. For systems of different dimensionalities, the function $\beta(g)$ is schematically plotted in Figure 1.4.

In Figure 1.4, we see that, for $d = 3$ (curve a), there is at least one zero of the function $\beta(g)$, which we denote by g_c. This zero corresponds to an unstable fixed point of Equation (1.3.6). Indeed, let a system of size L_0 have a conductance $g(L_0) \equiv g_0 > g_c$ and let the system size $L > L_0$ be monotonically increased. Since $\beta(g) > 0$ for $g > g_c$, the conductance g increases with L and the function $\beta(g)$ also increases, moving away from the point g_c and asymptotically approaching the value $\beta = 1$ at large g. This means that, for large systems, the conductance is proportional to L, i.e., the conductivity tends to a constant value σ_0 as $L \to \infty$, as expected for extended states. On the contrary, if initially the conductance g_0 is smaller than g_c, then we obtain that the conductivity exponentially decreases as $L \to \infty$, i.e., it corresponds to localized states. Thus, the fixed point g_c separates the regions of the initial conditions corresponding to localized and extended states. The quantity $(g_0 - g_c)/g_c$ may be considered as a parameter that determines the energy separation from the mobility edge.

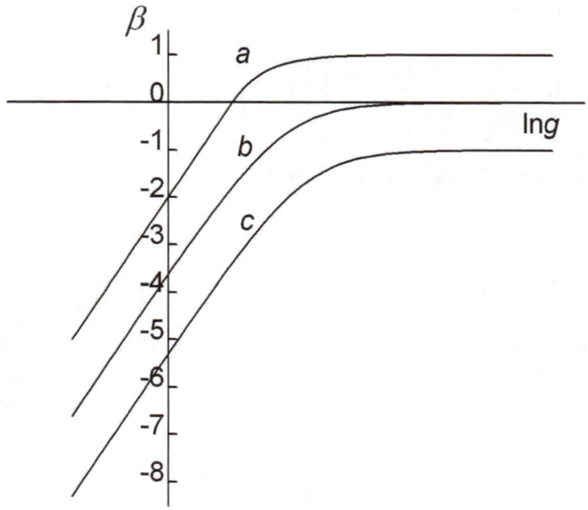

Figure 1.4 The shape of the scaling function $\beta(g)$ for systems of different dimensionalities: (a) $d = 3$; (b) $d = 2$; (c) $d = 1$

1.3.3 Model solutions

Since, by assumption, the function $\beta(g)$ is smooth, near the fixed point g_c it may be approximated by

$$\beta(g) = (g - g_c)/(\nu g_c), \tag{1.3.11}$$

where the parameter ν determines the slope of the function $\beta(g)$ at the point $g = g_c$. However, increasing L, we inevitably leave the vicinity of the point g_c where the linear approximation (Equation 1.3.11) is valid. For a qualitative discussion, we may use the model interpolation expression

$$\beta(g) = \frac{(g - g_c)}{g - (1-\nu) g_c}, \tag{1.3.12}$$

which gives the correct slope of $\beta(g)$ at $g = g_c$ and the correct asymptotic behavior at large L (if $\nu g_c = A$). Using this expression, we can explicitly perform the integration and find the macroscopic conductivity corresponding to a given initial value g_0. The final result for the conductivity is only slightly affected by the detailed behavior of the function $\beta(g)$ in the region of intermediate g.

Equation (1.3.6) with the function $\beta(g)$, defined by Equation (1.3.12), is easily integrated and we obtain

$$\left(\frac{g - g_c}{g_M - g_c}\right)^{\nu} \left(\frac{g}{g_M}\right)^{1-\nu} = \frac{L}{l}. \tag{1.3.13}$$

Here, the 'initial' (in L) condition was used corresponding to the lower 'cutoff' at the microscopic length by setting $g(l) = g_M > g_c$. From Equation (1.3.13), we find for the conductivity at large L (for $g >> g_c$)

$$\sigma = \frac{g}{L} \simeq \frac{e^2}{\hbar l}\left(\frac{g_M - g_c}{g_c}\right)^{\nu}\left(1 + \frac{\nu g_c}{g}\right) = \frac{e^2}{\hbar}\left(\frac{1}{\xi} + \frac{\nu g_c}{L}\right), \tag{1.3.14}$$

where we have introduced the *correlation length*

$$\xi = l\left(\frac{g_M - g_c}{g_c}\right)^{-\nu} \tag{1.3.15}$$

(here $(g_M - g_c)/g_c << 1$). It follows from Equation (1.3.13) that, for a system of size $L >> \xi$, g depends linearly on L (this is sometimes called the Ohmic regime), and the conductivity in the limit $L \to \infty$ is

$$\sigma = \frac{e^2}{\hbar \xi}, \tag{1.3.16}$$

the dependence of g on L is nonlinear for finite $L < \xi$.

Since charge transport is due to electrons with the Fermi energy ε_F, g_M is the conductance at this energy. By assumption, the quantity g_M is a smooth function of the Fermi energy ε_F and the conductivity vanishes for $g_M = g_c$ corresponding to the mobility edge ε_c. Expanding $g_M(\varepsilon_F)$ near ε_c, we obtain

$$g_M(\varepsilon_F) = g_c + \mathrm{d}g_M(\varepsilon_F)/\mathrm{d}\varepsilon_F|_{\varepsilon_F=\varepsilon_c}(\varepsilon_F - \varepsilon_c). \quad (1.3.17)$$

Thus, according to Equation (1.3.15), ν describes the divergence of the correlation length and the continuous decrease in the conductivity at $T = 0\,\mathrm{K}$ (1.3.16) as the Fermi level ε_F approaches the mobility edge ε_c. Using Equation (1.3.17), we obtain

$$\sigma = \frac{e^2}{\hbar l}\left(\frac{\varepsilon_F - \varepsilon_c}{\varepsilon^*}\right)^{\nu}, \quad (1.3.18)$$

where we have introduced the characteristic energy $\varepsilon^* = (\mathrm{d}\ln g_M(\varepsilon_F)/\mathrm{d}\varepsilon_F|_{\varepsilon_F=\varepsilon_c})^{-1}$.

The exponent ν can be calculated in the case of $(2 + \varepsilon)$ space dimensions assuming ε to be small and using perturbation theory. The $(2 + \varepsilon)$ perturbation theory gives $\nu = \varepsilon^{-1}$; extending this to three dimensions, one obtains $\nu = 1$ [8]. The same value of ν is also obtained in the self-consistent theory of localization for noninteracting electrons [47].

The behavior of the conductivity in the region of localized states, i.e., for $g(L_0) = g_0 < g_c$, is also easily investigated using the asymptotic expression for $\beta(g)$ at small g and the expansion of $\beta(g)$ in the neighborhood of g_c. Integrating Equation (1.3.6), we find

$$\int_g^{g_0} \frac{\mathrm{d}g}{g\beta(g)} = -\ln\left(\frac{L}{L_0}\right). \quad (1.3.19)$$

Separating the most divergent terms (as $g \to 0$ and $g \to g_c$) in the integral on the left-hand side in Equation (1.3.14), we may write it in the form

$$\int_g^{g_0} \frac{\mathrm{d}g}{g\beta(g)} = -\ln\ln\left(\frac{g_1}{g}\right) + \nu\ln\left(\frac{g_0 - g_c}{g_c}\right) + \varphi(g_0, g). \quad (1.3.20)$$

The function $\varphi(g_0, g)$ determined by Equations (1.3.19) and (1.3.20) is small compared with the divergent terms explicitly written out in Equation (1.3.20). Using Equations (1.3.19) and (1.3.20), we can find the asymptotic behavior of the conductance $g(L)$ as $L \to \infty$, i.e., as $g \to 0$, if the initial value is close to g_c. We have

$$g = g_c \exp(-L/\alpha), \quad (1.3.21)$$

where

$$\alpha = B^{-1}L_0\left(\frac{g_c - g_0}{g_c}\right)^{-\nu} \quad (1.3.22)$$

is the localization length on the insulator side of the transition, $B = \varphi(g_c, 0)$, and $L_0 \sim l$. This approach predicts that the localization length α diverges at the mobility edge with the same exponent ν as that for the correlation length.

For two-dimensional systems, assuming the function $\beta(g)$ to be smooth and monotonic, we can schematically plot it (Figure 1.4, curve *b*). We see that this function is everywhere negative and nonzero. It results from Equation (1.3.6) that $dg/dL < 0$ for any L so that increasing the system size L, we always get to the region of small g, where the asymptotic expression (Equation 1.3.7), corresponding to localized states, is valid. Thus, the one-parameter scaling theory of localization leads to the conclusion that for potential scattering *in an infinite two-dimensional system, all states are localized* [43].

We consider now such Fermi energies for which there exists an interval of L, where expression (1.3.9) is valid, i.e., for these ε_F and at least for some L, we have $g_0 >> \pi^{-2}\ln(L/l)$. Then the constant of integration is determined by the requirement that Equation (1.3.19) coincides with (1.3.9) in the region of large g. It follows that $L_0 = l$, and Equation (1.3.19) becomes

$$\int_{g_0}^{g} \frac{dg}{g\beta(g)} = \ln(L/l). \tag{1.3.23}$$

We note that writing the solution in the form of Equation (1.3.23) with $L_0 = l$ does not mean that we use Equation (1.3.6) for sizes $L_0 \sim l$, where the assumption of the one-parameter scaling theory, generally, fails; Equation (1.3.23) is obtained by matching the solution of Equation (1.3.6) with the correct asymptotic expression (1.3.9) in the region $L >> l$.

For two-dimensional degenerate electron gas, the Boltzmann conductivity (Equation 1.2.5) is proportional to the Fermi energy, $\sigma_0/\sigma_{\min}^{(2)} = \pi^2 g_0 = \varepsilon_F/\bar{\varepsilon}$, where $\sigma_{\min}^{(2)} = e^2/\pi^2\hbar$, ε_F is measured from the edge of the intrinsic conduction band in the absence of the random potential, and $\bar{\varepsilon} \simeq 0.3\hbar/\tau$. At $\tau = \text{const}$ the mean free path l depends on the Fermi energy according to

$$l = l(g_0) = l_1(\varepsilon_F/\bar{\varepsilon})^{1/2}, \tag{1.3.24}$$

where $l_1 = (0.6\hbar\tau/m)^{1/2}$

Equation (1.3.23) determines conductance of the system as a function of its size L and the position of the Fermi level. Explicitly separating in Equation (1.3.26) the most divergent terms as $g \to 0$ and $g_0 \to \infty$, we may write [48]

$$\ln\ln((g_1 + g)/g) + \pi^2 g_0 + \varphi(g, g_0) = \ln(L/l). \tag{1.3.25}$$

Here g_1 is determined by the asymptotic form of $\beta(g)$ at small g and the function $\varphi(g, g_0)$ determined by Equations (1.3.23) and (1.3.25) is small compared with the first two terms on the left-hand side in Equation (1.3.25) as $g \to 0$ and $g_0 \to \infty$. From Equation (1.3.25) we obtain

$$g = g_1\{\exp[L/\alpha(g, g_0)] - 1\}^{-1}, \tag{1.3.26}$$

where

$$a(g, g_0) = \exp[\pi^2 g_0 + \varphi(g, g_0)]l. \tag{1.3.27}$$

For small g, the quantity $\alpha(g, g_0)$ does not depend on g and becomes equal to the localization radius of strongly localized states. Generally, the function $\alpha(g, g_0)$ is the *localization length* that characterizes the rate of conductance decay with increasing L. According to Equation (1.3.27), it depends exponentially on $\pi^2 g_0 = \varepsilon_F/\bar{\varepsilon}$; for large ε_F and at a fixed L, $\exp[L/\alpha(g, g_0)]$ is no longer large compared with unity so that the asymptotic behavior at large L is not reached. The critical energy ε_c (sometimes called *the apparent mobility edge*) may be defined by the condition $L/\alpha(g, g_0) = 1$; taking account of Equation (1.3.27), we may write it in the form

$$\varepsilon_c/\bar{\varepsilon} = \ln[L/l(\varepsilon_c)] - \varphi[g_1/(e-1), \varepsilon_c/\pi^2\bar{\varepsilon}]. \qquad (1.3.28)$$

As the Fermi level crosses ε_c, we pass from the region where the exponential asymptotic behavior of the conductance is reached and the conductance is very small to the region where this asymptotic behavior is not reached. Because of the strong exponential Fermi-energy dependence of $\alpha(g, g_0)$, the energy interval, in which the conductance varies very quickly, though continuously, is very narrow. According to Equation (1.3.28), the position of the apparent mobility edge depends logarithmically on L or on the corresponding inelastic cutoff length. We note that the mobility edge determined by Equation (1.3.28), is located in the region of large $\pi^2 g_0$, where the use of (1.3.9) as the boundary condition to Equation (1.3.6) is justified.

We can illustrate this general argument by a model example, explicitly specifying the function $\beta(g)$. We choose this function in a simple interpolation form

$$\beta(g) = \ln\left(\frac{g}{g+\pi^{-2}}\right). \qquad (1.3.29)$$

this function has correct asymptotic behavior at large and small g.

In this case, straightforward integration of Equation (1.3.26) gives

$$\psi\left(\frac{g}{g+\pi^{-2}}\right) - \psi\left(\frac{g_0}{g_0+\pi^{-2}}\right) = \ln(L/l), \qquad (1.3.30)$$

where $\psi(z)$ is the function defined by the expression

$$\psi(z) = \ln\ln(1/z) - \int_0^z \frac{du}{(1\text{-}u)\ln(1/u)}. \qquad (1.3.31)$$

For $g >> \pi^{-2}$, we have

$$\psi\left(\frac{g}{g+\pi^{-2}}\right) \simeq -\pi^2 g - 0.5\ln(\pi^2 g) - C, \qquad (1.3.32)$$

where $C \simeq 0.13$; for $g << \pi^{-2}$, the main contribution comes from the first term on the right-hand side in Equation (1.3.31). Accordingly, at small g we obtain an expression of the type (1.3.29), in which

$$\alpha(g_0) = 2(\pi^2 g_0)^{1/2} \exp(\pi^2 g_0 + C) l(g_0). \qquad (1.3.33)$$

In this case, the energy ε_c is approximately determined by the equation

$$\varepsilon_c/\bar{\varepsilon} = \ln[L/l_1] - \ln(\varepsilon_c/\bar{\varepsilon}) - C. \qquad (1.3.34)$$

It follows from Equation (1.3.33) that the conductance falls sharply to very small values in an energy interval of width of about $\bar{\varepsilon}$ near the threshold ε_c. We note that $\sigma(\varepsilon)$ is the energy-dependent conductance that determines the extended-state component of conductance at finite temperatures. If the Fermi level lies within the mobility gap, $\varepsilon_F < \varepsilon_c$, then it follows from Equation (1.1.3) that the temperature dependence of the extended-state conductance has an activated form. The estimation of the integral in Equation (1.1.3) gives an activated behavior of the conductivity with the preexponential factor $\sigma_a = \sigma_a^{(2)}$ determined by the form of the function $\sigma(\varepsilon)$ near the apparent mobility edge ε_c. For $\varepsilon - \varepsilon_c >> \bar{\varepsilon}$, the function $\sigma(\varepsilon)$ is almost linear

$$\sigma(\bar{\varepsilon}) \simeq \sigma_{\min}^{(2)} (\varepsilon - \varepsilon_c)/\bar{\varepsilon}, \qquad (1.3.35)$$

and for $\varepsilon < \varepsilon_c$

$$\sigma(\bar{\varepsilon}) \simeq \sigma_{\min}^{(2)} \exp[-L/\alpha(\varepsilon)], \qquad (1.3.36)$$

where $\alpha(\varepsilon)$ is defined by Equation (1.3.33), in which $\pi^2 g_0$ is replaced by $\varepsilon/\bar{\varepsilon}$. For $kT >> \bar{\varepsilon}$, the main contribution to the integral in Equation (1.1.3) comes from the region of energies in which the approximation (1.3.35) is valid, and $\sigma_a \simeq \sigma_{\min}^{(2)}(kT/\bar{\varepsilon})$. For $kT << \bar{\varepsilon}$, we obtain $\sigma_a \simeq \sigma_{\min}^{(2)}(\bar{\varepsilon}/kT)\Gamma(kT/\bar{\varepsilon})$, where $\Gamma(z)$ is the Gamma function. In this case, the temperature dependence is related to the shift of the energy region giving the main contribution to the conductance.

Thus, the scaling theory of localization predicts (see Equations 1.3.27, 1.3.33) that the localization length increases very rapidly with $\pi^2 g_0 = \varepsilon_F/\bar{\varepsilon}$ and may easily attain very large (macroscopic) values [49]. For realistic system parameters, the condition that the localization length $\alpha(\varepsilon)$ is equal to the sample size L or the dephasing (inelastic scattering) length L_φ can be easily satisfied. If the localization length $\alpha(\varepsilon)$ is greater than L or L_φ, the system behaves as if the states were extended. This behavior is, however, different from that of standard metals. Indeed, in this case, the conductivity is expected to increase logarithmically with temperature; this logarithmic dependence comes from the weak localization corrections, as discussed in Section 1.2.2. Near the apparent mobility edge, the scaling approach predicts a gradual transition (in a narrow energy interval) from the slow logarithmic temperature dependence (for states above the apparent mobility edge) to the exponential temperature dependence in the region of localized states.

1.3.4 Some predictions of the scaling theory

In the above discussion, we considered mostly the case of very low temperatures. Thouless has argued that at nonzero temperatures inelastic scattering breaks quantum interference that produced the localization corrections. Accordingly, for three-dimensional systems, in expression (1.3.14) for the conductivity, one should replace L by the cutoff length

$$L_{\varphi} = \sqrt{D\tau_{\varphi}}, \tag{1.3.37}$$

where D is the diffusion coefficient and τ_{φ} is the dephasing time (if $L_{\varphi} < L$). Thus, we have

$$\sigma(T) = \frac{e^2}{\hbar}\left(\frac{1}{\xi} + \frac{v}{L_{\varphi}(T)}\right). \tag{1.3.38}$$

For electron–electron scattering, the characteristic dephasing time is $\tau_{ee} \simeq \hbar / T^p$. In this case, $L_{\varphi} = \sqrt{D\tau_{ee}} \simeq \sqrt{\hbar D} T^{p/2}$. In the region close to the transition, the first term on the right-hand side in Equation (1.3.38) tends to zero, and the second term can become dominant.

For two-dimensional systems, at finite temperatures inelastic scattering leads to a cutoff in the divergence, and the cutoff length is L_{φ}. It follows from Equations (1.3.30), (1.3.32), or (1.2.21) that, for $\xi > L_{\varphi}$, the conductivity acquires a correction, which depends logarithmically on temperature,

$$\sigma = \sigma_0 + \frac{pe^2}{2\pi^2\hbar}\ln(T/T_0). \tag{1.3.39}$$

Some of the available data on the temperature dependence of the conductivity will be discussed in Sections 1.4 and 1.5 for three-dimensional and two-dimensional systems, respectively.

The scaling theory of localization was extremely successful and was used as a basis for interpretation of transport properties of disordered solids for several decades. This is somewhat surprising, since it was developed for noninteracting charge carriers, whereas the estimations show (see Section 1.2.3) that the energy of Coulomb interaction could be much greater than the kinetic energy so that the interaction effects are expected to be important (especially for the two-dimensional systems corresponding to the critical dimensionality). Accordingly, the effect of charge carrier interactions may not be reduced to the effect of the inelastic cutoff length L_{φ}, and substantial effort was directed to generalizing the theory to systems of interacting electrons [50–52]. It was suggested [53] that one can incorporate interaction effects phenomenologically into the scaling approach. Indeed, the conclusion of the scaling theory for noninteracting electrons about the localization of all electronic states at arbitrarily weak disorder is based on the asymptotic behavior of the scaling function at large g, in particular, on the sign of the constant A in expression (1.3.9). The quantum interference weak-localization corrections give $A > 0$. On the other hand, as noted in Section 1.2.2, in the presence of spin–orbit scattering, it is possible that $A < 0$, giving $\beta(g) > 0$ at large g [35]. In this case, a fixed point $g = g_c$ appears ($\beta(g_c) = 0$) implying the possibility of the existence of a metal–insulator transition in two-dimensional systems. However, it was argued that in real systems, in addition to interactions of the spin–orbit universality class, one can find different universality classes (depending on the experimental situation) that result in a change of sign of the quantum correction in expression (1.3.9) [8]. With regard to electron–electron interactions (see Section 1.2.3), it was the shown that strong interactions can, in principle, result in a transition to a metallic state [50, 51]. However, there are certain difficulties in the theory, since the interaction strength appears to be divergent at scaling towards strong coupling; this precludes the limiting transition to large scales and low temperatures. In [53], the implication of formal extension of the one-parameter scaling theory for two-dimensional systems was discussed under the assumption that the interactions can be described by setting $A < 0$ (i.e., $\beta(g) > 0$ at large g). Using expression (1.3.9) with $A < 0$, we can easily integrate Equation (1.3.6) in the metallic region ($g > g_c$) and obtain

$$g(L) = g_0 + |A| \ln(L/l). \tag{1.3.40}$$

Setting the cutoff length $L_\varphi \sim T^{-p}$ and assuming that L_φ is less than the system size, we obtain, at low temperatures

$$g(L) \sim \ln(T_0/T), \tag{1.3.41}$$

i.e., we obtain a diverging conductance (vanishing resistance) as $T \to 0$. This is a *non-Fermi-liquid state*, since the nature of the electronic states is quite different from that of the states in the absence of the interactions which would be localized in this case.

1.3.5 Minimum metallic conductivity

Mott has extended the approach of the weak-scattering transport energy to the region near the mobility edge using the following argument [1, 54]. For a degenerate electron gas ($T = 0\,\mathrm{K}$), the value of the Fermi momentum k_F decreases as we approach the mobility edge, the corresponding characteristic de Broglie's wavelength λ_F increases, and if it becomes comparable to the mean free path l, the weak scattering Condition (1.1.2) is violated. As disorder becomes stronger, the states at the Fermi level become Anderson localized; this happens as $k_\mathrm{F} l \sim 1$. Using the weak scattering theory, one may write expression (1.2.5) in the form

$$\sigma = \frac{ne^2}{\hbar k_F^2}(k_F l). \tag{1.3.42}$$

For a spherical Fermi surface, in the three-dimensional case we have

$$n = (3\pi^2)^{-1} k_\mathrm{F}^3. \tag{1.3.43}$$

Ioffe and Regel argued that, in order to apply the standard scattering picture, the electron wavelength k_F^{-1} must be shorter than the mean free path l and both k_F^{-1} and l should exceed the interatomic spacing a (the Ioffe–Regel criterion [55]). Using Equations (1.3.42) and (1.3.43) along with the Ioffe–Regel criterion, we obtain $\sigma = \frac{ne^2}{\hbar k_\mathrm{F}^2}(k_F l) = \frac{e^2}{3\pi^2\hbar} k_\mathrm{F}^2 l \geq \frac{e^2}{3\pi^2\hbar} k_\mathrm{F} \geq \frac{e^2}{3\pi^2\hbar a}, \frac{e^2}{3\pi^2\hbar l}$.

Mott has concluded that the conductivity due to electrons in extended states cannot be smaller than the value

$$\sigma_{\min} = C\frac{e^2}{\hbar l}, \tag{1.3.44}$$

where C is constant (in this case $C = (3\pi^2)^{-1}$). The quantity determined by Equation (1.3.44) was called *the minimum metallic conductivity.* Different values of C suggested in the literature are of the same order of magnitude; a value suggested by Mott is $C = 0.026$. Using this value and $l = 3$ Å, one obtains

$$\sigma_{\min} \simeq 200\ \Omega^{-1}\mathrm{cm}^{-1}. \tag{1.3.45}$$

Mott has argued that, since the conductivity cannot be less than that determined by Equation (1.3.44), the conductivity at $T = 0$ K jumps abruptly from the value $\sigma_{\min}$ to zero, i.e., the metal–insulator transition is discontinuous (the discontinuity of the function $\sigma(\varepsilon)$ corresponding to dashed line a in Figure 1.5). Mott's concept of the minimum metallic conductivity and discontinuous metal–insulator transition disagrees with the scaling theory of localization, considered above, that predicts a continuous transition (curve b in Figure 1.5) as the Fermi level crosses the mobility edge ε_c for the three-dimensional systems at $T = 0$. In the following, we discuss some of the relevant experimental results and some developments of the scaling theory of localization.

Mott's arguments predict the existence of the mobility edge and a discontinuous metal–insulator transition, which can be also applied to two-dimensional systems [1, 54]. These arguments imply that the conductivity abruptly drops from the value

$$\sigma_{\min}^{(2)} = C_2 \frac{e^2}{\hbar} \tag{1.3.46}$$

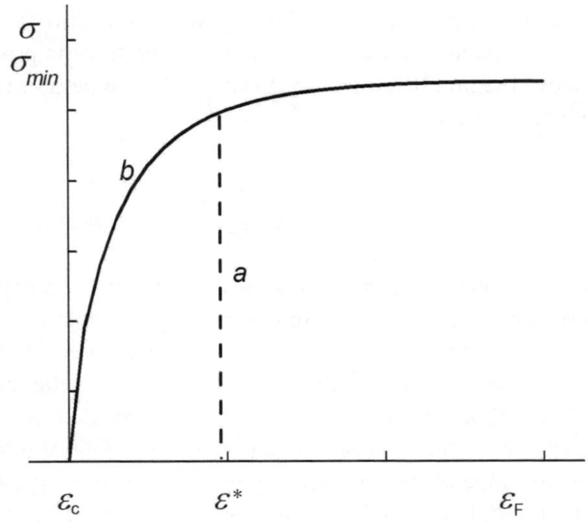

Figure 1.5 Behavior of the conductivity at $T = 0$K near the mobility edge ε_c (schematic). (*a*) Discontinuous transition implied by Mott's concept of the minimum metallic conductivity; (*b*) continuous variation predicted by the scaling theory of localization; the critical region corresponds to the energies $\varepsilon_c < \varepsilon < \varepsilon^*$, where the boundary of the critical region ε^* is the energy at which $\sigma(\varepsilon^*) = \sigma_{\min}$

to zero; according to Mott's argument and to numerical calculations, $C_2 \simeq 0.1$ [1]. It corresponds to the value of the two-dimensional minimum metallic conductivity $\sigma_{\min}^{(2)} \simeq 2.5 \times 10^{-5}\,\Omega^{-1}$ (we recall that, in the two-dimensional case, the dimensions of the conductivity and conductance coincide). In the two-dimensional case, the quantity $\sigma_{\min}^{(2)}$ is universal; it contains no length scale.

As noted above, Mott's argument contradicts the conclusion of the one-parameter scaling theory of localization for noninteracting charge carriers, which states that, for two-dimensional electron systems, all states are localized and no mobility edge is expected. In a sense, the energy ε_c separating strongly and weakly localized states might play the role of the mobility edge, since for the states with $\varepsilon > \varepsilon_c$, the localization length α exceeds the system size (or the relevant phase-breaking length) and these states are virtually extended. However, even above the apparent mobility edge, the 'insulator' features of the conductivity behavior corresponding to the behavior of weak-localization corrections with a negative temperature coefficient of the resistivity are expected to persist. On the other hand, a signature of a metallic-type behavior (as, in particular, implied by Mott's argument) is a positive temperature coefficient of the resistivity at low temperatures. In the following sections, we discuss in more detail the results of the experimental studies of the conductivity near the mobility edge for both three-dimensional and two-dimensional systems that seem to require a substantial modification of the traditional approaches, based not only on the concept of the mobility edge, but also on the one-parameter scaling theory.

1.4 EXTENDED-STATE CONDUCTION IN THREE DIMENSIONS

1.4.1 Activated conduction

For amorphous semiconductors, where the Fermi level lies within the mobility gap, the main contribution to the conductivity at not too low temperatures is provided by electrons in extended states above the mobility edge. In this case, the temperature dependence of the conductivity has an activation form

$$\sigma = \sigma_a \exp\left(-\frac{\varepsilon_a}{kT}\right), \tag{1.4.1}$$

where σ_a is the the preexponential factor and ε_a is the activation energy. If we assume, in accordance with Mott's concept of the minimum metallic conductivity, that the variation of the conductivity $\sigma(\varepsilon)$ is step-like and the conductivity is zero below ε_c and $\sigma(\varepsilon) \simeq \sigma_{\min}$ above it (at least within the layer of width of the order of kT near the mobility edge), then from Equation (1.1.3) we obtain $\sigma_a = \sigma_{\min}$ and $\varepsilon_a = \varepsilon_c - \varepsilon_F$, provided that the position of the Fermi level ε_F with respect to the mobility edge ε_c is independent of temperature.

A temperature dependence of the conductivity of the activation type (Equation 1.4.1) is a characteristic property both of crystalline (ordered) and disordered semiconductors. Such temperature dependence is related to extended-state conduction due to charge carriers activated to the mobility edge. The preexponential factor provides information about the electronic states and conduction mechanism near the mobility edge. However, it is often problematic to extract information about the behavior of the energy-dependent conductivity $\sigma(\varepsilon)$ from the experimental observations of the activated conduction for several reasons.

For disordered materials an activated behavior of the type specified by Equation (1.4.1) is usually observed at not too low temperatures, so that electrons in a relatively broad range of energies above the mobility edge contribute to the preexponential factor. With lowering temperature, substantial deviations from Equation (1.4.1) occur, related to the onset of hopping conduction over localized states. To find the behavior of the function $\sigma(\varepsilon)$, one needs to know the conductivity at $T = 0\,\mathrm{K}$, i.e., the measurements must be performed at very low temperatures, which is possible only if the Fermi level lies very close to or above the mobility edge.

Next, the preexponential factor can be affected by the variation of the position of ε_F with respect to the band edge ε_c with temperature (e.g., due to the statistical shift of the Fermi level or to the shift of ε_c produced by temperature-dependent disorder). Setting $\varepsilon_c - \varepsilon_F = \varepsilon_a - \zeta T$, we obtain that the activation energy ε_a is equal to the value obtained by the linear extrapolation of the temperature-dependent difference $\varepsilon_c - \varepsilon_F$ to $T = 0\,\mathrm{K}$ and the preexponential factor is multiplied by $\exp(\zeta/k)$; once $\sigma(\varepsilon) \simeq \sigma_{\min}$ near the mobility edge, we obtain

$$\sigma_a = \sigma_{\min} \exp(\zeta/k). \tag{1.4.2}$$

Experimentally, one can often perform measurements on the same or similar samples with activation energies that differ due to different preparation conditions, doping, or external treatments; e.g., in a-Si:H, the activation energy can be varied in a wide range by prolonged preliminary illumination (*the Staebler–Wronski effect*) [56]. If the preexponential factor were independent from the activation energy, the series of linear $\ln \sigma - T^{-1}$ plots would extrapolate to the same value as $T^{-1} \to 0$. For three-dimensional systems, however, the situation, where the conductivity is activated, but the preexponential factors are different from $\sigma_{\min}$ and from each other, is quite common. One of the reasons is that $\sigma_{\min}$ contains the length l (see Equation 1.3.28) that can change when the Fermi level is shifted. Moreover, for numerous disordered semiconductors, the empirical relation known as *the Meyer–Neldel rule* has been established, relating the preexponential factor σ_a and the conductivity activation energy ε_a

$$\ln \sigma_a = \ln \sigma_{a0} + G\varepsilon_a. \tag{1.4.3}$$

Equation (1.4.3) has been observed for semiconductor oxides [57], amorphous hydrogenated silicon (a-Si:H) [58], chalcogenide glasses [59], polymers [60], and ionically conducting crystals and glasses [61]. The Meyer–Neldel rule applies to chemically closely related semiconductors and to semiconductors where Fermi level positions vary due to preparation conditions, annealing, etc. Detailed studies of the Meyer–Neldel rule were performed for a-Si:H, where the activation energy could be varied due to the Staebler–Wronski effect. The ubiquitous value of G is about $G \simeq 25\,\mathrm{eV}^{-1}$.

Although the Meyer–Neldel rule seems to be an almost universal characteristic of semiconductors and is widely studied, to date there is no universal interpretation of this rule for all materials. Equation (1.4.3) is obtained if the temperature coefficient ζ is proportional to the activation energy. However, such a dependence does not follow from the conventional picture of the activated conduction; it can be obtained using some specific assumptions on the density of localized states required to produce a linear activation energy dependence of the temperature shift of the Fermi level (the statistical shift) [62]. The calculations of the statistical shift of the Fermi energy for some realistic appropriately chosen density-of-states models for a well documented case of a-Si:H were performed [63]. These calculations showed that, indeed, the Meyer–Neldel rule, observed in light-soaking experiments for a-Si: H, can be related to the statistical shift of the Fermi level and the resulting temperature dependence of the difference $\varepsilon_c - \varepsilon_F$.

An implication of the Meyer–Neldel rule is that the preexponential factor for the activated extended-state conductivity is determined not only by the properties of electronic states at the mobility edge, but also by the shape of the density of states in the mobility gap. One more complicating factor is that the parameters of the activated conduction may depend on the presence of large-scale fluctuations. In this case, the band edge ε_c corresponds to the classical percolation level in a random large-scale fluctuation potential.

Thus, the experimental data on the activated conductivity in disordered semiconductors are in general agreement with the concept of the mobility edge and the minimum metallic conductivity. However, the measurements of the activated conduction of disordered semiconductors, where the Fermi level lies in the region of localized states, do not provide straightforward information on the behavior of the conductivity $\sigma(\varepsilon)$ for energies ε lying close to the transition. To check Mott's assumption on the minimum metallic conductivity and to obtain reliable information about the features of electronic states and the behavior of the conductivity near the mobility edge, one should use low-temperature measurements in the region of the metal–insulator transition under the conditions where the Fermi level is located in the immediate vicinity of the mobility edge.

1.4.2 Extended-state conduction near the metal–insulator transition

The metal–insulator transition was observed in numerous disordered solids where the position of the Fermi level with respect to the mobility edge could be varied. This can be achieved, for example, in doped crystalline semiconductors by varying the impurity concentration and compensation. There also exists a possibility to finely tune the mobility edge by applying magnetic field [64], uniaxial stress [65], or by transmutation doping [66, 67]. This can be also done for amorphous semiconductors (a-Si:Nb [68]) and conducting polymers [69]. Magnetic field causes shrinkage of the impurity ground-state wavefunctions thus increasing localization and shifting the mobility edge upwards. Applying stress admixes more extended impurity excited states, thus stimulating delocalization. The change in impurity concentration N can affect the electronic properties of the system in different ways. For doped semiconductors, the quantity l is expected to be the distance between uncompensated impurities $N^{-1/3}$ (see, e.g., [70–72]), and Equation (1.3.44) assumes the form

$$\sigma_{\min} = C\frac{e^2}{\hbar}N^{1/3}. \tag{1.4.4}$$

As predicted by Mott, transition from an insulator to metallic state may be due to electron correlations (*Mott's transition*) and is expected to occur as the average spacing between impurities $N^{-1/3}$ becomes smaller than the critical spacing $N_c^{-1/3}$, where N_c is the critical impurity concentration determined by the condition

$$N_c^{-1/3}a_B \cong 0.25 \tag{1.4.5}$$

and a_B is the Bohr radius [2]. Condition (1.4.5) implies that delocalization occurs as the overlap energy becomes comparable to the energy of repulsion between electrons localized on the same impurity. On the other hand, a change in impurity concentration N affects the random potential in the material, thus resulting in a disorder-driven transition (*Anderson transition*). Generally, a change in N also affects the electron concentration n (related to a shift in the mobility edge ε_c), so that the metal–insulator transition is an *Anderson–Mott transition*, where the effects both of disorder and electron–electron interaction can play an important role.

In accordance with Equation (1.3.18), the variation of the zero-temperature conductivity with impurity concentration can by described by

$$\sigma(0) = \sigma_M \left(\frac{N - N_c}{N_c} \right)^{\nu}, \tag{1.4.6}$$

where the exponent ν is the same as in Equation (1.3.18) and the preexponential factor is $\sigma_M \sim \sigma_{min}$.

To compare the theoretical predictions on the variation of the zero-temperature conductivity near the metal–insulator transition, one needs to extrapolate the results of finite-temperature measurements of the conductivity to $T = 0\,K$. Such extrapolation is not trivial, in spite of the fact that very low temperatures are used (down to tens of mK), since some specific features of the conductivity appear in this temperature range and the results sometimes depend on the extrapolation method used. The standard criterion for a metal–insulator transition implies that the metallic state corresponds to a finite resistivity at $T = 0\,K$, whereas in The insulator state, the resistivity diverges as $T \to 0\,K$. As discussed in Section 1.2.1, the classical Boltzmann theory predicts that, for metallic conduction, the temperature coefficient of the resistivity $d\rho/dT$ at low temperatures is usually positive, since scattering is enhanced with increasing temperature. Therefore, the condition $d\rho/dT > 0$ is often considered as a 'signature' of a metal and the condition $d\rho/dT < 0$ as a signature of an insulating state; accordingly, the condition $d\rho/dT|_{T \to 0K} = 0$ is considered as a criterion for a metal–insulator transition. Actually, in the low-temperature range, the temperature dependence of the conductivity near the metal–insulator transition may be nonmonotonic and special care should be taken when describing the conductivity behavior. Moreover, it should be noted that, generally, the $d\rho/dT|_{T \to 0K} = 0$ criterion for a metal–insulator transition is not equivalent to the criterion, based on the appearance of the resistivity divergence as $T \to 0\,K$ and these criteria give different critical parameters for the metal–insulator transition.

Thus, in order to perform a reliable extrapolation at $T \to 0\,K$, one must know the law of the temperature variation of the conductivity in the low-temperature range. For three-dimensional systems, in the classical metallic region (for weak disorder, $k_F l >> 1$), where a description based on the Boltzmann equation may be used, the low-temperature conductivity has the form

$$\sigma = \sigma_0 - AT^s, \tag{1.4.7}$$

where σ_0 is the residual conductivity due to scattering by structural disorder and impurities and A is constant. Additional scattering (by phonons or electron–electron collisions) is usually enhanced with increasing temperature so that both A and s are positive (for electron–electron scattering, $s = 2$). Thus, the conductivity σ exhibits a typically metallic behavior, decreasing with increasing temperature.

With decreasing $k_F l$, the disorder becomes stronger and the conductivity decreases, approaching the value σ_{min} determined by Equation (1.4.4). In contrast to Mott's idea, the scaling theory of localization predicts that the conductivity varies continuously in the *critical region* where $\sigma < \sigma_{min}$, where the conventional weak-scattering picture does not apply. In this region, quantum interference effects become important. The boundary of the critical region is defined by $\sigma \simeq \sigma_{min}$ or $\xi \simeq l$ (up to a factor of order unity). Inside the critical region, the conductivity is given by Equation (1.3.38). The temperature-dependent correction describing the quantum interference can be rewritten, taking into account that the diffusion constant D is related to the conductivity by the Einstein relation, $\sigma = e^2 \mathrm{d}n/\mathrm{d}\varepsilon_F D$ (for interacting particles, $\mathrm{d}n/\mathrm{d}\varepsilon_F$ may not coincide with the one-particle density of states at the Fermi level). Accordingly, we obtain

$$\sigma = \frac{e^2}{\hbar}\left(\frac{1}{\xi} + \nu\sqrt{\frac{e^2(\mathrm{d}n/\mathrm{d}\varepsilon_F)T^p}{\hbar\sigma}}\right). \tag{1.4.8}$$

Thus, the temperature coefficient of the metallic conductivity is expected to change its sign as we pass from the classical to the critical region; as noted above, this happens as the value of the conductivity becomes close the minimum metallic conductivity σ_{min}, given by Equation (1.3.44).

If the quantum correction is small (in the region of the applicability of the weak-localization theory) and $p = 1$, then σ is weakly dependent on temperature and, therefore, varies according to the law

$$\sigma = a_1 + b_1 T^{1/2}, \tag{1.4.9}$$

where $b_1 > 0$. Since the correlation length ξ diverges as we approach the mobility edge and a_1 decreases, the second term becomes dominant near the transition, and the temperature dependence of the conductivity in the second term on the right-hand side of Equation (1.4.8) becomes appreciable. If the first term is small compared with the second, this dependence can be found by solving Equation (1.4.8) with respect to σ [73]

$$\sigma = \frac{e^2}{\hbar}\left\{\frac{2}{3\xi} + \nu^{2/3}\left(\frac{\mathrm{d}n}{\mathrm{d}\varepsilon_F}\right)^{1/3} T^{p/3}\right\}. \tag{1.4.10}$$

This is a temperature dependence of the form

$$\sigma = a_2 + b_2 T^{1/3} \tag{1.4.11}$$

(for $p = 1$).

The conductivity near the mobility edge was experimentally studied for numerous disordered solids. We summarize some of the results of the experimental studies of the conductivity near the metal–insulator transition by the example of doped germanium (see, e.g., [74, 75]).

1. Figure 1.6 shows the temperature variation of the conductivity for a series of Ge:As samples at varying impurity concentration [75]. A procedure allowing for a reliable extrapolation of the finite-temperature conductivity to $T = 0\,\text{K}$ has been described [75, 76]. The temperature dependence of the low-temperature conductivity for barely metallic samples obeys Equation (1.4.11) with temperature independent a_2.

The $T^{1/3}$ dependence was observed for different doped semiconductors: InSb [64], GaAs [64, 66], and Ge [75]. The metal–insulator transition in these experiments was approached by changing both the impurity concentration N and the magnetic field. For samples with concentrations farther from the critical one, the variation of the conductivity can be described by the $T^{1/2}$ dependence (Equation 1.4.9); the change from $T^{1/2}$ to $T^{1/3}$ was observed when the particle energy approaches the mobility edge [64, 66, 77].

2. As the impurity concentration is increased, the temperature coefficient of the resistivity changes sign at some concentration $N_c^* > N_c$. According to the theoretical arguments above, this occurs as $\sigma(0) = \sigma_{\min}$ and corresponds to the transition from the critical to the classical region, where $\sigma > \sigma_{\min}$. Thus one can estimate the minimum metallic conductivity by identifying it with the characteristic value of the conductivity at which its temperature

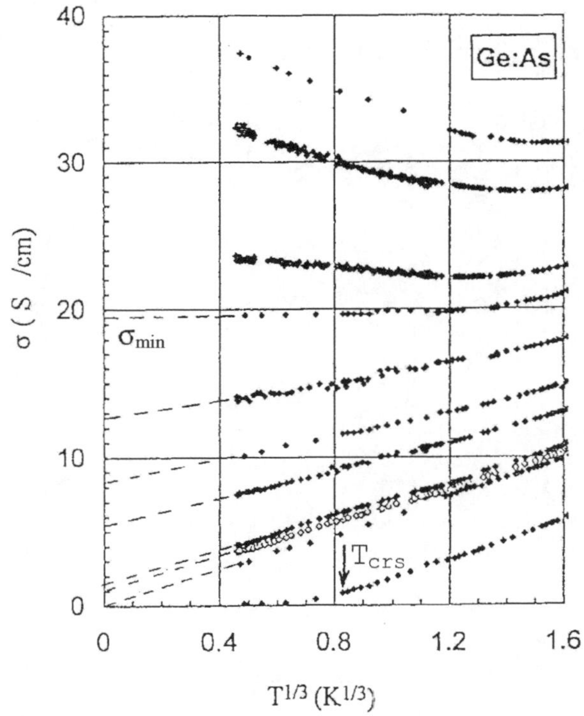

Figure 1.6 Temperature dependence of the conductivity for a series of Ge:As samples plotted as a function of $T^{1/3}$. The impurity concentrations are (from top to bottom): $5.38 \times 10^{17}\,\text{cm}^{-3}$, $5.15 \times 10^{17}\,\text{cm}^{-3}$, $4.60 \times 10^{17}\,\text{cm}^{-3}$, $4.45 \times 10^{17}\,\text{cm}^{-3}$, $4.17 \times 10^{17}\,\text{cm}^{-3}$, $3.91 \times 10^{17}\,\text{cm}^{-3}$, $3.82 \times 10^{17}\,\text{cm}^{-3}$, $3.58 \times 10^{17}\,\text{cm}^{-3}$, $3.56 \times 10^{17}\,\text{cm}^{-3}$, $3.50 \times 10^{17}\,\text{cm}^{-3}$, $3.00 \times 10^{17}\,\text{cm}^{-3}$. The arrow shows the temperature of the temperature-induced metal–insulator transition (reproduced from [75] with permission from Wiley-VCH)

coefficient changes sign [75]. The results of this procedure of estimation of the minimum metallic conductivity for different materials with different critical impurity concentrations N_c are shown in Figure 1.7. The order-of-magnitude estimate of σ_{min} agrees with Mott's value and the dependence $\sigma_{min}(N_c)$ is well described by Equation (1.4.4) (with different slopes C for n- and p-type materials).

3. The experimental studies have shown that a transition from metallic to insulator state is a continuous transition; the values of the zero-temperature conductivity $\sigma(0)$ in the critical region above the mobility edge can be substantially smaller than the minimum metallic conductivity σ_{min}, but remain finite, clearly indicating a metallic state. However, it is seen in Figure 1.6 that the $T^{1/3}$ dependence is obeyed not only in the critical region $N_c < N < N_c^*$ $(0 < \sigma < \sigma_{min})$, but for $N < N_c$ in the insulator region close to the transition for $T > T_{crs}$. Below T_{crs}, the conductivity is low and its temperature dependence is exponential, corresponding to variable range hopping; above T_{crs} its temperature variation becomes the same as in the critical region on the metallic side of the transition. Thus a crossover to metallic behavior or a temperature-induced *insulator–metal transition* must occur at $T = T_{crs}$ as we increase the temperature [78].

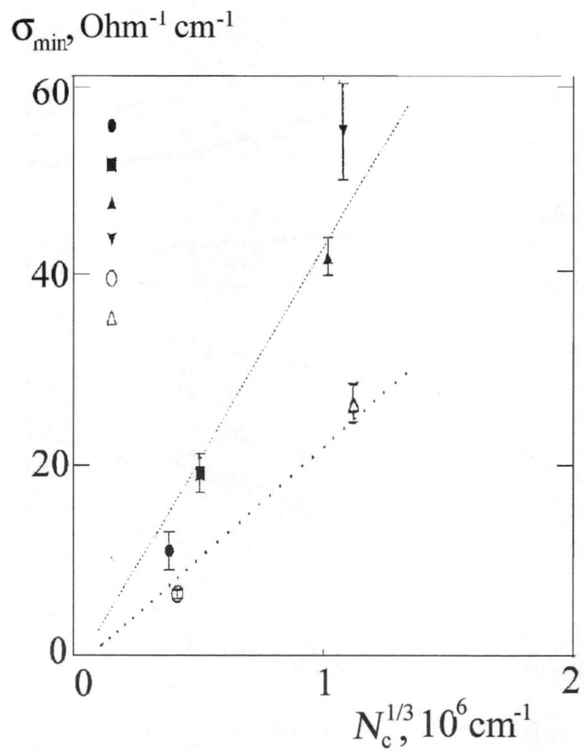

Figure 1.7 Mott's minimum metallic conductivity estimated from the temperature dependence of the conductivity as a function of $N_c^{1/3}$ (N_c is the critical impurity concentration). The slopes are $C = 0.12$ for n-type and $C = 0.06$ for p-type materials (reproduced from [75] with permission from Wiley-VCH)

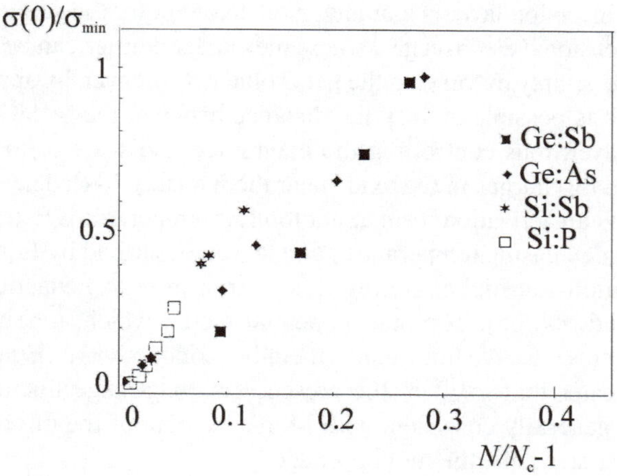

Figure 1.8 Extrapolated values of the normalized zero-temperature $\sigma(0)/\sigma_{min}$ as a function of impurity concentration for different doped semiconductors (reproduced from [75] with permission from Wiley-VCH)

The variation of $\sigma(0)$ below σ_{min} is usually well described by Equation (1.4.6). It is seen in Figure 1.8 that the data for differently doped Si and Ge samples fall on a universal curve. This shows the normalization role of σ_{min} in the scaling behavior of zero-temperature conductivity. It is worth noting that theories for interacting charge carriers predict the value $\nu = 1/2$ [79, 80].

It should be noted that the experimental data on the critical exponent ν are controversial; different values have been reported for different materials, ranging from 1/2 to 2. Sometimes, even the values reported for the same material were different (for instance, in Si:P the values 1/2 [81], 1.3 [82], and 1 [75] were derived). Such discrepancies might be due to the sample properties (in particular, the inhomogeneity in the impurity distribution) and, especially, due to the method of extrapolating the conductivity to zero temperature. The errors may arise due to using the data for samples with concentrations outside the critical region. In [75], the data for a number of Ge and Si samples doped with different impurities (that yielded admittedly different values of ν) were reprocessed using the method described in [76]. The problem of the critical indices does not seem to be completely resolved; indeed, the data obtained for neutron transmutation doped germanium samples with controlled disorder and homogeneity [83] indicate that the critical indices (at least on the insulator side of the metal–insulator transition) may be different, depending on the degree of disorder.

1.5 APPARENT MOBILITY EDGE AND EXTENDED-STATE CONDUCTION IN TWO-DIMENSIONAL SYSTEMS

1.5.1 Experimental studies of the mobility edge in low-mobility two-dimensional systems

Early experimental studies of the temperature dependence of the two-dimensional conductivity were performed on inversion layers in metal oxide–silicon field-effect transistors

(MOSFETs). A Si inversion layer is a unique model system for the study of electronic processes in a two-dimensional electron gas, where the carrier density can be varied over several orders of magnitude simply by varying the gate voltage. Moreover, by applying an additional substrate voltage, it is possible to vary the distance between the Si/SiO_2 interface and the two-dimensional layer, thus controlling the magnitude and scale of the random potential fluctuations created by charges in the oxide near the interface [84]. The studies showed that the conductivity has an activation form at not too low temperatures (a transition to hopping was observed with decreasing temperature) and is well described by Equation (1.4.1) at low electron concentrations; at higher concentrations transition to nonactivated metallic-type conduction occurred [84, 85]. The preexponential factor, which, for short-range potential fluctuations, was close to the minimum metallic conductivity, (Equation 1.3.46), and appeared to be greater than $\sigma_{\min}^{(2)}$ in the presence of long-range fluctuations [85]. These observations were generally consistent with Mott's concept of the minimum metallic conductivity and the existence of the mobility edge.

Subsequent studies of the conductivity behavior in two-dimensional disordered systems (thin metallic films [86], inversion layers in Si MOSFETs [87, 88] have demonstrated the existence of a logarithmic temperature dependence of the conductance in the metallic region above the mobility edge, with a positive temperature coefficient of resistivity, characteristic of the insulating state. Such a dependence agreed with the prediction of the scaling theory of localization and with the concept of the apparent mobility edge separating the regions of strong and weak localization of electronic states (Section 1.3.3); experiments in a magnetic field showed that a substantial logarithmic contribution comes from the correction to the Drude conductivity related to electron–electron interactions in the diffusive regime (Section 1.2.3). To realize the diffusive regime (i.e., to ensure the realization of the condition $l \ll L_\varphi$ at accessible temperatures), low-mobility samples were used.

1.5.2 Evidence for a true metal–insulator transition in high-mobility two-dimensional systems

The measurements of the temperature dependence of the conductivity in high-mobility Si MOSFETs evidence a real metal–insulator transition in two-dimensional systems [89, 90]. This evidence was corroborated by studies of the temperature dependence of the conductivity in similar Si MOSFETs [91], as well as by measurements on *p*-type Si/SiGe [92] and *n*-type Si/SiGe [93] heterostructures, *p*-AlGaAs [94, 95], *n*-AlGaAs [96], and *n*-AlAs [97] structures.

We briefly summarize the experimental observations for two-dimensional high-mobility systems.

1. A metallic behavior (positive temperature coefficient of the resistivity) is observed down to the lowest accessible temperatures at charge carrier concentrations n exceeding some critical concentration n_c. Below this critical concentration, the behavior of the resistance is insulating, thus indicating that a metal–insulator transition occurs in two dimensions (Figure 1.9). Near the critical concentration, the temperature variation of the resistivity may be nonmonotonic; at $n \simeq n_c$ the resistivity is of the order of the quantum unit of resistance, $h/e^2 \sim 25.6\,\mathrm{k}\Omega$ and is almost independent of temperature (the separatrix in Figure 1.9).

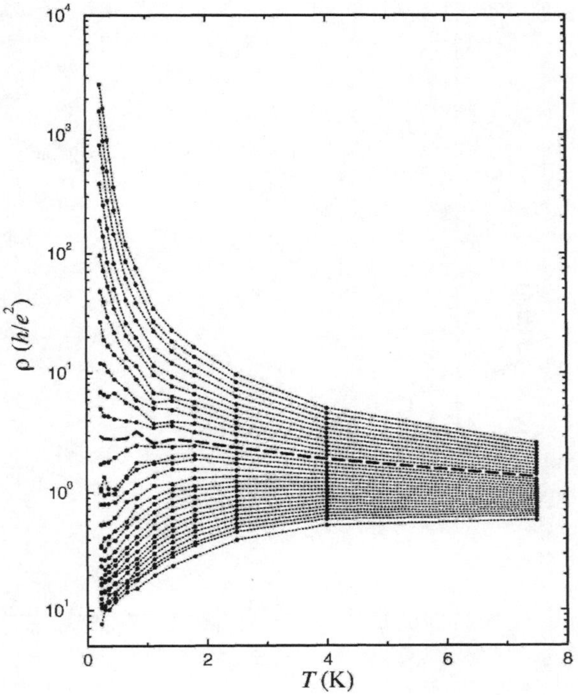

Figure 1.9 Resistivity (in units of h/e^2) of Si MOSFET as a function of temperature for electron concentration varying from $7.12 \times 10^{10}\,cm^{-2}$ to $13.7 \times 10^{10}\,cm^{-2}$ (from top to bottom). The critical density for the metal–insulator transition is $9.6 \times 10^{10}\,cm^{-2}$, indicated by the dashed line (reproduced with permission from [90]; Copyright 1995 by the American Physical Society)

2. Below a temperature of about 2 K, the resistivity exponentially decreases with temperature according to the law

$$\rho(T) = \rho_0 + \rho_1 \exp\left(-(T_0/T)^p\right), \tag{1.5.1}$$

where $p \simeq 1$ [98]. The parameter ρ_1 varies linearly with concentration and vanishes at the transition. The temperature T_0 is sample dependent and increases with concentration [95], $T_0 \propto |\delta n|^q$, where $\delta n = n - n_c$ and $q \simeq 1$. For high-mobility structures, the resistivity decreases with temperature by about an order of magnitude, whereas in the insulating region the resistivity increases sharply with decreasing T.

3. The resistivity for each particular sample (at not too low temperatures) may be scaled using a single scaling parameter T_0 (Figure 1.10). The resistivity data are reduced into two branches, insulating for $n < n_c$ and metallic for $n > n_c$. The scaling parameter T_0 has a critical behavior around a critical concentration n_c and decreases upon approaching the critical electron concentration.

4. The magnetoresistance in a weak perpendicular field is negative [91, 98], which indicates the quantum interference contribution to the conductivity of the two-dimensional metallic state.

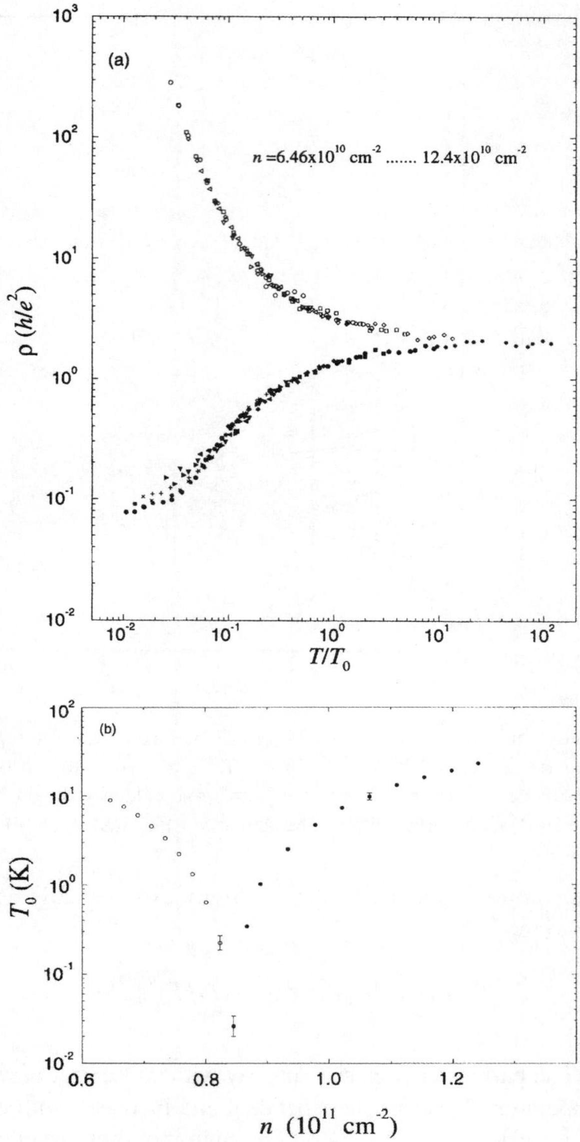

Figure 1.10 Temperature scaling of the resistivity (a) and scaling parameter T_0 vs electron density (b). Open symbols correspond to the insulating side of the transition and closed symbols to the metallic one (reproduced with permission from [90]; Copyright 1995 by the American Physical Society)

These features are in obvious disagreement with the predictions of the scaling theory of localization and are difficult to explain using the standard Fermi-liquid theory. Kravchenko *et al.* [90] argued that these results are evidence for the existence of a true metal–insulator transition in two dimensions. Indeed, scaling behavior is one of the signatures of a phase transition [99], suggesting the existence of a true metallic state in a high-mobility two-dimensional system. A substantial drop of resistivity on the metallic side of the transition (by an order of magnitude) at low temperatures is, however, difficult to explain using a conventional theory. The difference between the results for high-mobility and low-mobility systems can be related to the higher mobility combined with higher charge carrier effective masses; under these conditions the effect of carrier–carrier interactions is enhanced.

The possibility that carrier–carrier interactions would lead to a qualitative modification of the behavior predicted by the theories for noninteracting electrons was discussed in [50–52], where it has been argued that for weak disorder and arbitrarily strong interactions, a two-dimensional system scales toward a state with finite nonzero conductivity with lowering temperature (as discussed in Section 1.3.4, the rigorous proof of this statement encounters some difficulties). A phenomenological introduction of the interaction effects into the scaling theory of localization also shows that a metallic state that can exist in this case in an unconventional non-Fermi-liquid-type state whose existence is due to electron–electron interactions. Anyhow, a consistent theory that describes the interplay of disorder and interactions near the metal–insulator transition is still lacking.

1.5.3 Evidence against a true metal–insulator transition in two-dimensional systems

A different point of view is that the specific features of the low-temperature transport that seem to indicate a true quantum metal–insulator transition (positive metallic-type temperature coefficient of the resistivity in the apparently metallic region, scaling, etc.) can basically be understood in the context of the conventional 'classical' Fermi-liquid-type theory. A number of experimental results confirm this point of view.

(i) Metallic-type temperature coefficient of the resistivity was observed for densities about 30 times greater than the critical density (for resistivities about 100 times smaller than the resistivity at the transition) [100]. In this region, quantum interference and interaction effects are known to be small compared with the large variation of the Drude resistivity; this suggests that the main $g(T)$ dependence in the metallic phase is not due to quantum interference and interaction effects.

(ii) A correlation between the existence of several conducting bands, i.e., of several parallel gases of charge carriers (e.g., holes of different bands in GaAs/AlGaAs heterostructures), and the metallic behavior was studied [101]. The observed metal–insulator transition in the hole gas depended only slightly on the concentration of charge carriers in the split-off band (the concentration in the parallel channel); if it were determined by the interactions, one would expect that increasing the parallel concentration would screen the interactions between holes and suppress the metallic phase in the hole gas. On the other hand, a clear correlation between the resistance and the measured interband carrier scattering was established for the structures under study [101].

(iii) Some experiments have been successfully interpreted in the framework of normal Fermi-liquid behavior of charge carriers on the metallic side of the transition (using the Hartree–Fock approximation), including weak-localization corrections [100, 102, 103]. In [102], magnetoresistance measurements were used to extract the logarithmic corrections to the Drude conductivity in the 'metallic' phase of a high-quality two-dimensional GaAs hole system at low temperatures. It has been shown that the phase coherence is preserved in the metallic regime with evidence for normal Fermi liquid behavior and that the anomalous exponential decrease in resistivity with decreasing temperature in the metallic phase was not due to quantum interference or strong interaction effects.

Thus, the experiments indicate that the fundamental difference in the behavior of the conductivity of a two-dimensional gas in the metallic and dielectric regions, corresponding to the metal–insulator transition and the 'anomalous' metallic-type behavior may not be due to the appearance of a new quantum state of a system of strongly interacting charge carriers. An alternative 'quasiclassical' explanation is that anomalous behavior of the resistivity can be described using a Fermi-liquid-type approach. To describe the metallic behavior in this approach, a number of studies used an assumption of temperature-dependent charge carrier scattering.

1.5.4 Temperature-dependent charge carrier scattering

(a) Origin of the temperature dependence of scattering at low temperatures

At low temperatures, phonon scattering may usually be neglected in the temperature range where the anomalous metallic behavior is observed; in Si MOSFET two-dimensional electron systems and two-dimensional *n*-GaAs systems, phonon scattering may be disregarded at $T < 5\,\mathrm{K}$, whereas in two-dimensional *p*-GaAs systems, it can play a role at $T > 1\,\mathrm{K}$, being responsible for the observed nonmonotonic variation of the conductivity at intermediate temperatures ($1\,\mathrm{K} < T < 5\,\mathrm{K}$) [104]. It is known that the main scattering mechanisms in two-dimensional inversion layers are scattering by interface roughness, which is known to be important at higher impurity concentrations (as the two-dimensional electron gas is located closer to the interface), and scattering by charged impurity centers (at the Si/SiO_2 interface in Si MOSFETs) [105]. Moreover, in some structures, additional effects affecting the conductivity and scattering are to be taken into account, such as scattering between the split heavy-hole bands (in GaAs/AlGaAs heterostructures), recharging of trap states at the interface [106], freeze-out of free carriers [107], etc.

In this approach, the main factor responsible for metallic conductivity in a two-dimensional electron gas is *temperature-dependent disorder*, arising, in particular, from the temperature dependence of the screening of static disorder. It appeared that the experimental observations of the 'anomalous' metallic properties of two-dimensional electron systems (a metallic variation of the conductivity, a strong decrease in conductivity with decreasing temperature in the metallic region) can be explained in the framework of the theory based on the quasi-classical Boltzmann approach, even disregarding the quantum interference effects [108]. The single-site approximation was used, since usually the condition $k_F l >> 1$ is satisfied [109], and electron–electron interactions were taken into account through screening described within the random phase approximation; indeed, screening of the interactions with charged impurities (and, generally, with surface roughness) is the key factor that determines such behavior of the system [107, 108, 110]. For elastic impurity scattering, the low-temperature conductivity $\sigma(T) = ne^2\tau(T)/m$ is described by [108, 110]

$$\sigma(T) = \sigma(0)\left\{1 - 1.39C(n)(kT/\varepsilon_F) - 0.81C^2(n)\,C_{3/2}(kT/\varepsilon_F)^{3/2} + O(T^2)\right\}, \quad (1.5.2)$$

where $0 < C(n) < 1$ is a density-dependent form factor and $\sigma(0) = ne^2\tau(0)/m$ is the zero-temperature conductivity within lowest-order coupling to the impurities. It has been shown that if higher-order electron–electron interaction terms are taken into account, in particular, in the ballistic regime for $kT << \varepsilon_f$, i.e., for $\hbar/\tau << kT << \varepsilon_f$ (see Section 1.2.3, Equation 1.2.25), the leading linear term in Equation (1.5.2) survives [42]. These conditions, as well the conditions for the validity of Equation (1.5.2), are quite difficult to satisfy, and the measured conductivity virtually never has a linear temperature dependence (except at high densities, where the semiclassical random-phase approximation transport theory is accurate).

To explain the main features of the apparent metal–insulator transition in two-dimensional systems, a model was suggested, taking into account localized states in the conduction band tail with regard to doubly occupied states in the upper Hubbard band. In this model, the metallic behavior of the resistance was related to the activation of localized electrons to the conduction band, leading to the suppression of nonlinear screening of the disorder potential [124].

Another possibility, discussed in the context of the metal–insulator transition in a two-dimensional hole gas in GaAs/AlGaAs heterostructures, is the effect of inelastic Coulomb hole scattering between the two spin–orbit-split heavy-hole bands on the conductivity in the metallic region. It appeared that the temperature dependence of the measured inelastic interband scattering rate obeyed the relation

$$S(T) = S_0 + S_1 \exp\left(-(T_0/T)^p\right), \tag{1.5.3}$$

i.e., had exactly the form of Equation (1.5.1) with almost the same value of T_0. This mechanism requires a certain band structure and is not universal for all structures but the result (Equation 1.5.3) indicates that the activated temperature dependence of the resistance in the metallic regime may reflect the increase in inelastic interband scattering with temperature and does not necessarily imply the existence of an interaction-induced true metallic state.

(b) Extension of the scaling theory

A more comprehensive theory would require a simultaneous description of the features of quasiclassical scattering and of the localization and interaction quantum corrections. A generalization of the scaling theory of localization to the case of temperature-dependent disorder and temperature-dependent scattering has been given [111]. Following [111], one can assume that the disorder potential generally consists of the two components so that the classical (Drude) conductivity σ_D can be represented as

$$\sigma_D^{-1} = \sigma_1^{-1} + \sigma_0^{-1}(T), \tag{1.5.4}$$

where σ_1 is the temperature-independent residual conductivity, whereas σ_0 is the temperature-dependent metallic contribution (such that $d\sigma_0/dT < 0$). A straightforward generalization of the approach suggested in [43] is the assumption that the conductance, in addition to the dependence on the system size L, may also explicitly depend on T due the corresponding dependence of the disorder. Now in Equation (1.3.19) of the scaling theory, we have $g_0 = g_D = \sigma_D$, $\sigma_D = \sigma[l(T),T]$ and the scaling function is same as in Equation (1.3.19). When $L_\varphi < L$, one has to replace $\ln(L/l)$ by $\ln(L_\varphi/l)$ on the right-hand side of Equation (1.3.23). Assuming for simplicity that $g_D = \sigma[l(T),T] \sim l(T)$ and differentiating Equation (1.3.19) with respect to T, we obtain

$$\frac{1}{\beta(g)}\frac{d\ln g}{d\ln T} = \frac{d\ln L_\varphi}{d\ln T} + \left(1 - \frac{1}{\beta(g_D)}\right)\frac{d\ln g_D}{d\ln T}. \tag{1.5.5}$$

Since the phase-breaking time τ_φ may depend on disorder, the authors of [106] have set $\tau_\varphi \sim T^{-p}\sigma^{2\gamma-1}$, where p and γ are some constants. In addition, $D \sim \sigma$ so that $L_\varphi = \sqrt{D\tau_\varphi} \sim \sigma^r T^{-p/2}$. Then Equation (1.5.5) takes the form

$$\left(\frac{1}{\beta(g)} - \gamma\right)\frac{d\ln g}{d\ln T} = \frac{p}{2} + \left(\frac{1}{\beta(g_D)} - 1\right)\frac{d\ln g_D}{d\ln T}. \tag{1.5.6}$$

Equation (1.5.6) is supplemented with the 'boundary' condition $g(T_0) = g_D(T_0)$, where T_0 is the cutoff temperature at which localization effects become negligible, $L_\varphi(T_0) = l(T_0)$. Thus one can account for the $g(T)$ dependence, arising both from the phase-breaking processes (the first term on the right-hand side) and the temperature dependence of the classical conductance $g_D(T)$ (the second term).

This approach provides a rather good description of the experimental data in high-mobility MOSFETs, including the metallic behavior of the conductivity above some critical concentration n_c, the temperature dependence of $\sigma(T)$ near n_c, and a nonmonotonic variation of $\sigma(T)$ in a range of concentrations near n_c: a maximum of the conductivity could be the result of a superposition of a temperature dependent scattering and weak localization and interaction effects.

The quasiclassical (Drude) resistivity was assumed to be temperature dependent due to the variation in carrier scattering. One of the reasons for the metallic behavior of the resistivity is the temperature-dependent screening of charged impurity scattering centers at the interface temperature-dependent concentration of charged defects (see Section 1.5.2.1).

A nonmonotonic temperature dependence may be explained by additionally taking account of some classical scattering mechanism giving rise to the insulator-type temperature-dependent contribution to the resistivity related, e.g., to the metallic freeze-out of the free carriers [107] or the quantum localization corrections. Even a simplified classical approach makes it possible to describe (at least qualitatively) the behavior of two-dimensional systems near the metal–insulator transition, including the nonmonotonic temperature dependence at concentrations close to the critical one, concentration dependence of the conductivity, scaling properties; a good agreement was attained for high-mobility low-density charge carrier gas in Si MOSFETs and GaAs heterostructures. The observed scaling properties of the conductance $\sigma(T,n) \approx \sigma(T/T_0)$, as well as the concentration dependence of the scaling parameter T_0 [90], can be reproduced using the Boltzmann model [111].

(c) Effects of a large-scale fluctuation potential

Up to now, we have discussed homogeneous systems in the absence of strong inhomogeneities in the concentration distribution. However, the presence of a fluctuation potential $V(\mathbf{x})$

is a characteristic feature of impurity semiconductors [112]; in two-dimensional systems, this potential can be created by ionized impurities (with concentrations of about 10^{12}–$10^{13}\,\mathrm{cm}^{-2}$), which are usually present at the heterointerfaces, creating potential fluctuations in the plane of the two-dimensional charge carrier gas. A large-scale potential can be also intentionally produced by introducing an array of quantum dots with random parameters [113].

The amplitude of the fluctuation potential increases with decreasing electron concentration, as the screening becomes weaker and strongly nonlinear. Due to this, the spatial distribution of electrons becomes highly inhomogeneous; the electrons are concentrated in the regions of lower potential energy (valleys). The conductivity problem for an electron gas in the presence of a fluctuation potential is known to be intimately related to the continuum percolation problem [5]. An electron with energy ε can move only in the classically accessible regions defined by the condition $\varepsilon > V(\mathbf{x})$. At $T = 0\,\mathrm{K}$, the electrons are located in the regions where $\varepsilon_F > V(\mathbf{x})$ forming *puddles* and the transport is controlled by electron transfer between the puddles. At low ε_F, such regions are isolated; they grow in size with increasing ε_F until, at some critical $\varepsilon_F = \varepsilon_c$ (*percolation threshold*), isolated puddles merge together to form an infinite classically accessible region (the infinite *percolation cluster*). For $\varepsilon_F > \varepsilon_c$, electrons at the Fermi level can travel classically (above the potential landscape) over the entire system. Thus, a percolation-type metal–insulator transition is expected at some critical average charge carrier concentration n_c, corresponding to the condition $\varepsilon_F = \varepsilon_c$. Such a transition was discussed by Efros for modulation-doped heterostructures [114, 115]. The real situation is somewhat more involved than the simple classical continuum percolation problem, since the form of the fluctuation potential depends on the carrier concentration due to *nonlinear screening* [5]. Nonlinear screening results in substantial variation (flattening) of the potential in the valley regions filled by electrons.

Experimental studies of conductance in low-carrier concentration MOSFETs indicate a substantial role that can be played by the fluctuation potential in such systems. Thus, the correlation between n_c and the quality of the sample and the nonlinearity of current–voltage characteristics in unexpectedly weak electric fields were observed [116, 117]. These results could be explained using the classical percolation approach. At concentrations below the critical concentration n_c, metallic puddles in a two-dimensional hole gas are separated by potential barriers and the energy ε_a of activation to the percolation level is equal to the difference between the percolation level ε_c and the Fermi level. In an electric field E, the barrier height is decreased by eEL, where L is the puddle dimension. If the energy eEL becomes equal to the activation energy, the barrier vanishes and the conductivity grows abruptly, leading to highly nonlinear $I - V$ curves ('breakdown'). Since the cluster dimension diverges near the threshold, the critical electric field for the breakdown appears to be low.

For energies only slightly exceeding the percolation level ε_c, large classically accessible regions are connected by constrictions corresponding to saddle points of the potential landscape. If the Fermi level lies close to the percolation level (either slightly above or below it), the conduction is expected to be controlled by the constrictions for which the saddle point energies ε_{QC} are close to ε_c. In the regions of constrictions, a purely classical picture of conduction may, however, be incomplete; it must be supplemented by including the possibility of quantum effects such as tunneling through the constrictions between the puddles.

Thus, tunneling through the constrictions can play an important role, even for a large-scale potential. For energies just above the percolation threshold, the width of the constrictions of the infinite cluster may be comparable to the de Broglie wavelength [118–121]. Such narrow channels are usually rather short and the constrictions represent *quantum point contacts* (QPC), corresponding to saddle points of the large-scale potential. Quantum tunneling through these saddle points is a ballistic process. For degenerate electron gas, the conductance of such a QPC can be expressed using the Landauer formula

$$G(\varepsilon_F, T) = \frac{2e^2}{h}\int d\varepsilon\left(-\frac{\partial f_F(\varepsilon)}{\partial\varepsilon}\right)\mathrm{T}(\varepsilon)$$
$$= \frac{2e^2}{h}\frac{1}{1+\exp[\varepsilon_{QC}-\varepsilon_F/kT]}. \quad (1.5.7)$$

Here, $f_F(\varepsilon)$ is the Fermi–Dirac distribution function and $\mathrm{T}(\varepsilon)$ is the transmission through the QPC. Expression (1.5.7) was used in [118, 119] with $\mathrm{T}(\varepsilon) = \theta(\varepsilon_{QC} - \varepsilon_F)$. It should be noted that using (1.5.7) implies that for $\varepsilon_{QC} < \varepsilon_F$, the temperature dependence of the conductance is such that $dG(\varepsilon_F,T)/dT < 0$, and the resistance of a QPC has an activated contribution.

Due to the randomness of the values of ε_{QC} for different QPCs, the problem can be reduced to the classical bond percolation problem [3, 5]. By Equation (1.5.7), at $T = 0\,\mathrm{K}$, the conductance $G(\varepsilon_F,0)$ of a QPC is close to $2e^2/h$ if $\varepsilon_{QC} < \varepsilon_F$ and $G(\varepsilon_F,0) = 0$ if $\varepsilon_{QC} > \varepsilon_F$. We identify the puddles with sites and QPCs with bonds and assume that two puddles (sites) are bonded if $\varepsilon_{QC} < \varepsilon_F$ (the bond is 'open') and they are not bonded (the bond is 'closed') if $\varepsilon_{QC} > \varepsilon_F$. For a given distribution of the values of ε_{QC}, on increasing ε_F, we increase the number of open bonds until at some $\varepsilon_F = \varepsilon_c$ the infinite percolation cluster of conjugated open bonds appears. Neglecting the resistances of the puddles and assuming that the conductivity is controlled by QPCs, we obtain that the conductivity of the system is determined by the conductivity $G(\varepsilon_c,0) = 0$ of the critical QPC (or of the QPCs if there are several QPCs that control the resistance in the percolation path). The percolation threshold $\varepsilon_F = \varepsilon_c$ corresponds to the metal–insulator transition at $T = 0\,\mathrm{K}$: for $\varepsilon_F > \varepsilon_c$, the density of the percolation cluster increases and the zero-temperature conductance varies as $(\varepsilon_F - \varepsilon_c)^t$, where $t \simeq 1.3$ is the conductivity index [5, 122]. The conductance as a function of the electron density was numerically calculated for a simple 20×20 square array of QPCs assuming a uniform distribution of the QPC energies. The results of the calculation appeared to be in good agreement with the behavior of the low-temperature resistivity observed near the transition [118].

At finite temperatures, an important feature of the model [118] is that, according to Equation (1.5.7), the resistance of a conducting QPC with $\varepsilon_{QC} < \varepsilon_F$ increases exponentially with temperature (saturating as $T \to 0\,\mathrm{K}$), whereas the conductance of an insulating QPC increases exponentially. At high temperatures near the transition ($|\varepsilon_{QC} - \varepsilon_F| << kT$), the resistances of the QPCs on both sides of the transition are almost equal. As the temperature is lowered, the difference between the QPC resistances with ε_{QC} lying on different sides of the transition increases exponentially. Indeed, the resistance of systems on the insulating side of the transition determined by the critical QPCs with $\varepsilon_F < \varepsilon_{QC}$ grows exponentially. On the other hand, the behavior of the resistance of metallic samples controlled by conducting QPCs with $\varepsilon_{QC} < \varepsilon_F$ is more complicated and is related to the structure of the percolation cluster. If we disregard the variation in the structure of the percolation cluster, the temperature dependence of the resistance of the system is controlled by the temperature dependence of the resistance of conducting QPCs. It is expected to increase (typical metallic behavior) due the increase in the resistance of a conducting QPC with temperature determined by Equation (1.5.7)

$$R(\varepsilon_F, T) = (h/2e^2)\{1 + \exp[\varepsilon_{QC} - \varepsilon_F/kT]\}. \qquad (1.5.8)$$

Equation (1.5.8) has the functional form of Equation (1.5.1), where the parameter T_0 varies linearly with $\varepsilon_{QC} - \varepsilon_F$, i.e., with density, and vanishes at the transition, in accordance with the experimental observations. Equation (1.5.8) predicts saturation of the resistance at low temperatures. However, when passing from the resistance of a QPC to the resistance of the system, one should also take into account the variation in the number of bonds that belong to the percolation cluster. This number increases with temperature, since the insulating QPCs with $\varepsilon_{QC} > \varepsilon_F$ lying close to ε_c join the percolation cluster. Therefore, at low temperatures, where the resistances of the percolation cluster are weakly temperature dependent, the increase in the density of the percolation cluster may result in the increase in the system conductance. Accordingly, in a range of concentrations near the critical concentration corresponding to the condition $\varepsilon_F = \varepsilon_c$, the temperature dependence of the conductivity may vary nonmonotonically; a monotonic metal-like behavior is regained with increasing ε_F above the critical value.

Additional confirmation of the important role of random QPCs comes from the studies of a quasi-two-dimensional electron gas in Si MNOS structures with an inversion n channel [120, 121]. In the mesoscopic structures under study, the percolation cluster consisted of independent parallel paths and the resistance of each path was controlled by a single or a few QPCs. In this case, the shape of the current–voltage characteristic appeared to be sensitive to the variation of the number of QPCs in the optimal current paths. Using the description of a QPC by a parabolic potential, it appeared to be possible to describe the temperature and gate voltage dependences of the resistivity with reasonable accuracy.

1.6 CONCLUSIONS

Extended-state transport in disordered solids is an important problem, attracting a lot of attention for many years. Generally, to describe transport processes, an approach that takes into account an intricate interplay of disorder and electron–electron interaction effects is needed. High structural disorder substantially complicates the understanding of transport properties, since the conventional methods and concepts of the transport theory based on the weak scattering concept cannot be directly applied to the materials in question. However, most of the theories proposed for the description of transport are based on different versions of a quasi-classical weak-scattering approach and Fermi-liquid description of electron–electron interaction effects (modified to take account of the correction describing quantum effects). Somewhat surprisingly, such approaches used for the description of transport in extended states are often quite efficient. One example is Ziman's theory of liquid metals, which was successfully applied to describe the resistivity of many simple metals, due, in particular, to the fact that the pseudopotentials for electron interaction with ions are strongly screened. For dirty metals, for which the Fermi level lies deeply in the conduction band, the temperature dependence of the low-temperature conductivity determined by quantum interference corrections to the result of the weak elastic scattering theory (Drude conductivity) appeared to be in agreement with experiment.

For extended-state transport near the mobility edge, the kinetic energy, Coulomb interaction energy, and the characteristic disorder energy are often of the same order of magnitude; moreover, as the Fermi level approaches the mobility edge (in the vicinity of the metal–insulator transition), the Wigner–Seitz parameter related to the ratio of the Coulomb interaction energy to the kinetic energy becomes large, thus making it necessary to take interaction effects into account. However, even in this case, the main concepts of the one-parameter scaling theory of localization initially developed for noninteracting electrons (with some modifications) often appear to be a reasonable foundation for the description of transport.

In spite of extensive effort, some of the basic important issues are still not completely resolved. In bulk materials, although the scaling theory of the metal–insulator transition (with regard to interaction effects) adequately describes low-temperature transport in the critical region for numerous doped semiconductors, problems still remain, such as the problem of the critical indices. Moreover, in [123], experimental indications are discussed that, in contrast to the predictions of the scaling theory, the metal–insulator transition in amorphous alloys is discontinuous, implying the existence of Mott's minimum metallic conductivity. For two-dimensional systems, the fundamental problem of the existence of a true metal–insulator transition still remains controversial and different models (discussed in Section 1.5) were suggested that can qualitatively describe the anomalous metallic behavior. Thus, further research, both experimental and theoretical, is needed to elucidate the state of the electron system near the apparent metal–insulator transition.

ACKNOWLEDGEMENT

This study was supported by the Russian Foundation for Basic Research.

REFERENCES

[1] N.F. Mott and E.A. Davis, *Electronic Processes in Non-crystalline Materials*, 2nd edn, Clarendon, Oxford, 1979.
[2] N.F. Mott, *Metal–Insulator Transition,* 2nd edn, Taylor and Francis, London, 1974.
[3] I.P. Zvyagin, *Transport Phenomena in Disordered Semiconductors*, Moscow State University Press, Moscow, 1984.
[4] Y. Imry, *Introduction to Mesoscopic Physics*, University Press, Oxford, 2002.
[5] B.I. Shklovskii and A.L.Efros, *Electronic Properties of Doped Semiconductors*, Springer, Heidelberg, 1984.
[6] V.L. Bonch-Bruevich, R. Enderlein, B. Esser, R. Keiper, A.G. Mironov, and I.P. Zvyagin, *Elektronentheorie ungeordneter Halbleiter*, Deutscher Verlag der Wissenschaften, Berlin, 1984.
[7] V.F. Gantmacher, *Electrons in Disordered Media*, Fizmatlit, Moscow, 2003.
[8] P.A. Lee and T.V. Ramakrishnan, *Rev. Mod. Phys.*, **57**, 287 (1985).
[9] N.F. Mott and W.D. Twose, *Adv. Phys.*, **10**, 107 (1961).

[10] E. Abrahams, S.V. Kravchenko, and M.P. Sarachik, *Rev. Mod. Phys.*, **73**, 251 (2001).
[11] B.L. Altshuler, D.L. Maslov, and V.M. Pudalov, *Physica E*, **9**, 209 (2001).
[12] P.W. Anderson, *Phys. Rev.*, **109**, 1492 (1958).
[13] N.F. Mott, *Phil. Mag.*, **22**, 7 (1970).
[14] J.M. Ziman, *Phil. Mag.*, **6**, 1013 (1961).
[15] J.M. Ziman, *Models of Disorder*, University Press, Cambridge, 1979.
[16] W.A. Harrison, *Pseudopotentials in the Theory of Metals*, Benjamin, Menlo Park, 1966.
[17] M.L. Cohen and V. Heine, in: *Solid State Physics*, vol. 24, p. 37, Academic Press, New York, 1970; Mott N.F. *Phil. Mag. B*, **44**, 265 (1981).
[18] V.L. Bonch-Bruevich, *The Electronic Theory of Heavily Doped Semiconductors*, Elsevier, New York, 1966.
[19] I.P. Zvyagin, *Fiz. Tverdogo Tela.*, **6**, 2972 (1964).
[20] J.S. Langer and T.S. Neal, *Phys. Rev. Lett.*, **16**, 984 (1966).
[21] M.J. Katz, *Phys. Rev. A*, **140**, 1323 (1965).
[22] J.B. Krieger and T. Meeks, *Phys. Rev. B*, **8**, 2780 (1973).
[23] D. Chattopadhyay and B.R. Nag, *Phys. Rev. B*, **12**, 5676 (1975).
[24] J.W. Harrison and J.R. Hauser, *Phys. Rev. B*, **13**, 5347 (1976).
[25] I.S. Shlimak, A.L. Efros, and I.V. Yanchev, *Sov. Phys. Semicond.*, **11**, 149 (1977).
[26] A.L. Efros and M.E. Raikh, in: *Optical Properties of Mixed Crystals*, R.J. Elliott and I.P. Ipatova (eds), Elsevier, New York, 1988, p. 133.
[27] S. Fahy and E.P. O'Reily, *Appl. Phys. Lett.*, **83**, 3731 (2003).
[28] V. Venkataraman, C.W. Liu, and J.C. Sturm, *Appl. Phys. Lett.*, **63**, 2795 (1993).
[29] C. Michel, P.J. Klar, S.D. Baranovskii, and P. Thomas, *Phys. Rev. B*, **69**, 165211 (2004).
[30] L.P. Gor'kov, A.I. Larkin, and D.E. Khmel'nitskii, *Pis'ma Zh. Eksp. Teor. Fiz.*, **30**, 248 (1979) [*JETP Lett.*, **30**, 228 (1979)].
[31] B.L. Altshuler, A.G. Aronov, and D.E. Khmel'nitskii, *Solid State Commun.*, **39**, 619 (1981).
[32] B.L. Altshuler, A.G. Aronov, and D.E. Khmel'nitskii, *J. Phys. C*, **15**, 7367 (1982).
[33] B.L. Altshuler, D.E. Khmel'nitskii, A.I. Larkin, and P.A. Lee, *Phys. Rev. B*, **22**, 5142 (1980).
[34] P.A. Lee, *J. Non-Cryst. Sol.*, **35–36**, 21 (1981).
[35] S. Hikami, A.I. Larkin, and Y. Nagaoka, *Prog. Theor. Phys.*, **63**, 707 (1980).
[36] B.L. Altshuler and A.G. Aronov, *Zh. Eksp. Teor. Fiz.*, **77**, 2028 (1979) [*Sov. Phys. JETP*, **50**, 968 (1979)].
[37] B.L. Altshuler and A.G. Aronov, *Solid State Commun.*, **30**, 115 (1979).
[38] W.L. McMillan and J.M. Mochel, *Phys. Rev. Lett.*, **46**, 556 (1981).
[39] R.C. Dynes and J. Garno, *Phys. Rev. Lett.*, **46**, 137 (1981).
[40] M. Lee, J.G. Massey, V.L. Nguyen, and B.I. Shklovskii, *Phys. Rev. B*, **60**, 1582 (1999).
[41] H. Fukuyama, *J. Phys. Soc. Japan*, **48**, 2169 (1980).
[42] G. Zala, B.N. Narozhny, and I.L. Aleiner, *Phys. Rev. B*, **64**, 214204 (2001); E.A. Galaktionov, A.K. Savchenko, S.S. Safonov, Y.Y. Proskuryakov, L. Li, M. Pepper, M.Y. Simmons, D.A. Ritchie, E.H. Linfield, and Z.D. Kvon, *cond-mat*/0402139 (2004).
[43] E. Abrahams, P.W. Anderson, D.C. Licciardello, and T.V. Ramakrishnan, *Phys. Rev. Lett.*, **42**, 673 (1979).
[44] D.J. Thouless, *Phys. Rev. C*, **13**, 93 (1974).
[45] D.J. Thouless, *Phys. Rev. Lett.*, **39**, 1167 (1977).
[46] L.P. Kadanoff, *Physica*, **2**, 263 (1966).
[47] D. Vollhardt and P. Wofle, *Phys. Rev. B*, **22**, 4666 (1980).
[48] I.P. Zvyagin, *Phil. Mag. B*, **46**, 111 (1983).
[49] A. MacKinnon, *Z. Phys. B*, **53**, 1 (1983).
[50] A.M. Finkel'stein, *Zh. Eksp. Teor. Fiz.*, **84**, 168 (1983) [*Sov. Phys. JETP*, **57**, 97 (1983)].
[51] A.M. Finkel'stein, *Z. Phys. B*, **56**, 189 (1984).

[52] C. Castellani, C. DiCastro, P.A. Lee, and M. Ma, *Phys. Rev. B*, **30**, 527 (1984).
[53] V. Dobrosavljević, E. Abrahams, E. Miranda, and S. Chakravarty, *Phys. Rev. Lett.*, **79**, 455 (1997).
[54] N.F. Mott, *Phil. Mag.*, **26**, 505 (1972).
[55] A.F. Ioffe and A.R. Regel, *Prog. Semicond.*, **4**, 237 (1960).
[56] D.L. Staebler and C.R. Wronskii, *Appl. Phys. Lett.*, **31**, 292 (1977).
[57] W. Meyer and H. Neldel, *Z. Techn. Phys.*, **12**, 588 (1937).
[58] W.E. Spear, D. Allan, P.G. LeComber, and A. Ghait, *Phil. Mag. B*, **41**, 419 (1980).
[59] K. Shimakawa and F. Abdel-Wahab, *Appl. Phys. Lett.*, **70**, 652 (1997).
[60] K. Miyairi, Y. Ohta, and M. Ieda, *J. Phys. D*, **21**, 1519 (1988).
[61] T. Dosdale and R.J. Brook, *Solid State Ionics*, **8**, 297 (1983).
[62] I.P. Zvyagin, *J. Non-Cryst. Sol.*, **97–98**, 83 (1987).
[63] H. Overhof, in: *Hydrogenated Amorphous Silicon*, H. Neber-Aeschbacher (ed.), *Solid State Phenomena*, **44–46**, Scitec Publications, Zurich, 1995, p. 535.
[64] D.J. Newson and M. Pepper. *J. Phys. C*, **19**, 3983 (1986).
[65] T.F. Rosebaum, R.F. Milligan, M.A. Paalanen, G.A. Thomas, R.N. Bhatt, and W. Lin, *Phys. Rev. B*, **27**, 7509 (1983).
[66] K.J. Friedland, A.N. Ionov, R. Rentzsch, Ch. Gladun, and H. Vinzelberg, *J. Phys. Cond. Mat.*, **2**, 3759 (1990).
[67] A.G. Zabrodskii, *Phil. Mag. B*, **81**, 1153 (2001).
[68] G. Hertel, D.J. Bishop, E.G. Spencer, J.M. Rowell, and R.C. Dynes, *Phys. Rev. Lett.*, **50**, 743 (1983).
[69] M. Reghu, C.O. Yoon, D. Moses, Y. Cao, and A.J. Heeger, *Phys. Rev. B*, **47**, 1758 (1993).
[70] H. Fritzsche, in: *Metal Non-metal Transitions in Disordered Solids*, L.R. Friedman and D.P. Tumhall (eds), SUSSP, Edinburgh, 1978, p. 190.
[71] G. Biscupski, H. Dubois, O. Laborde, and X. Zotos, *Phil. Mag. B*, **42**, 19 (1980).
[72] E.M. Gershenson, L.B. Litvak-Gorskaya, and G. Ya. Lugovaya, *Fiz. Tech. Poluprov.*, **7**, 1284 (1981).
[73] B.L. Altshuler and A.G. Aronov, *Pis'ma Zh. Eksp. Teor. Fiz.*, **37**, 349 (1983) [*JETP Lett.*, **37**, 410 (1983)].
[74] P. Dai, Y. Zhang, and M.P. Sarachik, *Phys. Rev. Lett.*, **66**, 1914 (1991).
[75] I. Shlimak, *Phys. stat. sol. (b)*, **205**, 287 (1998).
[76] I. Shlimak, M. Kaveh, R. Ussyshkin, V. Ginodman, and L. Resnick, *Phys. Rev. Lett.*, **77**, 1103 (1996).
[77] M.C. Malliepaard, M. Pepper, R. Newbury, and G. Hill, *Phys. Rev. Lett.*, **61**, 369 (1988).
[78] I. Shlimak, M. Kaveh, R. Ussyshkin, V. Ginodman, L. Resnick, and V.F. Gantmakher, *J. Phys. Cond. Mat.*, **9**, 9873 (1997).
[79] S. Hikami, *Phys. Rev. B*, **24**, 2671 (1981).
[80] W. Sasaki, in: *Anderson Localization*, T. Ando and H. Fukuyama (eds), Springer, Berlin, 1988.
[81] M.A. Paalanen, T.F. Rosebaum, G.A. Thomas, and R.N. Bhatt, *Phys. Rev. Lett.*, **48**, 1284 (1982).
[82] H. Stupp, M. Hornung, M. Lakner, O. Madel, and H.V. Lohneysen, *Phys. Rev. Lett.*, **71**, 2634 (1993).
[83] R. Rentzsch, A.N. Ionov, Ch. Reich, V. Ginodman, I. Shlimak, P. Fozooni, and M.J. Lea, *Fiz. Tverdogo Tela*, **41**, 837 (1999).
[84] M. Pepper, *Proc. Roy. Soc. Lond. A*, **353**, 225 (1977).
[85] D.C. Tsui and S.J. Allen, *Phys. Rev. Lett.*, **32**, 1200 (1974).
[86] G.J. Dolan and D.D. Osherhoff, *Phys. Rev. Lett.*, **43**, 721 (1979).
[87] D.J. Bishop, D.C. Tsui, and R.C. Dyne, *Phys. Rev. Lett.*, **44**, 1153 (1980).
[88] M.J. Uren, R.A. Davies, M. Kaveh, and M. Pepper, *J. Phys. C*, **14**, 5737 (1981).

[89] S.V. Kravchenko, G.V. Kravchenko, J.E. Furneaux, V.M. Pudalov, and M. D'Iorio, *Phys. Rev. B*, **50**, 8039 (1994).

[90] S.V. Kravchenko, W.E. Mason, G.E. Bowker, J.E. Furneaux, V.M. Pudalov, and M. D'Iorio, *Phys. Rev. B*, **51**, 7038 (1995).

[91] D. Popovic, A.B. Fowler, and S. Washburn, *Phys. Rev. Lett.*, **79**, 1543 (1997).

[92] P.T. Coleridge, R.L. Williams, Y. Feng, and P. Zawadzki, *Phys. Rev. B*, **56**, R12764 (1997).

[93] T. Okamoto, K. Hosoya, S. Kawaji, and A. Yagi, *Phys. Rev. Lett.*, **82**, 3875 (1999).

[94] M.Y. Simmons, A.R. Hamilton, M. Pepper, E.H. Linfield, P.D. Rose, D.A. Ritchie, A.K. Savchenko, and T.G. Griffiths, *Phys. Rev. Lett.*, **80**, 1292 (1998).

[95] Y. Hanein, U. Meirav, D. Shahar, C.C. Li, D.C. Tsui, and H. Shtrikman, *et al.*, *Phys. Rev. Lett.*, **80**, 1288 (1998).

[96] Y. Hanein, D. Shahar, J. Yoon, C.C. Li, D.C. Tsui, and H. Shtrikman, *Phys. Rev. B*, **58**, R7520 (1998).

[97] S.J. Papadakis and M. Shayegan, *Phys. Rev. B*, **57**, R15068 (1998).

[98] V.M. Pudalov, *JETP Lett.*, **66**, 175 (1997).

[99] Sh. Ma, *Modern Theory of Critical Phenomena*, Benjamin, Massachusets, 1976.

[100] V.M. Pudalov, G. Brunthaler, A. Prinz, and G. Bauer, *Phys. Rev. B*, **60**, R2154 (1999).

[101] Y. Yaish, O. Prusa, E. Buchstab, Sh. Shapira, G.B. Yoseph, U. Sivan, and A. Stern, *cond-mat*/9904324 (1999).

[102] M.Y. Simmons, A.R. Hamilton, M. Pepper, E.H. Linfield, P.D. Rose, and D.A. Ritchie, *cond-mat*/9910368 (1999).

[103] V. Senz, T. Heinzel, T. Ihn, and K. Ensslin, G. Dehlinger, D. Grützmacher, and U. Gennser, *Phys. Rev. B*, **61**, R5082 (2000).

[104] S. Das Sarma and E.H. Hwang, *Phys. Rev. B*, **61**, R7838 (2000).

[105] T. Ando, A.B. Fowler, and F. Stern, *Rev. Mod. Phys.*, **54**, 437 (1982).

[106] B.L. Altshuler and D.L. Maslov, *Phys. Rev. Lett.*, **82**, 145 (1999).

[107] S. Das Sarma and E.H. Hwang, *Phys. Rev. Lett.*, **83**, 164 (1999).

[108] S. Das Sarma and E.H. Hwang, *Phys. Rev. B*, **69**, 195305 (2004).

[109] V.M. Pudalov, M.E. Gershenson, H. Kojima, G. Brunthaler, A. Prinz, and G. Bauer, *Phys. Rev. Lett.*, **91**, 126403 (2003).

[110] A. Gold and V.T. Dolgopolov, *Phys. Rev. B*, **33**, 1076 (1986).

[111] B.L. Altshuler, D.L. Maslov, and V.M. Pudalov, *Phys. stat. sol. (b)*, **218**, 193 (2000).

[112] V.A. Gergel and R.A. Suris, *JETP*, **84**, 719 (1983).

[113] E. Ribeiro, R.D. Jäggi, T. Heinzel, K. Ensslin, G. Medeiros-Ribeiro, and P.M. Petroff, *Phys. Rev. Lett.*, **82**, 996 (1999).

[114] A.L. Efros, *Solid State Commun.*, **65**, 1281 (1988).

[115] A.L. Efros, *Solid State Commun.*, **70**, 253 (1989).

[116] V.T. Dolgopolov, A.A. Shashkin, G.V. Kravchenko, C.J. Emeleus, and T.E. Whall, *Pis'ma Zh. Eksp. Teor. Fiz.*, **62**, 162 (1995).

[117] S.V. Kravchenko, D. Simonian, M.P. Sarachik, W. Mason, and J.E. Furneaux, *Phys. Rev. Lett.*, **77**, 4938 (1996).

[118] Y. Meir, *Phys. Rev. Lett.*, **83**, 3506 (1999).

[119] Y. Meir, *Phys. Rev. B*, **61**, 16470 (2000).

[120] B.A. Aronzon, D.A. Bakaushin, A.S. Vedeneev, A.B. Davydov, E.Z. Meilikhov, and N.K. Chumakov, *Fiz. Tech. Poluprov.*, **35**, 448 (2001).

[121] A.B. Davydov, B.A. Aronzon, D.A. Bakaushin, and A.S. Vedeneev, *Fiz. Tech. Poluprov.*, 36, 1241 (2002).

[122] D. Stauffer and A. Aharony, *Introduction to Percolation Theory*, Taylor and Francis, London, 1992.

[123] A. Möbius and C.J. Adkins, *cond-mat*/0111017 (2001).

[124] V.I. Kozub and N.V. Agrinskaya, *Phys. Rev. B*, **64**, 245103 (2001).

2 Description of Charge Transport in Amorphous Semiconductors

Sergei Baranovski and Oleg Rubel

Faculty of Physics and Material Sciences Center, Philipps University Marburg, Germany

2.1 INTRODUCTION

In this chapter and in Chapter 6 [1] we present basic ideas for description of charge carrier transport in disordered inorganic and organic materials. Charge transport in disordered

Charge Transport in Disordered Solids with Applications in Electronics Edited by S. Baranovski.

materials via extended states analogous to that in crystalline semiconductors is discussed in detail in Chapter 1 of this book [2]. Here we will focus on the transport properties determined by the presence of localized electron states in disordered materials. Localized states can either play the role of traps terminating charge carrier transport via extended states or they can be used by charge carriers in the so-called hopping transport mode, in which the carriers move via direct tunneling between the localized states.

The development of the study of disordered materials is rather curious. The field was, to a great extent, stimulated by the discovery of the semiconducting glasses, such as amorphous selenium, a-Se, and other chalcogenide glasses, for example, a-As_2Se_3 [3]. These materials are usually obtained by quenching from the melts. Such glassy semiconductors first gave rise to the hope that in various device applications one would be able to replace rather expensive crystalline semiconductors by much cheaper and better manufacturable semiconductor glasses. In the 1960s and 1970s this caused a real burst in experimental and theoretical study of glassy semiconductors, reflected in several monographs. To the interested reader we can recommend the book by Mott and Davis [4]. At the same time much research interest was devoted to the problems of hopping transport in doped crystalline semiconductors, where localized states for charge carriers (electrons and holes) are created by donors and acceptors. The latter systems provide a really valuable test field for theoretical description of hopping transport, since the electronic structure of individual localized states in doped semiconductors is well known. For shallow impurities these states are simple hydrogen-like electron states with renormalized Bohr radius. Due to the screening of the core Coulomb potential of the impurity atom by the semiconductor matrix, the wavefunction of the valence electron or hole on shallow donors and acceptors has a much larger spatial extension than that in a hydrogen atom. Nevertheless the structure of the electron state is very similar to that in a hydrogen atom. This well-known electron structure of localized states allows one to develop theoretical description of hopping transport in full detail. A perfect description of the theory can be found in the book by Shklovskii and Efros [5]. Unfortunately, chalcogenide semiconductor glasses appear extremely resistent against doping. This makes such systems unfavorable for device applications, in which doping effects play a decisive role as, for example, in transistors. Chalcogenide materials are now mostly used for rewritable optical memory storage devices, where the pronounced difference in optical and electrical properties between the amorphous and crystalline phases serves for data storage [6, 7]. These materials were, however, not able to replace expensive crystalline systems in routine semiconductor device applications.

A new era in the study of disordered materials for applications in semiconductor electronics began in the 1970s with another class of disordered systems: inorganic amorphous semiconductors, such as amorphous silicon a-Si, amorphous germanium, a-Ge, and their alloys. These materials are usually prepared as thin films by the deposition of atomic or molecular species. Particularly, hydrogenated amorphous silicon, a-Si:H, has been attracting much research attention, since incorporation of hydrogen essentially improves conducting properties favorable for device applications of amorphous semiconductors. In 1975 Spear and LeComber [8] showed that a-Si can be efficiently doped by donors and acceptors. This opened a possibility for using amorphous semiconductors in traditional semiconductor electronics. Many other disordered materials, such as hydrogenated amorphous carbon, a-C:H, and its alloys, polycrystalline and microcrystalline silicon, are rather close to a-Si:H with respect to their charge transport properties. These materials are used in various device applications as described in detail in Chapter 3 of this book [9] on the example of a-Si:H. Several good monographs are devoted to optoelectronic phenomena in amorphous semiconductors. Among others we can recommend the book by Overhof and Thomas [10] and that of Street [11].

In recent years much research has also been devoted to the study of various other disordered materials such as organic semiconductors and dye-sensitized amorphous semiconductors used in particular for photovoltaic applications. While the scientific community dealing with chalcogenide glasses, doped crystalline semiconductors and amorphous semiconductors usually followed the traditions of research based on the deep theoretical concepts developed by Mott, Anderson, Pollak, Shklovskii, Efros, Thomas, Overhof, Zvyagin and others, many researchers working with more modern systems, as for example, organic disordered materials and dye-sensitized solar cells, sometimes perform rather obscure theoretical treatments of the optodynamic phenomena in these materials, being unaware of the classical results, known for decades for treatment of the analogous phenomena in inorganic amorphous materials. It might therefore be instructive for researchers working in the widespread fields of disordered materials to learn more about the basic concepts developed for description of the charge transport effects in inorganic amorphous semiconductors, such as a-Si:H, where the theoretical concepts are already well established. In this chapter we present some of these basic concepts. In Chapter 6 of this book [1], we show how these concepts can be easily extended for description of charge transport in organic disordered materials, such as conjugated and molecularly doped polymers. We would like to emphasize that in various chemically very different organic and inorganic disordered materials the charge transport phenomena look very similar and can be described by similar theoretical concepts. Unfortunately, communities working with different materials are rather separated from each other. They organize scientific meetings and publish their research results, often being unaware of the achievements in parallel fields. Therefore the exchange of ideas between scientific communities dealing with similar charge transport phenomena in chemically different materials might be very useful. We claim that ideas presented below for the theoretical treatment of transport phenomena in inorganic amorphous materials may be of use for researchers working with other disordered materials.

2.2 GENERAL REMARKS ON CHARGE TRANSPORT IN DISORDERED MATERIALS

Although the literature on the transport phenomena in amorphous semiconductors is enormously rich, there are still many open questions in this field due to many problems specific to such materials. In contrast to ordered crystalline semiconductors with well-defined electronic energy structure consisting of energy bands and energy gaps, the electronic energy spectrum of disordered materials can be treated as quasi-continuous. Instead of bands and gaps, one can only distinguish in disordered materials between extended and localized states. In the former states, the charge carrier wavefunction is spread over the whole volume of a sample, while in the latter states the wavefunction of a charge carrier is localized in a spatially restricted region, and a charge carrier in such a state cannot spread as a plane wave, like in ordered materials. Localized electron states are known in ordered systems as well. In crystalline materials electrons and holes can be spatially localized, occupying donor and acceptor states or some other impurity states and structural defects. However, in the energy spectrum of such materials, the localized states usually appear in the form of δ-like discrete energy levels. In the essentially disordered semiconductors, on the contrary, the energy levels related to the spatially localized states usually fill the energy spectrum continuously. In disordered materials there exists some energy level that separates the extended states from the localized ones. This energy level is called the *mobility edge*, as described in detail in Chapter 1 of this book [2]. We will consider in the following mostly the energy states for electrons rather than for holes. Electron states above the mobility edge are extended and below the edge the states are localized. For holes the localized states lie energetically above the extended states. The energy region between the mobility edges for holes and electrons is called the *mobility gap*. The latter is analogous to the bandgap in ordered systems, albeit, contrary to the bandgap, the mobility gap contains energy states, namely the spatially localized states. Since the density of states (DOS) defined as the concentration of states per unit energy, per unit volume usually decreases when the energy moves away from the mobility edges towards the center of the mobility gap, the energy regions of localized states in the vicinity of the mobility edges are called *band tails*. We would like to emphasize that the charge transport properties depend essentially on the energy spectrum in the vicinity and below the mobility edge, i.e., in the band tails. Unfortunately the precise structure of the energy spectrum, even in the band tails, is not known for almost all disordered materials. The whole variety of optical and electrical investigation techniques have not yet proven able to determine this spectrum. Some of the experimental techniques directed to determining DOS are described in Chapter 3 [9]. Although these techniques provide some estimates for the DOS, the problem is still far from its solution. Since the information on the energy spectrum provided by experimental study is rather vague, it is difficult to develop a consistent theoretical description of charge transport from first principles. The absence of reliable information on the energy spectrum and on the structure of the wavefunctions in the vicinity and below the mobility edges can be considered as the main problem for researchers in their attempts to describe quantitatively the charge transport properties of disordered materials.

The general view of the energy spectrum in a disordered inorganic semiconductor is schematically shown in Figure 2.1. An analogous picture can be found in Chapter 3. The energy levels ε_v and ε_c denote the mobility edges for the valence and conduction bands, respectively. Electron states in the mobility gap between these energies are spatially localized. The states below ε_v and above ε_c can be occupied by delocalized holes and electrons. Some peaks in the DOS are shown in the mobility gap, which can be caused by some defects with particularly high concentration. These parts of the spectrum are discussed in more detail in Chapter 3. Although there is a consensus between researchers on the general view of the DOS in disordered materials, the particular structure of the energy spectrum is not known for most disordered systems. From the theoretical point of view the problem of calculating this spectrum is enormously difficult. There have been many attempts to deduce the shape of the DOS in amorphous semiconductors by fitting various experimental data using some particular assumptions on the energy spectrum. We consider in the subsequent sections some of these attempts based on the study of thermally stimulated currents or on the study of the so-called dispersive transport. Several complimentary results can be found in Chapter 3. Here we first discuss some general transport properties of amorphous semiconductors established experimentally.

Particular attention of researchers is usually given to the temperature dependence of the electrical conductivity since this dependence can indicate the underlying transport mecha-

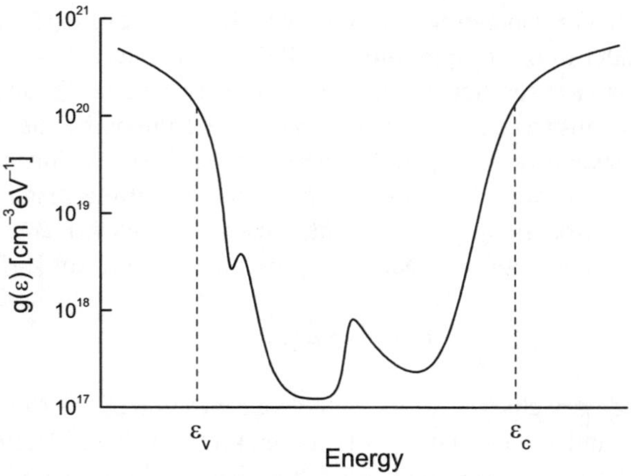

Figure 2.1 Density of states of a noncrystalline semiconductor (schematic); ε_c and ε_v correspond to mobility edges in the conduction band and valence band, respectively

nism. In a rather broad temperature range, the direct current (DC) conductivity in disordered materials has the form

$$\sigma = \sigma_0 \exp\left[-\left(\frac{\Delta}{kT}\right)^{\beta}\right], \tag{2.1}$$

where the preexponential factor σ_0 depends on the underlying system and the power exponent β depends on the material and also sometimes on the temperature range in which the conductivity is studied; Δ is the activation energy and k is the Boltzmann constant. In many inorganic disordered materials, such as vitreous and amorphous semiconductors, σ_0 is of the order of $10^2 - 10^4\ \Omega^{-1}\,\text{cm}^{-1}$. In such materials the power exponent β is close to unity at temperatures close to and higher than room temperature, while at lower temperatures, β can be essentially smaller than unity. In organic disordered materials considered in Chapter 6, values of β larger than unity also have been reported. For such systems the value $\beta \simeq 2$ is usually considered appropriate [12].

Another important characteristic of the electrical properties of disordered materials is their alternate current (AC) conductivity measured under application of an external alternating electric field with frequency ω. It has been established in numerous experimental studies that the real part of the AC conductivity in most disordered semiconductors depends on frequency according to the power law

$$\mathrm{Re}\,\sigma(\omega) = C\omega^{s}, \tag{2.2}$$

where C is constant and the power s is usually slightly smaller than unity. This power law has been observed in numerous materials at different temperatures in a wide frequency range. This frequency dependence differs drastically from that predicted by the standard kinetic theory developed for quasi-free charge carriers in crystalline systems. Equation (2.2) indicates the decisive role of electron transitions between localized states in disordered semiconductors. The corresponding theory is described in detail in Chapter 9 of this book [13]. We will consider in our chapter only the DC conductivity.

In order to develop a theoretical picture for transport properties of any material, the first questions one should answer are the question on the spectrum of the energy states for charge carriers and the question on the spatial structure of such states. Since these two central questions are not yet answered appropriately for noncrystalline materials, the theory of charge transport in disordered systems should be still considered as a phenomenological one. The DC conductivity can be generally represented in the form [14]

$$\sigma = e\int d\varepsilon\, \mu(\varepsilon)n(\varepsilon), \tag{2.3}$$

where e is the elementary charge, $n(\varepsilon)d\varepsilon$ is the concentration of electrons in the states with energies between ε and $\varepsilon + d\varepsilon$ and $\mu(\varepsilon)$ is the mobility of these electrons. Integration in Equation (2.3) is carried out over all energies ε. Under equilibrium conditions, the concentration of electrons $n(\varepsilon)d\varepsilon$ is determined by the density of states $g(\varepsilon)$ and the Fermi function $f(\varepsilon)$ dependent on the position of the Fermi energy ε_F (or on a quasi-Fermi energy in the case of stationary excitation of electrons):

$$n(\varepsilon) = g(\varepsilon)f(\varepsilon), \tag{2.4}$$

where

$$f(\varepsilon) = \frac{1}{1+\exp\left(\dfrac{\varepsilon-\varepsilon_F}{kT}\right)}. \tag{2.5}$$

The Fermi level in almost all known inorganic disordered semiconductors in realistic conditions is situated in the mobility gap, i.e., in the energy range which corresponds to spatially localized electron states. The charge carrier mobility $\mu(\varepsilon)$ in the localized states below the mobility edge is much smaller than that in the extended states above the mobility edge. Therefore at high temperatures, at which a considerable fraction of electrons can be found in the delocalized states above the mobility edge, these states dominate the electrical conductivity of the system. The corresponding transport mechanism under such conditions is similar to that in ordered crystalline semiconductors. Electrons in the states within the energy range of the width of the order kT above the mobility edge dominate the conductivity. In such a case the conductivity can be estimated as

$$\sigma \simeq e\mu_c n(\varepsilon_c)kT, \tag{2.6}$$

where μ_c is the electron mobility in the states above the mobility edge ε_c, and $n(\varepsilon_c)kT$ is their approximate concentration, where

$$n(\varepsilon_c) \simeq f(\varepsilon_c)g(\varepsilon_c) = \frac{g(\varepsilon_c)}{1+\exp\left(\dfrac{\varepsilon_c-\varepsilon_F}{kT}\right)}. \tag{2.7}$$

This equation is valid under the assumption that the typical energy scale of the DOS function $g(\varepsilon)$ above the mobility edge is larger than kT. The position of the Fermi level in disordered materials usually depends only slightly on temperature. Combining Equations (2.6) and (2.7), one obtains the temperature dependence of the DC conductivity described by Equation (2.1) with $\beta \simeq 1$ observed in most inorganic disordered semiconductors at high temperatures.

In order to obtain the numerical value of the conductivity in this high-temperature regime, one needs to know the density of states in the vicinity of the mobility edge $g(\varepsilon_c)$ and also the magnitude of the electron mobility μ_c in the delocalized states above ε_c. While the magnitude of $g(\varepsilon_c)$ is usually believed to be close to the DOS value in the vicinity of the band edge in crystalline semiconductors, there is no consensus among researchers on the magnitude of μ_c. Unfortunately, there are no reliable theoretical calculations for this quantity in most disordered materials. The only exception is provided by the so-called mixed crystals, as described in Chapter 1 of this book. In amorphous semiconductors μ_c is usually estimated to be in the range between $1\,\text{cm}^2/\text{V s}$ and $10\,\text{cm}^2/\text{V s}$ [4, 10].

At lower temperatures, at which fewer electrons occupy extended states above ε_c and many more electrons are in localized states, hopping electron transitions between localized states can play an essential and even a decisive role in the charge transport. In the next section we consider the hopping transport mode in more detail. The hopping transport mechanism determines transport phenomena in inorganic semiconductors in the range of temperatures well below room temperature, while in organic disordered materials this mechanism is believed to dominate charge transport at all temperatures. The latter topic is addressed in Chapter 6. The next section on the general features of hopping transport can be therefore also considered as an introduction to Chapter 6.

2.3 HOPPING CHARGE TRANSPORT IN DISORDERED MATERIALS VIA LOCALIZED STATES

Electron transport via delocalized states above the mobility edge dominates electrical conductivity of disordered materials only at high enough temperatures, at which an essential fraction of charge carriers fill these states. With decreasing temperature, the concentration of such electrons described by Equation (2.7) decreases exponentially and consequently their contribution to electrical conductivity diminishes. Under such circumstances, tunneling transitions of electrons between localized states in the band tails dominate charge transport in disordered semiconductors. This transport regime is called *hopping conduction*, since an incoherent sequence of tunneling transitions of charge carriers resembles a series of their hops between randomly distributed sites. Each site in this picture provides a spatially localized electron state with some energy ε. In the following we will assume that the localized states for electrons with concentration N_0 are randomly distributed in space and their energy distribution is described by the DOS function $g(\varepsilon)$:

$$g(\varepsilon) = \frac{N_0}{\varepsilon_0} G\left(\frac{\varepsilon}{\varepsilon_0}\right), \tag{2.8}$$

where ε_0 is the energy scale of the DOS distribution.

A tunneling transition probability of an electron from a localized state i to a lower in energy localized state j depends on the spatial separation r_{ij} between sites i and j as

$$\nu_{ij} = \nu_0 \exp\left(-\frac{2r_{ij}}{\alpha}\right), \tag{2.9}$$

where α is the localization length which we assume equal for sites i and j. This length determines the exponential decay of the electron wave function in the localized states as shown in Figure 2.2. The preexponential factor ν_0 in Equation (2.9) depends on the electron interaction mechanism that causes the transition. Usually it is assumed that electron transitions contributing to charge transport in disordered materials are caused by interactions of electrons with phonons. Often the coefficient ν_0 is simply assumed to be of the order of the phonon frequency ~$10^{13}\,\mathrm{s}^{-1}$, although a more rigorous consideration is in fact necessary to determine ν_0. Such a consideration should take into account the particular structure of the electron localized states and also the details of the interaction mechanism [15, 16]. Often values of ν_0 larger than $10^{13}\,\mathrm{s}^{-1}$ are necessary to reach agreement between theoretical results and experimental data [17].

When an electron performs a transition upward in energy from a localized state i to a higher in energy localized state j, the transition rate depends on the energy difference between the states. This difference should be compensated, for example, by absorption of a phonon with the corresponding energy [18]:

$$\nu(r_{ij}, \varepsilon_i, \varepsilon_j) = \nu_0 \exp\left(-\frac{2r_{ij}}{\alpha}\right) \exp\left(-\frac{\varepsilon_j - \varepsilon_i + |\varepsilon_j - \varepsilon_i|}{2kT}\right). \tag{2.10}$$

Equations (2.9) and (2.10) were written for the case in which electron occupies site i whereas site j is empty. If the system is in thermal equilibrium, the occupation probabilities of sites with different energies are determined by the Fermi statistics. This effect can be taken into account by modifying Equation (2.10) and adding the terms which account for the relative energy positions of sites i and j with respect to the Fermi energy, ε_F. Taking into account these occupation probabilities one should write the transition rate between sites i and j in the form [18]

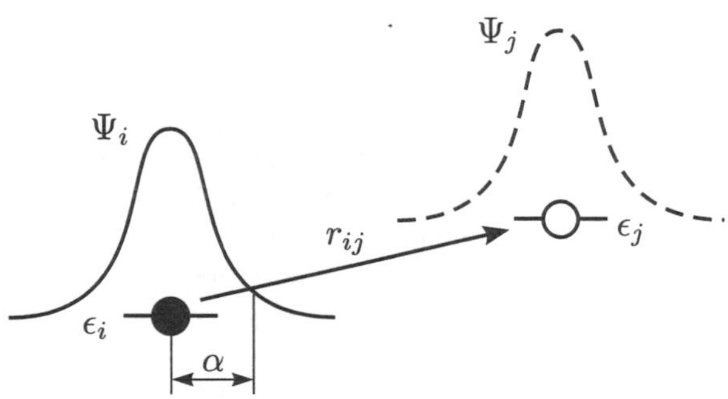

Figure 2.2 Hopping transition between two localized states i and j with energies ε_i and ε_j, respectively. The solid and dashed lines depict the carrier wavefunctions on sites i and j, respectively; α is the localization radius

$$\nu_{ij} = \nu_0 \exp\left(-\frac{2r_{ij}}{\alpha}\right)\exp\left(-\frac{|\varepsilon_i - \varepsilon_F| + |\varepsilon_j - \varepsilon_F| + |\varepsilon_j - \varepsilon_i|}{2kT}\right). \tag{2.11}$$

With the help of these formulas the problem of the theoretical description of hopping conduction can be easily formulated. One has to calculate the conductivity which is provided by transition events with the rates described by Equation (2.11) in the manifold of localized states with the DOS described by Equation (2.8).

2.3.1 Nearest-neighbor hopping

Before presenting the correct solution of the hopping problem we would like to emphasize the following. The style of the theory for electron transport via localized states in disordered materials essentially differs from that of the theories for electron transport in ordered crystalline materials. While in crystalline systems the description is usually based on various averaging procedures, in disordered materials the averaging procedures can in many cases lead to extremely erroneous results. We find it instructive to first analyze some of such approaches in order to illustrate the difference in the description of charge transport for ordered and disordered materials. Treating the scattering rates of electrons in ordered crystalline materials, one usually proceeds by averaging the scattering rates over the ensemble of scattering events. A similar procedure is often tried for disordered systems as well, although it is known from textbooks (see, for instance [5]) how erroneous such an approach can be in the case of disordered materials.

Let us consider the simplest example of hopping processes, namely, hopping of an electron through a system of isoenergetic sites randomly distributed in space with some concentration N_0. It will be assumed that electron states are strongly localized and the strong inequality $N_0\alpha^3 << 1$ is fulfilled. In such a case electrons prefer to hop between the spatially nearest sites and therefore this transport regime is often called the nearest-neighbor hopping (NNH). This type of hopping transport takes place in many real systems at high enough temperatures when the thermal energy kT is lager than the energy scale ε_0 of the DOS. In such situation the energy-dependent terms in Equations (2.10) and (2.11) do not play any essential role and the hopping rates are determined solely by the spatial terms. The rate of the transition of an electron between two sites i and j is described in such a case by Equation (2.9). The average transition rate is easily obtained by weighting this expression with the probability to find the nearest neighbor at some particular distance r_{ij} and by integrating over all possible distances:

$$\langle \nu \rangle = \nu_0 \int_0^\infty dr \exp\left(-\frac{2r}{\alpha}\right) 4\pi r^2 N_0 \exp\left(-\frac{4\pi}{3} r^3 N_0\right) \simeq \pi \nu_0 N_0 \alpha^3. \tag{2.12}$$

Assuming that this average hopping rate describes the mobility, diffusivity and conductivity of charge carriers, one apparently comes to the erroneous conclusion that these quantities are linearly proportional to the concentration of localized states N_0. However, experiments evidence an exponential dependence of transport coefficients on N_0 [5].

Let us look therefore at the correct solution of the problem. This solution is provided in the case considered, $N_0\alpha^3 << 1$, by the percolation theory [5]. In order to find the transport path, one treats a pair of sites as connected if the relative separation between the sites is smaller than some given distance R and checks whether there is a continuous path through the system via such connected sites. If such path is not possible the magnitude of R is increased and the procedure is repeated. At some particular value $R = R_c$ a continuous path through the infinite system via sites with relative separations $R \leq R_c$ arises. Various mathematical considerations give for R_c in three-dimensional space the relation [5, 19, 20]

$$\frac{4\pi}{3} N_0 R_c^3 = B_c, \tag{2.13}$$

where $B_c = 2.7 \pm 0.1$ is the average number of neighboring sites available within the distance smaller than R_c. The corresponding value of R_c should be inserted into Equation (2.9) in order to determine such kinetic coefficients as the mobility, diffusivity and conductivity [5]. The idea behind this procedure is the following. Due to the exponential dependence of the transition rates on the distances between the sites, the rates of electron transitions over distances $r < R_c$ are much larger than those over distances R_c. Such fast transitions do not play any essential role as a limiting factor for electron transport and hence they can be neglected in the calculations of the resistivity of the system. Transitions over distances R_c are the slowest among those which are still necessary for the DC transport and hence such transitions determine the conductivity. The structure of the percolation cluster responsible for charge transport is shown schematically in Figure 2.3.

The transport path consists of the quasi-one-dimensional segments each containing a 'difficult' transition over the distance $\sim R_c$. Inserting R_c determined by Equation (2.13) into Equation (2.9) one obtains the dependence of the conductivity on the concentration of localization sites proportional to $\exp(-2R_c/\alpha)$ in the form

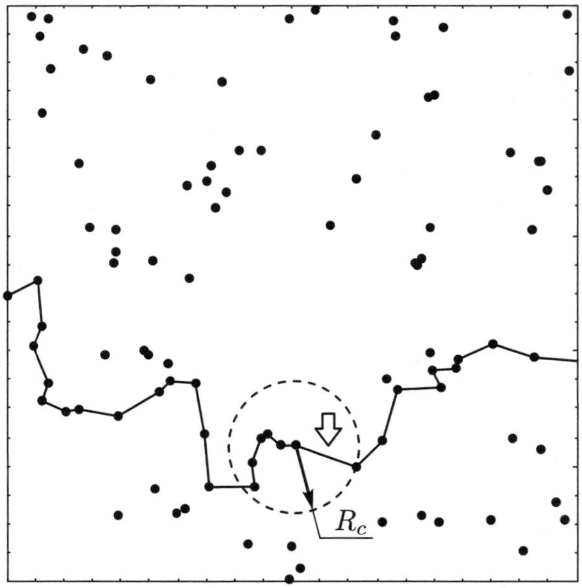

Figure 2.3 A schematic typical transport path with the lowest resistance. Circles depict localized states. The arrow points out the most 'difficult' transition over the distance R_c

$$\sigma = \sigma_0 \exp\left(-\frac{\gamma}{\alpha N_0^{1/3}}\right), \tag{2.14}$$

where σ_0 is the concentration-independent preexponential factor and $\gamma = 1.73 \pm 0.03$. Such arguments do not allow one to determine the exponent in the kinetic coefficients with accuracy better than a number of the order of unity [5]. One should however note that the quantity in the exponent in Equation (2.14) is much larger than unity for a system with strongly localized states under validity of the inequality $N_0\alpha^3 << 1$. This inequality justifies the above derivation. In numerous experimental studies of the hopping conductivity via randomly placed impurity atoms in doped crystalline semiconductors the dependence described by Equation (2.14) has been confirmed [5]. The drastic difference between this correct result and the erroneous one based on Equation (2.12) is apparent. Unfortunately the belief of many researchers in the validity of the procedure based on the averaging of hopping rates is so strong that the agreement between Equation (2.14) and the experimental data is sometimes called occasional. We would like to emphasize once more that the ensemble averaging of hopping rates leads to erroneous results. The magnitude of the average rate in Equation (2.12) is dominated by rare configurations of very close pairs of sites with separations of the order of the localization length α. Of course, such pairs allow very fast electron transitions, but electrons cannot move over considerable distances by using only such close pairs. Therefore, the magnitude of the average transition rate is irrelevant for calculations of the hopping conductivity [5, 21, 22]. The correct concentration dependence of the conductivity in the NNH regime is given by Equation (2.14).

So far, Equation (2.14) was obtained under the assumption that only spatial factors determine transition rates of electrons via localized states. This assumption is valid only at rather high temperatures. If the temperature is not as high and the thermal energy kT is comparable to or smaller than the energy spread of the localized states involved into the charge transport process, the problem of calculating the hopping conductivity becomes much more complicated. In such a case, the interplay between the energy-dependent and the distance-dependent terms in Equations (2.10) and (2.11) determines the conductivity. The lower the temperature, the more important become the energy-dependent terms in the expressions for transition probabilities of electrons in Equations (2.10) and (2.11). If the spatially nearest-neighboring sites have very different energies as shown in Figure 2.4,

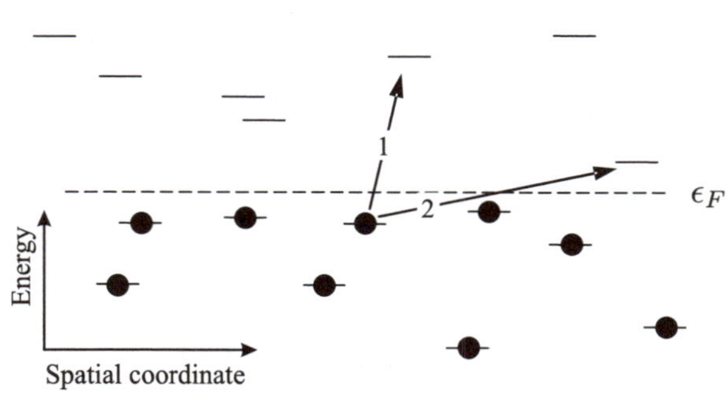

Figure 2.4 Two alternative hopping transitions between occupied states (full circles) and unoccupied states. The dashed line depicts the position of the Fermi level. Transitions (1) and (2) correspond to the nearest-neighbor hopping and to the variable range hopping regimes, respectively

the probability of electron transition between these sites upward in energy can be so low that it would be more favorable for the electron to hop to a more distant site if this site has energy closer to the initial one than to the nearest neighbor. Hence the typical length of electron transitions increases with decreasing temperature. This transport regime was therefore named variable-range hopping. We describe this transport mode in the following section.

2.3.2 Variable-range hopping

The concept of variable-range hopping (VRH) was put forward by Mott [4] who considered electron transport via a system of randomly distributed localized states at low temperatures. We start by presenting Mott's arguments. At low temperatures, the most efficient transitions for transport are electron transitions between states with energies in the vicinity of the Fermi level, since only in this energy range can filled and empty states with close energies be found. Consider the hopping conductivity resulting from energy levels within a narrow energy strap with the width $2\Delta\varepsilon$ symmetric with respect to the Fermi level as shown in Figure 2.5. The energy width of the efficient strap for electron transport can be estimated from the relation

$$g(\varepsilon_{\mathrm{F}})\Delta\varepsilon r^3(\Delta\varepsilon) \simeq 1. \tag{2.15}$$

This criterion is similar to that used in Equation (2.13) for NNH, though we do not care here about numerical coefficients. Here we have to do with the percolation problem in four-dimensional space since, in addition to the spatial terms considered in Section 2.3.1, we

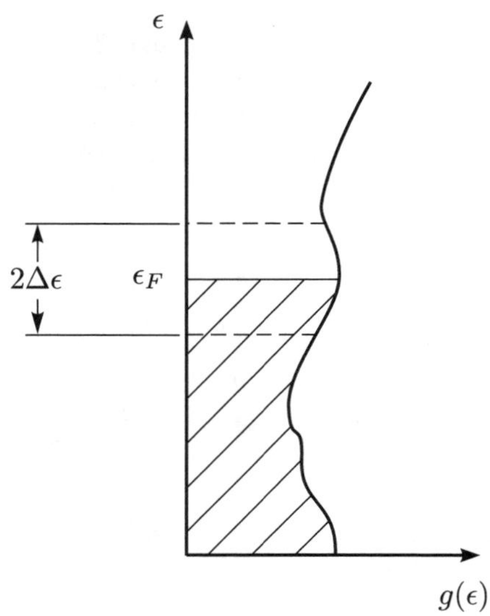

Figure 2.5 Effective region in the vicinity of the Fermi level, where the charge transport takes place at low temperatures

have now to consider also the site energies. For the transition rates described by Equation (2.11) the corresponding percolation problem has not yet been solved precisely. In Equation (2.15) it is assumed that the energy width $2\Delta\varepsilon$ is rather small and the DOS function $g(\varepsilon)$ is nearly constant in the range $\varepsilon_F \pm \Delta\varepsilon$. One can obtain the typical hopping distance from Equation (2.15) as a function of the energy width $\Delta\varepsilon$ in the form

$$r(\Delta\varepsilon) \simeq [g(\varepsilon_F)\Delta\varepsilon]^{-1/3} \tag{2.16}$$

and substitute it into Equation (2.10) in order to express the typical hopping rate

$$\nu = \nu_0 \exp\left\{-\frac{2}{[g(\varepsilon_F)\Delta\varepsilon\alpha^3]^{1/3}} - \frac{\Delta\varepsilon}{kT}\right\}. \tag{2.17}$$

The optimal energy width $\Delta\varepsilon$ that provides the maximum of the hopping rate can be determined from the condition $d\nu/d\Delta\varepsilon = 0$. The result reads

$$\Delta\varepsilon = \left[\frac{2kT}{3g^{1/3}(\varepsilon_F)}\right]^{3/4}. \tag{2.18}$$

After substitution of Equation (2.18) into Equation (2.17) one obtains the famous Mott formula for the temperature-dependent conductivity in the VRH regime

$$\sigma = \sigma_0\left[-\left(\frac{T_0}{T}\right)^{1/4}\right], \tag{2.19}$$

where T_0 is the characteristic temperature:

$$T_0 = \frac{\beta}{kg(\varepsilon_F)\alpha^3}. \tag{2.20}$$

Mott gave only a semi-quantitative derivation of Equation (2.19), from which the exact value of the numerical constant β cannot be determined. Various theoretical studies in three-dimensional systems suggest for β values in the range from 10.0 to 37.8 [23]. According to our computer simulations, the appropriate value is close to 17.6 [24].

The Mott law implies that the density of states in the vicinity of the Fermi level is independent of energy. However it is known that due to the long-range electron–electron interactions between localized electrons a gap (the so-called Coulomb gap) in the DOS arises in the vicinity of the Fermi energy [25, 26]. The gap is shown schematically in Figure 1.6. Using simple semi-quantitative arguments, Efros and Shklovskii [26] suggested a parabolic shape for the DOS function in the vicinity of the Fermi level

$$g(\varepsilon) = \frac{\eta\kappa^3}{e^6}(\varepsilon - \varepsilon_F)^2, \tag{2.21}$$

where κ is the dielectric constant, e is the elementary charge and η is a numerical coefficient. This result has been later confirmed by numerous computer simulations (see, for

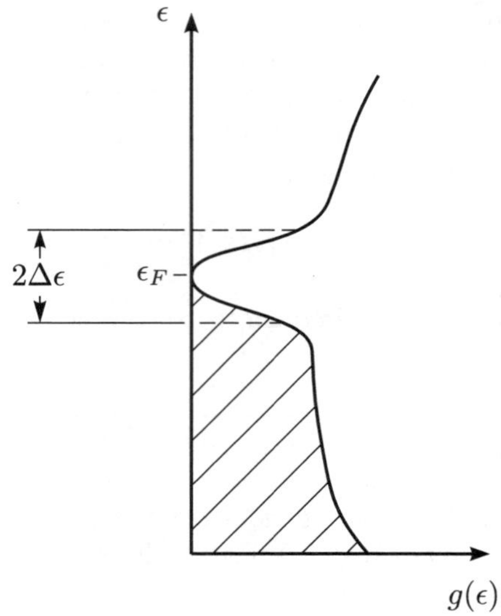

Figure 2.6 Schematic view of the Coulomb gap

example, [27]). At low temperatures, the density of states near the Fermi level has a parabolic shape, and it vanishes exactly at the Fermi energy. With rising temperature, the gap disappears (see, for example, [28]).

As we have seen above, localized states in the vicinity of the Fermi energy are the most efficient ones for transport at low temperatures. Therefore the Coulomb gap essentially modifies the temperature dependence of the hopping conductivity in the VRH regime at low temperatures as compared with Equation (2.19). The formal analysis of the T-dependence of the conductivity in the presence of the Coulomb gap is similar to that for Mott's law discussed above. Using the parabolic energy dependence of the DOS function one comes to the result [5]

$$\sigma = \sigma_0 \left[-\left(\frac{\tilde{T}_0}{T} \right)^{1/2} \right] \tag{2.22}$$

with $\tilde{T}_0 = \tilde{\beta} e^2 / \kappa \alpha k$, where $\tilde{\beta}$ is a numerical coefficient.

Equations (2.19) and (2.22) are among the most famous theoretical results in the field of variable-range hopping conduction. However these formulas are usually of little help for researchers working with noncrystalline materials, such as amorphous, vitreous or organic semiconductors. The reason is the following. The above formulas were derived for the case of either constant DOS (Equation 2.19) or a parabolic DOS (Equation 2.22) in the energy range essential for hopping conduction. These conditions can be usually met in the impurity band of a lightly doped crystalline semiconductor. However, in noncrystalline materials, the energy distribution of localized states is described by the DOS function which has a much stronger energy dependence than the parabolic one. In amorphous, vitreous and microcrystalline semiconductors the energy dependence of the DOS function is believed to be exponential, while in organic materials it is usually assumed as Gaussian. In such cases, new concepts are necessary in order to describe the hopping conduction. In the next section we present such concepts and calculate the dependences of the conductivity on temperature and on the concentration of localized states in inorganic disordered materials.

2.4 DESCRIPTION OF CHARGE-CARRIER ENERGY RELAXATION AND HOPPING CONDUCTION IN INORGANIC NONCRYSTALLINE MATERIALS

In most inorganic noncrystalline materials, such as vitreous, amorphous and polycrystalline semiconductors the localized states for electrons are distributed in a rather broad energy range with the width of the order of an electron volt. The DOS function that describes this energy distribution in such systems is believed to have an exponential shape [9]

$$g(\varepsilon) = \frac{N_0}{\varepsilon_0} \exp\left(-\frac{\varepsilon}{\varepsilon_0}\right), \tag{2.23}$$

where energy ε is counted positive from the mobility edge towards the center of the mobility gap, N_0 is the total concentration of localized states in the band tail, and ε_0 determines the energy scale of the tail. We consider here electrons as charge carriers. The results for holes can be obtained in an analogous way. The magnitudes of ε_0 in inorganic noncrystalline materials are believed to vary between ~0.025 and ~0.05 eV, depending on the system under consideration [9].

It is worth noting, that the arguments in favor of the purely exponential shape of the DOS in the band tails of inorganic noncrystalline materials described by Equation (2.23) cannot be considered as well justified. They are usually based on rather ambiguous interpretation of experimental data. One of the strongest arguments in favor of Equation (2.23) is the experimental observation of the exponential decay of the light-absorption coefficient for photons with an energy deficit ε with respect to the energy width of the mobility gap (see, for example, [4, 9]). One should mention however that this argument is valid only under the assumption that the energy dependence of the absorption coefficient is determined solely by the energy dependence of the DOS. In many cases, in particular in doped semiconductors, the matrix element for the electron excitation by a photon does strongly depend on energy [5, 29, 30]. Hence the argument on the shape of the DOS based on the energy dependence of the light absorption coefficient should be taken cautiously.

Another argument in favor of Equation (2.23) comes from measurements of the dispersive transport in time-of-flight experiments. In order to interpret the observed time dependence of the mobility of charge carriers, it is convenient to assume the DOS for the band tail in the form of Equation (2.23) (see, for example, [31]). In the following section we analyze the dispersive transport and show how the conclusion on the exponential shape of the band tail DOS in the inorganic disordered materials has been derived in the study of this phenomenon.

2.4.1 Dispersive transport in disordered materials

Nonstationary processes in noncrystalline semiconductors differ sometimes essentially from such processes in the crystalline counterparts. As an example, we consider the time-of-flight study of the charge-carrier mobility. In this technique electrons and holes are usually generated by a short pulse of well-absorbed light in the vicinity of a sample surface. The sample is affected by the external constant electric field that causes motion of charge carriers towards the opposite sample surface. Which carriers move—electrons or holes—depends on the field polarity. Assuming the density of charge carriers to be low, the initial spatial width of the carrier packet small, the sample macroscopically homogeneous, the recombination in the sample volume negligible, one can easily interpret experimental observations. The key quantity in the time-of-flight measurements is the electrical current density as a function of time. If the injection of charge carriers from the contacts can be neglected, the current is [32]

$$I(t) = \frac{\mathrm{d}P(t)}{\mathrm{d}t} - \frac{\mathrm{d}Q(t)}{\mathrm{d}t}, \tag{2.24}$$

where $P(t)$ is the dipole moment of the system due to moving carriers and $Q(t)$ is the total charge of moving carriers. In crystalline semiconductors the curve $I(t)$ has usually a pronounced plateau followed by a rapid fall of the current, as shown schematically in Figure 2.7. The typical time τ_{tr} of the carrier transit through the sample is called the *transit time*. Knowing the length of the sample L and the strength of the applied electric field F, one can easily determine the carrier drift velocity $v_{\mathrm{d}} = L/\tau_{\mathrm{tr}}$ and the carrier mobility:

$$\mu = \frac{L}{F\tau_{\mathrm{tr}}}. \tag{2.25}$$

In crystalline semiconductors with broad plateaus in the $I(t)$ curve, the shape of the carrier packet can be described by a Gaussian curve with width $w = \sqrt{D\tau_{\mathrm{tr}}}$ determined by the diffusion coefficient of carriers D. Therefore this type of transit is called *Gaussian* and the transport regime with such transits typical for ordered crystalline semiconductors is called

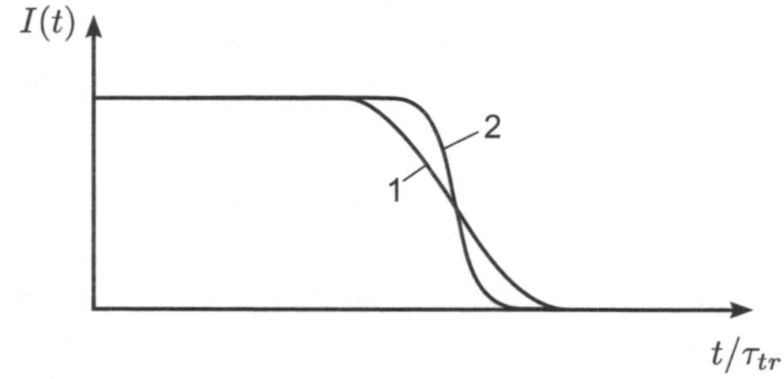

Figure 2.7 Electrical current as a function of the normalized time. Curve 2 corresponds to a longer transit time τ_{tr} than curve 1

Gaussian transport. The decay time of the current in such regime is $\tau_d \sim \upsilon_d^{-1}\sqrt{D\tau_{tr}}$. The latter is determined by the charge carriers leaving the sample after their transits.

In the noncrystalline inorganic semiconductors, the shape of the current transits in the time-of-flight measurements drastically differs from that shown in Figure 2.7. In the disordered materials the plateau in the curves $I(t)$ usually does not show up and the after-transit kink in the dependence $I(t)$ seen in Figure 2.7 is replaced by a pronounced featureless tail as shown in the insert in Figure 2.8. This behavior is caused by a very wide dispersion of transit times for charge carriers. Therefore this transport regime is called *dispersive transport* in contrast to the Gaussian transport with narrow distribution of transit times caused solely by diffusive broadening of the carrier packet. It is difficult to determine a single transit time τ_{tr} in the time-of-flight measurements for dispersive transport from the featureless curves $I(t)$. However if one plots the $I(t)$ curve on a double-logarithmical scale, as shown in Figure 2.8, it is usually possible to observe the time corresponding to the change in the slope of the current decay curves $I(t)$. This time is identified as a transit time τ_{tr} for dispersive transport. Approximately, one can describe $I(t)$ in the form

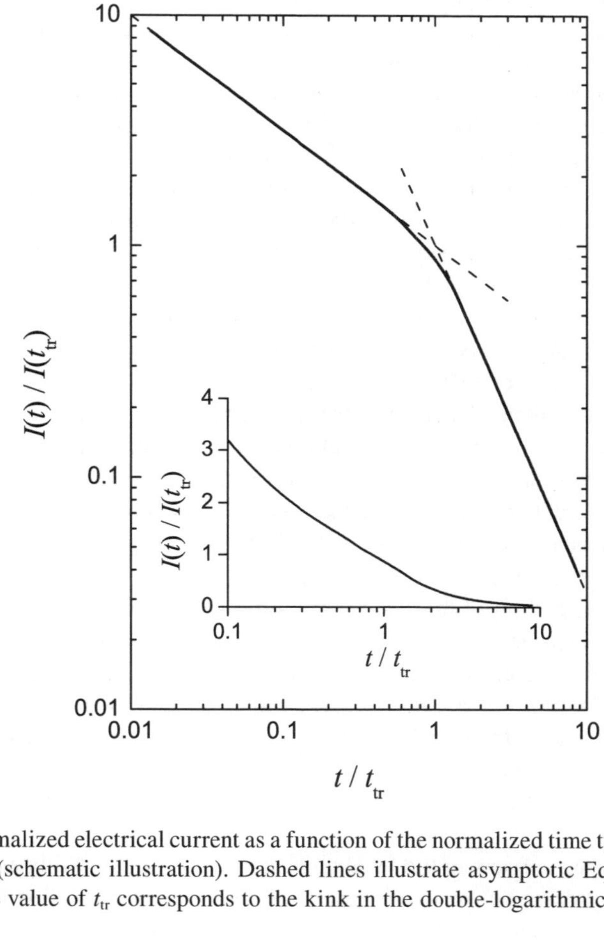

Figure 2.8 Normalized electrical current as a function of the normalized time typical for dispersive transport regime (schematic illustration). Dashed lines illustrate asymptotic Equations (2.26) with $\alpha_i = \alpha_f = 0.5$. The value of t_{tr} corresponds to the kink in the double-logarithmic curve

$$I(t) \propto \begin{cases} t^{-(1-\alpha_i)}, \mathrm{t} < \tau_{\mathrm{tr}} \\ t^{-(1+\alpha_f)}, \mathrm{t} > \tau_{\mathrm{tr}} \end{cases}, \tag{2.26}$$

where α_i and α_f are smaller than unity and often $\alpha_i \simeq \alpha_f$ [32].

The anomalous time dependence of the transient current can be understood if one assumes that the motion of the center of mass of the charge carrier packet $\bar{x}$ slows down in course of time so that

$$\bar{x}(t) \propto t^{\alpha} \tag{2.27}$$

and the spatial width of the carrier packet is described not by the macroscopic diffusion, but rather it is of the order $\bar{x}$ [32]

$$\omega(t) \sim \bar{x}(t). \tag{2.28}$$

Combining Equations (2.24), (2.27), and (2.28) one obtains Equation (2.26) [32]. Assuming that at low electric fields the shift of the center of mass of the carrier packet is linearly proportional to the field strength, $\bar{x} \sim Ft^{\alpha}$, one obtains the field dependence of the transit time in the form $\tau_{\mathrm{tr}} \sim (L/F)^{1/\alpha}$, which agrees well with experimental observations [32].

A more sophisticated mathematical description of the dispersive transport in the framework of a continuous random walk model leading to time dependences described above was suggested by Scher and Montroll [33]. A microscopic transport mechanism behind this phenomenological treatment should be further clarified. Scher and Montroll [33] suggested that the continuous random walk of a charge carrier with the necessary properties can arise in a hopping regime, in which charge carriers move via isoenergetic localized states with transition rates exponentially dependent on separations between such states as described in Section 2.3. According to this picture a carrier will be confronted with successively more and more difficult transitions in course of its random hopping motion, which would slow down its motion continuously, as described phenomenologically by Equation (2.27). However, Pollak [34] showed that the distribution of local transition times in a random walk would lead to the time dependences necessary to explain the broad dispersion of transient times only if there are traps, which catch charge carriers within times much smaller than the times of the release of carriers out of these traps. Otherwise carriers would be captured at time scales comparable to the release times and no dispersion would arise [34]. Hopping via randomly distributed isoenergetic sites would not lead to the broad dispersion of transient times since carriers always can leave the traps at a time scale at which they are captured [34]. Pollak also showed that a system with a sufficiently broad energy distribution of traps can, on the contrary, provide the necessary dispersion of transit times since carriers can be trapped rapidly from the transport states into energetically deep traps where they spend a long time before they are released back into transport states. The energy distribution of traps would lead to the broad distribution of the release times due to the exponential dependence of the release time on the trap energy [34]. Below we consider this model of dispersive transport in more detail following the very transparent approach of Orenstein and Kastner [31].

If one assumes that carriers in the localized states below the mobility edge ε_c are immobile, one can easily describe the carrier transport and energy relaxation (also called thermalization) in the framework of the so-called *multiple-trapping* model. In this model hopping transitions of electrons between localized states are neglected and charge carriers can move only via extended states above ε_c. This assumption can be true if the cutoff of the mobility is sharp compared with energy dependence of the occupation numbers of states determined by temperature and described by the Fermi distribution. Let us consider the energy distribution of electrons generated by a short pulse in the extended states above ε_c. Assuming a constant capture crossection of carriers into localized band-tail states, the states below ε_c will be first uniformly populated and the distribution of carriers after the first trapping event is determined solely by the density of states (DOS) in the tail. As time goes on, carriers from shallow states will be released into conducting states above ε_c, while the trapping process remains random. This will lead to the redistribution of trapped carriers from shallow states into deeper energy states in the band tail. A demarcation energy, $\varepsilon_d(t)$, separates the deep states whose occupation remains uniform from shallow states which are in thermal equilibrium with ε_c [31]. This demarcation level corresponds to the energy at which the trap-release time, $\tau(\varepsilon)$, is of the order of the delay time after the laser pulse, t. Counting energies ε of localized states positive from the mobility edge ε_c towards the center of the mobility gap, one can express the release time as

$$\tau(\varepsilon) = \nu_0^{-1}\exp\left(\frac{\varepsilon}{kT}\right) \tag{2.29}$$

and obtain the demarcation energy, determined by the condition $\tau(\varepsilon_d) = t$, in the form

$$\varepsilon_d(t) = kT\ln(\nu_0 t), \tag{2.30}$$

where ν_0 is the attempt-to-escape frequency, usually assumed equal to the frequency of phonons (lattice vibrations) of the order 10^{12}–$10^{13}\,\mathrm{s}^{-1}$ [31]. Thus $\varepsilon_d(t)$ moves with time downward in the band tail away from ε_c as depicted in Figure 2.9. The carriers will be distributed

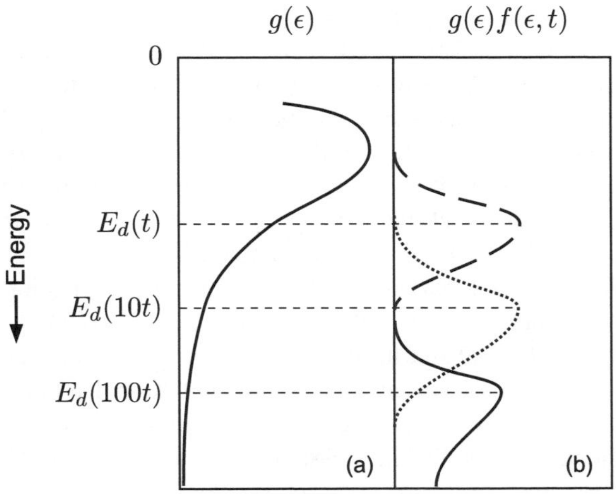

Figure 2.9 Density of states (a) in which electrons are rapidly trapped after excitation. The time evolution of the electron distribution after excitation (b) is shown for three different times: t, $10t$, and $100t$ (reproduced from [31] with permission from Elsevier)

in energy according to the product of the occupation probability and the density of states, $g(\varepsilon)$. For cases where $g(\varepsilon)$ decreases rapidly with energy away from ε_c, most of photoexcited carriers reside near $\varepsilon_d(t)$ as illustrated in Figure 2.9.

The time-dependent current, $I(t)$, is proportional to the carrier density, $n(t)$ in high-mobility states above ε_c at time t. Due to the definition of the demarcation energy $\varepsilon_d(t)$ there is a thermal equilibrium between the states in the vicinity of the mobility edge ε_c and the states in the vicinity of $\varepsilon_d(t)$. Therefore one can easily estimate the concentration of electrons in the conducting states at ε_c as [31]

$$\frac{n(t)}{N} \cong \frac{g(\varepsilon_c)}{g(\varepsilon_d)} \exp\left(\frac{\varepsilon_c - \varepsilon_d}{kT}\right), \tag{2.31}$$

where N is the total concentration of photoexcited charge carriers. Keeping in mind that we count energies positive downward from the mobility edge $\varepsilon_c \equiv 0$ and using Equations (2.23) and (2.30), one obtains from Equation (2.31)

$$\frac{n(t)}{N} \cong \left(\frac{1}{\nu_0 t}\right)^{1-\alpha}, \tag{2.32}$$

with $\alpha = kT/\varepsilon_0$. This analysis applies directly to the time-of-flight measurements where the current density, proportional to the density of mobile carriers $n(t)$, decreases with time according to Equation (2.26). Moreover, the linear temperature dependence of the dispersion parameter α derived above is often observed experimentally in inorganic noncrystalline semiconductors [32]. This agreement between experimental data on dispersive transport and the very transparent theoretical picture based on the energy dependence of the DOS, $g(\varepsilon)$, described by Equation (2.23) is often used as a reason to assume the validity of such a DOS in inorganic noncrystalline semiconductors such as a-Si:H.

Before finishing the description of the dispersive transport we would like to emphasize one important feature of the energy relaxation of charge carriers in the exponential DOS described by Equation (2.23). The values of the energy scale ε_0 of the DOS function in almost all known inorganic noncrystalline materials lie above ~0.025 eV, which means that, at temperatures below room temperature, thermal energy kT is smaller than ε_0. Under such conditions the demarcation energy in the above picture moves with time away from the mobility edge without a limit. However in real systems the concentration of localized states in the band tail is not infinite and the above picture loses its validity as soon as the total amount of photogenerated carriers N becomes comparable to the number of localized states in the band tail below the demarcation energy. Using the above equations one can estimate the corresponding time delay $t_{\lim}$ from the condition $\nu_0 t_{\lim} \simeq (N_0/N)^{1/\alpha}$. At times t larger than $t_{\lim}$ carriers achieve thermal equilibrium and the transport is no longer dispersive, since the concentration of mobile carriers cannot decrease with time due to the energy relaxation described above. It usually makes not much sense to study the behavior of a single charge carrier and to calculate its mobility in a system with the exponential DOS described by Equation (2.23) since this carrier infinitely relaxes downward in energy and its mobility continuously decreases. The effect of the occupation of localized states in the band tail will be described in Section 2.8.

So far we have considered the trivial version of the multiple-trapping model for the energy relaxation of charge carriers in the exponential band tail allowing only transitions between localized and extended states and neglecting direct tunneling transitions between localized states in the band tail. In the next section we remove this restriction and consider the same problem of the energy relaxation of charge carriers in the exponential band tail, including tunneling hopping transitions between localized states. The conclusion from this study will be the following. Almost all statements of the simplified consideration above remain valid. One should simply replace the mobility edge ε_c by some energy level in the band tail, called the transport energy.

2.4.2 The concept of the transport energy

In the following we continue to assume that the DOS in a band tail of a noncrystalline material has the energy dependence described by Equation (2.23). This simple function will allow us to introduce in the most transparent analytical form some valuable concepts developed so far for description of electrical conduction in noncrystalline semiconductors. We first present the concept of the so-called transport energy, which in our view, provides the most transparent description of the charge transport and energy relaxation of electrons in such materials. The crucial role of a particular energy level in the hopping transport of electrons via localized band-tail states with the DOS described by Equation (2.23) was first recognized by Grünewald and Thomas [35] in their numerical analysis of equilibrium variable-range hopping conductivity. This problem was later considered by Shapiro and Adler [36], who came to the same conclusion as Grünewald and Thomas, namely, that the vicinity of some particular energy level dominates the hopping transport of electrons in the band tails. In addition, they achieved an analytical formula for this level and showed that its position does not depend on the Fermi energy.

Independently, a rather different problem of nonequilibrium energy relaxation of electrons by hopping through the band tail with the DOS described by Equation (2.23) was solved at the same time by Monroe [37]. He showed that an electron, starting from the mobility edge, most likely makes a series of hops downward in energy. This character of the relaxation process changes drastically at some particular energy ε_t, which Monroe called the *transport energy* (TE). The hopping process near and below TE resembles a multiple-trapping type of relaxation with the TE playing the role of the mobility edge. In the multiple-trapping relaxation process [31] only electron transitions between delocalized states above the mobility edge and the localized band-tail states are allowed, while hopping transitions between the localized tail states are neglected. Hence, every second transition brings electron to the mobility edge. The TE of Monroe [37] plays the role of this edge for the hopping relaxation. It coincides exactly with the energy level discovered by Grünewald and Thomas [35] and by Shapiro and Adler [36] for equilibrium hopping transport. Shklovskii *et al.* [38] have shown that the same energy level ε_t also determines recombination and transport of electrons in the nonequilibrium steady state under continuous photogeneration in a system with the DOS described by Equation (2.23).

We see, therefore, that the TE determines both equilibrium and nonequilibrium and both transient and steady-state transport phenomena. The question then arises as to why this energy level is so universal that hopping of electrons in its vicinity dominates all transport phenomena. We give below a derivation of the TE based on consideration of a single hopping event of an electron localized deep in the band tail. It is the transport energy that maximizes the hopping rate as a final electron energy in the hop, independently of its initial energy [39]. All derivations below are carried out for the case $kT < \varepsilon_0$.

Consider an electron in a tail state with energy ε_i. According to Equation (2.9), the typical rate of a downward hop of such an electron to a localized state deeper in the tail energy $\varepsilon_j \geq \varepsilon_i$ is

$$\nu_{\downarrow}(\varepsilon_i) = \nu_0 \exp\left[-\frac{2r(\varepsilon_i)}{\alpha}\right], \tag{2.33}$$

where

$$r(\varepsilon) = \left[\frac{4\pi}{3}\int_{\varepsilon}^{\infty} \mathrm{d}x\, g(x)\right]^{-1/3}. \tag{2.34}$$

The typical rate of an upward hop of such an electron to a localized state shallower in the tail with energy $\varepsilon_j \leq \varepsilon_i$ is

$$\nu_{\uparrow}(\varepsilon_i, \delta) = \nu_0 \exp\left[-\frac{2r(\varepsilon_i - \delta)}{\alpha} - \frac{\delta}{kT}\right], \tag{2.35}$$

where $\delta = \varepsilon_i - \varepsilon_j \geq 0$. This expression is not exact. The average nearest-neighbor distance, $r(\varepsilon_i - \delta)$, is based on all states deeper than $\varepsilon_i - \delta$. For the exponential tail this is equivalent to considering a slice of energy with the width of the order ε_0. This works for a DOS that varies slowly compared with kT, but not in general. It is also assumed for simplicity that the localization length α does not depend on energy. The latter assumption can be easily released on the cost of a somewhat more complicated form of the equations. We will analyze the hopping rates at a given temperature T, and try to find the energy difference δ which provides the fastest typical hopping rate for an electron placed initially at energy ε_i. The corresponding energy difference δ is determined by the condition

$$\frac{\mathrm{d}\nu_{\uparrow}(\varepsilon_i, \delta)}{\mathrm{d}\delta} = 0. \tag{2.36}$$

Using Equations (2.23), (2.34), and (2.35), we obtain that the hopping rate in Equation (2.35) has its maximum at

$$\delta = \varepsilon_i - 3\varepsilon_0 \ln \frac{3\varepsilon_0 (4\pi/3)^{1/3} N_0^{1/3} \alpha}{2kT}. \tag{2.37}$$

The second term in the right-hand side of Equation (2.37), but with the opposite sign, determines the energy level ε_t called, after Monroe [37], the transport energy:

$$\varepsilon_t = 3\varepsilon_0 \ln \frac{3\varepsilon_0 (4\pi/3)^{1/3} N_0^{1/3} \alpha}{2kT}. \tag{2.38}$$

We see from Equation (2.37) that the fastest hop occurs to a state with energy in the vicinity of the TE, independently of the initial energy ε_i, provided ε_i is deeper in the tail than ε_t, i.e., if $\delta \geq 0$. This result coincides with that of Monroe [37]. At low temperatures, the TE ε_t is situated deep in the band tail, while with rising temperature it moves upward towards the mobility edge. At some temperature T_c, the TE merges with the mobility edge. At higher temperatures, $T > T_c$, the hopping exchange of electrons between localized band-tail states becomes inefficient and the dynamic behavior of electrons is well accounted for by the multiple-trapping model described in Section 2.4.1. At low temperatures, $T < T_c$, the TE replaces the mobility edge in the multiple-trapping process [37], as shown in Figure 2.10. The width W of the maximum of the hopping rate is determined by the requirement that near ε_t the hopping rate, $\nu_\uparrow(\varepsilon_i, \delta)$, differs by less than a factor of e (the base of natural logarithms) from the value $\nu_\uparrow(\varepsilon_i, \varepsilon_i - \varepsilon_t)$. One finds [38]

$$W = \sqrt{6\varepsilon_0 kT}. \tag{2.39}$$

For shallow states with $\varepsilon_i \leq \varepsilon_t$ the fastest hop, on the average, is a downward hop to a nearest spatially localized state in the band tail with the rate determined by Equations (2.33) and (2.34). We recall that the energies of electron states are counted positive downward from the mobility edge towards the center of the mobility gap. This means that electrons in the shallow states with $\varepsilon_i \leq \varepsilon_t$ normally hop into deeper states with $\varepsilon > \varepsilon_i$, whereas electrons in the deep states with $\varepsilon_i > \varepsilon_t$ usually hop upward into states with energies near ε_t in the energy interval W, determined by Equation (2.39).

This shows that ε_t must play a crucial role in those phenomena which are determined by hopping of electrons in the band tails. Let us consider here for illustration the hopping

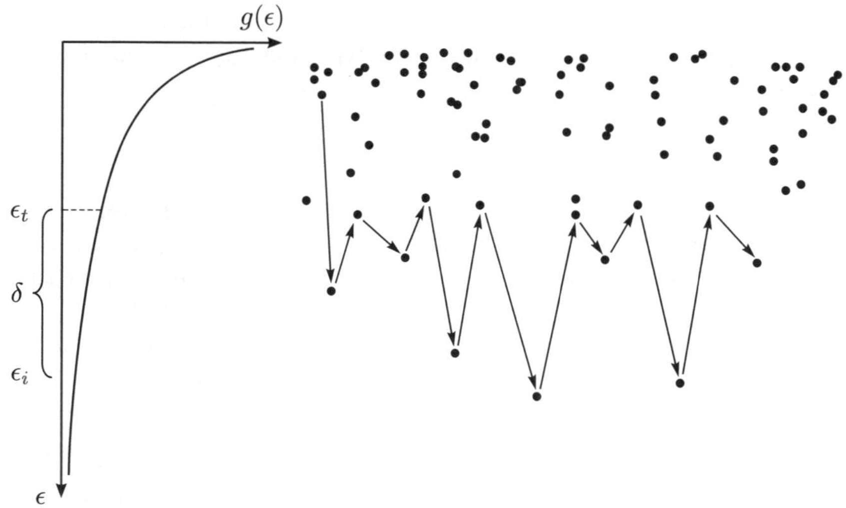

Figure 2.10 Hopping path via the transport energy. In the left frame, exponential DOS is shown schematically. The right frame depicts the transport path constructed from upward and downward hopes. The upward transitions bring a charge carrier to the sites with energies in the vicinity of the transport energy ε_t

energy relaxation of electrons in a system with the DOS described by Equation (2.23). This problem was studied initially by Monroe [37].

Consider an electron in some shallow localized energy state with the energy close to the mobility edge. Let the temperature be low, $T < T_c$, so that the TE, ε_t, lies well below the mobility edge, which we consider here as a reference energy $\varepsilon_c = 0$. The problem is to find the typical energy $\varepsilon_d(t)$ of the electron as a function of time, t. At early times, as long as $\varepsilon_d(t) < \varepsilon_t$, the relaxation is governed by Equations (2.33) and (2.34). The depth $\varepsilon_d(t)$ of an electron in the band tail is determined by the condition

$$\nu_{\downarrow}[\varepsilon_d(t)]\, t \simeq 1. \tag{2.40}$$

This leads to the double logarithmic dependence $\varepsilon_d \propto \varepsilon_0 \ln[\ln(\nu_0 t)] + C$ with constant C dependent on ε_0, N_0, and α. Indeed, Equations (2.33) and (2.40) prescribe the logarithmic form for the time dependence of the hopping distance $r(t)$ and Equations (2.23) and (2.34) lead to another logarithmic dependence $\varepsilon_d[r(t)]$ [37]. At the time

$$t_c \simeq \nu_0^{-1} \exp\left(\frac{3\varepsilon_0}{kT}\right) \tag{2.41}$$

the typical electron energy $\varepsilon_d(t)$ approaches the TE, ε_t, and the style of the relaxation process changes. At $t > t_c$, every second hop brings the electron into states in the vicinity of the TE, ε_t, from where it falls downward in energy to the nearest in space localized state. For the latter relaxation process the typical electron energy analogous to the demarcation energy in Equation (2.30) is determined by the condition [37]

$$\nu_{\uparrow}[\varepsilon_d(t), \varepsilon_t]\, t \simeq 1, \tag{2.42}$$

where $\nu_{\uparrow}[\varepsilon_d(t), \varepsilon_t]$ is the typical rate of electron hop upward from the demarcation energy towards the TE [37]. This condition leads to the typical energy position of the relaxing electron at time t determined as

$$\varepsilon_d(t) \simeq 3\varepsilon_0 \ln[\ln(\nu_0 t)] - 8\varepsilon_0/(N_0\alpha^3). \tag{2.43}$$

This is a very important result, showing that in a system with purely exponential energy dependence of the DOS, described by Equation (2.23), the typical energy of a set of independently relaxing electrons would continuously dive with time deeper and deeper into the mobility gap, as described in Section 2.4.1.

We would like to emphasize once more that one should be cautious with application of theoretical methods traditional for crystalline materials to the description of charge transport phenomena in disordered systems. For example, in some theoretical studies based on the Fokker–Planck equation, it has been claimed that the maximum of the energy distribution of electrons coincides with the TE, ε_t, and hence it is independent of time (see, for example [40]). Such statements are in contradiction to the above result with the maximum of the distribution at energy $\varepsilon_d(t)$ given by Equation (2.43). One should realize that the Fokker–Planck approach presumes a diffusive style of charge carrier energy relaxation. Hence, it is invalid for description of the energy relaxation in the exponential tails, in which electron can jump over the full energy range of the DOS (from a very deep energy state towards the TE) in a single hopping event [41].

2.5 EINSTEIN'S RELATIONSHIP FOR HOPPING ELECTRONS

2.5.1 Nonequilibrium charge carriers

In this section we would like to consider a very interesting problem related to the nonequilibrium energy relaxation of charge carriers in the band-tail states. It is well known that at low temperatures, $T \leq 50\,\mathrm{K}$, the photoconductivity in various inorganic noncrystalline materials, such as amorphous and microcrystalline semiconductors, does not depend on temperature [42–44]. At low temperatures, the TE, ε_t, lies very deep in the band tail and most electrons hop downward in energy. Concomitantly, the photoconductivity is a temperature-independent quantity determined by the energy-loss hopping of electrons via the band-tail states [45]. In such hopping relaxation, neither diffusion coefficient D, nor mobility of carriers μ, depends on temperature and the conventional form of the Einstein relationship $\mu = eD/kT$ cannot be valid. The question arises then as to what is the relation between μ and D for the energy-loss hopping relaxation. We answer this question below.

Let us start considering a system of nonequilibrium electrons in the band-tail states at $T = 0$. The only process that can happen with an electron is its hop downward in energy (upward hops are not possible at $T = 0$) to a nearest localized state in the tail. The rate of such process is described by Equations (2.33), (2.34), and (2.23). If the spatial distribution of localized tail states is isotropic, the probability of finding the nearest neighbor is also isotropic in the absence of the external electric field. In such a case, the process of the hopping energy relaxation of electrons leads to the spatial movement of electrons that resembles the diffusion in space. However, the median length of a hop (the distance r to the nearest neighbor available), as well as the median time $\tau = \nu_{\downarrow}^{-1}(r)$ of a hop (see Equation 2.33) increases in the course of relaxation, since the hopping process brings electrons deeper into the tail. Nevertheless one can ascribe to such a process a diffusion coefficient [45]

$$D(r) = \frac{1}{6}\nu_{\downarrow}(r)r^2. \tag{2.44}$$

Here $\nu_{\downarrow}(r)r^2$ replaces the product of the 'mean free path' r and the 'velocity' $r\nu_{\downarrow}(r)$ and the coefficient 1/6 accounts for the spatial symmetry of the problem. According to Equations (2.33), (2.34), (2.23), and (2.44), this diffusion coefficient decreases exponentially with increasing r and hence with the number of successive electron hops in the relaxation process.

In order to calculate the mobility of electrons in their energy-loss hopping relaxation under the influence of the electric field, one should take into account the spatial asymmetry of the hopping process due to the field [21, 45]. Let us consider an electron in a localized state at energy ε. If the external electric field with a strength F is applied along direction x, the concentration of tail states available for a hop of this electron at $T = 0$ (i.e., those, which have energies deeper in the tail than ε) is [45]

$$N(\varepsilon, x) = N(\varepsilon)\left(1 + \frac{eFx}{\varepsilon_0}\right), \tag{2.45}$$

where

$$N(\varepsilon) = \int_{\varepsilon}^{\infty} \mathrm{d}\varepsilon g(\varepsilon) = N_0 \exp\left(-\frac{\varepsilon}{\varepsilon_0}\right). \tag{2.46}$$

The strong inequality $eFx << \varepsilon_0$ was assumed in the derivation of Equation (2.45).

Due to the exponential dependence of the hopping rate on the hopping length r, the electron predominantly hops to the nearest tail state among available states if $r \gg \alpha$, which we assume to be valid. Let us calculate the average projection $\langle x \rangle$ on the field direction of the vector r from the initial state at energy ε to the nearest available neighbor among sites with concentration $N(\varepsilon, x)$ determined by Equation (2.45). Introducing spherical coordinates with the angle θ between r and x-axis, we obtain [21]

$$\begin{aligned}\langle x \rangle = &\int_0^{2\pi} \mathrm{d}\phi \int_0^{\pi} \mathrm{d}\theta \sin\theta \int_0^{\infty} [\mathrm{d}r r^3 \cos(\theta) N(\varepsilon, r, \cos\theta)] \\ &\times \exp\left[-\int_0^{2\pi} \mathrm{d}\phi \int_0^{\pi} \mathrm{d}\theta \sin\theta \int_0^{r} \mathrm{d}r' r'^2 N(\varepsilon, r', \cos\theta)\right].\end{aligned} \tag{2.47}$$

Substituting Equation (2.45) for $N(\varepsilon, r', \cos\theta)$, calculating the integrals in Equation (2.47) and omitting the second-order terms

$$\left(\frac{eN^{-1/3}(\varepsilon)F}{\varepsilon_0}\right)^2 \ll 1 \tag{2.48}$$

we obtain

$$\langle x \rangle = \frac{eFN^{-2/3}(\varepsilon)\Gamma(5/3)}{3\varepsilon_0(4\pi/3)^{2/3}}, \tag{2.49}$$

where $\Gamma(z) = \int_0^{\infty} \mathrm{d}y \exp(-y) y^{z-1}$ is the gamma-function and $N(\varepsilon)$ is determined by Equation (2.46). Equation (2.49) gives the average displacement in the field direction of an electron that hops downward from a state at energy ε to the nearest available neighbour in the band tail. The average length $\langle r \rangle$ of such a hop is

$$\langle r \rangle = \int_0^{\infty} \mathrm{d}r 4\pi N(\varepsilon) \exp\left[-\frac{4\pi}{3} N(\varepsilon) r^3\right] = \Gamma(4/3)\left[\frac{4\pi N(\varepsilon)}{3}\right]^{-1/3}. \tag{2.50}$$

One can ascribe to the hopping process the mobility

$$\mu = \frac{v}{F} = \frac{\langle x \rangle \nu(\langle r \rangle)}{F} = \frac{\Gamma(5/3) e N^{-2/3}(\varepsilon) \nu(\langle r \rangle)}{3(4\pi/3)^{2/3} \varepsilon_0} \tag{2.51}$$

and the diffusion coefficient

$$D = \frac{1}{6}\langle r\rangle^2 \nu(\langle r\rangle) = \frac{1}{6}N^{-2/3}(\varepsilon)\nu(\langle r\rangle)\frac{\Gamma^2(4/3)}{(4\pi/3)^{2/3}}. \tag{2.52}$$

Expressions (2.51) and (2.52) lead to the relationship between μ and D in the form

$$\mu = \frac{2\Gamma(5/3)e}{\Gamma^2(4/3)\varepsilon_0}D \simeq 2.3\frac{e}{\varepsilon_0}D. \tag{2.53}$$

This formula replaces the Einstein relationship $\mu = eD/kT$ in the case of the energy-loss hopping of electrons in the exponential band tail. One should realize that Equation (2.53) was derived in the linear regime with respect to the applied field under the assumption that $eFx \ll \varepsilon_0$. According to Equation (2.49), the quantity $\langle x\rangle$ is proportional to $N^{-2/3}(\varepsilon) = N_0^{-2/3}$ $\exp[2\varepsilon/(3\varepsilon_0)]$, i.e., it increases exponentially in the course of the relaxation towards larger localization energies ε. This means that, for deep localized states in the band tail, the condition $eFx \ll \varepsilon_0$ breaks down. The energy border for application of the linear theory depends on the strength of the electric field, F. The smaller the value of F, the deeper in the tail is this border. However, for any F, there is always a border energy in the tail below which the condition $eFx \ll \varepsilon_0$ can no longer be fulfilled and nonlinear effects play the decisive role for hopping conduction of charge carriers. In Section 2.9 we show how one can describe the nonlinear effects with respect to the applied electric field.

The diffusivity-to-mobility ratio for the nonequilibrium transport in amorphous semiconductors has been tested experimentally by Gu *et al.* [46]. It was claimed that the hole diffusion coefficient in the experimental study is not more than twice as large as predicted by the classical Einstein formula $D = kT\mu/e$, if one uses the experimental value of the mobility μ. This result was considered as the upper bound for any true failure of the Einstein relation. It is worth noting, however, that the measurements of the drift mobility were carried out by Gu *et al.* for temperatures above 200 K, while Equation (2.53) has been derived for the limit of the infinitesimal temperature [21]. Computer simulations [47] show that, at such high temperatures, the relation between μ and D differs very little from the classical Einstein formula. Therefore it is not surprising that no essential deviations from this formula were manifested in the experiments of Gu *et al.* [46]. In order to check any considerable deviation from the conventional Einstein formula caused by nonequilibrium conditions, one should perform experiments at much lower temperatures $T < 50$ K, at which transport of charge carriers is dominated by the energy-loss hopping.

2.5.2 Equilibrium charge carriers

One should clearly realize that Equation (2.53) is valid for the nonequilibrium energy-loss relaxation, in which only downward in energy electron transitions between localized states can occur. This regime is valid at low temperatures, at which the transport energy ε_t is very deep in the band tail. With increasing temperature, the upward in energy hops become more and more effcient for electron relaxation. Under these circumstances, the relation between μ and D evolves gradually with rising temperature from its temperature-independent form at $T = 0$ to the conventional Einstein relationship, $\mu = eD/kT$ [47] at $kT > \varepsilon_0$.

It is, however, worth noting that, even in the case of thermal equilibrium, the Einstein formula in its classical form $\mu = eD/kT$ is valid only for nondegenerate systems, in which most carriers are distributed in the states with energies much higher than the Fermi level ε_F. In such a case, the Fermi distribution can be approximated by the Boltzmann function, for which one easily obtains the relation between the drift mobility and the diffusion constant in the classical form $\mu = eD/kT$ [48]. In the opposite case of the degenerate system, a generalized Einstein formula should be used [49, 50]

$$\mu = eD\frac{1}{n}\frac{\partial n}{\partial \varepsilon_F}, \tag{2.54}$$

where n is the total concentration of charge carriers that determines the position of the Fermi level ε_F.

Ritter *et al.* [51] were the first who applied Equation (2.54) to the degenerate system of charge carriers distributed in the exponential DOS, and obtained a remarkable result that the Einstein formula in such a systems reads as

$$\mu = \frac{e}{\varepsilon_0}D. \tag{2.55}$$

We would like to emphasize that this formula is valid for the system of charge carriers in thermal equilibrium, while Equation (2.53) describes the relation between μ and D for essentially nonequilibrium process of the energy-loss hopping.

Whether the equilibrium system of charge carriers in the exponential DOS should be considered as degenerate or a nondegenerate depends solely on the relation between the energy scale of the DOS function ε_0 and the thermal energy kT. Remarkably, it does not depend on the concentration of charge carriers n. Straightforward calculations [51, 52] and computer simulations [47] show that at $kT < \varepsilon_0$, the relation between μ and D is described by Equation (2.55) and the system of equilibrium carriers should be considered as a degenerate one. In the case $kT > \varepsilon_0$, the equilibrium system of carriers is always nondegenerate and the relation between μ and D is described by the classical Einstein formula $\mu = eD/kT$.

2.6 STEADY-STATE PHOTOCONDUCTIVITY

Specific photoconducting properties of amorphous semiconductors, in particular of a-Si : H, are responsible for device applications of amorphous materials in solar cells and in field transistors, and therefore these properties are most often studied [9]. The literature with experimental data on the photoconductivity (PC) of amorphous semiconductors is enormously rich. The phenomenon has been studied for more than 30 years. Nevertheless, the theoretical interpretation of the effect is still a subject of controversy. In Figure 2.11 a temperature dependence of the steady-state PC $\sigma_p(T)$ in a-Si : H is shown [43]. This shape of the $\sigma_p(T)$ curve has been well known for 20 years [42]. Moreover, almost all different amorphous materials, including a-BCx : H, a-B : H, a-Se, alloy glasses, a-Si, a-Ge, intrinsic a-Si : H as well as *p*-doped and *n*-doped a-Si : H demonstrate such a dependence, as shown in Figure 2.12 [42, 43, 53–55]. Being common for different amorphous materials, this

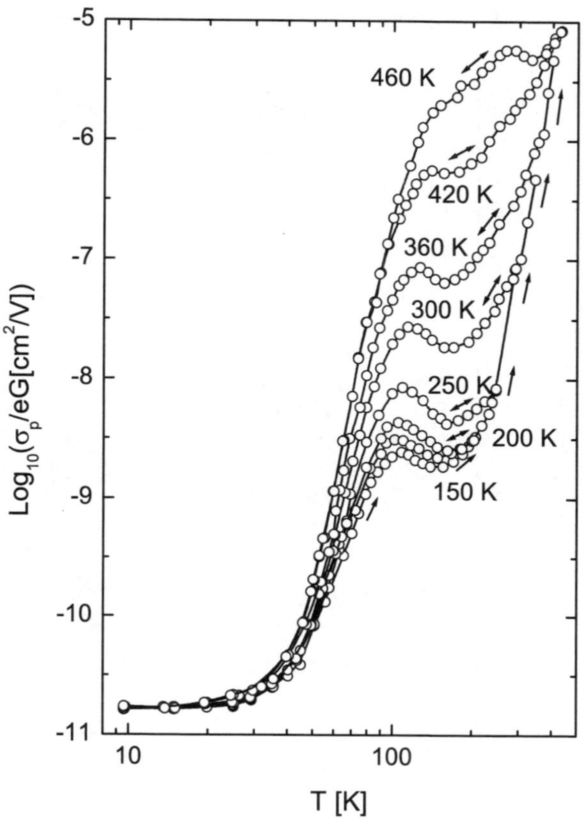

Figure 2.11 Temperature dependences of the normalized photoconductivity measured with the generation rate $G = 5.3 \times 10^{17}\,cm^{-3}s^{-1}$. The samples were cycled to successively higher temperatures 150, 200, 250, 300, 360, 420, and 460 K as indicated. Arrows show direction of temperature change (reproduced from [43] with permission from Taylor & Francis Ltd.)

dependence $\sigma_p(T)$ needs a general theoretical interpretation. Experiments show that transport of photoexcited carriers at low temperatures is fundamentally different from that near and above room temperature. Below 50 K the PC becomes temperature independent and it is essentially independent of the doping level and of the concentration of defects [42, 53, 54] in contrast to the strong dependence on the material parameters and on temperature above ~100 K [9]. In Section 2.6.1 we describe a theory of the PC in a-Si : H at temperatures $T < 50$ K, at which the PC is temperature independent. In Section 2.6.2 we address the steep rise of the PC at higher temperatures.

2.6.1 Low-temperature photoconductivity

The relatively simple and universal behavior of the PC at low temperatures suggests that the conduction process is related to very general features of amorphous semiconductors.

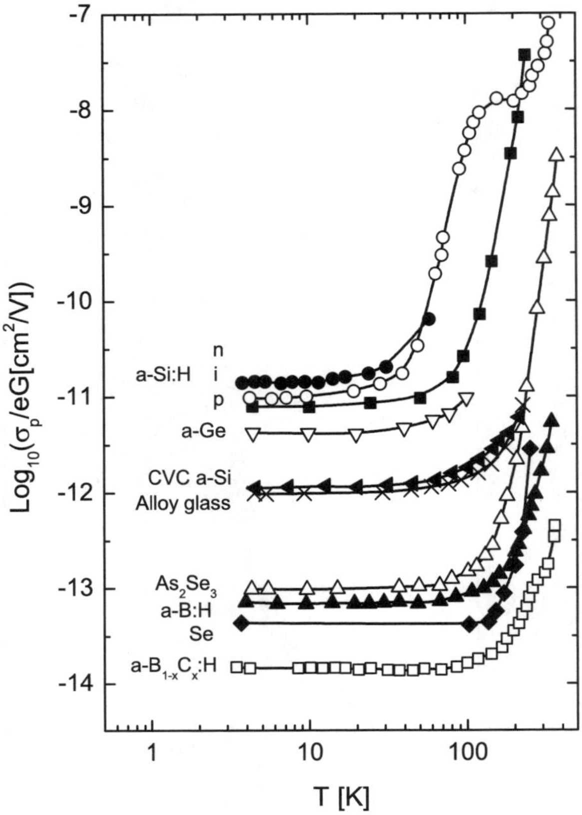

Figure 2.12 Temperature dependence of the normalized photoconductivity for several amorphous semiconductors, illustrating the universality of the shape of the $\sigma_p(T)$ curves. The alloy glass is sputtered film of the composition $As_{35}Te_{28}S_{21}Ge_{15}S_1$ (reproduced from [54] with permission from Elsevier)

Hoheisel *et al.* [42] suggested that the low-T photoconductivity $\sigma_p(0)$ is due to photoexcited carriers passing through the extended states before they become trapped into localized band-tail states. With the mobility in the extended states about $10\,cm^2V^{-1}s^{-1}$ and a capture time of free carriers $\tau \simeq 10^{-13}$ s the authors found good agreement with their experimental value $\sigma_p/eG \approx 10^{-12}\,cm^2V^{-1}s^{-1}$. Here the photoconductivity σ_p is normalized by the electron charge and by the bulk generation rate G. It is difficult, however, to explain with this model the observation of Vanecek *et al.* [56] that σ_p/eG is independent of the photon energy $\hbar\omega$ above the mobility gap energy and decreases only slowly when $\hbar\omega$ is decreased to values smaller than the gap. Moreover, σ_p begins to rise with temperature at temperatures sufficiently low, that reexcitation of charge carriers from tail states into extended states cannot play an essential role. Furthermore, time-of-flight measurements of the carrier mobility indicate that, at low temperatures, the mobility is dominated by hopping processes between localized states in the band tails [57–60].

In order to understand electronic properties of a system under steady-state generation, one should first find the appropriate recombination mechanism that provides the balance to the continuous photogeneration of charge carriers. Recombination between electrons and holes can be either radiative or nonradiative. Numerous studies of the PC and of the photoluminescence (PL) in a-Si : H with concentration of dangling bonds below $10^{16}\,\mathrm{cm}^{-3}$ revealed that, at least at low temperatures, recombination is dominated by radiative processes [61–65]. Therefore we describe below the theory of the PC at low temperatures, taking into account the radiative recombination mechanism [38, 45, 66].

As first mentioned by Hoheisel *et al.* [42], electrons and holes that recombine geminately do not contribute to the steady-state PC. We recall that in geminate recombination, electron recombines with the hole created with this electron in the same generation event. Therefore, in order to calculate the steady-state PC, one should focus on the carriers recombining nongeminately. Nongeminate recombination arises when the electron–hole pair succeeds in separating to a distance $\sim\frac{1}{2}n^{-1/3}$, where n is the steady-state concentration of electrons (or holes) under the chosen generation intensity G. Hence the first step in the calculation of the PC is to determine n as a function of G.

Let us consider a fate of a geminate electron–hole pair created by a photon with energy close to the width of the mobility gap. Electron and hole are quickly captured into localized states in the band tails. From these states they can either recombine with the rate

$$\nu_R(R) = \tau_0^{-1}\exp\left(-\frac{2R}{\alpha}\right), \tag{2.56}$$

or make a jump into localized states with lower energies (upward hops are not possible at low temperatures). For simplicity we consider hopping of electrons only, assuming holes immobile. An electron makes a hop to a localized state at distance r with the rate

$$\nu_{\downarrow}(r) = \nu_0\exp\left(-\frac{2r}{\alpha}\right). \tag{2.57}$$

In Equations (2.56) and (2.57) R is the electron–hole separation and α is the localization length of the electron. The preexponential factor in Equation (2.56) is of the order of the typical dipole radiative rate $\tau_0^{-1} \simeq 10^8\,\mathrm{s}^{-1}$, while the preexponential factor for nonradiative transition in Equation (2.57) is of the order of the phonon frequency $\nu_0 \approx 10^{13}\,\mathrm{s}^{-1}$. As shown by Shklovskii *et al.* [38, 45, 66], the huge difference in the magnitudes between ν_0 and τ_0^{-1} introduces a characteristic length scale

$$R_c = \frac{\alpha}{2}\ln(\nu_0\tau_0). \tag{2.58}$$

At electron–hole separations r smaller than R_c, recombination is improbable since $\nu_R(R) << \nu_{\downarrow}(r)$ due to the large difference in the preexponential factors. At separations r larger than R_c, recombination can compete with hopping diffusion and the probability of an electron–hole pair avoiding radiative recombination up to separations $r >> R_c$ is determined by the universal function [38, 45, 66]

$$\eta(r) = \xi\left(\frac{R_c}{r}\right)^{\beta} \tag{2.59}$$

with $\xi = 3.0 \pm 0.1$ and the critical index $\beta = 1.16 \pm 0.01$ [38]. Below, we use the approximate value $\beta \simeq 1$.

The steady-state concentration of electrons n at low temperatures is determined by the equation

$$G\eta\left(\frac{n^{-1/3}}{2}\right) = \frac{n}{\tau_0}\exp\left(-\frac{n^{-1/3}}{\alpha}\right) \tag{2.60}$$

with the solution

$$n(G) = [\alpha L(G)]^{-3}, \tag{2.61}$$

where quantity $L(G)$ satisfies the equation

$$L = \ln\left\{\left[\tau_0 G \alpha^3 L^2 \ln(\tau_0 \nu_0)\right]^{-1}\right\}. \tag{2.62}$$

The left-hand side of Equation (2.60) represents the generation rate of electrons, which survive through hopping to the distance $\frac{1}{2}n^{-1/3}$ and therefore avoid the geminate recombination. The right-hand side is the recombination rate of such electrons. Equations (2.61) and (2.62) determine the dependence of the steady-state concentration of electrons on the generation rate. This dependence is to be compared with the experimental one obtained by measurements of the light-induced electron spin resonance. The latter show the dependence $n(G) \propto G^{\delta}$ with $\delta \simeq 0.2$ for generation rates in the range $10^{16} < G < 10^{21}\,\text{cm}^{-3}\text{s}^{-1}$ [67]. This result is in good agreement with the result of Equations (2.61) and (2.62) for parameters $\alpha \simeq 1\,\text{nm}$ and $\nu_0\tau_0 \simeq 10^4$, usually assumed for a-Si:H [11].

These arguments provide the solution of the apparent contradiction between the geminate character of the photoluminescence and nongeminate character of the electron spin resonance signal at $G < 10^{19}\,\text{cm}^{-3}\text{s}^{-1}$ [9]. As shown in [38, 45, 66] the photoluminescence is determined by the majority of generated electron–hole pairs recombining at distances of the order of R_c, determined by Equation (2.58). Only a small proportion of electron–hole pairs generated per second in a cm^3 survive to distances $n^{-1/3} \gg R_c$, and hence recombine nongeminately, as described by the survival probability $\eta(n^{-1/3})$ in Equation (2.59). However, the lifetime $\tau(n^{-1/3})$ of these nongeminate pairs is so large that they accumulate in the system and determine the electron spin resonance signal. In order to calculate the magnitude of the photocurrent density j, one should multiply the generation rate of electron–hole pairs which recombine nongeminately (the expression in the left-hand side of Equation 2.60) with the typical field-induced dipole moment along the field direction d attained by a nongeminate pair before it recombines. The latter quantity depends on the strength of the applied electric field F and on the energy scale of the DOS ε_0. Straightforward calculations [38, 45, 66] give

$$d(F) = n^{-1/3}\frac{e^2 F n^{-1/3}}{12\varepsilon_0} \tag{2.63}$$

leading to the expression for the photoconductivity $\sigma_p = j/F$ in the form

$$\sigma_{\mathrm{p}} = G\eta\left(\frac{n^{-1/3}}{2}\right)\frac{e^2 F n^{-1/3}}{12\varepsilon_0}. \tag{2.64}$$

Using the solution of Equation (2.60) one obtains

$$\sigma_{\mathrm{p}} = G\frac{e^2\alpha^2}{12\varepsilon_0}\ln(\nu_0\tau_0)L. \tag{2.65}$$

Since L depends logarithmically on the generation rate G, we find from Equations (2.65) and (2.62)

$$\sigma_{\mathrm{p}} \propto G^y \tag{2.66}$$

with

$$y = 1 - \frac{1}{L}. \tag{2.67}$$

Equations (2.65) and (2.67) give the values $\sigma_{\mathrm{p}}/eG = 4 \times 10^{-12}\,\mathrm{cm^2/V}$ and $\gamma = 0.93$ at $G = 10^{20}\,\mathrm{cm^{-3}s^{-1}}$ for the reasonable choice of material parameters $\alpha \simeq 1\,\mathrm{nm}$, $\nu_0\tau_0 \simeq 10^4$, and $\varepsilon_0 = 0.025\,\mathrm{eV}$ [38]. Experimentally one finds $\sigma_{\mathrm{p}}/eG \simeq 10^{-11}\,\mathrm{cm^2/V}$ and $\gamma = 0.95 \pm 0.02$ [42, 68]. So far we have considered the low-temperature PC in an intrinsic amorphous semiconductor. The detailed theory of the low-temperature PC in doped amorphous semiconductors can be found in [69]. In the next section we discuss the temperature dependance of the PC.

2.6.2 Temperature dependence of the photoconductivity

Let us first estimate the temperature $\tilde{T}$, at which photoconductivity in amorphous semiconductors becomes dependent on temperature, and above which the picture of the energy-loss hopping is no longer valid. This happens when the transport energy $\varepsilon_t(T)$ (TE) crosses the low-temperature quasi-Fermi level $\varepsilon_{\mathrm{F}}^0$. The latter quantity can be also considered as the typical recombination energy of electrons in the energy-loss hopping, which for the exponential DOS can be determined as [38]

$$\varepsilon_{\mathrm{F}}^0 = \varepsilon_0 \ln\left(\frac{N_0}{n}\right), \tag{2.68}$$

where n is the steady-state concentration of electrons. The energy-loss hopping described in Section 2.6.1 dominates the transport and recombination of carriers at such low temperatures that the TE is deeper in the tail than the quasi-Fermi level $\varepsilon_{\mathrm{F}}^0$. With rising temperature, the TE moves upward into more shallow states. Equating $\varepsilon_{\mathrm{F}}^0$ determined by Equation (2.68) to $\varepsilon_t(T)$ determined by Equation (2.38) one obtains

$$\tilde{T} = \frac{3\varepsilon_0}{2k}\left(\frac{4\pi}{3}n\alpha^3\right)^{1/3}. \tag{2.69}$$

Substituting Equation (2.61) into this expression one finds [38]

$$\tilde{T} = \frac{3\varepsilon_0}{2k}\left(\frac{4\pi}{3}\right)^{1/3}\frac{1}{L(G)}. \tag{2.70}$$

Using the values of L from Section 2.6.1, one obtains $\tilde{T} \simeq 45\,\mathrm{K}$ for $G \simeq 10^{20}\,\mathrm{cm}^{-3}\mathrm{s}^{-1}$. This value of $\tilde{T}$ agrees well with the experimental data that show the PC almost temperature independent at temperatures below approximately 50 K. Since L depends on the generation rate only logarithmically, the value of $\tilde{T}$ is not very sensitive to the magnitude of the generation rate.

At temperatures higher than $\tilde{T}$, the TE lies above the low-temperature recombination level ε_F^0, and the picture of the energy-loss hopping described in Section 2.6.1 is no longer valid for calculations of the PC. At such temperatures the upward-energy hops of charge carriers towards the vicinity of the TE determine both transport and recombination. The calculations of the PC at such elevated temperatures are described in detail in [38]. In these calculations one first derives the balance equation that determines the steady-state concentration of charge carriers n under continuous generation with excitation density G, in which recombination is due to the thermally stimulated diffusion of electrons via states in the vicinity of the TE. Possessing the value of the concentration, one determines the temperature-dependent quasi Fermi level ε_F for the chosen DOS determined by Equation (2.23). Then it is easy to calculate the magnitude of the PC, $\sigma_p(T, G)$, via Equations (2.3)–(2.5) using the value of the quasi-Fermi level for ε_F and the easily calculated value of the mobility via the localized states in the vicinity of the TE [38]. The results of such calculations describe the steep increase of the PC at temperatures above $\tilde{T}$ seen in Figures 2.11 and 2.12. The interested reader can find all necessary details of the theory in [38]. Here we would like to discuss only one interesting phenomenon not considered in [38].

In Figure 2.11 it is clearly seen that the steep increase of the PC with temperature that begins at $T \simeq 50\,\mathrm{K}$ is followed by the decrease of the PC at $T \simeq 100\,\mathrm{K}$. This is a rather universal behavior for the PC in amorphous semiconductors, which can be easily interpreted in the following picture. Equations (2.69) and (2.70) show that the temperature $\tilde{T}$, at which the slow energy-loss hopping is replaced by the much faster motion of charge carriers via the states in the vicinity of the TE, is proportional to the energy scale of the band tail ε_0. Since this scale for the valence band in a-Si:H is approximately twice as large as that for the conduction band [9], holes become mobile at temperature $\tilde{T}_h$ approximately twice as large as the corresponding temperature $\tilde{T}_e$ for electrons. Therefore the general picture of the temperature-dependent PC in amorphous semiconductors is as follows. At temperatures below $\tilde{T}_e \simeq 50\,\mathrm{K}$ the PC is almost temperature independent, being determined by the slow energy-loss hopping and recombination of charge carriers. At $T > \tilde{T}_e \simeq 50\,\mathrm{K}$, but with $T < \tilde{T}_h \simeq 100\,\mathrm{K}$, electrons become more mobile and their thermal activation towards localized states in the vicinity of the TE (for electrons) strongly enhances the PC due to the enormous increase of the electron mobility that overcompensates the decrease in their concentration due to the enhanced recombination compared with the low-temperature case.

At temperature $T \simeq \tilde{T}_h \simeq 100\,K$, holes also become much more mobile compared with their mobilities in the energy-loss hopping regime at low temperatures. These mobile holes can drastically enhance the recombination efficiency, which leads to the drastic decrease of the steady-state carrier concentration at $T \geq \tilde{T}_h \simeq 100\,K$ and concomitantly to the decrease of the PC. At higher temperatures the TE merges with the mobility edge and the description based on the transport energy concept is no longer valid. At $T \geq 200\,K$ the increase of the PC with temperature can be well described by analogy with the same behavior in crystalline semiconductors, since the PC at such high temperatures is determined by transport via delocalized conduction band states above the mobility edge. This general picture of the PC is valid for various amorphous semiconductors and it can be used for interpretation of numerous experimental data [42, 43, 53–55].

2.7 THERMALLY STIMULATED CURRENTS—A TOOL TO DETERMINE DOS?

Measurements of the thermally stimulated conductivity (TSC) belong to the routine methods to study the electronic properties of semiconductors. In these measurements a sample is cooled to a low temperature T_0, then photoexcited for a time period t_e and, after a delay period t_0, the sample is heated the dark at a constant rate γ while the thermostimulated current is measured. The TSC technique has attracted increasing attention in studies of disordered semiconductors, in the hope of obtaining important information on the energy distribution of the density of states (DOS) in the mobility gap. It is worth noting, however, that the information on the DOS extracted from the TSC measurements essentially depends on the interpretation of the TSC phenomenon.

The first theory of the TSC in a system with a continuous energy distribution of localized states below the band edge was suggested by Simmons *et al.* [70]. In this theory only carriers in the extended states above the mobility edge were considered as mobile, and it was assumed that the carriers thermally released from the traps are swept out of the sample before they can recombine. Such a theory might be valid, perhaps at very high electric fields, at which carriers are swept out very efficiently and also at high temperatures at which hopping via localized band-tail states can be neglected. However, measurements of the TSC are usually carried out at low fields and at low temperatures. In such conditions hopping and recombination processes cannot be neglected. The theory of the TSC taking at least recombination into account was suggested by Fritzsche and Ibaraki [71]. This theory is still widely used to interpret modern experiments. Fritzsche and Ibaraki assumed that after a long delay t_0 (about 1 h) the distribution of photoexcited carriers in the trapping states corresponds to the thermal equilibrium at temperature T_0. Only carriers in the extended states above the mobility edge were considered as contributing to the current. The TSC was assumed to arise from a balance between the thermal emission and recombination. As the sample is heated in the dark, the thermal emission occurs from progressively lower-lying traps, and the magnitude of the TSC σ_{TSC} is therefore expected to be proportional to the product of the DOS $g(\varepsilon)$ and the occupation probability $f_0(\varepsilon, T_0)$ at the end of the delay time [71]

$$\sigma_{TSC} \propto e\mu_c \tau_0 g(\varepsilon_m) f_0(\varepsilon_m, T_0), \tag{2.71}$$

where e is the electron charge, $\mu_c\tau_0$ is the free electron mobility–lifetime product and ε_m is the time-dependent energy, from which the thermal emission is most efficient. This energy moves towards the middle of the mobility gap with time and temperature [72]. The crucial assumptions of this approach are the following: (i) the retrapping processes were not taken into account at all; (ii) the exact balance between the thermal emission and recombination was assumed.

Generally the TSC is determined by the interplay between the thermal emission of electrons from the traps into conducting states and their retrapping and recombination. The concentration of the conducting electrons n obeys the equation

$$\frac{\mathrm{d}n}{\mathrm{d}t} = \frac{\mathrm{d}n_+}{\mathrm{d}t} - \frac{\mathrm{d}n_-}{\mathrm{d}t} - \frac{n}{\tau_0}, \tag{2.72}$$

where $\mathrm{d}n_+/\mathrm{d}t$ and $\mathrm{d}n_-/\mathrm{d}t$ are the rates of the increase of n due to thermal release of electrons from the traps and its decrease due to retrapping, respectively, and τ_0 is the lifetime of the conducting electrons that depends on n, T and also on the whole concentration of trapped carriers in the system. Fritzsche and Ibaraki [71] neglected the term $\mathrm{d}n_-/\mathrm{d}t$ and they assumed the 'steady state' $\mathrm{d}n/\mathrm{d}t = 0$, to be valid, which automatically leads to Equation (2.71).

Predictions of this approach were verified experimentally. Zhou and Elliott [73] were the first to claim inconsistencies between this theory and experimental results at low temperatures. Analyzing the similarity between the temperature dependence of the TSC and that of the PC, Zhou and Elliott came to the conclusion that the TSC is controlled by recombination rather than by thermal emission of trapped electrons. Under such circumstances Equation (2.71) is not valid. Besides this inconsistency, it is not possible to justify the neglect of the retrapping processes in the TSC. Even at room temperature the retrapping time is by many orders of magnitude smaller than the recombination time [74]. At lower temperatures at which the TSC is usually measured the relation between these times in favor of the fast retrapping should be even more drastic. Therefore the retrapping processes must be undoubtedly taken into account in the theoretical interpretation of the TSC.

A theory of the TSC taking into account trapping, retrapping and recombination has been suggested by Gu *et al.* [75], but under the assumption that only delocalized carriers in the states above the mobility edge are mobile. This is correct at rather high temperatures while the TSC is usually studied at low temperatures, at which transport is via hopping of electrons in the band tails. A comprehensive theory of the TSC in the hopping regime taking into account trapping, retrapping and recombination was suggested in [76]. This theory is based on the concept of the transport energy (TE) introduced in Section 2.4.2. We will not present this theory here in detail. Instead we draw below a qualitative picture of the TSC in the hopping regime.

Thermally stimulated conductivity σ_{TSC} is determined by the expression

$$\sigma_{\mathrm{TSC}} = e\mu(T)n(T), \tag{2.73}$$

where $n(T)$ is the concentration of mobile electrons in the states with energies in the vicinity of the TE and $\mu(T)$ is their hopping mobility. The mobility of electrons in the vicinity of the TE sharpy increases with T, since the TE moves with rising temperature towards the energy states with higher DOS, making the tunneling of electrons between localized states much more efficient. In the correct description, the low-temperature TSC is determined by the interplay between the hopping mobility $\mu(T)$ increasing with T and the concentration of carriers in the vicinity of the TE, decreasing in the course of time due to recombination and also due to the higher energy distance between the Fermi level and the TE with rising temperature. There is no necessity to consider the steady-state condition $dn/dt = 0$.

In Figure 2.13 typical temperature dependences of the TSC are shown for different initial temperatures T_0, which were obtained at the heating rate $\gamma = 2$ K/min. For $T_0 < 65$ K a pronounced maximum at $T_m \simeq 90$ K is observed, which is independent of T_0. Such curves can be found in numerous publications [73, 76, 77], albeit with different values of T_m, always independent of T_0. In order to get a better insight into the physics behind these results, let us compare the values of σ_{TSC} at $T \simeq 55$ K in Figure 2.13 for $T_0 = 20$ K and $T_0 = 50$ K. In the latter case $\sigma_{TSC}(55\,K)$ is much lower than in the former. The mobility determined just by the actual temperature ($T \simeq 55$ K) is the same in both cases. Therefore, in accord with Equation (2.73), the only reason for the difference in the values of $\sigma_{TSC}(55\,K)$ can be the difference in the values of $n(T)$ in these two cases. As long as the light intensity G and the excitation duration t_0 were the same [76], the total amount of generated carriers was also the same in these two cases. Hence the only reason why there are more carriers at $T = 55$ K in the case of $T_0 = 20$ K than in the case of $T_0 = 50$ K is that the recombination process at $T_0 = 50$ K is more efficient than at $T_0 = 20$ K. Therefore, after the delay period $t_0 = 30$ min, which was the same in both cases [76], there were many more electrons at the start of heating at $T_0 = 20$ K than at $T_0 = 50$ K. During the slow heating process the carriers continue to recombine and their number disappearing from the system per second is higher for higher total amount of carriers, i.e., recombination at $T = 55$ K is more intensive in the case of $T_0 = 20$ K than in the case of $T_0 = 50$ K. Therefore, the TSC curve increases more steeply at $T_0 = 50$ K than at $T_0 = 20$ K since the concentration of mobile carriers $n(T)$ decreases less at $T_0 = 50$ K than at $T_0 = 20$ K, while the mobility increase with temperature in both cases is the same. After the curve of σ_{TSC} for $T_0 = 50$ K merges into that for $T_0 = 20$ K there is no longer any difference between these two curves. Curves for higher values of T_0 merge into

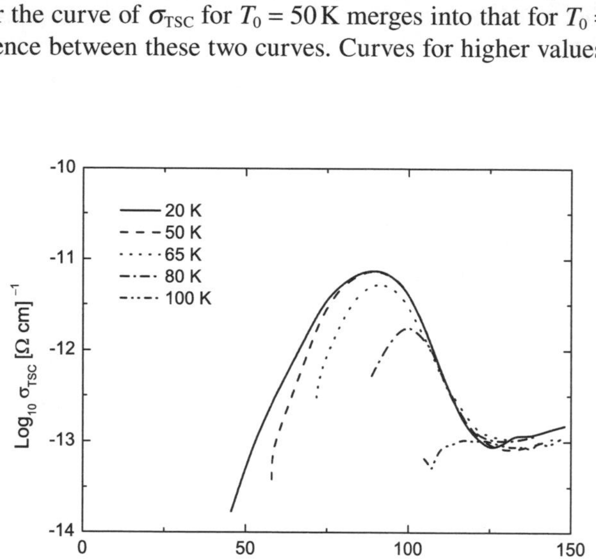

Figure 2.13 The TSC curves for a-Si:H with different starting temperatures T_0 in the range 20–100 K (reproduced with permission from [76]; Copyright 1997 by the American Physical Society)

the curve for $T_0 = 20\,\text{K}$ at higher temperatures, respectively. It is clearly seen in Figure 2.13 that there is some universal TSC curve (for given values of G, t_e, and t_0) and all data for different values T_0 merge into it. The higher is T_0 the lower is the starting concentration of carriers, because more of them recombine within the delay period t_0 at higher T_0. On the other hand, recombination is less efficient for lower concentration of carriers. The carrier mobility, depending only on T, does not depend on the starting temperature T_0. Therefore, the increase of σ_{TSC} with T is always steeper for higher T_0, in good agreement with experimental data (Figure 2.13). In the frequently used steady-state description of Fritzsche and Ibaraki [71] the tendency should be just the opposite: the initial rise of σ_{TSC} with T should be steeper for lower T_0, being determined by the function f_0 at $T = T_0$ in Equation (2.71). The maximum in the σ_{TSC} dependence in our picture is determined by the interplay between the mobility of charge carriers rising with temperature and their concentration decreasing due to recombination in the course of time. In order to check this conclusion, measurements of the TSC were carried out at different heating rates γ. According to our picture, the maximum should shift towards higher temperatures, and it should increase in the amplitude at higher heating rates, since recombination is less pronounced at shorter times needed to achieve the given value of T at larger γ. Experimental data presented in Figure 2.14 confirm the expected trend. They also clearly show that the temperature corresponding to the maximum of the TSC curve σ_{TSC} and the amplitude of the maximum itself are determined by the experimental conditions (for example, by the heating rate γ) and they should not be interpreted as a direct indication of some particular structure of the DOS of localized states in the band tails of disordered semiconductors. Therefore, we conclude that the TSC measurements should be interpreted with much more caution than is sometimes the case in the scientific literature. The DOS function does play an essential role in the calculation of the TSC. However, in order to make any conclusions on this important function, one should use the appropriate theory and compare its results with experimental data instead of claiming the shape of the DOS directly from the experimental curve using unjustified expressions such as Equation (2.71).

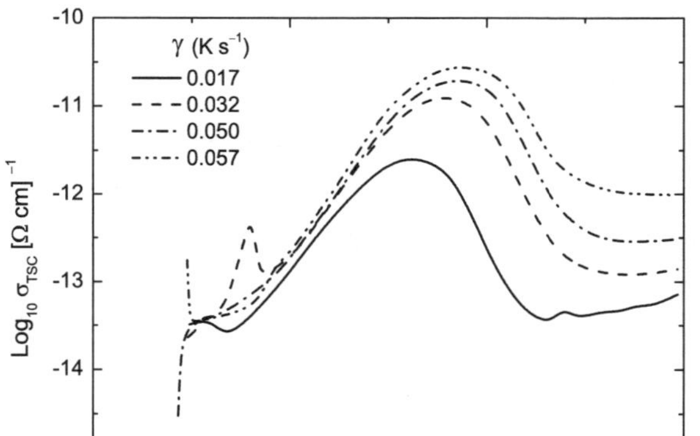

Figure 2.14 The TSC spectra in a-Si:H for different heating rates γ (reproduced with permission from [76]; Copyright 1997 by the American Physical Society)

2.8 DARK CONDUCTIVITY IN AMORPHOUS SEMICONDUCTORS

As discussed in Section 2.2, the experimentally observed dark conductivity $\sigma_d(T)$ in amorphous semiconductors at temperatures around room temperature can be well described by the activated Arrhenius law (Equation 2.1 with $\beta = 1$). The interpretation of this dependence is similar to that in crystalline semiconductors. Charge carriers are activated from the Fermi energy to the delocalized states above the mobility edge, where they move as quasi-free particles with some effective mass. The conductivity in such a case is well described by Equations (2.6) and (2.7). More interesting and complicated for description from the theoretical point of view is the behavior of the dark conductivity at lower temperatures, at which conduction is dominated by hopping transitions of charge carriers between the localized band-tail states. This transport process is a variable-range hopping (VRH) in the localized states with an exponential energy distribution. One way to calculate $\sigma_d(T)$ is to use the general Equation (2.3) and the concept of the transport energy (TE) described in Section 2.4.2. One can easily show that integrand in Equation (2.3) with concentration of charge carriers $n(\varepsilon)$ described by Equations (2.4), (2.5), and (2.23), and the hopping mobility $\mu(\varepsilon)$ approximated by Equation (2.51), has a sharp maximum at the energy ε equal to the TE determined by Equation (2.38). In such an approach one easily arrives at the expression for $\sigma_d(T)$ in the form [78]

$$\sigma_d = \frac{e^2 \nu_0 r_t^2}{kT} \exp\left(-\frac{2r_t}{\alpha}\right) \int_{-\infty}^{\varepsilon_t} \frac{d\varepsilon g(\varepsilon)}{1+\exp[(\varepsilon-\varepsilon_F)/kT]} \exp\left(\frac{\varepsilon-\varepsilon_t}{kT}\right) + \frac{Ce^2\nu_0}{\varepsilon_0} \int_{\varepsilon_t}^{0} \frac{d\varepsilon g(\varepsilon)}{1+\exp[(\varepsilon-\varepsilon_F)/kT]} \exp\left[-\frac{2r(\varepsilon)}{\alpha}\right], \tag{2.74}$$

where ε_F is the Fermi level and the typical hopping distance $r(\varepsilon)$ at energy ε can be estimated as

$$r(\varepsilon) = \left[\left(\frac{4\pi}{3B_c}\right)\int_{-\infty}^{\varepsilon} g(\varepsilon) d\varepsilon\right]^{-1/3} \tag{2.75}$$

with $r_t = r(\varepsilon_t)$ being the hopping distance at the TE. The coefficient $C = 2.3$ in Equation (2.74) is due to the modified form of the Einstein relation for energy-loss hopping given by Equation (2.53). The coefficient $B_c = 2.7$ in Equation (2.75) takes into account the necessity to fulfil the percolation criterion given by Equation (2.13).

Although the above approach, based on the concept of the transport energy, is transparent, it is not the most accurate for calculations of $\sigma_d(T)$. There is a more accurate, though less transparent approach based on the percolation theory. The details of the percolation approach can be found, for example, in [5]. Its application to the variable-range hopping in exponential band tails was first studied by Grünewald and Thomas [35] and later the theory was developed by Vissenberg and Matters [79]. In the percolation approach one treats the transition rates between all pairs of sites i and j, taking into account both the quantum mechanical transition probabilities and also the probabilities that one of the sites in a pair is filled by a charge carrier, while the other site is empty, as described by Equation (2.11). The hopping rate can be expressed as

$$\nu_{ij} = \nu_0 \exp\left(-\frac{2r_{ij}}{\alpha}\right)\exp\left(-\frac{|\varepsilon_i - \varepsilon_F| + |\varepsilon_j - \varepsilon_F| + |\varepsilon_j - \varepsilon_i|}{2kT}\right) \equiv \nu_0 \exp(-\xi_{ij}). \tag{2.76}$$

Connecting only sites with $\xi_{ij} < \xi$, and increasing ξ, one determines the value $\xi = \xi_c$, at which the infinite percolation cluster of interconnected sites first appears. This cluster is responsible for the hopping motion of charged carriers in the DC regime [5]. The critical value ξ_c determines the exponential temperature and concentration dependences of transport coefficients

$$\sigma_d = \sigma_0 \exp(-\xi_c), \tag{2.77}$$

where σ_0 is a prefactor not given by the percolation theory. The problem is to calculate the exponent ξ_c. Vissenberg and Matters [79] calculated $\sigma_d(T)$ using the classical percolation criterion that B_c, determined as the number of valid bonds per site at the percolation threshold, is equal to 2.8.

For a given total concentration of electrons in the system n Vissenberg and Matters [79] obtained the following expression for $\sigma_d(T)$

$$\sigma_d(n,T) = \sigma_0 \left[\frac{\pi n (\varepsilon_0/kT)^3}{(2\alpha)^3 B_c \Gamma(1 - kT/\varepsilon_0)\Gamma(1 + kT/\varepsilon_0)}\right]^{\varepsilon_0/kT}, \tag{2.78}$$

where $\Gamma(z) = \int_0^\infty dy \exp(-y) y^{z-1}$

It is worth noting that the temperature dependence of the dark conductivity provided by Equation (2.78) for the VRH in the exponential DOS has nothing to do with Mott's law described by Equation (2.19). Unfortunately, the opposite is often claimed in the scientific literature [80, 81]. Mott's law (Equation 2.19) is valid for the VRH process in the vicinity of the Fermi level in a system with energy-independent DOS. As we have seen in Section 2.3.2, even a slight energy dependence of the DOS, such as a parabolic one in the case of the Coulomb gap, drastically changes the temperature dependence of the VRH conductivity as compared with Mott's formula. For the parabolic gap, $\sigma_d(T)$ is described by Equation (2.22) instead of Equation (2.19). In the case of the exponential DOS described by Equation (2.23), the temperature dependence $\sigma_d(T)$ has an Arrhenius character $\sigma_d \propto \exp(-\Delta/kT)$, according to Equation (2.78) with an activation energy Δ that is weakly (logarithmically) temperature dependent [79]. Only in the limit of extremely low and practically unachievable temperatures can Mott's law be observed in systems with exponential DOS.

In order to verify the analytical percolation approach used by Vissenberg and Matters [79] and also in order to check the accuracy of the analytical approach based on the TE, we performed a series of straightforward computer simulations of the percolation problem for the dark hopping conductivity in the exponential DOS [78]. The simulation was based on a straightforward Monte Carlo algorithm directed to finding the percolation threshold in a system of randomly placed localization sites with energy distribution described by Equation (2.23). The sites were placed in a cube with a length chosen to provide a number of sites m for a given concentration N_0. No correlation between the spatial positions of sites and their energies was assumed. The hopping parameter ξ_{ij} was calculated for each pair of sites according to Equation (2.76) and the percolation threshold $\xi_c(m)$ was found, which provides a continuous path through the system via states with $\xi_{ij} < \xi_c(m)$. Five runs of the algorithm were performed for each set of the system parameters with different random generator seed numbers and the results were averaged over these runs. The accuracy in determination of $\xi_c(m)$ for each run was $\xi_c(m) \pm 0.1$. The standard deviations of the $\xi_c(m)$ values for single runs never exceeded 0.2. Calculations for each set of parameters were performed for $m = 10^5$, 2×10^5, and 5×10^5 and the value of ξ_c for the infinite system was determined via linear extrapolation of $\xi_c(m)$ as a function of $1/m$.

In Figure 2.15 the results of the computer simulation for ξ_c as a function of temperature are shown (squares) in comparison with the results of the percolation analytical approach of Vissenberg and Matters [79] (dashed-dotted line) and with the results obtained by the TE approach according to Equation (2.74) (dashed line). The simulation parameters, such as N_0, n, α, and ε_0, were chosen equal to those given in [79]. Since the analytical approach based on the TE concept does not allow one to determine the preexponential factor σ_0 in Equation (2.77), we adjusted this factor in order to equate ξ_c in the TE description to that obtained in the simulation for the lowest considered temperature $T = 55\,\text{K}$. Hence only the slope in the temperature dependence of ξ_c can be found in the framework of the TE approach. Furthermore, the TE approach for the chosen material parameters is self-consistent only at rather low temperatures $T < 70\,\text{K}$, since for higher temperatures the assumptions used for the derivation of the TE are no longer valid. Comparison with the computer simulation in Figure 2.15 shows that, in the range of low temperatures, the TE approach gives a correct temperature dependence $\sigma_d(T)$. The same can be generally claimed for the analytical approach based on the percolation theory [79]. This theory is self-consistent in a much broader temperature range than the TE description. The analytical percolation approach is supposed to be accurate at $kT < \varepsilon_0$ and $|\varepsilon_F| \gg \varepsilon_0$ [79]. These conditions are fulfilled in a very broad temperature range. Some deviations between the results of the analytical percolation approach and those of computer simulations at temperatures above ~100 K seen in Figure 2.15 are not of high importance since at temperatures above ~150 K, conduction in amorphous semiconductors is dominated by charge carriers in delocalized

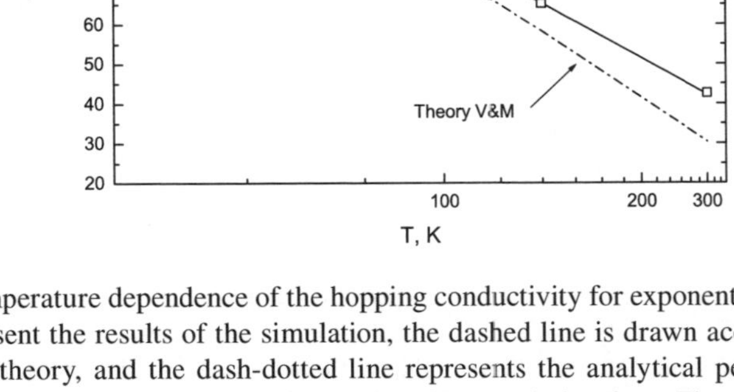

Figure 2.15 Temperature dependence of the hopping conductivity for exponential density of states. The squares represent the results of the simulation, the dashed line is drawn according to Equation (2.74) for the TE theory, and the dash-dotted line represents the analytical percolation theory of Vissenberg and Matters [79] (reproduced from [78] with permission from Elsevier)

states above the mobility edges and hopping conduction via localized states in the band tails plays a secondary role.

2.9 NONLINEAR FIELD EFFECTS

Transport phenomena under the influence of high electric fields in inorganic noncrystalline materials, such as amorphous semiconductors, has been the object of intensive experimental and theoretical study for decades. This is implied by observations of strong nonlinearity in the field dependences of the dark conductivity [82, 83], of the photoconductivity [84] and of the charge carrier drift mobility [82, 85, 86] at high electric fields. This effect is most pronounced at low temperatures, when charge transport is determined by electron hopping via localized band-tail states (Figure 2.16).

Whereas the field-dependent hopping conductivity at low temperatures was always a challenge for theoretical description, the theories for the temperature dependence of the hopping conductivity at low electric fields were successfully developed for all transport

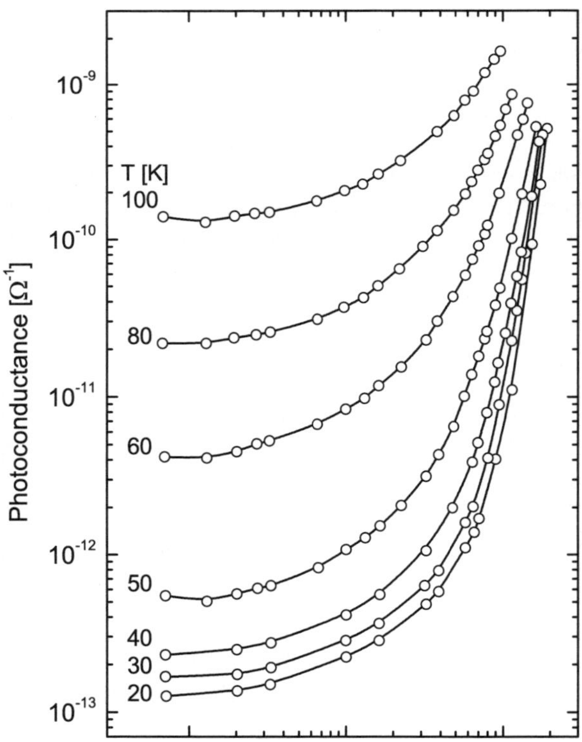

Figure 2.16 Dependence of the photoconductivity in a-Si:H on the electric field at different temperatures (reproduced from [84] with permission from Taylor & Francis Ltd.)

regimes discussed: for the dark conductivity [35, 79], for the drift mobility [37], and for the photoconductivity [38]. In all these theories, hopping transitions of electrons between localized states in the exponential band tails play a decisive role, as described above.

Shklovskii was the first who recognized that strong electric field plays, for hopping conduction, a role similar to that of temperature [87]. In order to obtain the field dependence of the conductivity $\sigma(F)$ at high fields, Shklovskii replaced the temperature T in the well-known dependence $\sigma(T)$ for low fields by a function $T_{\text{eff}}(F)$ of the form [87]

$$T_{\text{eff}} = \frac{eF\alpha}{2k}, \tag{2.79}$$

where e is the elementary charge, k is the Boltzmann constant, and α is the localization length of electrons in the band-tail states. A very similar result was obtained later by Grünewald and Movaghar in their study of the hopping energy relaxation of electrons through band tails at very low temperatures and high electric fields [88]. The same idea was also used by Shklovskii *et al.* [38], who suggested that, at $T = 0$, one can calculate the field dependence of the stationary photoconductivity in amorphous semiconductors by replacing the laboratory temperature T in formulas for the low-field finite-temperature theory by an effective temperature $T_{\text{eff}}(F)$ given by Equation (2.79).

It is easy to understand why electric field plays a role similar to that of temperature for the energy relaxation of electrons. In the presence of the field, the number of sites available for charge transport at $T = 0$ is essentially enhanced in the direction prescribed by the field, as shown in Figure 2.17. Hence, electrons can relax faster at higher fields. From the figure it is seen that an electron can increase its energy with respect to the mobility edge by the amount $\varepsilon = eFx$ in a hopping event over a distance x in the direction prescribed by the electric field. The process resembles thermal activation. The analogy becomes closer when we express the transition rate for the hop as

$$\nu = \nu_0 \exp\left(-\frac{2x}{\alpha}\right) = \nu_0 \exp\left(-\frac{2\varepsilon}{eF\alpha}\right) = \nu_0 \exp\left[-\frac{\varepsilon}{kT_{\text{eff}}(F)}\right], \tag{2.80}$$

where $T_{\text{eff}}(F)$ is provided by Equation (2.79).

This effective activation induced by electric field produces at $T = 0$ a Boltzmann tail in the energy distribution function of electrons via localized states, as shown by numerical calculations [89, 90]. In Figure 2.16, a field-dependent photoconductivity in a-Si:H is shown for several temperatures [84]. If we compare the values of the photoconductivity at the lowest measured temperature, $T = 20\,\text{K}$ in Figure 2.16, with the values of the low-field photoconductivity at $T = T_{\text{eff}} = eF\alpha/2k$ measured by Hoheisel *et al.* [42] and by Stradins and Fritzsche [43], we come to the conclusion that the data agree quantitatively if one assumes the value for the localization length $\alpha = 1.05\,\text{nm}$ [38], which is very close to the value $\alpha = 1.0\,\text{nm}$ known for a-Si:H from independent estimates [11]. This comparison shows that the concept of the effective temperature based on Equation (2.79) provides a powerful tool to estimate the nonlinearity of transport coefficients with respect to the electric field using the low-field results for the temperature dependences of such coefficients.

However, experiments are usually carried out not at $T = 0$, but at finite temperatures, and the question arises on how to describe the transport phenomena in the presence of both

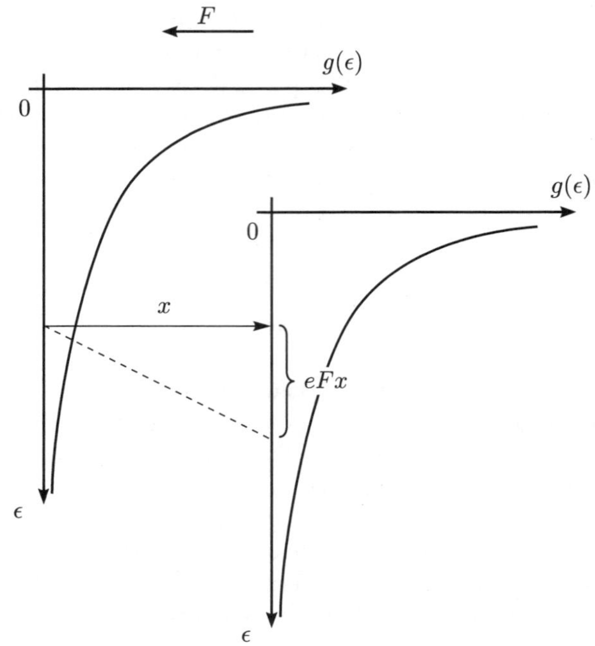

Figure 2.17 Tunnelling transition of a charge carrier in the band tail affected by a strong electric field. Traveling the distance x, the carrier acquires energy eFx, where F is the strength of the electric field, and e is the elementary charge (reproduced with permission from [91]; Copyright 1995 by the American Physical Society)

factors, finite T and high F. By studying the steady-state energy distribution of electrons in numerical calculations and computer simulations [89, 90] and by straightforward computer simulations of the steady-state hopping conductivity and the transient energy relaxation of electrons [91] the following result has been found. The whole set of transport coefficients can be represented as a function of a single parameter $T_{\text{eff}}(F, T)$

$$T_{\text{eff}}(F,T) = \left[T^{\beta} + \left(\gamma \frac{eF\alpha}{k} \right)^{\beta} \right]^{1/\beta} \tag{2.81}$$

with $\beta \simeq 2$ and values of γ in the range 0.5–0.9, depending on which transport coefficient is considered [91]. We are aware of no analytical theory that can support this numerical result.

In the end of this section we would like to make the following remark. Very often in the scientific literature it is claimed that transport coefficients in the hopping regime should have a purely exponential dependence on the applied electric field (see, for instance, [86]). The idea behind such statements seems rather transparent. Electric field diminishes potential barriers between localized states by the amount $\Delta\varepsilon = eFx$, where x is the projection of the hopping radius on the field direction. At a first glance, this should diminish the activation energies in Equations (2.10) and (2.11) by this amount $\Delta\varepsilon = eFx$, and lead to the term $\exp(eFx/kT)$ in the expressions for the charge carrier mobility, diffusivity and conductivity. One should, however, take into account that hopping transport in all real materials is the essentially variable-range hopping process. In such a process the interplay between spatial and energy-dependent terms in the exponents of the transition probabilities determines the conduction path, as discussed above. Therefore it is not enough to take into account the influence of the strong electric field only on the activation energies of single hopping transitions. One should consider the modification of the whole transport path due to the effect of the strong electric field. It is this VRH nature of hopping process that leads to a more complicated field dependence of the transport coefficients expressed by Equations (2.80), (2.81).

2.10 CONCLUDING REMARKS

In this chapter we have presented several basic concepts developed for the description of charge carrier transport in ionorganic disordered semiconductors, such as amorphous and microcrystalline materials. Although these concepts are based on the rather old ideas of the variable-range hopping conduction, they unfortunately have not yet become the working tool for many researchers dealing with amorphous materials. Furthermore, these concepts are even less known by researchers working with other kinds of disordered materials, such as for example organic materials or dye-sensitized structures, which are presently intensively studied for various device applications, particularly, in photovoltaics. We would like to emphasize that transport phenomena in chemically different disordered materials look rather similar, being determined by the presence of localized electron states. In such a case one can use the concepts described in this chapter for theoretical interpretation of the observed charge transport effects in chemically different materials. Here we have given a mathematical description, assuming that the density of localized states has the exponential shape described by Equation (2.23). Of course, the particular dependences of the kinetic charge transport coefficients on material parameters and temperature are specific for the particular shape of the DOS. However, the mathematical apparatus presented here is rather universal and it can be applied to the theoretical description of charge transport phenomena in materials with other shapes of the DOS than the exponential one. For example, in Chapter 6 of this book we show the application of the concepts described in the current chapter to organic disordered materials, in which the DOS is believed to be Gaussian.

ACKNOWLEDGEMENTS

The authors are indebted to numerous colleagues for stimulating and enlightening discussions. Among those are Boris Shklovski (University of Minnesota), Alexei Efros and Michael Raikh (Utah University), Hellmut Fritzsche (Chicago University), Peter Thomas, Walther Fuhs and Heinz Bässler (Philipps-University Marburg), Igor Zvyagin (Moscow State University) and many other colleagues. Financial support of the Deutsche Forschungsgemeinschaft and European Community (IP 'FULLSPECTRUM', N: SES6-CT-2003-502620) is gratefully acknowledged.

REFERENCES

[1] S.D. Baranovski and O. Rubel, Description of charge transport in disordered organic materials. In: *Charge Transport in Disordered Solids with Applications in Electronics*, S.D. Baranovski, (ed), John Wiley & Sons, Ltd, Chichester, (2006).

[2] I.P. Zvyagin, Charge transport via delocalized states in disordered materials. In: *Charge Transport in Disordered Solids with Applications in Electronics*. S.D. Baranovski, (ed), John Wiley & Sons, Ltd, Chichester, (2006).
[3] N.A. Goryunova and B.T. Kolomiets, *Izv. Akad. Nauk. SSSR Ser. Fiz.*, **20**, 1496 (1956).
[4] N.F. Mott and E.A. Davis, *Electronic Processes in Non-Crystalline Materials*. 2d edn. Clarendon Press, Oxford (1979).
[5] B.I. Shklovskii and A.L. Efros, *Electronic Properties of Doped Semiconductors*. Springer, Heidelberg (1984).
[6] A.V. Kolobov, P. Fons, A.I. Frenkel, A.L. Ankudinov, J. Tominaga, and T. Uruga, *Nature Materials*, **3**, 703 (2004).
[7] M. Wuttig, *Nature Materials*, **4**, 265 (2005).
[8] W.E. Spear and P.G. LeComber, *Solid State Commun.*, **17**, 565 (1975).
[9] W. Fuhs, Hydrogenated amorphous silicon—material properties and device applications. In: *Charge Transport in Disordered Solids with Applications in Electronics*. S.D. Baranovski, (ed), John Wiley & Sons, Ltd, Chichester, (2006).
[10] H. Overhof and P. Thomas, *Electronic Transport in Hydrogenated Amorphous Semiconductors*. Springer, Heidelberg (1989).
[11] R.A. Street, *Hydrogenated Amorphous Silicon*. Cambridge Solid State Science Series, Cambridge University Press, Cambridge (1991).
[12] H. Bässler, *Phys. Stat. Sol. (b)*, **175**, 15 (1993).
[13] I.P. Zvyagin, AC hopping transport in disordered materials. In: *Charge Transport in Disordered Solids with Applications in Electronics*. S.D. Baranovski, (ed), John Wiley & Sons, Ltd, Chichester, (2006).
[14] N.F. Mott, *Phil. Mag.*, **19**, 835 (1969).
[15] T. Holstein, *Phil. Mag. B*, **37**, 49 (1978).
[16] H. Scher and T. Holstein, *Phil. Mag. B*, **44**, 343 (1981).
[17] U. Siegner, D. Weber, E.O. Göobel, D. Benhardt, V. Heuckeroth, R. Saleh, S.D. Baranovskii, P. Thomas, H. Schwab, C. Klingshirn, J.M. Hvam, and V.G. Lyssenko, *Phys. Rev. B.*, **46**, 4546 (1992).
[18] A. Miller and E. Abrahams, *Phys. Rev.*, **120**, 745 (1960).
[19] N.W. Dalton, C. Domb, and M.F. Sykes, *Proc. Phys. Soc.*, **83**, 496 (1964).
[20] C. Domb and N.W. Dalton, *Proc. Phys. Soc.*, **89**, 859 (1966).
[21] S.D. Baranovski, T. Faber, F. Hensel, P. Thomas, and G.J. Adriaenssense, *J. Non-Cryst. Solids*, **198–200**, 214 (1996).
[22] S.D. Baranovski, H. Cordes, K. Kohary, and P. Thomas, *J. Non-Cryst. Solids*, **81**, 955 (2001).
[23] C.H. Seager and G.E. Pike, *Phys. Rev. B*, **10**, 1435 (1974).
[24] O. Rubel and S.D. Baranovski, unpublished work (2003).
[25] M. Pollak, *Disc. Faraday Soc.*, **50**, 13 (1970).
[26] A.L. Efros and B.I. Shklovskii, *J. Phys. C*, **8**, L49 (1975).
[27] S.D. Baranovski, A.L. Efros, B.L. Gelmont, and B.I. Shklovskii, *J. Phys. C*, **12**, 1023 (1979).
[28] I. Shlimak, M. Kaveh, R. Ussyshkin, V. Ginodman, S.D. Baranovski, H. Vaupel, P. Thomas, and R.W. van der Heijden, *Phys. Rev. Lett.*, **75**, 4764 (1995).
[29] S.D. Baranovski, F. Hensel, K. Ruckes, P. Thomas, and G.J. Adriaenssens, *J. Non-Cryst. Solids*, **190**, 117 (1995).
[30] P. Bozsoki, S.D. Baranovski, P. Thomas, and S.C. Agarwal, *Phys. Stat. Sol. (c)*, **1**, 113 (2004).
[31] J. Orenstein and M.A. Kastner, *Solid State Commun.*, **40**, 85 (1981).
[32] I.P. Zvyagin, *Kinetic Phenomena in Disordered Semiconductors (in Russian)*. Moscow University Press, Moscow (1984).
[33] H. Scher and E.W. Montroll, *Phys. Rev. B*, **12**, 2455 (1975).

[34] M. Pollak, *Phil. Mag.*, **36**, 1157 (1977).
[35] M. Grünewald and P. Thomas, *Phys. Stat. Sol.* (*b*), **94**, 125 (1979).
[36] F.R. Shapiro and D. Adler, *J. Non-Cryst. Solids*, **74**, 189 (1985).
[37] D. Monroe, *Phys. Rev. Lett.*, **54**, 146 (1985).
[38] B.I. Shklovskii, E.I. Levin, H. Fritzsche, and S.D. Baranovski, Hopping photoconductivity in amorphous semiconductors: dependence on temperature, electric field and frequency. In: *Advances in Disordered Semiconductors*. H. Fritzsche, (ed), Volume 3. World Scientific, Singapore 161 (1990).
[39] S.D. Baranovski, P. Thomas, and G.J. Adriaenssens, *J. Non-Cryst. Solids*, **190**, 283 (1995).
[40] A.V. Zherzdev, V.G. Karpov, A.B. Pevtsov, A.G. Pilatov, and N.A. Feoktistov, *Sov. Phys. Semicond.*, **26**, 421 (1992).
[41] S.D. Baranovski, T. Faber, F. Hensel, and P. Thomas, *J. Non-Cryst. Solids*, **198–200**, 222 (1996).
[42] M. Hoheisel, R. Carius, and W. Fuhs, *J. Non-Cryst. Solids* **63**, 313 (1984).
[43] P. Stradins and H. Fritzsche, *Phil. Mag. B.*, **69**, 121 (1994).
[44] J.H. Zhou, S.D. Baranovski, S. Yamasaki, K. Ikuta, K. Tanaka, M. Kondo, A. Matsuda, and P. Thomas, *Phys. Stat. Sol.* (*b*), **205**, 147 (1998).
[45] B.I. Shklovskii, H. Fritzsche, and S.D. Baranovskii, *Phys. Rev. Lett.*, **62**, 2989 (1989).
[46] Q. Gu, E.A. Schiff, S. Grebner, F. Wang, and R. Schwarz, *Phys. Rev. Lett.,* **76**, 3196 (1996).
[47] S.D. Baranovskii, T. Faber, F. Hensel, and P. Thomas, *Phys. Stat. Sol.* (*b*), **205**, 87 (1998).
[48] R.P. Feynman, R.B. Leighton, and M. Sands, *The Feynman lectures on physics*. Volume 1. Addison-Wesley Publishing, Palo Alto (1963).
[49] R.A. Smith, *Semiconductors*. Cambridge University Press, Cambridge (1978).
[50] N.W. Ashcroft and N.D. Mermin, *Solid state physics*. Holt, Rinehart & Winston, New York (1988).
[51] D. Ritter, E. Zeldov, and K. Weiser, *Phys. Rev. B*, **38**, 8296 (1988).
[52] N. Tessler and Y. Roichman, *Organic Electronics*, **5–6**, 200 (2005).
[53] H. Fritzsche, M.Q. Tran, B.G. Yoon, and D.Z. Chi, *J. Non-Cryst. Solids*, **137–138**, 467 (1991).
[54] H. Fritzsche, B.G. Yoon, D.Z. Chi, and M.Q. Tran, *J. Non-Cryst. Solids.*, **141**, 123 (1992).
[55] A. Ilie, N.M.J. Conway, B. Kleinsorge, J. Robertson, and W.I. Milne, *J. Appl. Phys.*, **84**, 5575 (1998).
[56] M. Vanecek, J. Kocka, P. Demo, E. Sipek, and A. Triska, *J. Non-Cryst. Solids*, **90**, 183 (1987).
[57] C. Cloude, W.E. Spear, P.G. LeComber, and A.C. Hourd, *Phil. Mag. B*, **54**, L113 (1986).
[58] W.E. Spear, C.S. Cloude, D. Goldie, and P. LeComber, *J. Non-Cryst. Solids*, **97–98**, 15 (1987).
[59] W.E. Spear and C.S. Cloude, *Phil. Mag. Lett.*, **55**, 271 (1987).
[60] W.E. Spear, Transport and Tail State Interactions in Amorphous Silicon. In: *Advances in disordered semiconductors*. H. Fritzsche, (ed), Volume 1. World Scientific, Singapore 721 (1988).
[61] R.A. Street, *Adv. Phys.*, **30**, 593 (1981).
[62] R.A. Street and D.K. Biegelsen, *Solid State Commun.*, **44**, 501 (1982).
[63] J.D. Joannopoulos and G. Locowsky, *The physics of hydrogenated amorphous silicon I, II*. Springer, Berlin (1984).
[64] K. Jahn, R. Carius, and W. Fuhs, *J. Non-Cryst. Solids*, **97–98**, 575 (1987).
[65] A. Vomwas and H. Fritzsche, *J. Non-Cryst. Solids*, **97–98**, 823 (1987).
[66] S.D. Baranovski, H. Fritzsche, E.I. Levin, I.M. Ruzin, and B.I. Shklovskii, *Sov. Phys. JETP*, **69**, 773 (1989).
[67] M. Bort, W. Fuhs, S. Liedtke, R. Stachowitz, and R. Carius, *Phil. Mag. Lett.*, **64**, 227 (1991).
[68] R.E. Johanson, H. Fritzsche, and A. Vomvas, *J. Non-Cryst. Solids*, **114** 274 (1989).

[69] S.D. Baranovski, P. Thomas, G.J., Adriaenssens, and Ö. Öktü, *Solid State Commun.*, **86**, 549 (1993).
[70] J.G. Simmons, G.W. Taylor, and M.C. Tam, *Phys. Rev. B*, **7**, 3714 (1973).
[71] H. Fritzsche and N. Ibaraki, *Phil. Mag. B*, **52**, 299 (1985).
[72] M. Zhu and H. Fritzsche, *Phil. Mag. B*, **53**, 41 (1986).
[73] J.H. Zhou and S.R. Elliott, *Phys. Rev. B*, **46**, 9792 (1992).
[74] G.J. Adriaenssens, S.D. Baranovski, W. Fuhs, J. Jansen, and Ö. Öktü, *Phys. Rev. B*, **51**, 9661 (1995).
[75] B. Gu, Z. Xu, and B. Dong, *J. Non-Cryst. Solids*, **97–98**, 479 (1987).
[76] S.D. Baranovski, M. Zhu, T. Faber, F. Hensel, P. Thomas, M.B. vonder Linden, and W.F. van der Weg, *Phys. Rev. B*, **55**, 16226 (1997).
[77] D.S. Misra, A. Kumar, and S.C. Agarwal, *Phys. Rev. B*, **31**, 1047 (1985).
[78] S.D. Baranovski, O. Rubel, and P. Thomas, *Thin Solid Films*, **487**, 2 (2005).
[79] M.C.J.M. Vissenberg and M. Matters, *Phys. Rev. B*, **57**, 12964 (1998).
[80] C. Godet, *J. Non-Cryst. Solids*, **299–302**, 333 (2002).
[81] C. Godet, *Phys. Stat. Sol. (b)*, **231**, 499 (2002).
[82] C.E. Nebel, R.A. Street, N.M. Johanson, and C.C. Tsai, *Phys. Rev. B*, **46**, 6803 (1992).
[83] A. Nagy, M. Hundhausen, L. Ley, G. Brunst, and E. Holzenkämpfer, *J. Non-Cryst. Solids*, **164–166**, 529 (1993).
[84] R. Stachowitz, W. Fuhs, and K. Jahn, *Phil. Mag. B*, **62** 5 (1990).
[85] H. Antoniadis and E.A. Schiff, *Phys. Rev. B*, **43**, 13957 (1991).
[86] K. Murayama, H. Oheda, S. Yamasaki, and A. Matsuda, *Solid State Commun.*, **81**, 887 (1992).
[87] B.I. Shklovskii, *Sov. Phys. Semicond.*, **6**, 1964 (1973).
[88] M. Grünewald and B. Movaghar, *J. Phys. C: Condens. Matter*, **1**, 2521 (1989).
[89] S. Marianer and B.I. Shklovskii, *Phys. Rev. B*, **46** 13100 (1992).
[90] S.D. Baranovski, B. Cleve, R. Hess, and P. Thomas, *J. Non-Cryst. Solids*, **164–166**, 437 (1993).
[91] B. Cleve, B. Hartenstein, S.D. Baranovski, M. Scheidler, P. Thomas, and H. Bässler, *Phys. Rev. B*, **51**, 16705 (1995).

3 Hydrogenated Amorphous Silicon—Material Properties and Device Applications

Walther Fuhs

Hahn-Meitner-Institut Berlin, Kekuléstraße 5, D-12489 Berlin, Germany

3.1 INTRODUCTION

Research in the field of amorphous semiconductors, from the beginning, has been driven by both the scientific interest in basic aspects of disorder in the properties of solids and technological applications. In the early years chalcogenide glasses were at the center of the interest due to thin-film applications in imaging, xerography, memory and switching devices. By then amorphous silicon and amorphous germanium, a-Si and a-Ge, were of more academic scientific interest. As simple elemental tetrahedrally bonded amorphous

Charge Transport in Disordered Solids with Applications in Electronics Edited by S. Baranovski.

semiconductors they served as model systems in which the disorder was less complicated, being defined not by chemical composition, but by the structural disorder only. The loss of long-range order is the structural characteristic of amorphous semiconductors. As a result, important theoretical concepts which are based on periodicity, fail, e.g., band structure, *k*-vector, Bloch states, effective mass and optical selection rules. The nature and structure of the local chemical bonding largely determine the physical properties.

The early forms of amorphous silicon had unacceptable electronic properties due to a high density of states in the energy gap which effectively pinned the Fermi level. The conductivity in this kind of amorphous silicon was very low and, below room temperature, in general determined by hopping transport among localized gap states at the Fermi level. Due to effective pinning of the Fermi level in a high density of gap states the conductivity could not be varied by doping, illumination or carrier injection. Such material properties prevented this material from being useful for electronic devices.

The situation changed when, at the beginning of the 1970s, new deposition techniques were discovered which enabled the preparation of amorphous silicon films with attractive semiconducting properties. It soon became clear that the improvement of the electronic quality of this form of amorphous silicon was due to the incorporation of hydrogen. This is why this material is referred to as hydrogenated amorphous silicon. Six milestones may be seen in the material and device development on the basis of this new material which led to a burst of research activities in this field and to exciting technological applications:

- introduction of plasma-enhanced chemical vapor deposition (PECVD) by Chittick, Alexander and Sterling in 1969 [1];
- discovery that the defect density in this new kind of amorphous silicon was low which resulted in high photoconductivity [2];
- demonstration of hydrogen passivation by the research group at Harvard University by studying sputtered a-Si:H and a-Ge:H [3]. Later on numerous studies gave proof that the superior semiconducting and photoelectric properties of this kind of amorphous silicon, a-Si:H, were due to the incorporation of hydrogen;
- successful substitutional *n*-type and *p*-type doping by addition of phosphine or diborane to the process gas. LeComber and Spear [4] demonstrated the control of the electrical conductivity over 10 orders of magnitude;
- first report on photovoltaic solar cells from a-Si:H by Carlson and Wronski in 1976 [5];
- first report on the fabrication and physics of a thin-film transistor by LeComber, Spear, and Gaith [6].

The preparation and properties of hydrogenated amorphous and microcrystalline semiconductors, a-Si:H and μc-Si:H and their alloys have been described in numerous review articles and monographs [7–11] and the development of this field of research is well documented in the proceedings of the biannual *International Conference on Amorphous and Microcrystalline Semiconductors* which are published in regular issues of the *Journal of Non-Crystalline Solids.* Today, hydrogenated amorphous silicon, a-Si:H, offers a mature material and device technology used for solar cells, thin-film transistors, sensors, imaging, radiation detectors and displays. Among the various material options for thin-film solar cells this is the only technology which so far has been able to overcome the barrier to mass production of large area modules and to occupy a reasonable share of the world market.

3.2 PREPARATION AND STRUCTURAL PROPERTIES OF AMORPHOUS SILICON

Amorphous silicon is prepared by deposition from the gas phase onto substrates which are held at temperatures far below the melting temperature. A large variety of techniques has been used: thermal evaporation, sputtering, chemical vapor deposition (CVD), photo-CVD, plasma-enhanced chemical vapor deposition (PECVD), and thermocatalytic hot-wire deposition (HWCVD). There is no principal difference in the structure of amorphous films prepared by the various methods. The difference lies in the deposition rate and the kind and concentration of defects (dangling bonds, voids). PECVD has led to the lowest defect densities and, therefore, is widely used now in research and in industrial applications.

Thermodynamically amorphous silicon, a-Si, is in a metastable state. The real structure can be varied experimentally in many ways and therefore the material properties strongly depend on the preparation conditions and on the treatment of the amorphous samples after deposition. Thermal annealing was shown to produce changes of practically all material properties (enthalpy, electrical properties, defect densities, optical properties) [12]. It has been found that the free energy of annealed a-Si (relaxed state) is about 0.11 eV/atom higher than that of crystalline silicon [13]. Heating above temperatures of about 500°C usually introduces a transition into the thermodynamically more stable crystalline phase. The kinetics of solid-phase crystallization (SPC) is characterized by nucleation and growth of crystal clusters at the expense of the surrounding amorphous material. Laser recrystallization of amorphous silicon has become an important technique for the fabrication of polycrystalline silicon thin-film transistors. Recently, SPC has attracted high interest by the fact that the presence of certain metals strongly modifies the nucleation and growth process. Such techniques enable the engineering of polycrystalline silicon thin films on foreign substrates at low temperatures for new device applications such as thin-film transistors or polycrystalline silicon thin-film solar cells. A particularly interesting example of this kind of processing is the creation of a polycrystalline silicon layer with an average grain size of more than 10 μm on glass substrates by an Al-induced layer exchange process [14].

In PECVD deposition, silane (SiH_4) or gas mixtures of silane with rare gases or hydrogen are decomposed in a glow discharge. In most cases parallel plate systems in a stainless steel reactor have been used. The initial process is the decomposition of silane following various pathways through electronic states excited by inelastic collisions with electrons in the plasma. The electron energy is broadly distributed between zero and some tens of eV. The excited states are dissociative states from which dissociation occurs spontaneously to SiH_3, SiH_2, SiH, Si and H. The reactive species undergo secondary reactions, mostly with parent molecules SiH_4 and H_2 which results in a steady-state distribution determined by the rate constants of the respective generation and annihilation rates. Therefore highly reactive species (SiH_2, SiH, Si) will have lower steady state concentrations than less reactive species (SiH_3). Steady-state densities have been measured using various gas-phase diagnostic tools. The result is that the SiH_3 radical is considered to be by far the most frequent species and is therefore considered to be the dominant precursor for growth of a-Si:H and μc-Si:H [15].

Matsuda [15] proposed the following model for the growth process: The SiH_3 radical reaches the film-growing surface and diffuses, eventually abstracting hydrogen bonded to a Si surface atom. The result is a SiH_4 molecule diffusing away and a Si dangling bond. Another SiH_3 may lock onto this site and form a Si–Si bond. The most important deposition parameters are: substrate temperature, base pressure, flow rates of the process gas, power density, and frequency. High-quality a-Si:H films are grown at deposition rates of typically 1 Å/s at substrate temperatures in the range 150–250°C.

Catalytic chemical vapor deposition (often termed hot-wire deposition HWCVD) is an alternative method for the preparation of device quality a-Si:H and μc-Si:H films at low process temperatures. With this method the reaction gases (SiH_4, H_2) are decomposed at a heated tungsten or tantalum wire. At wire temperatures above 1600°C SiH_4 decomposes by a catalytic surface reaction into Si and 4H. Secondary gas phase reactions with SiH_4 and H_2 lead to different kinds of radicals. Since the radical with the highest concentration is SiH_3 it is believed that the growth mechanisms are about the same in HWCVD and PECVD deposition. However, differences may arise from the fact that the concentration of atomic hydrogen is larger by an order of magnitude in HWCVD than in PECVD [16].

The simplest model for the structure of tetrahedrally bonded amorphous semiconductors is the continuum random network (CRN) with an average coordination number of four [17]. X-ray diffraction in general exhibits a diffused circular pattern. The Fourier transform of the intensity distribution is the radial distribution function (RDF) which describes the probability of finding another atom at a distance r from the reference atom. The example shown in Figure 3.1 demonstrates that the tetrahedral short-range order is maintained and that the long-range order is completely lost after only few neighbor shells [18]. The first peak describes the nearest neighbor shell with a Si–Si bond length of 0.236 nm which is very close to the crystalline value of 0.24 nm. The second peak is broadened and located at 0.386 nm, as compared with 0.384 nm in the crystalline material. The average bond angle estimated from such data is 109°. In crystalline Si the third peak is found at 0.465 nm. It is a characteristic of the tetrahedral amorphous structures that this peak is completely absent. Such RDF curves have been fitted by CRN models which allowed for changes in the bond

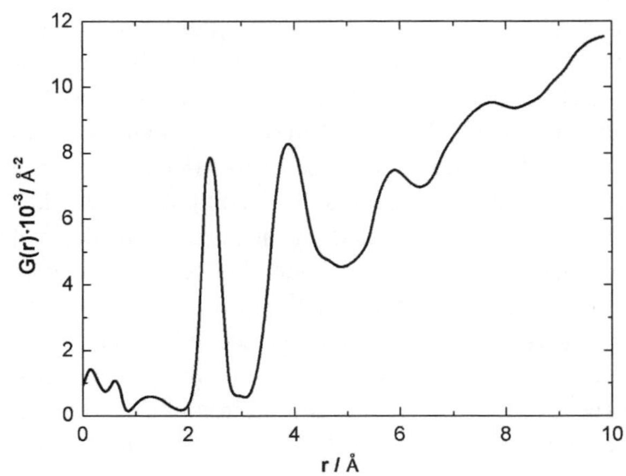

Figure 3.1 Reduced radial distribution function $G(r)$ of a-Si:H (reproduced from [65] with permission of Taylor & Francis)

lengths of up to 1% and changes in the bond angles of less than 10%. These observations are more or less the same for amorphous silicon and hydrogenated amorphous silicon. Of course the CRN is an idealized structure of a tetrahedrally bonded amorphous semiconductor. In real films (a-Si or a-Si:H) there may be inhomogeneities, columnar structure or void formation. In addition there are numerous defects such as vacancies, dangling bonds and impurities, in particular hydrogen, oxygen, carbon or nitrogen.

The progress attained using the plasma deposition method was that the density of deep defects (Si-dangling bonds) was tremendously reduced owing to the incorporation of hydrogen. The effect of hydrogen is to saturate dangling bond defects and to lower the average coordination number of the network which allows the construction of a more relaxed disordered structure. The hydrogen content can vary widely with the deposition conditions (5–40 at%). In optimized films the hydrogen concentration amounts to 5–15 at% and the density of neutral Si-dangling bonds is less than $10^{15}\,cm^{-3}$. Infrared spectroscopy reveals the absorption bands due to various bonding configurations of hydrogen. The spectra of optimized films show that the hydrogen is bonded predominantly in the Si–H configuration and the spectra are dominated by the absorption band of the Si–H stretching mode at $2000\,cm^{-1}$. Under less favorable deposition conditions in particular the absorption band at $2100\,cm^{-1}$ suggests the formation of dihydrate SiH_2 or the formation of clustered hydrogen. Figure 3.2 shows an example of the dependence of the intensity of the IR absorption bands on the deposition temperature T_S [19]. At low T_S the hydrogen content is high and the mode at $2100\,cm^{-1}$ clearly dominates. When T_S increases the intensity of this band decreases and the SiH-mode is enhanced. Concomitantly the defect density, measured by the spin density N_S of the Si-dangling bonds, decreases. The lowest defect density is obtained at a substrate temperature of about 250°C. Low defect concentration and preferential hydrogen bonding in monohydride configuration are often quoted as measures of film quality.

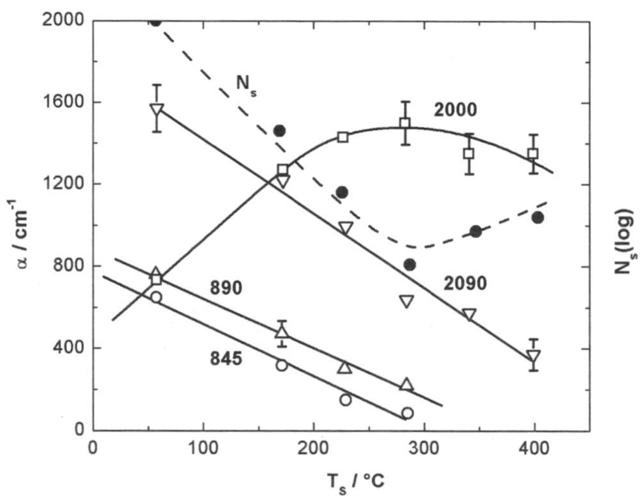

Figure 3.2 Absorption coefficient α of hydrogen stretching and bending modes (in cm^{-1}) and spin density N_S ($g = 2.0055$) in a-Si:H as a function of the deposition temperature. Deposition parameters: 5% SiH_4 in argon, flow rate 130 sccm, $P = 25\,W$ (reproduced from [19] with permission of the American Physical Society)

Annealing of the films at temperatures above the deposition temperature leads to the evolution of hydrogen which is accompanied by an enhancement of the defect density. Hydrogen evolution spectroscopy has frequently been used to explore the morphology of hydrogenated amorphous films. In such experiments the specimen is heated at a constant heating rate in ultra-high vacuum and the partial pressure of hydrogen is measured as a function of temperature. Figure 3.3 presents, as an example, results obtained on films prepared at different substrate temperatures [20]. In films deposited at around 250°C the evolution rate has a single broad peak at 550°C whereas films made at lower temperatures with a higher hydrogen content exhibit an additional structure at lower temperature near 350°C. The occurrence of the low-temperature peak points to the existence of a porous structure which enables rapid hydrogen diffusion. At very high substrate temperatures (above 400°C) the incorporation of hydrogen is reduced considerably.

According to the pronounced role of hydrogen in plasma-deposited hydrogenated silicon and the high concentration of hydrogen involved, such materials may be considered as hydrogen–silicon alloys. Practically all film properties depend on the hydrogen content. This is, for instance, demonstrated very impressively by the pronounced increase of the energy gap with the hydrogen content shown in Figure 3.4 [21]. Mobile hydrogen is believed to be the cause of reversible changes of electronic properties (see Section 3.3.3).

A particular attraction of the plasma deposition technique is the high-flexibility in the choice of deposition parameters and process gases which allows one to easily modify the properties of the deposited films. Doping can be achieved by adding controlled amounts of B_2H_6 or PH_3 to the process gas. Hydrogen dilution and high radiofrequency power result in the formation of microcrystalline silicon films (μc-Si:H). This material is often viewed as a two-phase material consisting of crystalline columns which are about 100–150 nm wide and which are formed by small nanocrystalline grains (10–20 nm) [22]. These columns are separated by strongly disordered regions which are considered as consisting of a-Si:H (Fig. 3.5). Hydrogen is present in high amounts (typically 5 at%) and considered to be

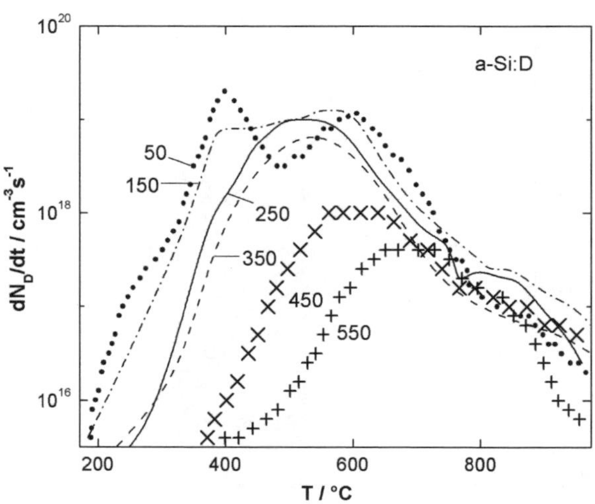

Figure 3.3 Hydrogen effusion spectra of deuterated a-Si:D films deposited at various substrate temperatures by radiofrequency PECVD (reproduced from [20] with permission of Elsevier)

mostly located in the disordered regions and on the surface of the columns or grains which results in excellent defect passivation. μc-Si:H decreases the value of the bandgap from 1.75 eV for a-Si:H to lower values as do a-$Si_{1-x}Ge_x$:H alloys which can be made by depositing from gas mixtures of SiH_4 and GeH_4. The extension to higher bandgaps has been achieved by alloying Si with C, O or N to form hydrogenated amorphous semiconductors a-$Si_{1-x}C_x$:H, a-$Si_{1-x}N_x$:H, and SiO_x:H. It is thus possible to control the energy gap in the range 0.7–3.6 eV by control of the gas-phase composition in the plasma discharge. For device

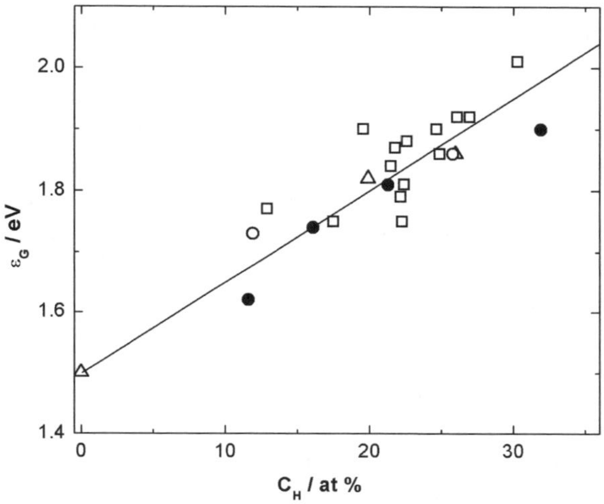

Figure 3.4 Optical energy gap ε_G as a function of the hydrogen content C_H (reproduced from [21] with permission of Elsevier)

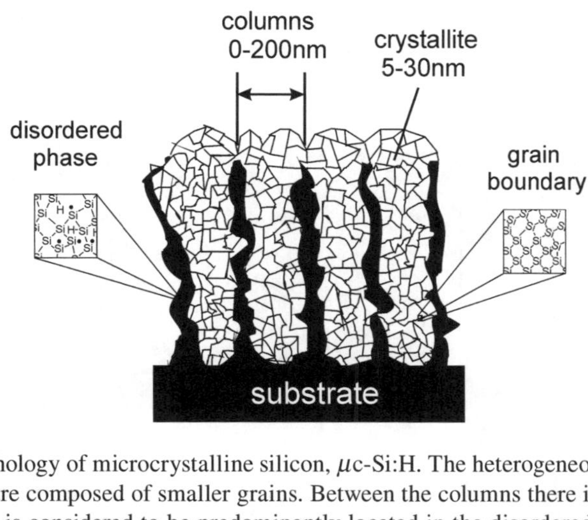

Figure 3.5 Morphology of microcrystalline silicon, μc-Si:H. The heterogeneous structure consists of columns which are composed of smaller grains. Between the columns there is a disordered phase (a-Si:H). Hydrogen is considered to be predominantly located in the disordered phase and on grain boundaries

development this ability of bandgap tuning is very attractive. However, the general experience is that alloying is always connected with the formation of additional defects such that the electronic properties, carrier lifetimes and doping efficiencies deteriorate with increasing concentration x of alloy constituents.

3.3 DENSITY OF STATES DISTRIBUTION IN THE ENERGY GAP

3.3.1 Model of the density of states distribution

The electronic properties of amorphous films are largely determined by the density and energy distribution $N(\varepsilon)$ of the localized states in the energy gap. Numerous experimental techniques have been applied for the investigation of intrinsic and extrinsic defect states, leading to the model depicted in Figure 3.6. Inside the bands the density of states distributions of the valence and conduction band of amorphous semiconductors differ only slightly from those of their crystalline counterparts. In the optical spectra therefore the main effect of disorder is a significant broadening of the spectra. The states deeper in the bands are considered to be extended, but of course they are no longer Bloch states. A characteristic of amorphous semiconductors is the disorder-induced localization of states near the band edges. Theoretical studies led to a model where the transition from extended to localized states occurs sharply at distinct energies ε_C and ε_V [23, 24, 25]. Since at ε_C and ε_V the carrier mobility is expected to drop by orders of magnitude these energies are called mobility edges. For transport in amorphous semiconductors the mobility edges play a similar role as the band edges in crystalline semiconductors (see Section 3.5).

Tails of localized states extend from both bands deep into the gap with an exponential energy distribution. It is assumed that these states arise from the short-range disorder.

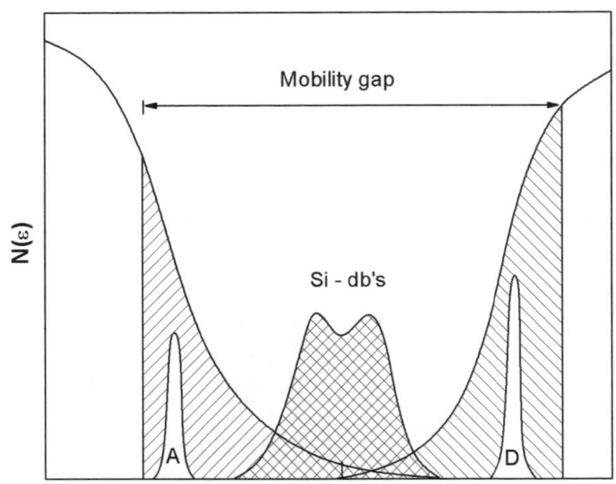

Figure 3.6 Model of the density of states distribution in the energy gap of hydrogenated amorphous silicon, a-Si:H, showing localized tail states, Si dangling bonds, donor (D) and acceptor (A) states. ε_C and ε_V denote the mobility edges

Model calculations, in fact, showed that variations of bond length, bond angles and dihedral angles result in tail states that are localized in the sense that they have a finite amplitude in limited spatial regions only [26]. Deep defect states are formed in a random network structure as a result of stress release, but may also arise from unfavorable deposition conditions. The simplest point defect is an unsaturated Si bond for which the term 'Si dangling bond' is generally used. The positive/neutral state of this defect is located around midgap in undoped material, the negatively charged state is shifted to higher energy by a positive correlation energy U of about 0.2 eV. Although one would expect to find a large variety of intrinsic defects, the Si dangling bond is the only one that so far has been identified microscopically.

In undoped device-grade a-Si:H the concentration of Si dangling bonds amounts to some $10^{14}\,cm^{-3}$. Besides these intrinsic defect states one expects extrinsic states that originate from impurities. Substitutional doping results in donor and acceptor states which are hidden underneath the broad tail. Although under normal conditions gas-phase-deposited films contain impurities such as oxygen, nitrogen or carbon of the order of $10^{19}\,cm^{-3}$, so far no gap states due to these impurities have been identified. In hydrogenated semiconductors hydrogen is the most frequent impurity. The beneficial role of hydrogen is to remove states from the gap by forming strong SiH bonds which are located deeper in the valence band.

For many years it was believed that the density of states distribution is a stable property of a-Si:H. This, however, is not the case: $N(\varepsilon)$ can vary with temperature, doping level, carrier injection, light absorption or voltage stress. As a result most of the electronic properties depend to some extend on sample treatment and history.

3.3.2 Band-tail states

Photoemission spectroscopy using UV light (UPS) or X-rays (XPS) allows a direct experimental determination of the density of states distribution in the bands [27]. The information obtained by this spectroscopy is strongly restricted to the surface region because of the small escape depth of the electrons. This is important because it is well known that the properties in the surface region of amorphous silicon may differ appreciably from the bulk. The escape depth is of the order of 10 nm below 10 eV and 1–2 nm at 1 keV. Photoemission measurements showed that the electronic structure of amorphous silicon is determined by the short-range order of the bonding which is tetrahedral, as in crystalline silicon, and that the disorder results in broadening of the spectra. The disorder energy (0.1–1 eV) appears to be small compared with the band width (about 10 eV) [27, 28]. It is found that SiH bonds introduce additional states in the valence band and induce a shift of the valence band edge to higher binding energy [29]. Since the conduction band is less affected by hydrogen bonding, the bandgap of hydrogenated amorphous silicon, a-Si:H increases. This may be the main reason why the energy gap of a-Si:H is larger than in crystalline silicon and increases with the hydrogen content (see Figure 3.4).

The influence of disorder is considerably more pronounced at the band edges where amorphous silicon has tails of localized states extending deep into the energy gap. Quantitative information on details of the tail state distribution at the valence band was obtained applying a more sensitive photoemission technique, total yield spectroscopy. The density of states distribution $N(\varepsilon)$ shown in Figure 3.7 was deduced from the total yield spectra by simply taking the energy derivative. $N(\varepsilon)$ extends over eight orders of magnitude, consisting

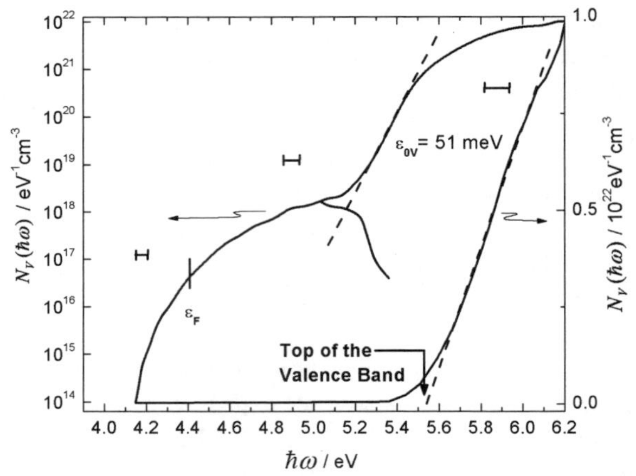

Figure 3.7 Photoemission yield measurement of the valence band density of states in a-Si:H in a logarithmic (left curve) and linear scale (right curve). ε_{0V} is the characteristic energy of the valence band tail and ε_F the position of the Fermi level. For details see text (reproduced from [30] with permission of World Scientific)

of a linear valence band edge at 5.65–6.05 eV, an exponential valence band tail, and a defect band centered near 5 eV. There is clearly a portion extending over some orders of magnitude which can be described by an exponential energy dependence:

$$N(\varepsilon) = N_0 \exp(-\varepsilon/\varepsilon_{0V}). \tag{3.1}$$

Here ε is the energy measured from the band edge at ε_V, $N_0 = 10^{21}\,\text{eV}^{-1}\,\text{cm}^{-3}$ and energy parameter ε_{0V} is the reciprocal slope of the exponent which here amounts to 51 meV.

The conduction band tail has been intensively investigated by transport experiments. In particular, time-of-flight experiments, as pioneered by W. E. Spear [31], have provided valuable information on the localized band tail states. In this experiment carriers are generated by short electron or photon pulses close to a blocking contact and the transit is recorded as the transient primary photocurrent under external bias. Depending on the external bias the transit of both electrons and holes can be studied. However, free carrier transport at the mobility edge cannot be observed directly because the carriers will experience multiple trapping in localized states (traps) during their transit. This is why the drift mobility μ_D determined from the transit time t_{tr} across the specimen is always smaller than the microscopic mobility μ_0 at the mobility edge by a factor which accounts for the time the carriers spend in traps:

$$\mu_D = \mu_0 \tau_f/(\tau_f + \tau_t). \tag{3.2}$$

In this expression τ_f and τ_t denote the times the carriers are free or spend in traps, respectively. A pronounced temperature dependence will be introduced by the thermal release in the multiple trapping process. The situation is simple when trapping occurs into a single trapping level at a certain energy. In this case one observes a well-defined transit time which is inversely proportional to the applied field. However, in amorphous materials the distribution of localized band tail states results in a distribution of release times and causes dispersive transport. Scher and Montroll [32] have shown that dispersive transport is a quite universal feature of amorphous semiconductors. In this case the drift mobility has an unusual field dependence and depends even on the thickness of the specimen. At room temperature electron transport in a-Si:H is nondispersive while the hole transport shows pronounced dispersion. The results from time-of-flight studies agree well with the prediction of exponential band edges. For high-quality a-Si:H the temperature dependence of the dispersion parameters resulted in energy parameters of the exponential band edges of $\varepsilon_{0V} \simeq 33$–$43\,\text{meV}$ for the valence band and $\varepsilon_{0C} \simeq 18$–$22\,\text{meV}$ for the conduction band [11, 31, 33]. Exponential band tail states are considered to be the reason for the exponential optical absorption edge (see Section 3.4). The Urbach parameter ε_0 derived from the slope of the spectra is generally associated with the slope of the valence band tail ε_{0V}. The actual values of the Urbach parameter varies strongly with the film quality, and increase with the defect density. Due to this experience ε_0 is quite generally taken as a measure of the film quality which can easily be determined from a measurement of the optical absorption edge.

The existence of exponential band tails is quite well established and there is reasonable agreement between the results from various techniques. However, it is experimentally unclear how far the exponential tails extend towards the mobility edge and how deep they reach into the energy gap. It is also an open question why the localized state distribution is exponential at all. From random disorder one would rather expect a Gaussian distribution. Thus the relationship between structural disorder and the exponentially distributed band tail states is still unclear. The distributions are unsymmetrical, the value of ε_{0V} being by almost a factor of two larger than ε_{0C}, which may be due to the different symmetry character of the states. The conduction band states are *s*-like and not so much influenced by bond angle disorder while the valence band states with their *p*-like character are expected to be much more sensitive.

3.3.3 Deep defect states

ESR-spectroscopy is the only method of defect characterization which allows, in principle, the microscopic identification of paramagnetic defects through the g-value of the observed resonance. Figure 3.8a presents ESR lines which have been observed in a-Si:H and assigned to states in different energy regions in the bandgap [34, 35, 36]. When the Fermi level is moved across the bandgap by doping these signals appear in different energy ranges (Figure 3.8b). The signal observed in undoped or weakly doped a-Si:H ($g = 2.0055$) has been assigned to the neutral three-fold coordinated Si atom which has a free sp^3 hybrid orbital (Si dangling bond, Si-db). When the Fermi level is shifted towards the conduction band by incorporation of P atoms these defect states will be negatively charged and therefore the signal intensity will decrease. At the higher doping levels a narrow resonance at $g = 2.0043$ emerges which has been attributed to conduction band-tail states which are singly occupied. When the Fermi level moves towards the valence band by boron doping the Si-dbs will be positively charged and a broad line emerges at $g = 2.012$ due to holes trapped in the valence band tail. The distribution of the electron and hole line intensities with energy reflects the exponential band-tail distributions. These assignments are supported by the observation of

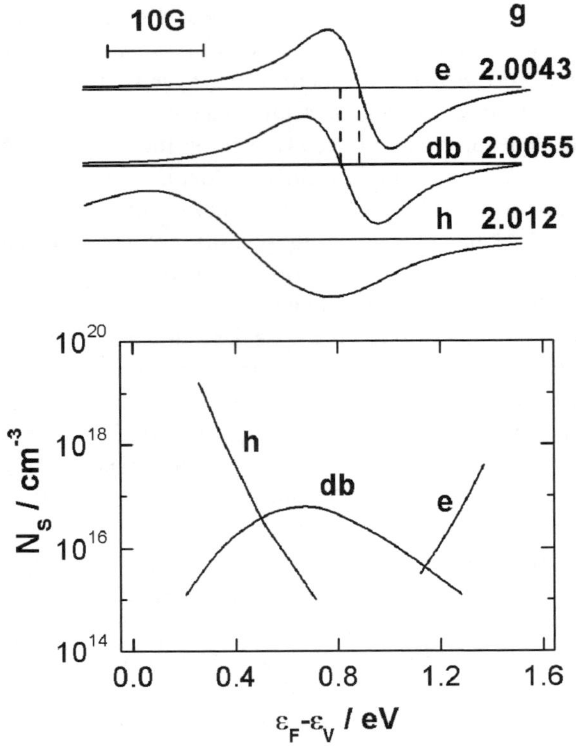

Figure 3.8 Electron spin resonance signals due to band-tail electrons ($g = 2.0043$), band-tail holes ($g = 2.012$) and dangling bonds ($g = 2.0055$) observed in a-Si:H, and spin density of these resonances as a function of the Fermi level (reproduced from [36] with permission of the American Physical Society)

the same signals in light-induced ESR spectra of undoped a-Si:H when the band tails are populated by the generation of excess carriers [34].

The assignment of the signal at $g = 2.0055$ to Si dangling bonds in the disordered network was at first based on the analogy with paramagnetic defect states at grain boundaries and surfaces/interfaces of crystalline silicon. The confirmation of this assignment came from studies of the ^{29}Si hyperfine spectra in undoped a-Si:H [37]. Figure 3.9 displays ESR spectra of a-Si:H which contained the natural abundance of the ^{29}Si isotope (about 4.5%). The central line corresponds to Si dangling bonds located at the ^{28}Si and ^{30}Si isotopes which have zero nuclear spin. With a sensitivity enhanced by a factor of 100 additional resonances emerge on both sides of this resonance. These two satellites are the hyperfine spectra of those Si-dbs which are located at the ^{29}Si nuclei with a nuclear spin $I = ½$. The careful analysis of these spectra led to the conclusion that the microscopic nature of the defect is indeed a Si dangling bond. The hybridization of about 10% *s*-character and 90% *p*-character indicates that the central Si atom has relaxed back towards the three nearest neighbor atoms, thus forming an almost planar structure. About half of the wavefunction appears localized at the central atom while the rest is distributed onto the nearest and next-nearest neighbor atoms. The structure of this defect thus is very similar to the well known P_b center at the Si/SiO_2 interface.

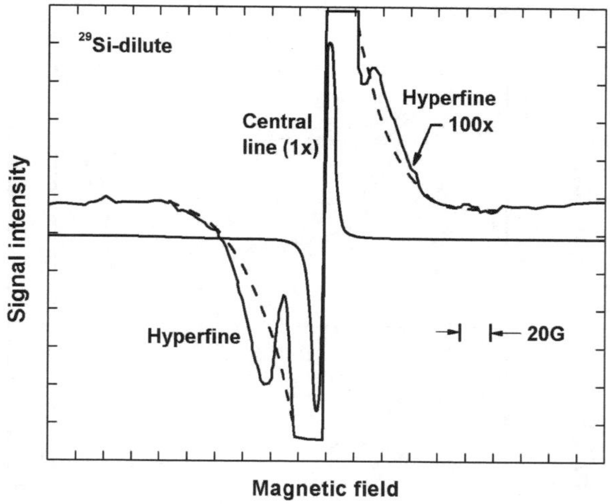

Figure 3.9 ESR derivative spectrum for an a-Si:H sample with naturally occurring isotopic abundance. The wings of the central line are amplified ×100 to show the structure arising from hyperfine satellites. The dashed line is a guide to the eye to indicate the background due to the central line (reproduced from [37] with permission of the American Physical Society)

The electronic properties of a-SiH are strongly determined by the concentration of the Si dangling bond defects. This defect essentially controls the carrier lifetimes and hence the electronic quality of the material. The spin density N_S of this resonance is determined by details of the preparation process. For instance, N_S decreases with increasing substrate temperature and increases when T_S exceed an optimum value which is close to about 250°C (see Figure 3.2). N_S is enhanced by particle bombardment and hydrogen effusion above T_S. In undoped films N_S can vary in the range 10^{14}–$10^{19}\,cm^{-3}$ [11]. The intensity of the ESR signal is a convenient measure of the electronic quality of the material.

Substitutional doping of a-Si:H is accompanied by a pronounced increase in the density of dangling bond defects. This phenomenon was explained by Street [38] in terms of defect compensation, assuming that a thermal equilibrium exists with the plasma at the surface of the film during the growth process. Later a similar increase of the defect density was observed in films that had been doped interstitially by indiffusion of Li [40] or ion implantation of group I elements [39] which suggested a more general defect compensation mechanism. Figure 3.10 displays as an example $N(\varepsilon)$ distributions of differently doped films derived from sub-bandgap absorption spectra measured by optical defect spectroscopy using photocurrent spectroscopy (constant photocurrent mode CPM) and photothermal deflection spectroscopy (PDS) [40]. In this figure only the full and dotted curves have been obtained from the experimental data, the broken curves are assumed supplements. These results clearly show the enhancement of the defect density with the shift of the Fermi level to either side (see arrows) and reveal in addition that the defect distributions differ considerably in n-type and p-type material. Similar results have been reported by other authors [41, 42]. It is widely accepted that these defect bands arise from positively charged dangling bonds (D^+) in p-type a-Si:H located in the upper half of the gap and to negatively charged dangling bonds (D^-) in n-type a-Si:H located in the lower half of the gap. It has to be noticed

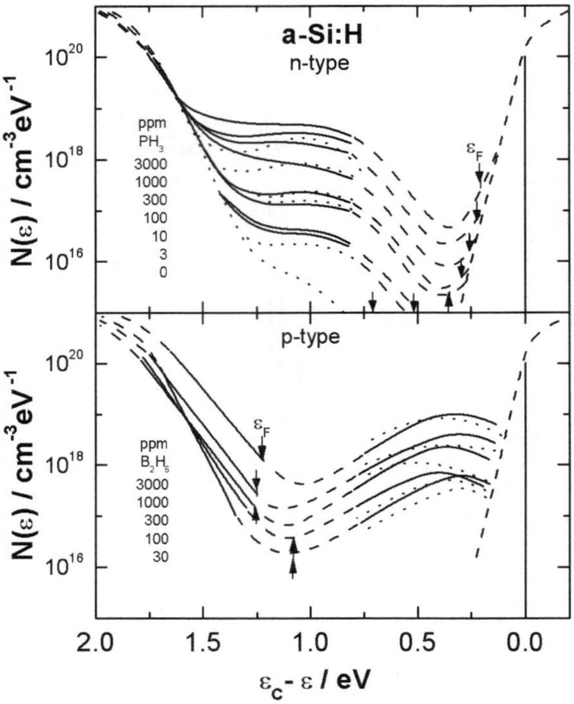

Figure 3.10 Density of states in the energy gap of phosphorus- and boron-doped a-Si:H derived from optical defect spectroscopy (dotted lines: photocurrent spectroscopy CPM, full lines: photo thermal deflection spectroscopy PDS). The broken lines are assumed supplements. The Fermi level position is indicated by the arrows (reproduced from [40] with permission of Taylor & Francis)

that this behavior is not related to correlation effects (negative correlation energy). The correlation energy of dangling bond defects in a-Si:H is positive. In undoped or weakly doped *n*-type films there may also be some contribution from neutral dangling bonds (D°) which are supposed to be located around midgap. The strong dependence of the defect density on the position of the Fermi level are interpreted as resulting from a decrease of the defect formation energy of charged defects by the charge transfer from the Fermi level ε_F to the defect.

There is a large body of information which shows that the defect structure of a-Si:H is not stable after film growth and fixed by the chosen substrate temperature, but is determined by an equilibrium process which is established after film growth at temperatures below T_S [43, 44]. Figure 3.11 shows an example which demonstrates to what extent the conductivity can be altered by a variation of the cooling rate when a sample is cooled from high to low temperatures. Above a temperature $T_G \simeq 410°C$ all curves merge into one curve which shows that the sample is in an equilibrium state. When the sample is cooled to temperatures below T_G the atomic and electronic structure is frozen into a slowly relaxing state which reminds one of the behavior of a glass. This behavior has been associated with the presence of hydrogen which becoming mobile at $T > T_G$ can act as a mediator of the equilibration

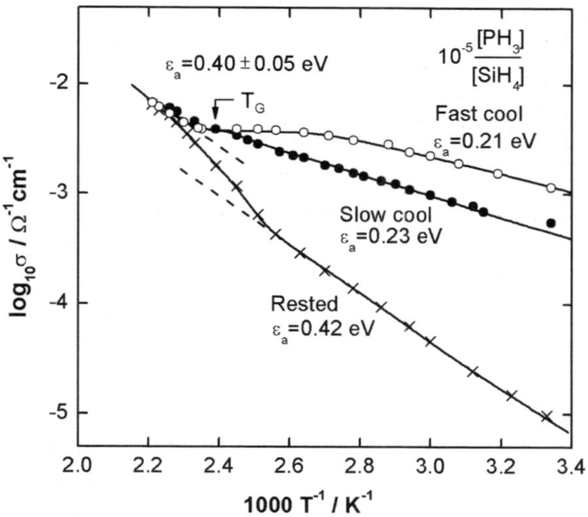

Figure 3.11 Temperature dependence of the dark conductivity after an anneal at 190°C for different cooling rates. ε_a denotes the activation energy of the conductivity. Rested: stored in the dark for several months (reproduced from [44] with permission of Elsevier)

process. Therefore the electronic properties at room temperature depend on the rate of cooling and are defined by the thermal equilibrium state at $T = T_G$. In the example shown in Figure 3.11 the conductivity at 300 K differs by more than two orders of magnitude, depending on the history of the sample. After the sample has been kept in the dark at 300 K for 7 months (slow cooling rate) the sample is almost in an equilibrium state showing the same activation energy as at $T > T_G$.

Such observations led to the conclusion that the density of states distribution in the energy gap of a-Si:H is the result of a chemical thermal equilibrium process. A detailed description of chemical reactions and theoretical concepts basing on thermodynamical arguments may be found in [11]. In order to explain the energetic position of the defect states (see Figure 3.10) it was proposed that in a disordered structure the energy of a Si dangling bond can take a range of values such that minimization of energy in the equilibration process will result in the formation of defects with the lowest energy. This is the basic idea of the defect pool model [45]. This idea was further developed by inclusion of a number of additional ingredients:

- It was proposed that the defect formation energy is different for the creation of dangling bonds in different charge states (–, 0, +) [46]. For instance, in *n*-type a-Si:H there is an energy gain by transfer of electronic energy in an amount of $\varepsilon_F - \varepsilon_{D-}$ when an electron is transferred from the Fermi level ε_F to the state of a negatively charged dangling bond at ε_{D-}. This is the effect which causes the different position of the dangling bond bands in the $N(\varepsilon)$ of *n*-type and *p*-type a-Si:H [40–42].
- It was suggested that the formation of dangling bonds occurs by breaking weak bonds which are related to the valence band tail states. This explains why the defect formation energy is very small [47, 48].

- It was argued that the defect density also depends on specific microscopic defect reactions [49].

These different aspects of defect formation were brought together in a theoretical formulation of the defect pool model [50] which allowed one to calculate the distribution of the defect states for undoped and doped a-Si:H and to show that such a model accounts for most of the observations. Further improvements of the model in particular dealt with the simultaneous formation of differently charged defects [51], and consideration of specific microscopic defect reactions [52].

The formation of metastable defects in a-Si:H which are created under external excitation such as carrier injection or illumination appears to be closely related to the defect equilibrium processes. The most famous of these observations is the light-induced defect creation. Figure 3.12 reproduces this observation from the historic publication by Staebler and Wronski [53]. The photoconductivity decreases as a function of the exposure time, and after illumination the dark conductivity has dropped by orders of magnitude, which indicates a pronounced shift of the Fermi level ε_F towards midgap. Annealing at temperatures of 120–150°C completely restores the original state. Due to its technological relevance this effect has been studied very intensively in the past, applying literally all methods that have been used to characterize physical properties of a-Si:H and many models have been put forward for interpretation (see, e.g., *Material Research Society Symposium Proceedings* **497**, 1997).

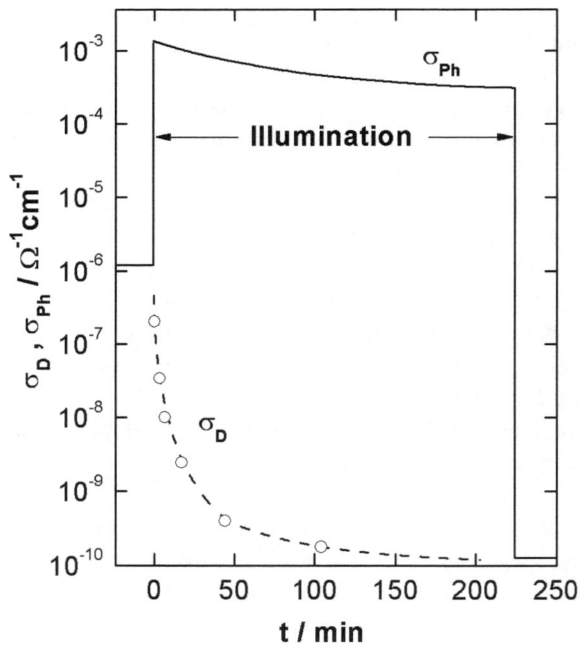

Figure 3.12 Photoinduced change of the photoconductivity σ_{PH} and the dark conductivity σ_D of a nominally undoped *n*-type a-Si:H film which had been deposited at 320°C. σ_D was measured by interrupting the illumination for short periods of time. The specimen was illuminated at 300 K with 200 mW/cm^2 at 600–900 nm (reproduced from [53] with permission of the American Institute of Physics)

Most authors agree that under external excitation silicon dangling bonds are created which decrease the carrier lifetime. Their density tends to saturate after long exposure times. Most results of gap state spectroscopies show that the light-induced defects cannot be distinguished from the defects created by unfavorable deposition conditions. The annealing process involves activation energies of 1–1.8 eV, and also the relaxation kinetics are very similar to those involved in the thermal relaxation process [11]. This behavior is taken as evidence that the defect creation and annealing involves the motion of hydrogen atoms. One proposal for the microscopic process is that the primary step is the trapping of excess carriers into band-tail states (weak bonds) which results in destabilization of the bond. When the bond is broken hydrogen is expected to move in from an adjacent SiH site to stabilize the defect [54]. This model explains the saturation of defect creation as arising from the competition of defect creation and annealing. The competition between trapping in band tails and recombination via the created defects accounts for the complicated nonlinear dependence of the density of light-induced defects on the generation rate and exposure time.

Metastability and defect equilibration, as formulated in the defect pool model, appear to be closely related. The formation energy of charged defects and therefore their concentration depends on the position of the Fermi energy ε_F. Therefore, any process which leads to a shift of ε_F results in a corresponding change of the defect density. The function of electronic devices (sensors, solar cells, transistors) is the result of a shift of ε_F by an external excitation. In operation this leads to degradation of the device performance which is often a serious challenge for device technology. Examples are the degradation of solar cells under extended exposure to light and the shift of the threshold voltage in the field effect transistor due to defect creation in the accumulation layer.

3.4 OPTICAL PROPERTIES

The optical spectra of amorphous silicon appear as broadened versions of the spectra of their crystalline counterpart. They differ in particular in two aspects [55]: (1) The fine structure of the crystalline spectra disappears since in a disordered structure there is no requirement of momentum conservation. (2) The imaginary part of the dielectric constant $\varepsilon_2(\omega)$ which depicts absorption processes, peaks at somewhat lower energy and the peak height depends on the preparation conditions. It has been suggested that disorder-induced weakening of the average bond strength is responsible for the observed red shift [55].

For applications of amorphous silicon in optoelectronic devices the position and shape of the absorption edge are of high relevance. Figure 3.13 compares absorption edges for a-Si:H, μc-Si:H and crystalline silicon c-Si. Whereas the spectra of the latter are quite similar, the absorption edge in a-Si:H is distinctively different. It is shifted to higher energy and in the visible spectral region the absorption in a-Si:H is larger by an order of magnitude than in c-Si. The latter has been assigned to the loss of momentum conservation in the disordered structure. The general behavior suggests the possibility of distinguishing three ranges in the absorption spectrum of the amorphous films.

(a) At $\alpha > 10^3\,\mathrm{cm}^{-1}$ absorption takes place between extended states and is often described by Tauc's expression [57]

$$(\alpha\hbar\omega)^{1/2} = B(\hbar\omega - \varepsilon_G). \tag{3.3}$$

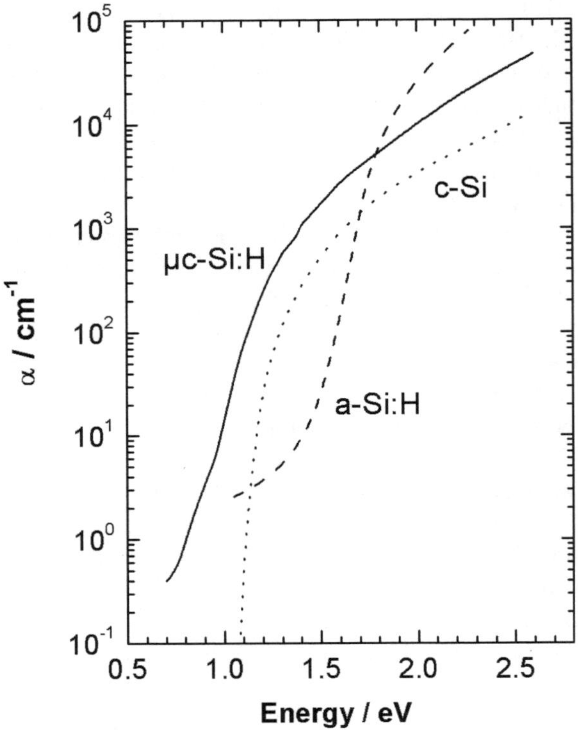

Figure 3.13 Optical absorption edges of hydrogenated amorphous and microcrystalline silicon, a-Si:H and μc-Si:H. For comparison the absorption edge of monocrystalline silicon (c-Si) is given. (reproduced from [56] with permission of the Materials Research Society)

In this expression $\hbar\omega$ is the photon energy, ε_G the energy gap and B an empirical constant. This relation has frequently been used to define an optical energy gap.

(b) For $\alpha < 10^3\,\mathrm{cm}^{-1}$ there is an exponential dependence of α on the photon energy often referred to as Urbach edge

$$\alpha \propto \exp[(\hbar\omega - \varepsilon_G)/\varepsilon_0], \tag{3.4}$$

ε_o is the Urbach parameter which is associated with the disorder in the amorphous film.

(c) In the low-energy range the curve levels off and forms a shoulder which has been assigned to defect absorption (see Section 3.3.3).

In an amorphous semiconductor an optical energy gap can be defined in various ways. In a rather arbitrary way the energies ε_{03} or ε_{04} where the absorption constant reaches values of $\alpha = 10^3$ or $10^4\,\mathrm{cm}^{-1}$ may be used to characterize the optical gap. Alternatively an optical gap may be defined on basis of Equation (3.3) which predicts that in a plot of $(\alpha\hbar\omega)^{1/2}$ versus $\hbar\omega$ the linear extrapolation of the straight line leads to a value of the optical gap (Tauc gap). The reasoning behind this expression is the concept of nondirect optical transitions which accounts for the loss of k-conservation in the disordered structure [57]. In this approach, one assumes that the optical transitions occur between the extended states of the valence and conduction bands whose density of states have a square root dependence on energy and that the momentum matrix element is constant. A critical discussion of these assumptions may be found in [8–11]. It has to be emphasized that there is no specific structure in $\alpha(\hbar\omega)$ at this energy. Nevertheless this definition of an optical gap has been quite commonly used for film characterization. In spite of the doubts about the validity of the underlying assumptions this definition is the only one with some physical significance. It may also be taken as a justification of the applied procedure that the value of $\varepsilon_G \simeq 1.75\,\text{eV}$ determined for optimized films compares favorably with values determined from transport measurements (see Section 3.5). The value of ε_G depends strongly on the hydrogen content C_H (see Figure 3.4).

Exponential absorption edges are a characteristic of amorphous semiconductors. In the best a-Si:H films this behavior has been found down to values of $\alpha \simeq 1\,\text{cm}^{-1}$. The value of the Urbach parameter ε_0 depends on temperature and in particular on the general disorder in the films. The Urbach edge in amorphous semiconductors can be related to the disorder-induced exponential band-tail states. Being determined by the broader of the two band-tail distributions, the Urbach parameter in a-Si:H is given by the slope of the valence band. In fact, ε_0 determined for optimally prepared films amounts to 45–50 meV which agrees well with the slope of the exponential valence band tail determined from photoemission or transport studies (see Section 3.3.2). The value of the Urbach parameter can easily be measured and therefore is a very convenient quantity to characterize the disorder in the amorphous films.

3.5 TRANSPORT PROPERTIES

Whereas in crystalline semiconductors the DC transport is fairly well understood and consistently described theoretically, the situation is far more complex in amorphous semiconductors. Hydrogenated amorphous silicon, a-Si:H, has attained the role of a model substance where, due to the broad information on material parameters and intensive careful transport experiments, a consistent picture based on theoretical concepts could be developed. But even in this case a variety of side effects influences the transport properties and often prevents the extraction of meaningful microscopic physical parameters from a particular experiment. The measurements are performed on thin films about 1 μm thick on glass or quartz substrates in lateral contact configurations. Surface and interface layers may have properties different from the bulk of the films, adsorbates may cause depletion or accumulation layers at the surface, the samples may be contaminated with impurities in differing amounts, impurities may diffuse in and hydrogen may diffuse out. A particular problem is the fabrication of stable ohmic contacts. Further complications arise from the observation that the density of states changes both with temperature [43, 44] and with extended illumination [53] (see Section 3.3). The defect equilibrium state is rapidly established at the high temperatures and is frozen in when the sample is cooled down below the critical temperature T_G. This is why all transport data are affected by the thermal history.

In an amorphous semiconductor three paths for transport are to be expected:

- In the high-temperature range transport can occur in extended states above the mobility edges ε_V and ε_C. In a-Si:H films the temperature dependence of the conductivity $\sigma(T)$ most often exhibits a well-defined activation energy which is associated with $\varepsilon_C - \varepsilon_F$ or $\varepsilon_F - \varepsilon_V$ for n-type or p-type films, respectively. In the more general case these activation energies are dependent on T. The existence of mobility edges has been questioned in theoretical work which considered the effect of electron–phonon coupling on disorder-induced localization [10]. It was shown that due to electron–phonon coupling at finite temperatures localization is lost at the mobility edges. The result is that the mobility edge should be substituted by the energy of a transport path which turns out to be located close to ε_C or ε_V and to depend only weakly on temperature. The carrier mobility cannot be deduced from steady-state experiments. Estimates lead in agreement with theoretical arguments to values of 1–10 cm^2/Vs for both electrons and holes [7, 10].

- When the temperature decreases, thermally activated tunneling among localized states in the energy gap with much lower mobility may prevail. In a-Si:H this is not observed in DC conductivity except at very low temperature under high optical excitation and very high electric fields [58] (see Figure 3.21 in Section 3.6.1).
- If the density of states near the Fermi level is high, hopping transport at the Fermi level can be the dominating process [7, 10]. Theoretical concepts for variable-range hopping have been developed and it appears that this mechanism has clearly been identified as the predominanting electronic transport process in highly disordered nonhydrogenated amorphous semiconductors. However, it turned out that in most cases the microscopic parameters deduced for instance from Mott's famous $T^{1/4}$ law were highly unrealistic.

Details of the theoretical description of charge transport in all these regimes can be found in the Chapter 2 of this book.

The observation that a-Si:H can effectively be doped from the gas phase in the PECVD deposition [4] strongly shifted the scientific and technological interest to transport in extended states. In the following, measurements of the DC conductivity and thermoelectric power will be described and it will be shown that two important aspects play an essential role in the interpretation of transport data in amorphous semiconductors: the statistical shift of the Fermi energy $\varepsilon_F(T)$ and the existence of long-range potential fluctuations.

When transport occurs at the mobility edge ε_C, the conductivity of n-type films can be written as

$$\sigma = \sigma_0 \exp[-(\varepsilon_C - \varepsilon_F)/kT]. \tag{3.5}$$

In this expression k denotes the Boltzmann constant and ε_F the Fermi energy. Similar expressions hold for hole conduction at ε_V. There has been much controversy about the magnitude of the prefactor σ_0. It sounds reasonable to identify the conductivity at the mobility edge with Mott's minimum metallic conductivity which leads to $\sigma_0 = 150\,\Omega^{-1}\,\text{cm}^{-1}$ [59]. More sophisticated theories have shown that the concept of minimum metallic conductivity may be inappropriate. However, several alternative theoretical approaches led to the conclusion that the prefactor should be close to this value, but with a different justification [10].

The thermoelectric power S is related to the Peltier coefficient Π by Onsager's relation

$$S = \Pi/eT. \tag{3.6}$$

Π/e is identical with the energy of the conducting electrons relative to the Fermi level which is given by the sum of $\varepsilon_C - \varepsilon_F$ plus a term which accounts for the kinetic energy of the carriers above the mobility edge which may be set to kT. This leads to the relation

$$e/k\,S(T) = (\varepsilon_C - \varepsilon_F)/kT + 1. \tag{3.7}$$

The problem in the analysis of experimental data is that the involved energies may depend on temperature, $\varepsilon_F(T)$ and $\varepsilon_C(T)$. This may arise, for instance, from a temperature dependence of the energy gap or of the transport path at ε_C or from a statistical shift of the Fermi energy. As a consequence, in an Arrhenius plot of Equation (3.5) the slope will give an apparent activation energy ε_σ which is different from $\varepsilon_C - \varepsilon_F$ and the extrapolation to $1/T = 0$ will result in a value for the prefactor σ_0^* which differs appreciably from the microscopic prefactor σ_0:

$$\sigma = \sigma_0^* \exp(-\varepsilon_\sigma/kT). \tag{3.8}$$

Similarly, a plot of S versus $1/T$ will give a slope ε_S which is different from the activation energy $\varepsilon_C - \varepsilon_F$ in the expression (3.7) for the thermoelectric power. ε_σ determined from Arrhenius plots according to Equation (3.8) can thus be markedly different from the actual value of $\varepsilon_C - \varepsilon_F$ and may not even have a simple physical meaning. As a very convenient method to determine the actual position of the Fermi energy one has to measure the DC conductivity at a given temperature $\sigma(T_o)$ and then to calculate $\varepsilon_C(T_o) - \varepsilon_F(T_o)$ from expression (3.5), inserting the microscopic prefactor of $\sigma_0 = 150\,\Omega^{-1}\,\text{cm}^{-1}$. Since this prefactor is universally valid, this procedure can be applied to all amorphous semiconductors which show activated behavior of the DC conductivity.

It was one of the most important breakthroughs in the physics of amorphous semiconductors when Le Comber and Spear [4] reported that amorphous silicon prepared by PECVD can effectively be doped by adding small amounts of phosphine PH_3 or diborane B_2H_6 to the silane SiH_4 in the discharge gas. In amorphous semiconductors the donors deliver their electrons to empty states at the Fermi level ε_F. Therefore, the resultant shift of ε_F depends on the density of states distribution. However a quantitative evaluation of the doping effect turns out to be rather difficult. In general, the concentration of the dopants in the films is different from that in the gas phase and widely different numbers between 0.5 and 5.2 are given for the incorporation ratio [11, 60]. Additional uncertainty arises from the lack of knowledge about the doping efficiency which describes the part of the incorporated impurity atoms located on substitutional sites. Figure 3.14 displays the variation of the dark conductivity with the gas phase concentration of the doping gases. For P-doping the maximum conductivity at 300 K, σ_{RT}, of $10^{-2}\,\Omega^{-1}\,\text{cm}^{-1}$ is attained at a concentration of 10^3–10^4 vppm PH_3. At higher doping levels, σ_{RT} decreases, presumably due to the generation of defect states. For boron doping, σ_{RT} decreases at low doping levels and attains, at higher doping levels, maximum values of about $10^{-3}\,\Omega^{-1}\,\text{cm}^{-1}$. Similar results have been obtained by ion implantation of various elements and also by indiffusion of elements which diffuse at low temperature below T_S.

The minimum of σ_{RT} at low B-doping concentrations is due to the change from n-type to p-type conduction when the Fermi level crosses midgap such that $\varepsilon_C - \varepsilon_F = \frac{1}{2}(\varepsilon_C - \varepsilon_V)$. Inserting $\sigma_{RT} = 10^{-12}\,\Omega^{-1}\,\text{cm}^{-1}$ and $\sigma_0 = 150\,\Omega^{-1}\,\text{cm}^{-1}$ in Equation (3.5) one obtains $\varepsilon_C - \varepsilon_V = 1.7$ eV for the magnitude of the mobility gap in a-Si:H at 300 K. A value of 1.75 eV was obtained from a similar analysis of compensated films [61].

Representative temperature dependences of the conductivity $\sigma(T)$ and the thermoelectric power $S(T)$ are displayed in Figure 3.15 for a series of phosphorus-doped a-Si:H films. The curves are roughly activated, the more strongly doped films show a characteristic kink at

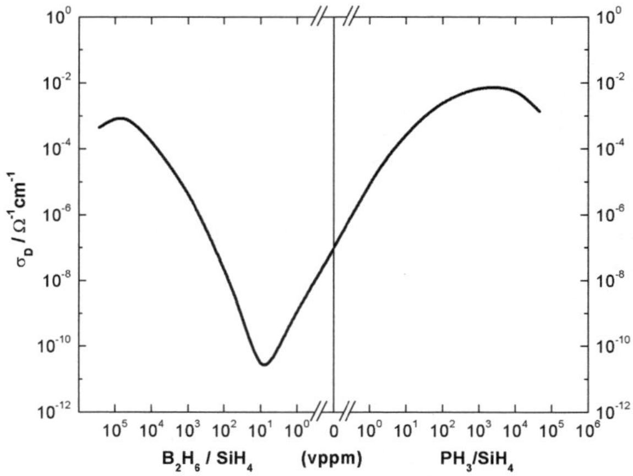

Figure 3.14 Room temperature conductivity σ_D of a-Si:H as a function of the dopant concentration in the gas phase. The curve is an average curve through data from various laboratories

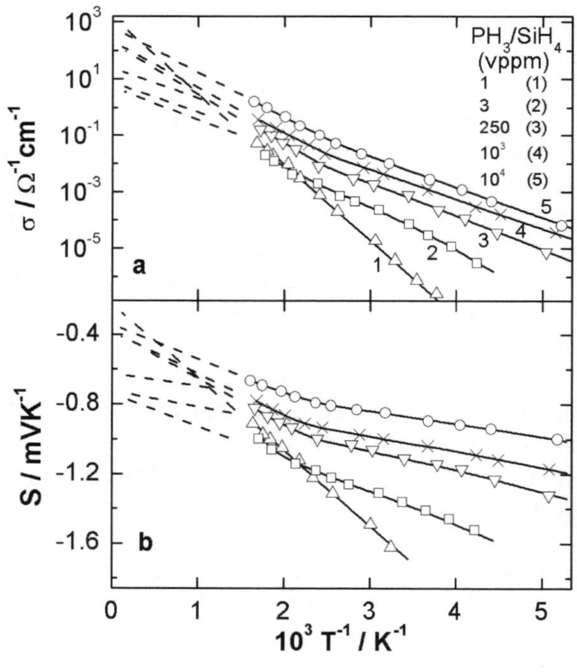

Figure 3.15 Temperature dependence of the conductivity σ and thermoelectric power S of a-Si:H at various doping levels (reproduced from [60] with permission of Elsevier)

about 400 K, which can be associated with the onset of defect equilibration taking place at $T > T_G$. The most striking feature is the wide spread of values for the apparent prefactor σ_0^* obtained from the extrapolation to $1/T = 0$. σ_0^* varies in the range 1–$10^4\,\Omega^{-1}\,cm^{-1}$ and depends on both temperature and doping level. This behavior is in accordance with observations by many other authors who also used other dopant atoms such as As, Sb, Bi, Li, K, and Na.

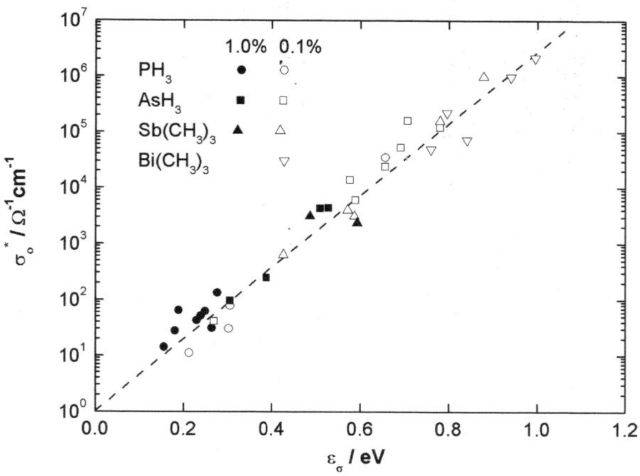

Figure 3.16 Preexponential factor σ_0^* as obtained from extrapolations of the temperature dependences of the conductivity of *n*-type a-Si:H (Equation 3.8, Arrhenius plots) to $1/T = 0$ as a function of the activation energy ε_σ at 300 K for various dopants (reproduced from [62] with permission of Springer)

The example shown in Figure 3.16 reveals that ln σ_0^* varies approximately linearly with the apparent activation energy ε_σ, thus following a Meyer–Neldel rule. This rule was first reported by Meyer and Neldel [63] in 1937 for baked metal oxide powders and has very frequently been observed in all kinds of systems. Beyer and Overhof [60] explained this behavior quantitatively by a statistical shift of the Fermi energy with temperature, $\varepsilon_F(T)$. Such a shift is required because, due to the neutrality condition, the total density of electrons is independent of temperature. This shift is also well known from crystalline semiconductors. For instance, in a singly doped *n*-type crystalline semiconductor, ε_F is close to the donor level at low temperatures and moves towards midgap with increasing temperature in the exhaustion regime where the electron concentration is constant. In an amorphous semiconductor, this shift will be large when the density of deep gap states is small compared with that in the band tails, which is the case in all hydrogenated semiconductors. Assuming for instance a linear dependence of the Fermi energy on temperature

$$\varepsilon_C - \varepsilon_F(T) = (\varepsilon_C - \varepsilon_{F^\circ}) + \delta T, \tag{3.9}$$

one easily finds

$$\sigma(T) = \sigma_0 \exp(-\delta/k)\exp[-(\varepsilon_C - \varepsilon_{F^\circ})/kT]. \tag{3.10}$$

Hence, from an Arrhenius plot (lnσ vs $1/T$) neither the actual value of $\varepsilon_C - \varepsilon_F(T)$, nor the microscopic prefactor σ_0 can be determined. There are very few simple cases in the literature where σ_0 can be estimated directly from experimental results. This is the case when the Fermi level is located at midgap in a region where the density of states is large such that the temperature dependence can be neglected, as realized in undoped electron-bombarded a-Si:H [36]. Similarly, in a perfectly compensated sample, the Fermi energy is located at midgap [61]. The prefactor observed in such samples is $\sigma_0^* \simeq 2300\,\Omega^{-1}\,\text{cm}^{-1}$.

Correcting for the temperature dependence of the energy gap, one obtains a value for the microscopic prefactor of $\sigma_0 \simeq 150\text{–}200\,\Omega^{-1}\,\mathrm{cm}^{-1}$, which is close to the theoretical value for the minimum metallic conductivity [61].

Beyer and Overhof [60, 64] have suggested discussing transport data by defining a Q-function which combines $\sigma(T)$ with $S(T)$

$$Q(T) = \ln(\sigma\Omega\mathrm{cm}) + e/k\, S(T). \tag{3.11}$$

The advantage of this Q-function is that the temperature dependences of ε_C and ε_F cancel. Hence Q does not depend on electron statistics, but contains information on the transport path at ε_C only. Equations (3.5) and (3.7) readily give a temperature-independent value of Q:

$$Q(T) = \ln(\sigma_0\Omega\mathrm{cm}) + 1. \tag{3.12}$$

Any change in the transport path should lead to a change of σ_0. Structure in $\sigma(T)$ and $S(T)$ which does not also show up in $Q(T)$, is due to the temperature dependence of ε_F. In particular, if there is a difference in the activation energies ε_σ and ε_S this will result in a nonzero slope ε_Q in a plot of Q vs $1/T$:

$$\varepsilon_Q = dQ/d(1/T) = \varepsilon_\sigma - \varepsilon_S. \tag{3.13}$$

In Figure 3.17, the results of the P-doped films are replotted in terms of the Q-function. The data fall nicely on straight lines, the kinks in the data of Figure 3.15 have disappeared. Two conclusions may be drawn from this result: (1) there is no indication of a second conduction path; (2) ε_σ and ε_S are different, the slope of the straight lines increases from 0.05 eV for the undoped film to 0.15 eV for higher doping levels. Such behavior has been observed for a large variety of doped and undoped, n-type and p-type films prepared under various conditions.

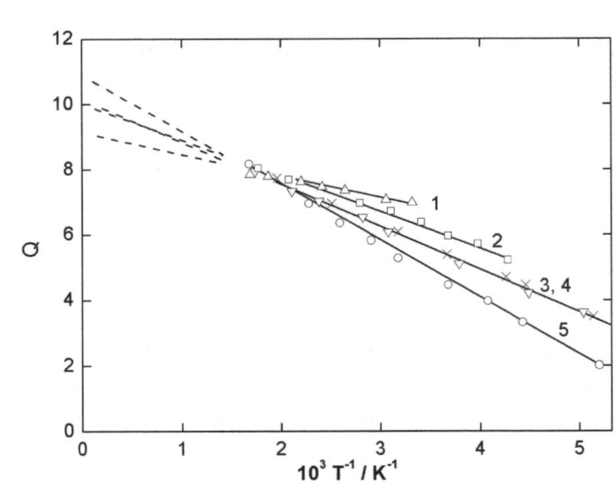

Figure 3.17 Q-function defined by Equation (3.11) of the phosphorus-doped films (same sample numbers as in Figure 3.15) (reproduced from [60] with permission of Elsevier)

The observations have consistently been explained by the proposal that long-range static potential variations may lead to $\varepsilon_Q \neq 0$ by causing long range fluctuations of the mobility edges which affect the thermoelectric power much more than the conductivity [64]. Potential fluctuations may arise from local density fluctuations, growth inhomogeneities or from electric fields due to charged defects. For the latter case, it was shown that in highly doped films, the random distribution of ionized donors results in internal potential fluctuations of the order of 0.15 eV on a length scale of typically 20 nm.

The DC transport properties of hydrogenated films appear to be quite well understood. The conductivity and the thermoelectric power can be described by the standard expressions, relevant microscopic parameters can be deduced and linked with the density of states distribution. As a universal feature, the value of the microscopic prefactor σ_0 is found experimentally to amount to 150–200 Ω^{-1} cm^{-1} and thus is close to what is expected from theory. The experimental results show that the temperature dependences of the transport parameters are strongly determined by the statistical shift of the Fermi energy and by long-range potential fluctuations. The understanding, however, is still incomplete. So far there is no explanation for the double sign anomaly of the Hall effect [10].

3.6 RECOMBINATION OF EXCESS CARRIERS

Photoconduction and photoluminescence have been studied intensively for a long time [9, 11, 65] since the recombination of excess carriers is of high relevance for device applications such as solar cells or sensors. Figure 3.18 presents a survey of the behavior of a-Si:H which demonstrates that photoconduction and photoluminescence have anticorrelated temperature dependences and thus appear to be competing processes. Clearly, two temperature ranges can be distinguished [66]:

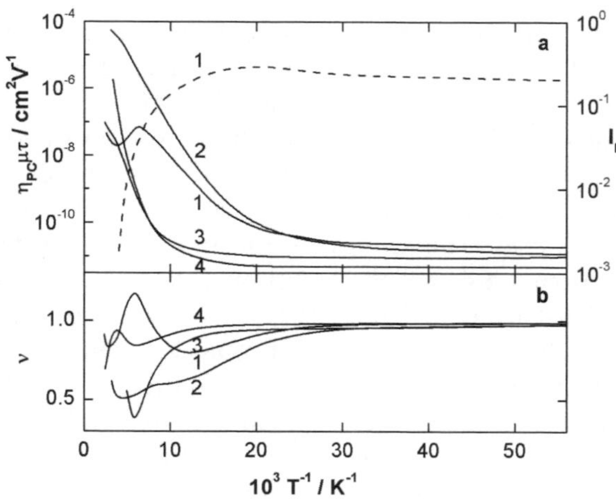

Figure 3.18 Temperature dependence of the normalized photoconductivity $\eta_{PC}\mu\tau$ (a) and of the exponent ν of the intensity dependence $\sigma_{PH} \simeq G^{\nu}$ of various a-Si:H films. PECVD-films: (1) undoped; (2) 100 vppm PH_3; (3) 1000 vppm B_2H_6; sputtered film: (4) undoped sputtered. The dashed line shows the temperature dependence of the photoluminescence intensity I_{PL} of sample (1) (reproduced from [66] with permission of Elsevier)

1. At $T < 60\,\text{K}$ the photoluminescence is very high with an efficiency close to unity, the photoconductivity (here divided by charge and generation rate σ_{PH}/eG) is very low and both quantities do not depend on temperature. σ_{PH} depends linearly on the generation rate and varies only slightly with the defect density or doping level. Such behavior suggests a mechanism of photoconduction which is pretty universal.

2. At $T > 60\,\text{K}$ the photoluminescence intensity is quenched and the photoconductivity increases strongly. In this high-temperature range the kinetics, the dependence on the generation rate (here shown by the exponent of the power law dependence $\sigma_{PH} \propto G^{\nu}$), the magnitude of σ_{PH} and the detailed kinetics are determined by the occupancy of the defect states.

The interpretation of such data requires information on the dominant recombination and transport paths, the density and energy distribution of the participating states. In addition, one has to consider the random location of the states in space. The intensive work on photoluminescence, photoconductivity and in particular spin-dependent recombination as studied by optically detected magnetic resonance (ODMR), and electrically detected magnetic resonance (EDMR), have provided evidence that Si dangling bonds are the most effective recombination centers in the entire temperature range.

3.6.1 Low-temperature regime ($T < 60\,\text{K}$)

The photoluminescence spectrum of a-Si:H of low defect density exhibits a single structureless emission band centered at 1.3–1.4 eV with a half-width of 0.25–0.3 eV. This emission has been associated with recombination between electrons and holes localized in the respective band tails. The quantum efficiency of the intrinsic emission is very high, the reported values amount to 0.3–1 for optimized films [65]. In defect-rich or highly doped films this intrinsic band is quenched and an additional structure appears at 0.8–0.9 eV which is somewhat broader (0.35–0.4 eV) and has a much lower efficiency. Figure 3.19 shows an example of such spectra for undoped and boron doped a-Si:H films.

The recombination kinetics are characterized by a broad distribution of lifetimes which extends over more than ten orders of magnitude from 10^{-8} up to $10^{2}\,\text{s}$ [65, 67–70]. This experimental experience led to the proposal of models which have in common that the carriers recombine by radiative tunneling between tail states such that the broad distribution of lifetimes results from a distribution of electron–hole separations. In the geminate pair model, it is assumed that diffusion during thermalization does not separate the photoexcited electrons and holes such that the recombination occurs between trapped electron–hole pairs which have been created in the same absorption event [11, 65]. In the distant pair model [71], it is supposed that the carriers are able to diffuse apart to large distances and recombine radiatively with the nearest available partner in a nongeminate process. When the generation rate G increases, there should be a transition from geminate to nongeminate recombination when the electron–hole separation decreases due to the increasing carrier concentrations. These models lead to different consequences for photoconduction because only those carriers can contribute to transport which escape geminate recombination. Therefore, the anticorrelated temperature dependences of the PL intensity and photoconductivity (Figure 3.18) have been taken as evidence for the predominance of geminate

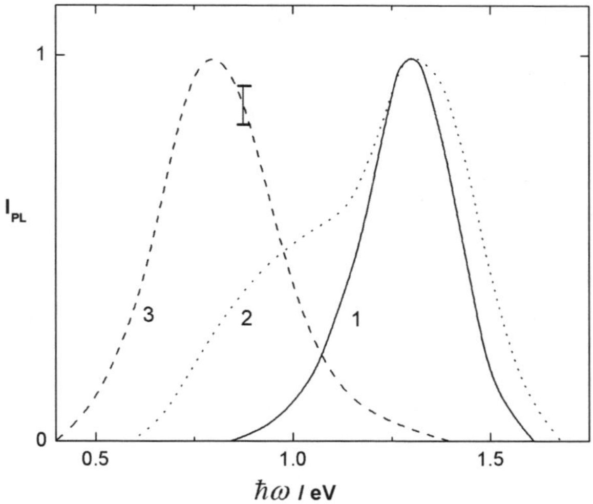

Figure 3.19 Normalized photoluminescence spectra at $T = 10\,K$ of a-Si:H films: (1) undoped; (2) doped 100 vppm B_2H_6; (3) doped 1000 vppm B_2H_6

recombination at low temperature. A similar anticorrelation has been reported to exist at high electric fields where it was found that the highly superohmic I–V dependence was connected with quenching of the photoluminescence [58] (see Figure 3.21).

The theoretical description starts from the assumption that the thermalization process in the band tails proceeds by hopping transitions. The rate for a hop over a distance r to a state of lower energy is given by

$$\nu = \nu_0 \exp(-2r/a) \tag{3.14}$$

with a prefactor $\nu_0 \simeq 10^{12}\,s^{-1}$ and the Bohr radius $a \simeq 10\,Å$ of the band tail electron. When the electron hops to a state which is by $\Delta\varepsilon$ higher in energy, this rate has to be multiplied by $\exp(-\Delta\varepsilon/kT)$. Thermalization competes with radiative tunneling, the rate of which is described by

$$\nu_r = \tau_0^{-1} \exp(-2R/a). \tag{3.15}$$

In this expression R denotes the electron–hole pair separation, and the prefactor is assumed to be $\tau_0 \simeq 10^{-8}\,s$ with a value typical for an allowed electrical dipole transition. Due to the difference of the prefactors of four orders of magnitude the final radiative transition is the rate-limiting step of the overall process and therefore the radiative lifetime is given by the inverse of the recombination rate

$$\tau_r = \tau_0 \exp(2R/a). \tag{3.16}$$

This model relates the broad distribution of radiative lifetimes with the distribution of electron–hole pair separations and is referred to as radiative tunneling model. The model appears to be supported by the characteristic dependence of the lifetime distribution on the generation rate G [67, 68]. The spectra measured by frequency-resolved spectroscopy (FRS) showed a major component at $\tau_1 \simeq 1$ ms and a minor contribution at $\tau_2 \simeq 1\,\mu$s. Both recombination channels are independent of G at excitation densities $G < G_0 \simeq 5 \times 10^{19}\,\mathrm{cm^{-3}\,s^{-1}}$ as one expects for predominant geminate recombination. The spectra shift to shorter times at $G > G_0$ when the carrier concentration becomes large such that τ_r decreases strongly due to the decreasing separation of the recombining carriers, and recombination occurs with the nearest available partner (distant pair recombination) [67, 71].

The radiative tunneling model, although accounting for many experimental observations, led to predictions which are difficult to reconcile with the experimental results, both at high and low generation rates. The first problem is that the model as sketched above does not allow for geminate recombination. Computer simulations on the basis of this model [72] which took into account the competition between thermalization and radiative recombination showed that geminate recombination and distant pair recombination can coexist. Details of these model calculations are presented in the Chapter 2 of this book. However, it turned out that the derived probability function $P(R)$ for geminate recombination at an electron–hole separation R was much broader than the experimental life time distribution. Later computer simulations using essentially the same model, but considering steady-state conditions [73] revealed that the carriers diffusing to large distances accumulate due to their large lifetime which finally results in prevailing distant pair recombination. The calculated lifetime distribution shifts monotonously with increasing excitation density to shorter times which is in clear contradiction with the experimental finding. At high generation rates, a number of experiments show that the total PL intensity increases sublinearly with G [67, 69] and concomitantly the FRS spectra shift to shorter times. These studies led unambiguously to the conclusion that the lifetime shortens at high values of G due to a nonradiative recombination channel [69]. Therefore it has been suggested to modify the model by assuming that only the geminate process is radiative while distant pair recombination is predominantly nonradiative [74]. The dependence of the lifetime spectra on the generation rate is then the result of a competition between radiative geminate recombination and nonradiative distant pair recombination, the lifetime distribution of which causes the strong shift to shorter times. With this assumption, both the shift of the spectra and the decrease of the quantum efficiency could be explained satisfactorily. However, the question remained unanswered what physically might be the difference between geminate and non-geminate pairs. It was proposed that the geminate pairs are identical with excitons and, indeed, the average lifetime of about 1 ms would be in the right order of magnitude for the recombination of triplet excitons [74]. Further insight was obtained from a recent study of frequency-resolved spectroscopy [70] which showed that at low generation rates $G < G_0$ the lifetime spectra consist in fact of three features. Besides the known structures at τ_1 and τ_2 which have been assigned to the geminate process there is a third component at longer times $\tau_3 \simeq 10^{-2}$–10^{2} s which moves to shorter times with increasing G until it merges into the τ_1 component. These experiments prove convincingly that at low G both geminate and non-geminate recombination do coexist. In fact this is what has been suggested for a long time. The steady-state carrier concentration in the band tail measured by light-induced spin resonance (LESR) exhibits, over the entire range of G values, a strongly sublinear dependence on the generation rate ($n_S \propto G^{0.17}$) [68]. Such behavior is characteristic of distant pair recombination [71]. This led to the suggestion that radiative recombination occurs predominantly by geminate kinetics (lifetimes τ_1 and τ_2) and the long-living pairs (lifetime τ_3) detected by LESR follow distant pair kinetics.

While time- or frequency-resolved measurements give valuable information on the kinetics of recombination, magnetic resonance studies aim at a microscopic identification of the recombining species. Various studies of optically detected magnetic resonance have led to the suggestion of excitonic recombination in a-Si:H. However, the interpretation of the traditional ODMR spectra turns out to be rather complicated. Often enhancing and quenching signals overlap and the influence of the experimental parameters (probing frequency, light intensity, microwave power, temperature) have led to conflicting observations. In particular, such measurements do not allow one to derive information on the specific coupling of the recombining electron–hole pairs. Recently, new methods for studies of spin-dependent recombination pulsed optically and electrically detected magnetic resonance (PEDMR and PODMR) have been introduced, and it was demonstrated that such techniques enable a convincing identification of recombination channels in a-Si:H [75]. The main difference between PODMR and the traditional ODMR is that in PODMR all spins are excited coherently by a high-power microwave pulse and can perform Rabi oscillations. The Rabi frequencies contain direct information on the spin state and coupling of the recombining species. The basic theory for the time-domain of spin-dependent recombination will be treated in Chapter 5 of this book. The results of such experiments on a-Si.H [75] is that there are clearly three different recombination channels: (1) nonradiative tunneling from band tail electrons into weakly coupled dangling bond states (see below); (2) recombination of exchange coupled electron–hole pairs which are assigned to excitons (single line at $g = 2.008$ involving a $S = 1$ triplet state); (3) recombination of strongly dipolar coupled electron–hole pairs with a minimum separation of about 9 Å while the average amounts to about 16 Å. Both exciton recombination and the recombination of strongly coupled electron–hole pairs are radiative processes. This is the first real proof of excitonic recombination and of radiative distant pair recombination in a-Si:H.

The kinetics of low-temperature radiative recombination in a-Si:H is dominated by geminate recombination at low and moderate generation rates and by nongeminate recombination at high generation rates. The geminate process can be assigned to excitonic recombination with a lifetime distribution consisting of two components τ_1 and τ_2. It is tempting to associate these components with singlet (τ_2) and triplet (τ_1) recombination of excitons. Even at the lowest generation rates, radiative recombination by distant pairs competes and finally becomes dominant at $G > 5 \times 10^{19}\,\mathrm{cm}^{-3}\,\mathrm{s}^{-1}$. In addition, at very high generation rates and very low temperatures ($T < 10\,\mathrm{K}$) a nonradiative channel becomes competitive which presumably is due to Auger recombination.

Si dangling bonds are known to be the most efficient centers for nonradiative recombination. When the density of these defects exceeds $10^{17}\,\mathrm{cm}^{-3}$, the PL intensity is strongly quenched due to nonradiative tunneling of band-tail electrons into defects (Figure 3.20). The microscopic nature of this process has been identified in many ODMR studies, most recently by PODMR [75]. The rate of this process can in analogy with relation (3.15) be written as

$$\nu_{\mathrm{nr}} = \nu_0 \exp(-2R_\mathrm{d}/a) \tag{3.17}$$

with $\nu_0 \simeq 10^{12}\,\mathrm{s}^{-1}$, the extension of the band tail state $a \simeq 10$ Å, and the tunneling distance R_d to the defect. Quenching of the PL intensity will take place when $\nu_{\mathrm{nr}} > \nu_\mathrm{r}$ which defines a critical radius $R_\mathrm{C} = 1/2a \ln(\nu_0\tau_\mathrm{r})$. For $R_\mathrm{d} < R_\mathrm{C}$ nonradiative recombination prevails. Street [65] using a random distribution of defects of density N_S, derived an expression for the quantum efficiency

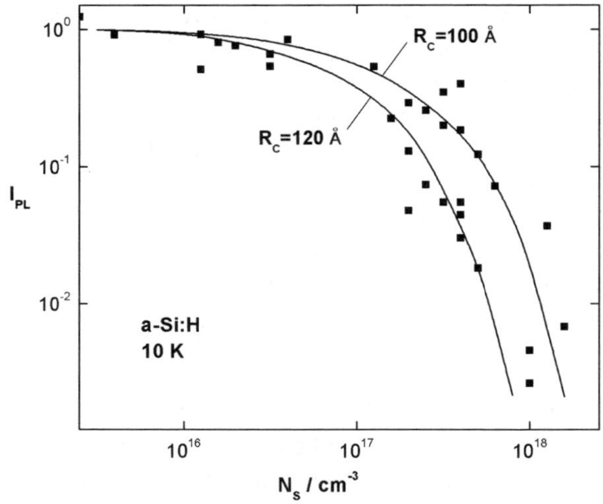

Figure 3.20 Photoluminescence intensity versus spin density of a variety of differently prepared a-Si:H films. The curves are fits to Equation (3.18) using the given values for R_C as parameters (reproduced from [18] with permission of Taylor & Francis)

$$\eta = \exp(-4\pi/3R_C^3 N_S). \tag{3.18}$$

The solid lines in Figure 3.20 for two R_C values of 100 and 120 Å show that this formula gives a reasonable description of the experimental results. $R_C = 100$ Å is consistent with the values $\tau_r = 1$ ms and $a = 10$ Å.

The low-temperature photoconductivity σ_{PH} has been associated with hopping in the band tail [58, 72]. The most important observations are [58, 66]: (1) $\sigma_{PH} \propto G$; (2) $\sigma_{PH}/eG \simeq 10^{-11}\,\text{cm}^2\,\text{V}^{-1}$, depending only slightly on the defect density; (3) decay of σ_{PH} on a time scale shorter than the average radiative lifetime; (4) pronounced field dependence of σ_{PH}. The latter is demonstrated by Figure 3.21 which shows *I–V* characteristics of a-Si:H under illumination at various temperatures. It is important to notice that the field dependence is connected with field quenching of the photoluminescence, which may arise from field-induced separation of geminate pairs. In case of band-tail hopping, a strong field dependence has to be expected since the applied field increases the number of available states in the field direction and thus tends to keep the carriers moving in a high density of states. A theory of hopping photoconductivity has been developed for finite temperatures and high electric fields which led to a satisfactory explanation of these observations [76–78]. These theories suggest that as a result of excitation to higher band-tail states a transport path develops in the band tail at an energy E_t which moves as a function of temperature and electric field strength to higher energy. This is connected with an enhancement of photoconductivity σ_{PH} and a concomittant quenching of the photoluminescence intensity. Finally at $T > 100$ K the transport channel is at the mobility edge and photoconduction arises predominantly from transport in the extended states.

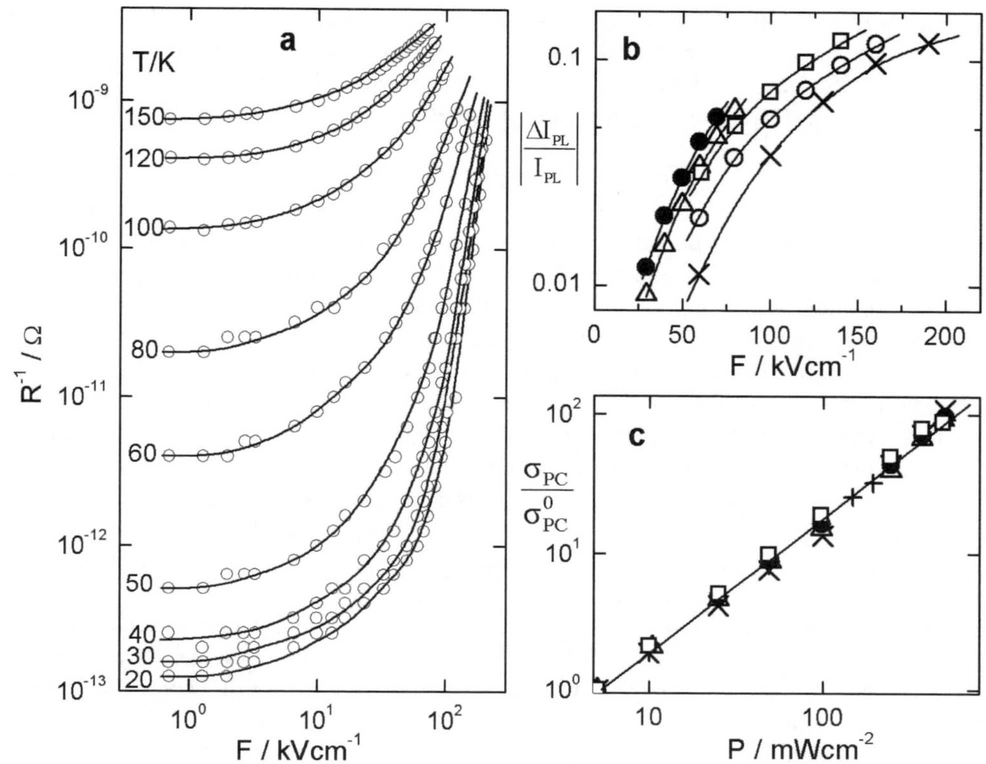

Figure 3.21 (a) I–V characteristics of illuminated a-Si:H (low defect density) plotted as photoconductance $1/R$ versus electric field strength F at various temperatures; (b) field quenching of the photoluminescence: (crosses) 30 K, (open circles) 50 K, (squares) 60 K, (triangles) 120 K, (full circles) 150 K; (c) dependence of the photocurrent I/I_0 (normalized to its value at $P = 5\,\mathrm{mWcm^{-2}}$) on the light intensity P at various values of the electric field strength F/kVcm^{-1}: (full circles) 6.7, (triangles) 50, (squares) 100, (crosses) 150, (plus signs) 200 (reproduced from [58] with permission of Taylor & Francis)

3.6.2 High-temperature regime ($T > 60\,\mathrm{K}$)

Photoconductivity measurements are performed in gap-like electrode arrangements. The strong influence of contact properties has often been ignored in the literature and may be the reason for many conflicting experiences in measurements of steady state and in particular transient photoconduction. In secondary photoconduction, electrons which are extracted at one electrode are to be replenished at the other contact. These secondary photocurrents can be described by an expression

$$j_{PH} = e\eta\Phi_{Ph}(1-R)[1-\exp(-\alpha d)]\tau/t_{tr}. \qquad (3.19)$$

In this expression Φ_{Ph} $(1 - R)$ is the photon flux corrected for the surface reflection, and η describes the quantum efficiency for the creation of free carriers. The gain g is defined by the number of carriers which cross the contact divided by the number of created free carriers and is given by $g = \tau/t_{tr}$, the ratio of the recombination lifetime τ and the transit time t_{tr} for electrons. The latter is given by $t_{tr} = L/\mu F$ where L is the electrode separation, F the electric field strength and μ the carrier mobility. Thus the gain can be written as

$$g = \mu\tau V/L^2. \tag{3.20}$$

In secondary photoconduction, the gain is determined by the $\mu\tau$ products and applied voltage and therefore can be much larger than one. Typical values for the $\mu\tau$ products are $(\mu\tau)_n \simeq 10^{-6}\,\mathrm{cm^2/V}$ and $(\mu\tau)_p \simeq 10^{-8}\,\mathrm{cm^2/V}$ [11]. This mode of photoconduction has to be distinguished from primary photoconduction which occurs when the electrodes are blocking, as realized in a Schottky barrier or pin diode under reverse bias (see Section 3.7). In general, the photocurrent will be the sum of electron and hole currents. However, the general finding in a-Si:H is that the current is carried predominantly by the majority carriers which in undoped and P-doped specimens are electrons and in B-doped samples are holes. As expected from expression (3.19), the spectral dependence of j_{PH} follows that of the absorption spectra (see Figure 3.13) exhibiting exponential edges. Most of the spectra show a shoulder at low photon energies which results from defect absorption and has frequently been used for defect characterization (see Sections 3.3.3 and 3.4).

The temperature dependence $\sigma_{PH}(T)$ is rather complex, being determined by both the trapping and thermal release from band-tail states and the temperature dependence of the recombination lifetime. Similarly complicated is the dependence on the generation rate G which follows to a good approximation a power law, $\sigma_{PH} \propto G^{\nu}$, with an exponent ν between 0.4 and 1.2 depending on sample quality, doping level and temperature (Figure 3.18). Of particular interest is the influence of doping. Above 80 K activated temperature dependences are often observed with activation energies of 0.1–0.15 eV for n-type and 0.2–0.3 eV for p-type films (see Figure 3.18). These energies may be attributed to an average trap depth for the respective majority carriers. This assignment is in agreement with results from time-of-flight studies, which also showed that the trap depth is about twice as large for holes as for electrons [11, 31, 33]. In undoped films of low defect density there is often a temperature range where $\sigma_{PH}(T)$ decreases with increasing temperature (Figure 3.18a, curve 1). This quenching effect is a result of a decrease of the recombination lifetime when minority carriers are excited either thermally or by optical excitation from deep trapping levels [79, 80]. The magnitude of the photoconductivity depends strongly on the position of the Fermi level. This is evident from Figure 3.18 and shown in more detail in Figure 3.22 where σ_{PH} at 300 K is plotted as a function of the position of the Fermi level, $\varepsilon_C - \varepsilon_F$, for a large number of undoped and doped a-Si:H samples. The most remarkable feature is the tremendous increase of σ_{PH} when the Fermi level moves from near midgap to about 0.4 eV below ε_C [79, 81]. Connected with this enhancement is a change in the recombination kinetics, which manifests itself in a systematic change of the intensity dependence. The exponent of the power law dependence decreases from 0.8 to about 0.5. Similar trends are observed for the boron-doped films ($\varepsilon_C - \varepsilon_F > 0.8$ eV), but with lower magnitude of σ_{PH} which may be due to the deeper trap depth in p-type a-Si:H. Qualitatively this behavior can be explained by considering the change of the occupancy of the defect states when the Fermi level is moved by doping.

Various recombination models have been put forward to explain the temperature and intensity dependences and also numerical simulations have been performed [9]. A lot of information has accumulated which shows that Si dangling bonds are the dominant

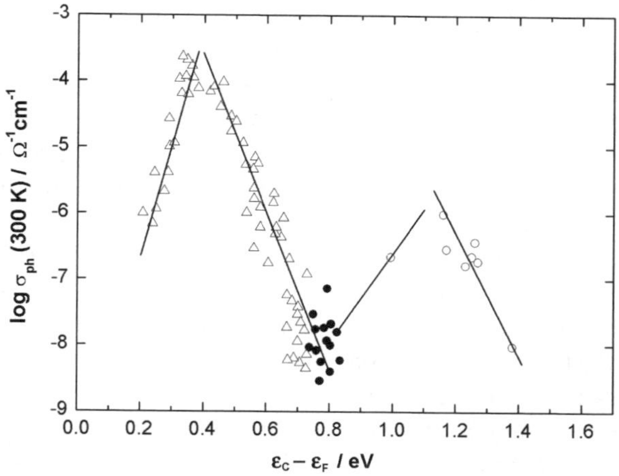

Figure 3.22 Photoconductivity at 300 K of undoped and doped a-Si:H prepared under different conditions for a photon flux of $10^{14}\,cm^{-2}\,s^{-1}$ as function of the activation energy of the conductivity ε_σ (reproduced from [79] with permission of the American Institute of Physics)

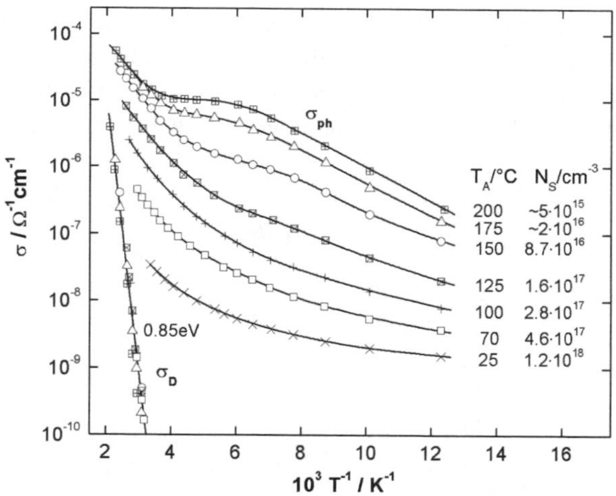

Figure 3.23 Photoconductivity and dark conductivity of an undoped a-Si:H film after electron bombardment and stepwise annealing at T_A which led to the given values of the spin density ($g = 2.0055$) (reproduced from [36] with permission of the American Physical Society)

recombination centers. For instance, this is evident from Figure 3.23, which displays temperature dependences of the dark and photoconductivity $\sigma_D(T)$ and $\sigma_{PH}(T)$ of an undoped film bombarded with 3 MeV electrons to create defects [36]. ESR measurements identified these defects as Si dangling bonds. By electron bombardment and subsequent annealing at the indicated temperatures the spin density N_S was varied between 1.2×10^{18} and about $5 \times 10^{15}\,cm^{-3}$. It is important to notice that in this particular film, this procedure did not affect $\sigma_D(T)$ which means that the dark Fermi level remained unchanged ($\varepsilon_C - \varepsilon_F = 0.85\,eV$).

The pronounced decrease of σ_{PH} with N_S is therefore not due to a change in the occupation of the states, but to a variation of the concentration of the relevant recombination centers. More detailed information has been obtained from studies of spin-dependent recombination by electrically detected magnetic resonance (EDMR). Such experiments give insight into the recombination dynamics and allow one to identify the participating centers by the *g*-values of the observed resonances. For spin-dependent recombination to occur, the participating states have to be paramagnetic such that the spin selection rules determine the transition probabilities. In a cw experiment such measurements are carried out by recording the change of σ_{PH} when the sample is brought into microwave resonance as in an ESR experiment. EDMR investigations led to the conclusion that the dominating recombination steps in a-Si:H are the tunneling transition from band-tail states into neutral dangling bonds and the hopping transport of holes among states in the valence band tail [36, 82]. The question whether these recombination steps are the dominant ones in the entire temperature range, even at 300 K and above, is still unresolved. Direct capture into Si dangling bonds may become dominant at the higher temperatures.

3.7 DEVICE APPLICATIONS

The reduction of the width of the tail state distribution and of the deep defect density in hydrogenated amorphous silicon was the key to the realization of device applications since these quantities determine the carrier lifetimes and the width of space-charge regions in the device. When for instance a metal–semiconductor contact is formed (Figure 3.24) the shape of the potential profile in the semiconductor can be calculated from the Poisson equation

$$V'' = -\rho/\kappa\kappa_0, \tag{3.21}$$

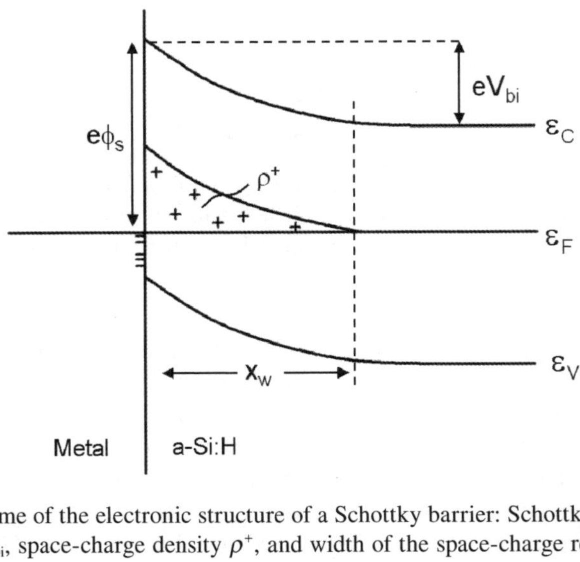

Figure 3.24 Scheme of the electronic structure of a Schottky barrier: Schottky barrier height $e\Phi_S$, built-in potential V_{bi}, space-charge density ρ^+, and width of the space-charge region x_W

where κ_0 and κ denote the absolute and relative dielectric constants, respectively, and $\rho(x)$ is the space charge density, which due to the energy distribution of states in the gap becomes a function of x. A simple solution can be obtained for the case of a homogeneous distribution of the density of states $N(\varepsilon) = N_0$ which results in

$$V = V_{bi}\exp(-x/x_w), \qquad x_w = (\kappa\kappa_0/e^2N_0)^{1/2}, \tag{3.22}$$

where x_w may be taken as a measure of the width of the space charge layer. Inserting $\kappa = 12$ and for the density of states distribution a high value of $N_0 = 10^{19}\,\text{cm}^{-3}\text{eV}^{-1}$, which may be taken as typical of nonhydrogenated a-Si, one obtains $x_w \simeq 8\,\text{nm}$. In such a thin barrier, transport will be dominated by tunneling which would result in low effective barrier heights and high saturation currents. The reduction of the density of states distribution to $N_0 = 10^{16}\,\text{cm}^{-3}\text{eV}^{-1}$ leads to a strong enhancement of the width of the space charge layer to $x_w \simeq 0.25\,\mu\text{m}$, as required for electronic devices.

Hydrogenated amorphous silicon a-Si:H is characterized by low carrier mobilities, low excess carrier lifetimes and therefore by a low minority-carrier diffusion length L_d. Whereas crystalline silicon devices in general are diffusion controlled, $L_d >> x_w$, the opposite is true for a-Si:H devices where $x_w > L_d$ is valid. Therefore the carrier collection in a-Si:H devices occurs predominantly by drift in the built-in field of the device. In addition, *p–n* junctions, which are the backbone of c-Si device technology, are not usable for a-Si:H due to the high defect densities in the doped layers. An important advantage of a-Si:H consists in the high optical absorption in the visible part of the spectrum (see Figure 3.13) which qualifies a-Si:H for applications in thin-film optoelectronic devices.

As far as performance is concerned, devices made from a-Si:H cannot compete with c-Si devices. However, a-Si:H can be deposited homogeneously on large-areas using inexpensive low-temperature deposition techniques. This offers advantages in applications where large area matters. Indeed, the most successful applications of a-Si:H are large-area sensors, displays or thin-film solar cells. In this section a short description will be given of the working principle of the devices which form the building blocks in such applications, namely Schottky barrier diodes, *p–i–n* diodes and thin-film transistors.

3.7.1 Schottky barrier diodes

Schottky barriers on a-Si:H can be made by evaporation of metals onto a-Si:H. In the simplest model, a barrier of height $e\Phi_S$ develops at the metal–a-Si:H interface which, according to Mott's rule [83] is given by the difference of the metal workfunction and the electron affinity of the semiconductor. In undoped a-Si:H films, the Fermi level position in the bulk is typically located at $(\varepsilon_C - \varepsilon_F)_b \simeq 0.7\,\text{eV}$ and a depletion layer for electrons is formed with a built-in potential eV_{bi} of

$$eV_{bi} = e\Phi_S - (\varepsilon_C - \varepsilon_F)_b. \tag{3.23}$$

The scheme of the electronic structure is shown in Figure 3.24. The behavior may be rather complex in detail due to different transport mechanisms across the interface such as thermionic emission, tunneling via gap states, or recombination currents. When the transport is limited by thermionic emission across the Schottky barrier, the *IV* characteristic can be written as [84]

$$j = j_0[\exp(eV/nkT) - 1],$$
$$j_0 = AT^2\exp(-e\Phi_S/kT). \tag{3.24}$$

In this expression A denotes the Richardson constant and n the quality factor which accounts for all deviations from ideal behavior that may arise from other transport channels. The height of the Schottky barrier $e\Phi_S$ has been determined for numerous metals from the temperature dependence of the saturation current density j_0 and from internal photoemission. The variation of $e\Phi_S$ with the metal workfunction differs appreciably from the behavior predicted by Mott's rule and is very similar to that reported for monocrystalline silicon [84]. The barriers on a-Si:H are in general higher than those in c-Si. The largest value of 1.15 eV was obtained for Pt on a-Si:H as compared with 0.8 eV on monocrystalline silicon [85]. It has been reported that the barrier height decreases with the doping level both for P-doping [86] and B-doping [87]. Such behavior has been interpreted by the shrinking of the space-charge region with increasing density of gap states and the resulting enhancement of tunneling currents which short out the barrier and thus lead to a lower effective barrier height.

It is an inherent problem of the fabrication of metal–semiconductor diodes that chemical reactions at the interface may modify the contact properties. Many metals are known to form silicides. Pd, Pt and Ni are examples for this group of metals for which a variety of surface reactions has been observed [86]. Al and Au have the tendency to interdiffuse and promote low-temperature crystallization of a-Si:H. Thus the properties of the diodes depend strongly on the preparational details, which may result in nonideal behavior, enhancing competing transport mechanisms. An enhancement of the density of states distribution by unfavorable deposition conditions may result in an enhancement of tunneling currents, interface states may cause enhancement of recombination currents. The relevance of these various mechanisms will also vary with temperature.

At room temperature, typical a-Si:H diodes have saturation currents of $j_0 = 10^{-11}$–10^{-12} A/cm^2 and the quality factor amounts to $n = 1.1$–1.25 [85]. Similar results have been reported for Schottky diodes formed between metals with low workfunction (rare earth metals) and p-type a-Si:H [87]. In spite of their relatively uncomplicated fabrication techniques the applications of Schottky barrier diodes appear rather limited. Most of the first solar cell structures were of the Schottky barrier type. However, the observed metastabilities and the expectation that p–i–n solar cells have a higher potential for high energy conversion efficiencies, soon disqualified Schottky diodes in photovoltaic applications. However, they may be used as photoconductive sensors in some types of array. Of course, such diodes are still important in research for certain types of studies of amorphous semiconductors which require space-charge layers, as for instance in the various kinds of capacitance spectroscopy.

3.7.2 *p–i–n* diodes

p–i–n diodes of a-Si:H are mostly used for photoconductive sensors arrays and thin-film solar cells [88]. The structure consists of an electronically active undoped a-Si:H layer, a-Si:H(i), contacted by thin highly P- or B-doped layers. Metals or transparent conducting oxides (TCO) may be used to provide contacts to the doped layers. Figure 3.25 displays a

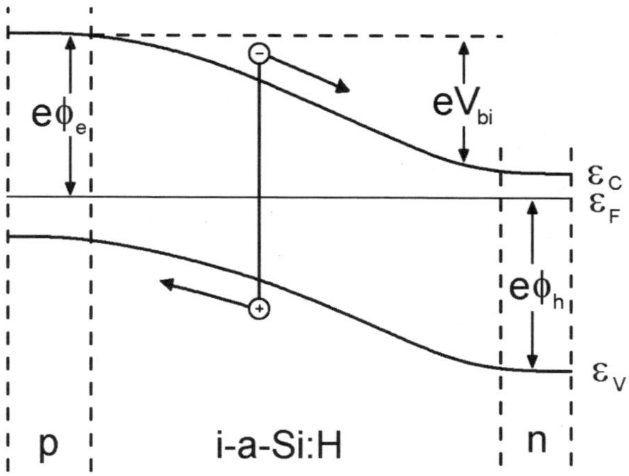

Figure 3.25 Scheme of the electronic structure of an a-Si:H *p–i–n* diode: V_{bi} built-in potential, $e\Phi_e$ and $e\Phi_h$ injection barriers for electrons and holes

scheme of the band profile of such a junction. The *n* and *p* layers provide the built-in potential V_{bi} of the junction, which is determined by the position of the Fermi level in the doped layers. Doping with 10^4 vppm P and B thus result in $eV_{bi} \simeq 1.35\,\text{eV}$. Since undoped a-Si:H is *n*-type with the Fermi level at 0.7–0.8 eV the potential in the *i*-layer changes from strong electron depletion at the *p*-side to accumulation at the *n*-side. The thickness of the doped layers is small, of the order of 100 Å, to minimize optical absorption losses, but large enough to prevent depletion. The barriers for carrier injection are high, $e\Phi_e \simeq 1.55\,\text{eV}$ for electrons at the *p*-layer and $e\Phi_h \simeq 1.65\,\text{eV}$ for holes at the *n*-layer, such that the doped layers form excellent blocking contacts under reverse bias. The optimum thickness of the *i*-layer depends on the actual application. For a solar cell a thickness of about 0.5 μm is typically used to avoid field-free regions which would reduce the carrier collection efficiency. The same argument applies to sensors which are operated in the photovoltaic mode without external bias. However, since the collection efficiency increases strongly with reverse bias, thicknesses up to some 10 μm may be used under high reverse bias in sensor applications.

When excess carriers are generated by light absorption they are collected by drift in the electric field in the *i*-layer (field collection), electrons are extracted into the *n*-layer and holes into the *p*-layer. Since no carriers are replenished at the blocking electrodes, both carriers have to be extracted. Operated in the photovoltaic mode or with reverse bias the *p–i–n* diode is an excellent example of a primary photoconductor with a gain that cannot exceed one. The response time is short, being given by the transit time of the carriers. The spectral distribution of the collection efficiency of a *p–i–n* diode prepared on a glass substrate (Figure 3.26) reaches maximum values of about 80–90% at wavelength 500–600 nm, which demonstrates that *p–i–n* diodes can be made with very little recombination losses. In this case, the collection length L_C defined by the distance that a carrier can move in the electric field before it recombines, $L_C = \mu\tau\, V/L$, has to be larger than the thickness of the *i*-layer $L_C > L$. The essential requirement for this is the ability to prepare *i*-layers with low density of silicon dangling bonds, which are the dominating centers for deep trapping and

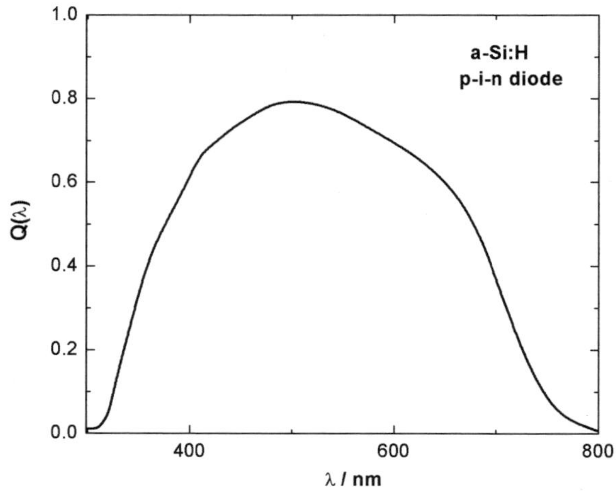

Figure 3.26 External collection efficiency $Q(\lambda)$ as a function of the wavelength λ of a representative *p–i–n* diode (unpublished measurement with permission from the Institute of Photovoltaics, Research Center Jülich)

recombination. The collection efficiency falls off above 600 nm due to the decrease of optical absorption in the *i*-layer. On the short wavelength side of the distribution, the collection efficiency decreases due to enhanced absorption in the doped layer through which the light enters in. This constitutes an important loss mechanism since the doped layers are dead layers in the sense that they do not contribute to charge collection due to the high defect densities which cause fast recombination.

3.7.3 Thin-film transistors

The small minority-carrier diffusion length in doped a-Si:H is the main obstacle to making bipolar transistors. The use of undoped a-Si:H in thin-film field effect transistors (TFT) was proposed as early as 1979 [6] and since then a tremendous development has led to industrialization in many very successful applications in display and sensor array technology [88, 89]. A standard TFT scheme is sketched in Figure 3.27. The transport channel is made from undoped a-Si:H (n-type). In the early days most of the TFTs were made by depositing the nitride Si_3N_4:H onto the active a-Si:H film. However, most of the TFT designs in the modern literature use the inverted structure where the dielectric is deposited first. Ohmic source and drain contacts are realized by highly P-doped a-Si(n^+) films. Metals for the source, drain and gate contact are deposited by vacuum evaporation. Typical channel geometries are 2–20 μm for the channel length L and >10 μm for the channel width W. These values keep decreasing as the technological development goes on. It is important that all process steps can be carried out at low temperature on any kind of substrate and that conventional photolithographic techniques can be used for the appropriate patterning. The devices work in the *n*-channel accumulation mode. When a positive voltage is applied to the gate, a highly conducting channel is formed in the a-Si:H by the accumulation layer induced close to the a-Si:H–a-Si_3N_4:H interface, which shorts out the

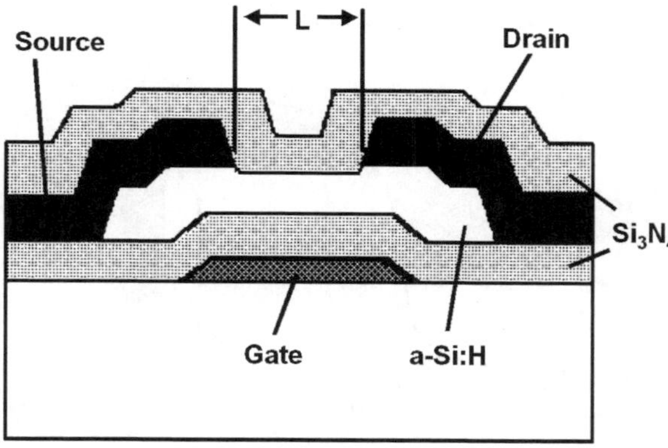

Figure 3.27 Scheme of an a-Si:H thin-film transistor

high resistance of a-Si:H between source and drain and thus switches the transistor into the conducting state.

The historic, but still representative, TFT characteristics from the publication by Snell, Mackenzie, Spear, LeComber and Hughes [90] are displayed in Figure 3.28. For the transfer characteristic the source drain current I_{SD} is plotted as a function of the gate voltage V_G and for the output characteristic I_{SD} is plotted as a function of the source drain voltage V_D. The dynamic range of seven orders of magnitude is very impressive: on-current 10^{-4} A, off-current 10^{-11}A. The description of such characteristics follows that of field-effect transistors made from crystalline silicon [84, 88, 89]. The induced accumulation charge of the channel Q_n (per unit area) is given by

$$Q_n = C_G[V_G - V_T - V(x)]. \tag{3.25}$$

In this expression C_G denotes the gate capacity per unit area, V_T the threshold voltage and $V(x)$ the local channel potential arising from the current flow I_{SD}. V_T is of the order of 1 V and arises from the fact that there is a small accumulation layer in the a-Si:H at the a-Si:H–a-Si_3N_4 interface. I_{SD} increases when the charge induced by V_G exceeds the charge in this accumulation layer. The channel current is determined by an effective mobility μ_{FE} (field-effect mobility), and the electric field $F(x)$ in the channel, $I_{SD} = W\, Q_n\, \mu_{FE}\, F(x)$. Integration across the channel leads to approximate solutions for the linear low-voltage region ($V_D < V_G$) and the saturation regime ($V_D > V_G$). These two cases differ in the role of the channel geometry. While in the first case the channel geometry remains unaltered by V_D, the channel narrows in the second case with increasing V_D, which finally results in pinch-off and saturation approximately at $V_D = V_G - V_T$.

$$I_{SD} = \mu_{FE} C_G (V_G - V_T) V_D\, W/L, \qquad V_D < V_G, \tag{3.26}$$

$$I_{SD} = \mu_{FE} C_G (V_G - V_T)^2\, W/2L, \qquad V_D > V_G. \tag{3.27}$$

These expressions can be used to determine μ_{FE}, either from the linear regime in Figure 3.28a or from a plot of $\sqrt{I_{SD}}$ versus V_G. For the data shown in Figure 3.28 these procedures

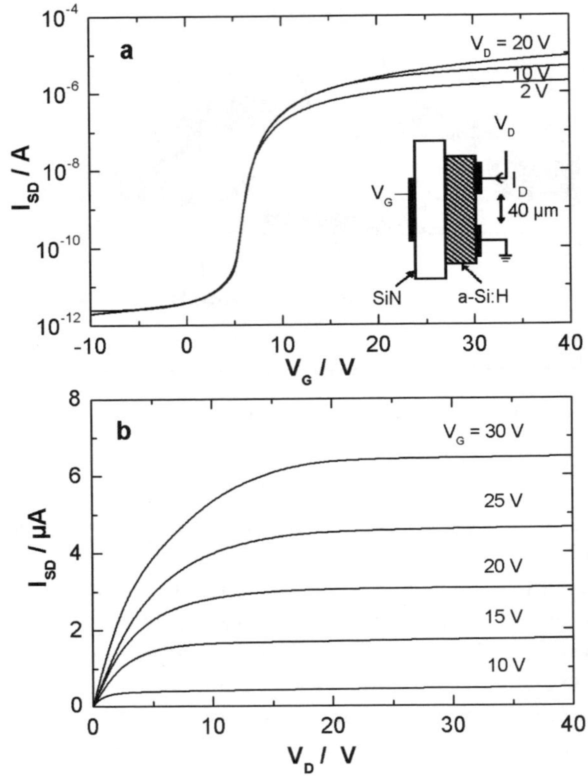

Figure 3.28 (a) Transfer characteristics I_{SD} vs V_G; (b) output characteristics I_{SD} vs V_D of an a-Si:H field effect transistor. The TFT channel is 500 μm wide and 40 μm long, and the thickness of the a-Si:H and Si_3N_4 layers are both 0.5 μm (reprinted from [90] with permission of Springer)

resulted in $\mu_{FE} = 0.4\,cm^2V^{-1}s^{-1}$. This value is very close to that of the room temperature drift mobility (see Section 3.3) as determined from time-of-flight experiments. As in case of the drift mobility, the field-effect mobility is expected to differ from the microscopic mobility in the extended states by a factor which accounts for the time that the carriers are trapped in localized states (Equation 3.2). The low value of μ_{FE} is the reason the channel conductance, defined by I_{SD}/V_D in a-Si:H TFTs, is smaller by a factor of about 10^3 than in crystalline Si TFTs. The channel conductance and thus the achievable currents would decrease further for *p*-type a-Si:H by another factor of about 100, which disqualifies *p*-type channel devices.

Undesired shifts of the threshold voltage V_T under high V_G when the device is held in strong accumulation have been observed. It turned out that it is very important to deposit the nitride with stoichiometric composition in order to prevent charge injection from the channel at the gate contact. Such trapping would cause unacceptable shifts of the threshold voltage. Additional degradation can result from interface states. Such effects were particularly severe in TFT structures where the gate dielectric was deposited onto the a-Si:H film, which was the conventional procedure in the early days. Proper control of the deposition conditions has resulted in the elimination of such deteriorating effects. An intrinsic effect is the creation of metastable defects in the accumulation layer [91, 92], which is related to the defect creation mechanisms described in Section 3.3.3 and which are very similar to light-induced degradation. It has been shown that such effects can be described in terms of a defect pool model [93]. When the voltage stress is applied at elevated temperatures these effects can be quite pronounced. At room temperature the defect creation rate is low and therefore the induced shifts of V_T in operation remain tolerable.

a-Si:H TFTs are now widely used in active matrix arrays for imaging and display such as liquid crystal displays, optical scanners, and radiation imaging arrays [88, 89]. Such arrays contain a high number of elements (*p–i–n* diodes, liquid crystal pixels) each combined with a TFT such that each pixel can be addressed or read out. A challenge for future work lies in the development of such thin-film technologies for flexible devices [94].

3.8 THIN-FILM SOLAR CELLS

The first publication on amorphous silicon solar cells in 1976 initiated a burst of research activity [5]. For this application it is most important that the absorption of a-Si:H in the visible spectral region is considerably larger than in c-Si (see Figure 3.13) thus enabling effective absorption of sunlight in a film which is only about 1 μm thick. In the early days, Schottky barriers, *p–n* junctions and *p–i–n* junctions were proposed. Due to their inherent advantages now only *p–i–n* structures (see Section 3.7.2) are used, which are based on field collection in the *i*-layer of the device. Such devices have been developed to a stable efficiency of about 13% for laboratory cells of small area [95]. This success, together with the development of low-cost manufacturing technology, has made amorphous silicon a viable choice for use in large-scale generation of electricity for terrestrial applications. This technology is attractive because of the reduction in material consumption due to the high optical absorption and because of the energy-saving production technologies with process temperatures not exceeding 300°C. Both arguments are the basis for the expectation that this technology will lead to substantial cost reduction for solar cells. The status, expectations and manufacturing technologies have been summarized in various publications [11, 88, 96]. At present various companies manufacture modules with stabilized efficiencies of 6–8%. So far, the technology based on hydrogenated amorphous silicon and its alloys is the only thin-film technology which has been able to surmount the barrier to mass production and occupies a 6% share of the rapidly expanding photovoltaic world market [97].

Two different device configurations have been developed (Figure 3.29). In the most commonly used superstrate cell, the structure is glass/TCO/p^+-a-Si/*i*-a-Si:H/n^+-a-Si/metal and the light enters through the glass. Instead of glass, other substrate material can be used, which is able to stand the process temperatures of up to 200°C. The substrate cell uses stainless steel (ss) as substrate with the inverted layer sequence ss/Ag-ZnO/a-Si(n^+)/a-Si(i)/a-Si(p^+)/TCO/metal grid. In both single-junction cells, the light enters through the p^+-layer which results in the highest collection efficiency because the collection length is smaller for holes than for electrons. The thickness of the active undoped absorber layer, a-Si:H(i), is about 0.5 μm to optimize carrier collection by drift in the electric field. The doped layers, which are photoelectronically dead layers, have to be as thin as possible, typically less than 10 nm, to avoid absorption losses. Due to the low lateral conductivity of such films both cell types need transparent conducting contacts (TCO), which are either zinc oxide (ZnO)

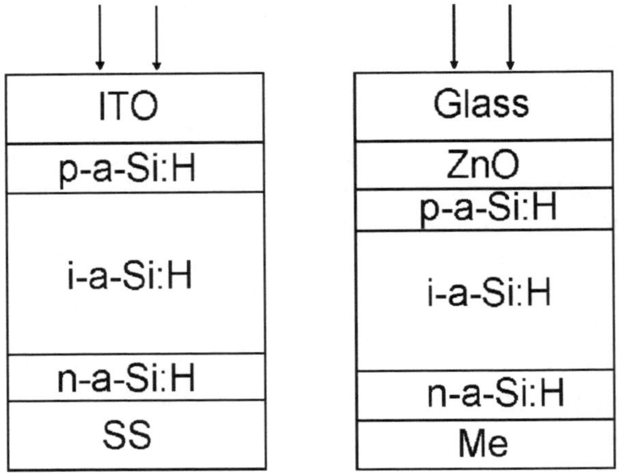

Figure 3.29 Scheme of the two types of single-junction *p–i–n* solar cells. Substrate cell: deposition on metal (stainless steel), superstrate cell: deposition on glass. In both cases illumination is through the *p*-layer

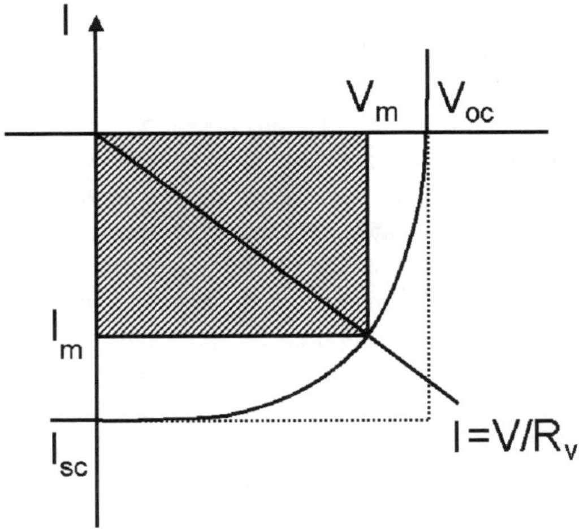

Figure 3.30 *I–V* characteristic of a solar cell under illumination: I_{SC} short-circuit current, V_{OC} open-circuit voltage, R_V load, maximum power point (I_m, V_m)

or indium–tin oxide (ITO). The spectral distribution $Q(\lambda)$ of a *p–i–n* structure has been discussed in Section 3.7.2 (see Figure 3.26).

The response of a solar cell under illumination is characterized by three parameters (Figure 3.30): short-circuit current I_{SC}, open-circuit voltage V_{OC} and the fill factor FF which is the ratio of the *IV* product at the maximum power point and $I_{SC}V_{OC}$. The efficiency of the solar cell is then defined by the ratio of the electrical output power P_{el} and the optical input power P_o:

$$\eta = P_{el}/P_o = I_{SC}V_{OC}\,FF/P_o. \tag{3.28}$$

As reference for P_o in general the spectral distribution of the solar spectrum is used after a 45° passage through the atmosphere (air mass AM1.5) normalized to a total power of 1 kW/m^2.

Single-junction cells have been developed to stable efficiencies in the range of 9–10% in the laboratory for small areas. Apparently there are intrinsic limitations for single *p–i–n* cells. In order to enhance the efficiency, one would have to enhance the optical absorption by increasing the thickness of the active *i*-layer. This, however, results in a decrease of the collection efficiency. The enhancement of optical absorption also relies on efficient light-trapping technology. Therefore the superstrate approaches use textured ZnO, which enhances light scattering such that the light path of weakly absorbed light is increased. In both types of solar cell the use of an efficiently reflecting back contact is essential, which enhanced the light path by reflecting the weakly absorbed light.

A major problem of the amorphous silicon technology has been the degradation of the amorphous films and devices under illumination. In spite of some progress with material stability this effect appears to be an intrinsic property of hydrogenated amorphous semiconductors (see Section 3.3.3). Considerable progress has been made concerning increased stability of solar cells by proper device engineering. The strategy is to make use of the experience that the metastability is more pronounced for thick than for thin *p–i–n* cells. It is therefore advantageous to use stacks of two or three thinner *p–i–n* cells. A difficult task in the engineering of such stack cells with two terminals is the fabrication of effective tunnel junctions and the current matching. In order to avoid losses, the thickness of the cells has to be carefully designed such that each of the cells delivers the same current. Such advanced device structures offer two advantages: (1) in a stack, the thickness of the single cells is reduced, which improves the stability by enhancing carrier collection through an increase of the internal electric field; (2) by using materials with different bandgaps a tandem structure can be made which leads to a better use of the solar spectrum. It is important that the degradation of the cells tends to saturate as a function of the exposure time (typically some 100 h illumination with sunlight) such that stabilized efficiencies can be guaranteed by the manufacturers. In Figure 3.31 the degradation of single *p–i–n* cells and stacked cells are compared [98]. The figure clearly shows the pronounced decrease of the efficiency of single-junction cells and the progress attained with tandem structures. Optimized tandem solar cells typically degrade by about 10% from the as-deposited state.

In the most successful triple-junction solar cell so far (Figure 3.32) of United Solar System Corporation (USSC) [95] the top cell uses a-Si:H with a bandgap of 1.8 eV for the intrinsic layer which absorbs the high energy part of the solar spectrum. The *i*-layer of the middle cell is an a-Si/Ge alloy which contains about 15% Ge, leading to a bandgap of 1.6 eV. The bottom cell that is designed to absorb the red part of the spectrum uses an *i*-layer of an a-Si/Ge alloy with 40–50% Ge and a bandgap of 1.4 eV. A textured Ag/ZnO back-reflector is used to enhance light trapping. Figure 3.33 shows the *I–V* characteristics and the collection efficiency of the best triple-junction cell so far [99]. The spectral response spans from 350 to 950 nm with a peak quantum efficiency exceeding 90% in the green spectral range. The high values of $Q(\lambda)$ above 800 nm give proof of an excellent back-reflector. In parallel with the preparation of the tandem stack, single cells have been made to optimize thickness and quantum efficiency. These results demonstrate the selective use of the solar spectrum, and the current values prove the quality of current matching. The total

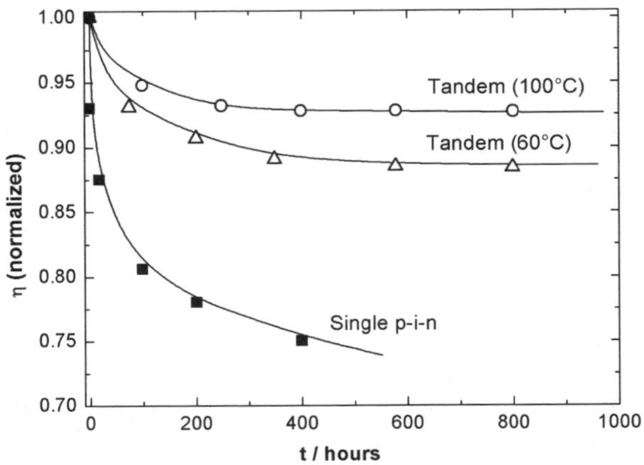

Figure 3.31 Degradation of a-Si:H solar cells: normalized conversion efficiency η as a function of the exposure time t to solar radiation (simulated AM1) for a single junction cell (squares) and tandem cells illuminated at 60°C (triangles) and 100°C (circles) (reproduced from [98] with permission of Elsevier)

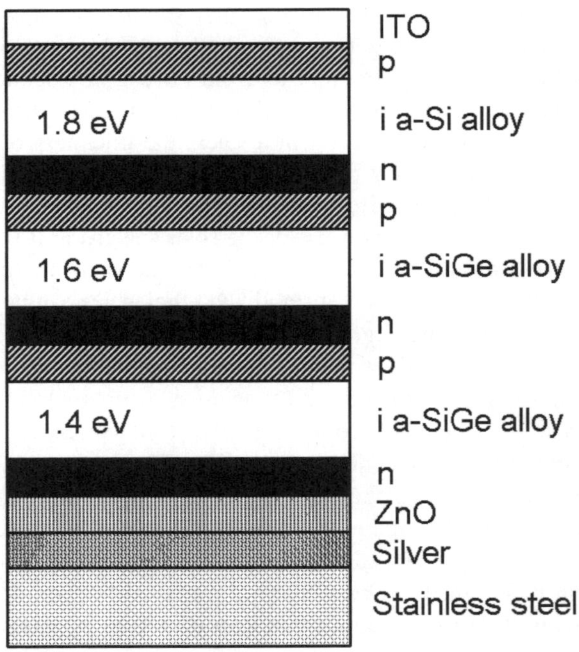

Figure 3.32 Scheme of the triple-junction cell of Unisolar Corporation

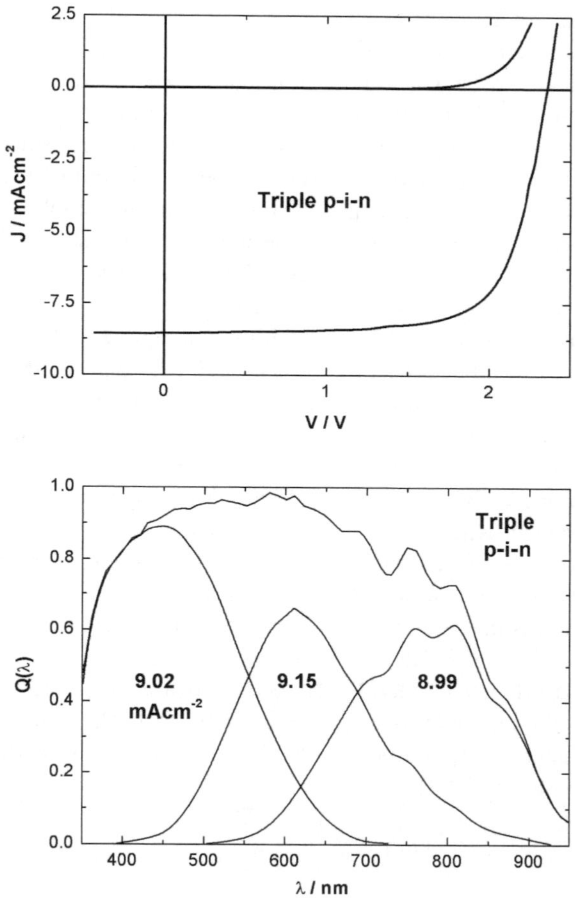

Figure 3.33 *I–V* characteristic and external quantum efficiency $Q(\lambda)$ of a triple-junction solar cell with an initial efficiency of 15.2% (area 0.25 cm^2). Electrical characteristics: V_{OC} = 2.344 *V*, j_{SC} = 8.99 mA/cm^2, FF = 0.722, η = 15.2%. The selective quantum efficiencies $Q_i(\lambda)$ were measured on single-component cells optimized for current matching in the triple stack (reproduced from [99] with permission of the Materials Research Society)

current density of the three component cells adds up to excellent 27.16 mA/cm^2. This cell has achieved the highest efficiency (active area) reported so far for laboratory cells (A = 0.25 cm^2): 15.2% initial, 13% stabilized. The use of high hydrogen dilution in the preparation of the intrinsic films appears to be one of the keys to these results. Hydrogen dilution is considered to lead to a more ordered structure of the material. a-Si:H deposited with high hydrogen dilution is characterized by a heterogeneous mix of an amorphous structure with small areas of ordered material with dimensions in the range of nm. When very high hydrogen dilution is used the films become microcrystalline. It turned out that the best results were obtained when the films were made with preparation parameters right at the threshold of microcrystalline silicon formation [95, 99].

Challenges for research at present are the development of materials which can be used in stacked tandem structures with a-Si:H, the enhancement of the deposition rate, improvement of the cell efficiencies and production yields. The a-Si/Ge alloys still have considerably poorer electronic properties than a-Si:H and in addition their deposition uses germane as process gas which constitutes an important cost factor. Therefore a number of research groups aim at replacing the a-Si/Ge alloy with microcrystalline silicon μc-Si:H [100–102]. μc-Si:H can be prepared by PECVD or HWCVD deposition techniques under conditions which, from a technological point of view, are compatible with those in the deposition of a-Si:H. The physics of this material is rather complex due to pronounced structural heterogeneity (see Figure 3.5). Single-junction *p–i–n* solar cells have been made with μc-Si:H absorbers (about 1.5–2 μm thick) with efficiencies of close to 10%, using both PECVD and HWCVD [103]. In this case, the growth rate of the radiofrequency PECVD deposition was enhanced to 0.5 nm/s by depositing in a regime with high pressure and discharge power. It is most interesting that the μc-Si:H with the highest crystallinity is not necessarily the most suitable material for solar cells. It has been demonstrated that the best efficiencies are obtained when the films grow close to the onset of amorphous growth at higher silane concentration and radiofrequency power [101].

The comparison of the absorption edges (Figure 3.13) suggests using μc-Si:H as bottom cell in a tandem concept with a-Si:H. Figure 3.34 shows the scheme of such a cell made up of a stack of two *p–i–n* structures with a-Si:H forming the top and μc-Si:H forming the bottom cell. The resulting quantum yield reveals again the advantage of such a device concept: The thin top cell, made of a-Si:H, uses the high-energy photons and the thicker bottom cell, made of μc-Si:H, uses the low-energy photons. Various laboratories have published stabilized efficiencies of 11–12% on small areas and values above 10% have already been reached on modules of larger areas [103–106]. The highest values so far have been reached by Kaneka Corporation, where also a production of modules on the basis of this technology has started [104].

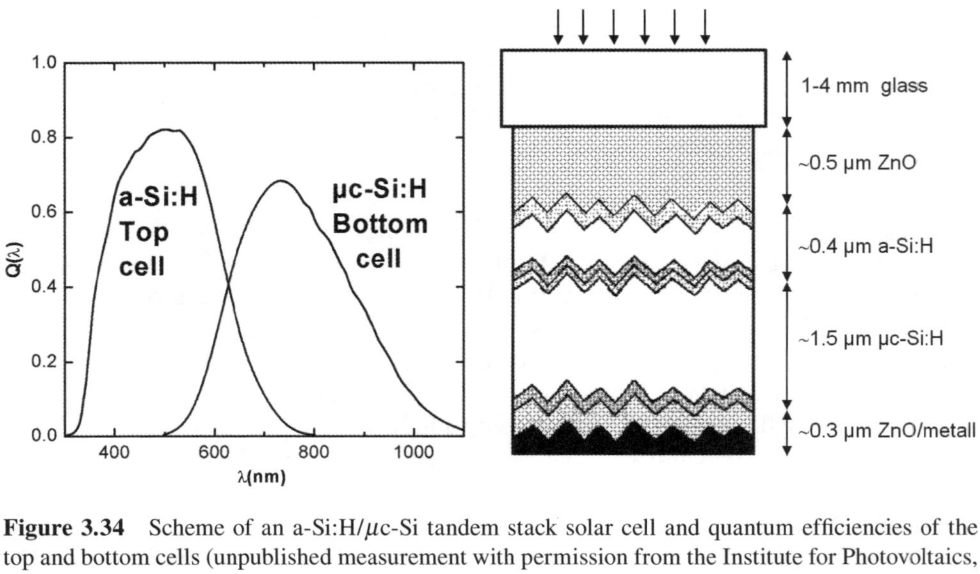

Figure 3.34 Scheme of an a-Si:H/μc-Si tandem stack solar cell and quantum efficiencies of the top and bottom cells (unpublished measurement with permission from the Institute for Photovoltaics, Research Center Jülich)

The industrial activity proves impressively that the technology which based on a-Si:H and related alloys has surmounted the barrier to commercialization. The progress achieved in this field is certainly based on the broad understanding of the physics of both materials and devices owing to 25 years of continuous research effort. Numerous applications of these amorphous semiconductors in photovoltaics, sensors, photodetectors and in particular displays has driven this research and established the mature technology available today.

REFERENCES

[1] R.C. Chittik, J.H. Alexander, and H.F. Sterling, The preparation and properties of amorphous silicon, *J. Electrochem. Soc.*, **116**, 77–81 (1969).

[2] R.J. Loveland, W.E. Spear, and A. Al-Shabaty, Photoconductivity and absorption in amorphous silicon, *J. Non-Cryst. Solids*, **13**, 55–68 (1973).

[3] A.J. Lewis, G.A.N. Connell, W. Paul, J. Pawlik, and R. Temkin, Hydrogen incorporation in amorphous Germanium, *AIP Conf. Proc.*, **20**, 27–33 (1974).

[4] P.G. LeComber and W.E. Spear, Substitutional doping of amorphous silicon, *Solid State Commun.*, **17**, 1193 (1975).

[5] D. Carlson and C. R. Wronsky, Amorphous silicon solar cells, *Appl. Phys. Lett.*, **28**, 671–673 (1976).

[6] P.G. LeComber, W.E. Spear, and A. Gaith, Amorphous silicon field effect device and possible applications, *Electronics Lett.*, **15**, 179–181 (1979).

[7] N.F. Mott and E.A. Davis, *Electronic Processes in Non-Crystalline Materials*, Oxford University Press, Oxford, 1979.

[8] J.D. Joannopoulos and G. Lucovsky, Hydrogenated Amorphous Silicon, in: *Topics in Applied Physics*, Springer, Heidelberg, 1984.

[9] J. Pankove (ed.), *Hydrogenated Amorphous Silicon,* in: *Semiconductor and Semimetals 21 A–D*, R.K. Willardson and A.C. Beer (eds), Academic Press, 1984.

[10] H. Overhof and P. Thomas, *Electronic Transport in Hydrogenated Amorphous Semiconductors*, Springer Tracts in Modern Physics Vol. 114, ed. H. Höhler, Springer Verlag Heidelberg, 1989.

[11] R.A. Street, *Hydrogenated Amorphous Silicon*, Cambridge University Press, Cambridge, 1991.

[12] C. Spinella, S. Lombardo, and F. Priolo, Crystal grain nucleation in amorphous silicon, *J. Appl. Phys.*, **84**, 5383–5414 (1998).

[13] E.P. Donovan, F. Spaepen, D. Turnbull, J.P. Poate, and D.C. Jacobson, Calorimetric studies of crystallization and relaxation of amorphous Si and Ge prepared by ion implantation, *J. Appl. Phys.*, **57**, 1795–1804 (1985).

[14] O. Nast and S.R. Wenham, Elucidation of the layer exchange mechanism in the formation of polycrystalline silicon by Al-induced crystallization, *J. Appl. Phys.*, **88**, 124–132 (2000).

[15] A. Matsuda, Microcrystalline silicon—growth and device applications, *J. Non-Cryst. Solids*, **338–340**, 1–12 (2004).

[16] H. Matsumura, H. Umemoto, and A. Masuda, CAT-CVD (hot-wire CVD): How different from PECVD in preparing amorphous silicon, *J. Non-Cryst. Solids*, **338–340**, 19–26 (2004).

[17] D.E. Polk, Structural model for amorphous silicon and germanium, *J. Non-Cryst. Solids*, **5**, 365–376 (1971).

[18] W. Schülke, Structural investigation of hydrogenated amorphous silicon by X-ray diffraction, *Phil. Mag. B*, **43**, 451–468 (1981).

[19] D.K. Biegelsen, R.A. Street, C.C. Tsai, and J.C. Knights, Hydrogen evolution and defect creation in amorphous Si:H alloys, *Phys. Rev. B*, **20**, 4839–4846 (1979).
[20] W. Beyer, Incorporation and stability of hydrogen in amorphous silicon and germanium, *J. Non-Cryst. Solids*, **198–200**, 40–45 (1996).
[21] H. Fritzsche, Characterization of glow-discharge deposited a-Si:H, *Solar Energy Materials*, **3**, 447 (1980).
[22] K. Tanaka, Recent progress in microcrystalline semiconductor thin films, *Mat. Res. Soc. Symp. Proc.*, **452**, 3–16 (1997).
[23] N.F. Mott, Electrons in disordered structures, *Adv. Phys.*, **16**, 49–144 (1967).
[24] L. Banyai, On the theory of electronic conduction in amorphous semiconductors, in: *Physique de Semiconducteurs*, M. Hulin (ed.), Dunod, Paris, 417–421 (1964).
[25] M.H. Cohen, H. Fritzsche, and S.R. Ovshinsky, Simple band model for amorphous semiconducting alloys, *Phys. Rev. Lett.*, **22**, 1065–1068 (1969).
[26] J. Dong and D.A. Drabold, Atomistic structure of band tail states in amorphous silicon, *Phys. Rev. Lett.*, **80**, 1928–1931 (1998).
[27] L. Ley, Photoemission and optical properties, in: *The physics of Hydrogenated Amorphous Silicon II*, Chap. 3, 61–168, J.D. Joannopoulos and G. Lucovsky (eds), Springer, Berlin, 1984.
[28] W.B. Jackson, S. Kelso, C.C. Tsai, J.W. Allen, and S.J. Oh, Energy dependence of the optical matrix elements in hydrogenated amorphous and crystalline silicon, *Phys. Rev. B*, **31**, 5187–5198 (1985).
[29] B. von Roedern, L. Ley, and M. Cardona, Photoelectron spectra of hydrogenated amorphous silicon, *Phys. Rev. Lett.*, **39**, 1576–1579 (1977).
[30] K. Winer and L. Ley, Near-surface electronic properties of a-Si:H, in: *Amorphous Silicon and Related Materials*, H. Fritzsche (ed.), World Scientific, Singapore, 365–408 (1989).
[31] W.E. Spear, Study of transport and related properties of a-Si:H by transient experiments, *J. Non-Cryst. Solids*, **59/60**, 1–8 (1983).
[32] H. Scher and E.W. Montroll, Anomalous transit time dispersion in amorphous solids, *Phys. Rev. B*, **12**, 2455–2477 (1975).
[33] T. Tiedje, Information on band tail states from time-of-flight experiments, in: reference [9], Vol. C, 207–237 (1984).
[34] R.A. Street, D.K. Biegelsen, and J.C. Knights, Defect states in doped and compensated a-Si: H, *Phys. Rev. B*, **24**, 969–984 (1981).
[35] H. Dersch, J. Stuke, and J. Beichler, Electron spin resonance of doped glow-discharge amorphous silicon, *phys. stat. sol. b*, **105**, 265–275 (1981).
[36] H. Dersch, L. Schweitzer, and J. Stuke, Recombination processes in a-Si:H: Spin dependent photoconductivity, *Phys. Rev. B*, **28**, 4678–4684 (1983).
[37] M. Stutzmann and D.K. Biegelsen, Microscopic nature of coordination defects in amorphous silicon, *Phys Rev. B*, **40**, 9834–9840 (1989).
[38] R.A. Street, Doping and the Fermi energy in amorphous silicon, *Phys. Rev. Lett.*, **49**, 1187–1190 (1982).
[39] S.K. Kalbitzer, G. Müller, P.G. LeComber, and W.E. Spear, The effects of ion-implantation on the electrical properties of amorphous silicon, *Phil. Mag. B*, **41**, 439–456 (1980).
[40] K. Pierz, W. Fuhs, and H. Mell, On the mechanism of doping and defect formation in a-Si: H, *Phil. Mag. B*, **63**, 123–141 (1991).
[41] J. Kocka, The density of states in undoped and doped amorphous hydrogenated silicon, *J. Non-Cryst. Solids*, **90**, 91–98 (1987).
[42] K. Winer, I. Hirabayashi, and L. Ley, Distribution of occupied near-surface band gap states in a-Si:H, *Phys. Rev. B*, **38**, 7680–7693 (1988).
[43] R.A. Street, J. Kakalios, C.C. Tsai, and T.M. Hayes, Thermal-equilibrium processes in amorphous silicon, *Phys. Rev. B*, **35**, 1316–1333 (1987).

[44] J. Kakalios and R.A. Street, Thermal equilibrium processes in doped amorphous silicon, *J. Non-Cryst. Solids*, **97–98**, 767–770 (1987).

[45] L. Ley and K. Winer, The defect pool model for amorphous hydrogenated silicon, *Proc. Int. Conf. Physics of Semiconductors* (Polish Institute of Physics, Warsaw 1989) 1633–1637.

[46] Y. Bar-Yam and J.D. Joannopoulos, Theories of defects in amorphous semiconductors, *J. Non-Cryst. Solids*, **97–98**, 467–474 (1987).

[47] M. Stutzmann, The defect density in amorphous silicon, *Phil. Mag. B*, **60**, 531–546 (1989).

[48] Z.E. Smith and S. Wagner, Band tails, entropy, and equilibrium defects in hydrogenated amorphous silicon, *Phys. Rev. Lett.*, **59**, 688–691 (1987).

[49] R.A. Street and K. Winer, Defect equilibria in undoped a-Si:H, *Phys. Rev. B*, **40**, 6236–6249 (1989).

[50] K. Winer, Chemical equilibrium description of the gap-state distribution in a-Si:H, *Phys. Rev. Lett.*, **63**, 1487–1490 (1989).

[51] G. Schumm and G.H. Bauer, Thermodynamical equilibrium gap-state distribution in undoped a-Si:H, *Phil. Mag. B*, **64**, 515–527 (1991).

[52] M.J. Powell and S.C. Deane, Improved defect-pool model for charged defects in amorphous silicon, *Phys. Rev. B*, **48**, 10815–10827 (1993).

[53] D.L. Staebler and C.R. Wronski, Optically induced conductivity changes in discharge-produced hydrogenated amorphous silicon, *J. Appl. Phys.*, **51**, 3262–3268 (1980).

[54] M. Stutzmann, W.B. Jackson, and C.C. Tsai, *Phys. Rev. B*, **32**, 23 (1985).

[55] D. Ewald, M. Milleville, and G. Weiser, Optical spectra of glow discharge deposited silicon, *Phil. Mag. B*, **40**, 291–303 (1979).

[56] R. Carius, F. Finger, U. Backhausen, M. Lysberg, P. Hapke, M. Otte, and H. Overhof, Electronic properties of microcrystalline silicon, *Mat. Res. Soc. Symp. Proc.*, **467**, 283–288 (1997).

[57] J. Tauc, *Amorphous and Liquid Semiconductors*, Plenum, New York, 1974.

[58] R. Stachowitz, K. Jahn, and W. Fuhs, Low-temperature transport and recombination in a-Si:H, *Phil. Mag. B*, **62**, 5–18 (1990).

[59] N.F. Mott, Conduction in non-crystalline materials, *Phil. Mag.*, **19**, 835–852 (1969).

[60] W. Beyer and H. Overhof, Doping effects in a-Si:H, in: reference [9], Vol. C, 257–304 (1984).

[61] W. Beyer, H. Mell, and H. Overhof, Transport properties in compensated a-Si:H films, *J. Physique,* **42 C 2**, 103–106 (1981).

[62] D.E. Carlson and C.R. Wronski, Amorphous silicon solar cells, in: *Amorphous Semiconductors, Topics in Applied Physics*, Vol. 38, M.H. Brodsky (ed.), Springer, Heidelberg, 287–329, 1979.

[63] W. Meyer and H. Neldel, Über die Beziehungen zwischen den Energiekonstanten ε und der Mengenkonstanten α in der Leitwerts-Temperaturformel bei oxydischen Halbleitern, *Zeitschrift f. Techn. Physik*, **12**, 588–593 (1937).

[64] H. Overhof and W. Beyer, A model for the electronic transport in hydrogenated amorphous silicon, *Phil. Mag. B*, **43**, 433–450 (1981).

[65] R.A. Street, Luminescence and recombination in hydrogenated amorphous silicon, *Adv. Phys.*, **30**, 593–676 (1981).

[66] M. Hoheisel, R. Carius, and W. Fuhs, Low temperature photoconductivity in a-Si:H films, *J. Non-Cryst. Solids*, **59–60**, 457–460 (1983).

[67] C. Tsang and R.A. Street, Recombination in plasma-deposited amorphous a-Si:H: Luminescence, *Phys. Rev. B*, **19**, 3027–3040 (1979).

[68] M. Bort, W. Fuhs, S. Liedke, R. Stachowitz, and R. Carius, Geminate Recombination in a-Si:H, *Phil. Mag. Lett.*, **64**, 227–233 (1991).

[69] Ambros, R. Carius, and J. Wagner, Lifetime distribution in a-Si:H: geminate-, nongeminate-, and Auger-processes, *J. Non-Cryst. Solids*, **137–138**, 555–558 (1991).

[70] T. Aoki, T. Shimizu, S. Komedoori, S. Kobayashi, and K. Shimakawa, Coexistence of geminate and non-geminate recombination in a-Si:H and a-Ge:H, *J. Non. Cryst. Solids*, **338–340**, 456–459 (2004).

[71] D.J. Dunstan, Kinetics of distant pair recombination–amorphous silicon luminescence at low temperature, *Phil. Mag. B*, **46**, 579 (1982).

[72] B.I. Shklovskii, H. Fritzsche, and S.D. Baranovski, Electronic transport and recombination in amorphous semiconductors at low temperatures, *Phys. Rev. Lett.*, **62**, 2989–2992 (1989).

[73] E.I. Levin, S. Marianer, and B.I. Shklovskii, Photoluminescence in amorphous silicon at low temperatures: Computer simulation, *Phys. Rev. B*, **45**, 5906–5918 (1992).

[74] R. Stachowitz, M. Schubert, and W. Fuhs, Non-radiative distant pair recombination in amorphous silicon, *J. Non-Cryst. Solids*, **227–230**, 190–196 (1998).

[75] K. Lips, C. Böhme, and T. Ehara, The impact of the electron spin on charge carrier recombination—the example of amorphous silicon, *J. Optoelectronics Adv. Mat.*, **7**, 13–24 (2005).

[76] B.I. Shklovskii, E.I. Levin, H. Fritzsche, and S.D. Baranovski, Hopping photoconductivity in amorphous semiconductors: dependence on temperature, electric field and frequency, in: *Advances in Disordered Solids*, Vol. 3, H. Fritzsche (ed.), World Scientific, Singapore, 161–191 (1990).

[77] S.D. Baranovski, P. Thomas, and G.J. Adriaenssens, The concept of transport energy and its application to steady-state photoconductivity in amorphous silicon, *J. Non-Cryst. Solids*, **190**, 283–287 (1995).

[78] B. Cleve, B. Hartenstein, S.D. Baranovski, M. Scheidler, P. Thomas, and H. Bässler, High-field hopping transport in band tail states of disordered semiconductors, *Phys. Rev. B*, **51**, 16705 (1995).

[79] P.E. Vanier, F.J. Kampas, R.R. Cordermann, and C. Rajeswaran, A study of hydrogenated amorphous silicon deposited by glow discharge in silane-hydrogen mixtures, *J. Appl. Phys.*, **56**, 1812–1820 (1984).

[80] W. Fuhs, Recombination of trapped carriers probed by sub-bandgap excitation, *J. Non-Cryst. Solids*, **77–78**, 593–602 (1985).

[81] D.A. Anderson and W.E. Spear, Photoconductivity and recombination in doped amorphous silicon, *Phil. Mag.*, **36**, 695–712 (1977).

[82] W. Fuhs and K. Lips, Recombination in a-Si:H films and pin structures studied by electrically detected magnetic resonance (EDMR*), J. Non-Cryst. Solids*, **164–166**, 541–546 (1993).

[83] W. Mönch, *Electronic Properties of Semiconductor Interfaces*, Springer Series in Surface Science, Springer, Heidelberg, 2004.

[84] S.M. Sze, *Physics of Semiconductor Devices*, John Wiley & Sons, Inc., New York, 1981.

[85] C.R. Wronski, D.E. Carlson, and R.E. Daniel, Schottky barrier characteristics of metal–amorphous-silicon diodes, *Appl. Phys. Lett.*, **29**, 602–605 (1976).

[86] R.J. Nemanich, Schottky barriers on a-Si:H, in: reference [9], Vol. C, 375–406 (1984).

[87] K.H. Greeb, W. Fuhs, H. Mell, and H.M. Welsch, Schottky barriers on *p*-type a-Si:H, *Solar Energy Materials*, **7**, 253–261 (1982).

[88] R.A. Street (ed.), *Technology and Applications of Amorphous Silicon*, Springer Heidelberg, 2000.

[89] T. Tsukada, *TFT/LCD Liquid Crystal Displays Addressed by Thin-film Transistors*, Japanese Technology Reviews (Section A: Electronics) Vol. 29, Gordon and Breach, Tokyo, 1996.

[90] A.J. Snell, K.D. Mackenzie, W.E. Spear, P.G. LeComber, and A.J. Hughes, Application of amorphous silicon field effect transistors in addressable liquid-crystal display panels, *Applied Physics*, **24**, 357–362 (1981).

[91] M.J. Powell, C. van Berkel, and J.R. Hughes, Time and temperature dependence of instability mechanisms in amorphous silicon thin-film transistors, *Appl. Phys. Lett.*, **54**, 1323–1325 (1989).

[92] N.H. Nickel, W. Fuhs, and H. Mell, Defect creation in the accumulation layer of a-Si:LH thin-film transistors, *Phil. Mag. B*, **61**, 251–261 (1990).

[93] M.J. Powell, S.C. Deane, I.D. French, J.R. Hughes, and W.I. Milne, A defect-pool model for near–interface states in amorphous thin-film transistors, *Phil. Mag. B*, **63**, 325–336 (1991).

[94] R.E.I. Schropp, B. Stannowski, and J. Rath, New challenges in thin-film transistor (TFT) research, *J. Non-Cryst. Sol.*, **299–302**, 1304–1310 (2002).

[95] S. Guha, J. Yang, and A. Banerjee, Amorphous silicon alloy photovoltaic research—Present and Future, *Prog. Photovoltaics: Res. Appl.*, **8**, 141–150 (2000).

[96] Y. Hamakawa (ed.), *Thin-Film Solar Cells*, Springer Series in Photonics, Springer V, Heidelberg, 2004.

[97] M. Schmela and J. Siemer, Wachstum um die Wette, *Photon 4/2005*, 28–38 (2005).

[98] D.E. Carlson, Monolithic amorphous silicon alloy solar modules, *Solar Energy Materials & Solar Cells,* **78**, 627–645 (2003).

[99] J. Yang and S. Guha, Amorphous silicon alloy materials and solar cells near the threshold of microcrystallinity, *Mat. Res. Soc. Symp. Proc.*, **557**, 239–250 (1999).

[100] J. Meier, P. Torres, R. Platz, S. Dubail, U. Kroll, J.A. Anna Selvan, N. Pellaton, Ch. Hof, D. Fischer, H. Keppner, A. Shah, K.-D. Ufert, P. Giannoules, and J. Koehler, On the way towards high efficiency thin-film silicon solar cells by the micromorph concept, *Mat. Res. Soc. Symp. Proc.*, **420**, 3 (1996).

[101] O. Vetterl, F. Finger, R. Carius, P. Hapke, L. Houben, O. Kluth, A. Lambertz, A. Mück, B. Rech, and H. Wagner, Intrinsic microcrystalline silicon: a new material for photovoltaics, *Solar Cell Materials & Solar Cells*, **62**, 97–108 (2000).

[102] K. Yamamotu, M. Yoshimi, Y. Tawada, S. Fukuda, T. Sawada, T. Meguro, H. Takata, T. Suezaki, K. Koi, K. Hayashi, T. Suzuki, and A. Nakajima, Large area thin-film silicon module, *Techn. Digest 12th Int. Photovoltaic Solar Energy Conf.*, 547–551 (2001).

[103] S. Klein, T. Repmann, and T. Brammer, Microcrystalline silicon films and solar cells deposited by PECVD and HWCVD, *Solar Energy*, **77**, 893–908 (2004).

[104] K. Yamamoto, A. Nakajima, M. Yoshimi, T, Sawada, S. Fukuda, T. Suezaki, M. Ichikawa, Y. Koi, M. Goto, and T. Meguro, A high efficiency thin film silicon solar cell and module, *Solar Energy*, **77**, 939–949 (2004).

[105] J. Meier, J. Spitznagel, U. Kroll, C. Bucher, S. Faÿ, T. Moriarty, and A. Shah, Potential of amorphous and microcrystalline silicon solar cells, *Thin Solid Films*, **451–452**, 518–524 (2004).

[106] T. Repmann, T. Kilper, W. Appenzeller, C. Zahren, H. Stiebig, and B. Rech, Up-scaling and optimization of thin-film solar cells and modules based on amorphous and microcrystalline silicon, *Conf. Rec. 31st Photovoltaic Specialist Conf.*, *IEEE-31*, USA, January 2005, 1383–1388 (2005).

4 Applications of Disordered Semiconductors in Modern Electronics: Selected Examples

Safa Kasap[1], John Rowlands[2], Kenkichi Tanioka[3] and Arokia Nathan[4]

[1]*University of Saskatchewan, Saskatoon, SK, S7N 5A9, Canada*
[2]*University of Toronto and Sunnybrook Health Sciences Centre, 2075 Bayview Avenue, Toronto, M4N 3N5, Canada*
[3]*NHK Science and Technical Research Laboratories, Tokyo, 175 Japan*
[4]*University of Waterloo, Waterloo, Canada, currently on a sabbatical leave at the Center for Advanced Photonics and Electronics, Electrical Division, 9 JJ Thomson Avenue, Cambridge University, Cambridge CB3 OFA, UK.*

4.1 PERSPECTIVES ON AMORPHOUS SEMICONDUCTORS

One of the biggest advantages of noncrystalline semiconductors is the ability of these materials to be prepared in *large areas* in a cost-effective way for what may be coined 'macroelectronics', such as displays, scanners, solar cells, image sensors, position sensors,

Charge Transport in Disordered Solids with Applications in Electronics Edited by S. Baranovski.

and similar large area applications. For example, high-quality amorphous and microcrystalline silicon layers can be easily deposited over large areas by plasma-enhanced chemical vapor deposition (PECVD), or hot-wire chemical vapor deposition (HWCVD), on inexpensive substrates to fabricate active matrix arrays for display and sensor applications. However, they all have low charge carrier drift mobilities compared with their crystalline counterparts. Charge carrier scattering from random periodic potentials in amorphous semiconductors, and from grain boundaries in microcrystalline semiconductors limits the band transport drift mobility, and limits the use of these materials in high-speed or high-gain applications. The mobility of electrons in hydrogenated amorphous silicon (a-Si:H) is typically 5–10 $cm^2 V^{-1} s^{-1}$, it is usually in the range 20–100 $cm^2 V^{-1} s^{-1}$ in nanocrystalline (nc-) and microcrystalline Si (μc-Si) and, for comparison, it is 1400 $cm^2 V^{-1} s^{-1}$ in single-crystal Si. In addition, the charge carrier transport in amorphous semiconductors is further limited by multiple trapping in shallow localized states inasmuch as these materials have a distribution of localized states in their mobility gap between E_v and E_c. The effective electron drift mobility in a-Si:H is ~1 $cm^2 V^{-1} s^{-1}$. One of the drawbacks of a-Si:H technology is that the hole drift mobility is two to three orders of magnitude smaller than the electron mobility which means that it is even more difficult to fabricate useful p-channel FETs and implement complementary circuits. Nonetheless, low-speed electronics is just as important as high-speed electronics in the electronics market. A low-speed flat panel display made from hydrogenated amorphous silicon (a-Si:H) TFTs (thin-film transistors) has a comparable cost to a high-speed crystalline Si microchip that runs the CPU.

The earliest large-scale application of amorphous semiconductors was in xerography (photocopying and printing) which relied on the photoconductivity of a-Se (amorphous selenium) and a-As_2Se_3 (arsenic triselenide) as described by Mort [1]. These xerographic photoconductors were called *photoreceptors*, and ushered in an era of amorphous semiconductor research. While a-Se photoreceptors have been replaced by organic photoreceptors, today a-Se finds use in two important applications. First, is the use of a-Se as an X-ray photoconductor material in recently developed and commercialized direct conversion X-ray image detectors [2, 3]. The second is the use of a-Se as an avalanche photoconductor in ultra-high-sensitive video tubes called HARP tubes that have been able to capture even starlight images [4]. (HARP is an acronym for 'high-gain avalanche rushing photoconductor' and the suffix *icon* makes it a vidicon tube.) Recently, Watanabe *et al.* have successfully incorporated a HARP layer, operating with avalanche gain, onto a CMOS image sensor to produce a prototype HARP-CMOS high-sensitivity image sensor [5].

Large-area electronics today has become almost totally dominated by a-Si:H and μc-Si:H (usually crystallized from a-Si:H) due to the discovery by LeComber and Spear in 1975 that it is indeed possible to dope a-Si:H and thereby change the conductivity by orders of magnitude in a way that resembles the doping of crystalline Si. It was possible to move the Fermi level in this amorphous semiconductor by doping; a very unusual and exciting discovery in the amorphous semiconductor field at the time. Once it was realized that it is possible to prepare high-quality films, dope a-Si:H and modify its electronic properties, the subsequent research very quickly led to the fabrication of various devices such as *pin* photodiodes, photoreceptors, thin-film field-effect transistors, active matrix arrays, displays, image sensors, position sensors, and multi-junction photovoltaic devices, and eventually to a number of successful commercial applications; Street's recent book [6] provides a comprehensive review of the technology and applications of a-Si:H and μc-Si:H.

The properties of both amorphous and microcrystalline silicon depend strongly on the preparation method and the final structure; defects and microcrystallinity (e.g., grain size and growth morphology). Indeed, this very dependence on the fabrication and material quality prevented the early researchers from identifying unifying characteristics. Improving the efficiency and stability of thin-film a-Si:H and μcSi:H photovolatic devices, and trying to understand the underlying physics of the material and the devices, have all been part of the ongoing research in the amorphous semiconductor field over the last ten years. There is currently much interest in developing useful high-quality nc-Si:H TFTs that have higher carrier mobilities than a-Si:H TFTs. An equally critical requirement is high-quality gate and passivation dielectric materials and associated interfaces, since the active channel region is extremely thin. Another topic of interest is the integration of a-Si:H and μc-Si:H TFTs with other technologies, most exciting of which is integration with organic light-emitting diodes (so called OLEDs) and the fabrication of TFT arrays on flexible substrates for the realization of flexible electronics whose applications range from nano- to giga-scale electronic circuits. Flexible electronics is a multidisciplinary area of research and its development requires unconventional design solutions for device and circuit integration and it paves the way for new manufacturing tools and techniques. More recent applications of a-Si:H TFT arrays have been in imaging detectors, including those for medical imaging.

4.2 DIRECT CONVERSION DIGITAL X-RAY IMAGE DETECTORS

A flat-panel X-ray image detector is essentially a large-area integrated circuit that is able to capture an X-ray image and convert it to a digital form for image display, analysis and storage as schematically illustrated in Figure 4.1. Such flat-panel detectors, or *imagers*, can replace the X-ray film/screen cassettes of today, and thereby provide a smooth transition to digital radiography. The flat panel image detector consists of millions of pixels, each of which acts as an individual detector, as shown schematically in Figure 4.2. Each pixel converts the radiation it receives to an amount of charge proportional to the incident radiation. In a *direct conversion* detector, a photoconductor such as a-Se is used to absorb the radiation incident at a pixel and convert it to a charge that is collected on the pixel electrode,

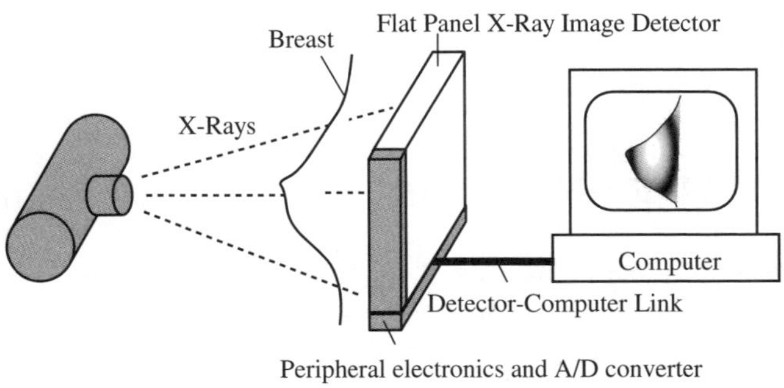

Figure 4.1 Schematic illustration of a flat-panel X-ray image detector, and its operation as an imager

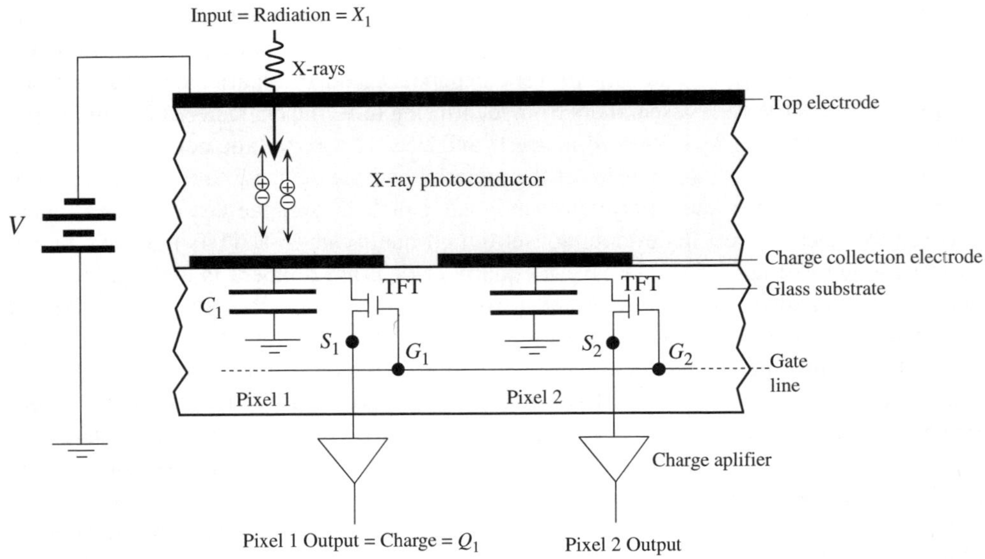

Figure 4.2 A highly simplified schematic diagram of the cross-sectional structure of two pixels of a direct conversion flat-panel X-ray image detector. TFT is a thin-film transistor switch (FET device) that is turned on when a voltage is applied to the gate (after [9] with permission of Elsevier)

then read out and converted to a digital signal. The electron–hole pairs (EHPs) generated in the photoconductor drift under the action of an applied field. Each pixel has a storage capacitance to receive and store the X-ray generated charge from the photoconductor. Figure 4.2 shows the cross-section of a flat-panel detector with two neighboring pixels, labeled 1 and 2. An incident radiation intensity X_1 on pixel 1, generates a charge Q_1 in the photoconductor which is stored on the capacitance C_1 at this pixel. When the gate of the pixel's transistor switch is activated, the charge Q_1 is read out and constitutes the signal for one of the pixels in the image. Direct conversion detectors have been extensively reviewed by a number of authors [2, 3, 7, 8]. There are also successful *indirect conversion* X-ray image detectors that first convert the X-rays incident at a pixel to light by using a phosphor, such as CsI, and then detect the light with a *pin* photodetector at the pixel. In this chapter we only consider direct conversion detectors, and examine the recent progress in a-Se based X-ray photoconductors.

4.3 X-RAY PHOTOCONDUCTORS

The performance of a direct conversion X-ray imager depends critically on the selection and design of the photoconductor. Given that it is not practical to focus X-rays and that the objective is to image body parts, it becomes necessary to find photoconductors that can be prepared in a large area. Due to the difficulties of growing suitable large-area single crystals, the research has focused on amorphous and polycrystalline photoconductive layers such as a-Se, ThBr, PbI_2, HgI_2, PbO, CdZnTe. Table 4.1 summarizes some of the most important

Table 4.1 Material properties of three selected photoconductors: a-Se, HgI_2 and $Cd_{0.95}Zn_{0.05}Te$. Data selectively combined from various sources [12]. *a* is at $F = 10\,V/\mu m$; and *b* is at $F = 20\,V/\mu m$. *d* is density. Resistivity ρ values are estimated

Photoconductor, state, preparation	d (g/cm^3)	δ (μm) at 20 keV	δ (μm) at 60 keV	E_g (eV)	$W_{\pm}$ (eV)	ρ (Ωcm)	Electron μ_e (cm^{2}V^{-1}s^{-1}) $\mu_e\tau_e$ (cm^2/V)	Hole μ_h (cm^{2}V^{-1}s^{-1}) $\mu_h\tau_h$ (cm^2/V)
Stabilized a-Se Vacuum deposition	4.3	48	976	2.1–2.2	45*a* 20*b*	10^{14}–10^{15}	$\mu_e = 0.003$–0.006 $\mu_e\tau_e = 0.3 \times 10^{-6}$–$10^{-5}$	$\mu_h = 0.12$ $\mu_h\tau_h \simeq 10^{-6}$–$6 \times 10^{-5}$
HgI_2 Polycrystalline, PVD	6.3	32	252	2.1	5	~4×10^{13}	$\mu_e = 88$ $\mu_e\tau_e \simeq 10^{-5}$–$10^{-4}$	$\mu_h \simeq 3$–4 $\mu_h\tau_h \sim 10^{-6}$
HgI_2 Polycrystalline, SP	6.3	32	252	2.1	5	~4×10^{13}	$\mu_e\tau_e \simeq 10^{-6}$–$10^{-5}$	$\mu_h\tau_h \sim 10^{-7}$
$Cd_{0.95}Zn_{0.05}Te$ Polycrystalline, Vacuum deposition	5.8	80	250	1.7	5	~10^{11}	$\mu_e\tau_e \simeq 2$–10^{-4}	$\mu_h\tau_h \simeq 3 \times 10^{-6}$

properties of a-Se, HgI_2 and CdZnTe, which are three of the photoconductors currently under intense study as large-area X-ray photoconductors [9, 10]. It is highly desirable that an X-ray photoconductor should have as high an X-ray sensitivity as possible, and should have a negligibly small dark current under operating conditions (e.g., when biased appropriately). In addition, it should exhibit long-term stability in its properties, insignificant X-ray damage and X-ray fatigue, that is, a loss of sensitivity upon prolonged exposure. The reduction in the sensitivity upon X-ray exposure is usually termed 'ghosting' inasmuch as it leads to a 'ghost' of a previous image.

The X-ray sensitivity can usually be addressed in terms of three controlling factors. The first factor is how much radiation is actually absorbed from the incident radiation that is useful in generating EHPs. The *quantum efficiency*, i.e., the fraction $\eta_Q(E) = [1 - \exp(-\alpha L)]$ of incident photons with photon energy E in the beam that are attenuated by the photoconductor, depends on the linear attenuation coefficient α of the photoconductor material and its thickness L. $\alpha = \alpha(E, Z, d)$ is a function of photon energy E, as well as the materials properties such as the atomic number Z and density d. The reciprocal of α, $1/\alpha$, is the *absorption depth* δ where the beam has been attenuated by 63%. The required photoconductor thickness is ideally several times the absorption depth δ.

The second factor is the generation of free EHPs, i.e., a quantity of charge ΔQ, from the absorbed radiation. The amount of radiation energy $W_\pm$ absorbed by a medium to create a single *free* electron and a single *free* hole (free EHP) is called the *ionization energy* or the EHP *creation energy*. This must be as low as possible because the free (or collectable) charge ΔQ generated from an absorbed radiation of energy ΔE is simply $q\Delta E/W_\pm$, where q is the electronic charge. $W_\pm$ depends on the bandgap E_g and, in some cases, as in a-Se, on the applied field F. (By Ramo's theorem, the total charge collected in the external circuit is one electron per free electron and free hole pair generated in the semiconductor.)

The third factor is how much of the photogenerated charge ΔQ is actually collected in the external circuit, which is characterized by the electron and hole drift mobilities (μ_e and μ_h) and lifetimes (deep trapping times, τ_e and τ_h), and their product $\mu\tau$, called the *carrier range*. (The subscripts e and h refer to electrons and holes.) The collection efficiency $\eta_{collect}$ is a function of the carrier *schubwegs* ($\mu\tau F$) per unit thickness ($\mu\tau F/L$), i.e., $x_e = \mu_e\tau_e F/L$ and $x_h = \mu_h\tau_h F/L$, where F is the field, and the normalized absorption depth δ/L [11].

4.4 STABILIZED AMORPHOUS SELENIUM (*a*-Se)

As a result of its commercial use as an electrophotographic photoreceptor, a-Se is one of the most highly developed photoconductors [1, 12]. It can be easily coated over a large area as a thick layer (e.g., 100–1000 μm). It can readily be coated onto any suitable substrate by conventional vacuum deposition techniques and without the need to raise the substrate temperature beyond 60–70°C. Thus, currently a-Se is one of the practical photoconductors for clinical medical X-ray imagers because it has an acceptable X-ray absorption coefficient, good charge transport properties for both holes and electrons and the dark current in a-Se is much smaller than many competing polycrystalline layers. The actual a-Se layer is called *stabilized* a-Se because pure a-Se crystallizes over time, resulting in a much higher dark conductivity, thus making it unsuitable as an X-ray photoconductor. Alloying pure a-Se with As (0.2–0.5% As) prevents crystallization, but introduces deep hole traps, and shortens the hole range. If the alloy is doped with 5–20 parts per million (ppm) of a halogen (e.g.,

Cl), the hole range is improved, but at the expense of electron range. The As alloyed and Cl doped material is called *stabilized* a-Se.

The drift of both electrons and holes in a-Se involves interactions with shallow and deep traps. Shallow traps reduce the drift mobility from its band value μ_0 to an effective drift mobility μ, and deep traps capture carriers and prevent the carriers from being collected, and hence reduce η_{collect}. The deep trap capture times or the lifetimes τ vary substantially between different samples and depend on various factors such as the source of the a-Se material, impurities, and the preparation method. The electron lifetime τ_e is particularly sensitive to impurities in the a-Se source material. The hole lifetime τ_h drops rapidly with decreasing substrate temperature whereas the electron lifetime τ_e is relatively insensitive to the substrate temperature [13]. Increasing the As concentration increases the electron range, whereas Cl doping increases the hole range [14]. Table 4.1 lists typical electron and hole ranges for stabilized a-Se photoconductors.

The EHP creation energy $W_\pm$ in a-Se has a strong dependence on the field F and weak dependence on the photon energy E [15–17]. $W_\pm$ decreases strongly with increasing field. In the case of a-Se, $W_\pm$ at a given X-ray photon energy E follows an empirical relation of the form $W_\pm \approx W_\pm^0 + BF^{-n}$ where B is a constant that depends on the energy, $W_\pm^0$ is the saturated EHP creation energy (at infinite field), and n is typically 0.7–1. The value of $W_\pm^0$ should follow the Que–Rowlands rule $2.2E_g + E_{\text{phonon}}$ [18], where E_{phonon} is a typical small phonon energy. With $E_g \simeq 2.2\,\text{eV}$, we would expect that $W_\pm^0 \approx 5$–$6\,\text{eV}$, but experiments indicate 6–8 eV [13]. The energetic primary electron generates many EHPs, but only a certain fraction of these are free to drift and the rest of the electrons and holes recombine before contributing to the photocurrent. Whether the field dependence of $W_\pm$ is dominated by *geminate* or *columnar* recombination has not been fully resolved and is currently an area of research [19]. Mah and Rowlands [20] interpreted their measurements of X-ray sensitivity change with X-ray energy in the range of 40 kVp to 18 MeV to indicate that at the lower energy the two mechanisms were comparable in importance, but at the higher energy only geminate recombination was acting due to the lower ionization efficiency of high energy electrons. $W_\pm$ decreases with increasing E, but its mechanisms have not yet been conclusively understood. At a typical operating field of $10\,\text{V}/\mu\text{m}$, $W_\pm$ is 35–55 eV over the diagnostic beam energy (12–120 keV). At $F = 20\,\text{V}/\mu\text{m}$ $W_\pm$ is ~20 eV, but such high fields in a thick a-Se photoconductor require a very large applied voltage (20 kV across a thickness of $1000\,\mu\text{m}$), which poses technical problems.

The dark current for metal/a-Se/metal single layer structures depends in a nonlinear fashion on the applied field and the nature of the metal/a-Se contact. Due to the high field necessary to operate the a-Se detector, the dark current in the simple metal/a-Se/metal structure must be reduced. This has been made possible by using a multilayer *p–i–n*-type structure (e.g., where the dark current is less than $0.1\,\text{nA/cm}^2$ at $F = 20\,\text{V}/\mu\text{m}$ [21].) The *p*-layer and *n*-layer terminologies are different from those customarily used for crystalline semiconductors. The *p*- and *n*-layers act as blocking layers by modifying the field at the metal semiconductor-metal. The *p*-layer has been appropriately doped to transport holes, but efficiently traps electrons; the opposite is true for the *n*-layer. The rate of emission of these deeply trapped carriers is so small that there is no significant current injection into the bulk a-Se layer. With these *n*- and *p*-layers, the electric field at the metal electrodes is sufficiently small to minimize charge injection from the contacts, which substantially reduces the dark current. The *i*-layer refers to a-Se that efficiently transports both holes and electrons.

The density of states (DOS) diagram for an amorphous semiconductor is the key to understanding its electrical and optical properties. Unfortunately, even though a-Se has been extensively studied over many years, there are still various uncertainties and controversies in its DOS function (the reader can compare models described and discussed in the literature [22–28]). A tentative schematic diagram is sketched in Figure 4.3, which highlights only the salient features in the DOS. The bandgap $E_c - E_v$ is about 2.1 eV. Over the time scale of typical transit times involved in a-Se X-ray photoconductor layers, the electron drift mobility μ_e is controlled primarily by the shallow traps at about $\Delta E'_1 = 0.30–0.35$ eV below E_c. μ_e is therefore thermally activated with an activation energy $\Delta E'_1$, and the measured activation energy is relatively field independent as borne out by experiments. There is a secondary smaller peak at about $\Delta E'_2 = 0.45–0.50$ eV below E_c, and a distribution of deep electron traps at about $\Delta E'_d \simeq 1.1–1.2$ eV below E_c. The concentration of deep traps depend on alloying, doping, and aging. The hole drift mobility μ_h is also thermally activated, but with an activation energy that depends on the field. New evidence points to a drift mobility μ_h that is likely to be controlled by a monotonically decreasing distribution of tail states, which explains the field dependence of the activation energy. There is a peak in the DOS at about $\Delta E_2 \simeq 0.45–0.50$ eV above E_v, and a distribution of deep hole traps at $\Delta E_d = 0.85–0.90$ eV above E_v. The Fermi level is about 0.95–1 eV above E_v, slightly below midgap (a-Se is p-type). $\Delta E'_2$ and ΔE_2 have been proposed to represent the positions of valence alternation pair (VAP) type defects (Se_3^+ and and Se_1^-) in the structure. The nature and the exact origin of the shallow traps at $\Delta E'_1$ and the deep traps at $\Delta E'_d$ from E_c and ΔE_d

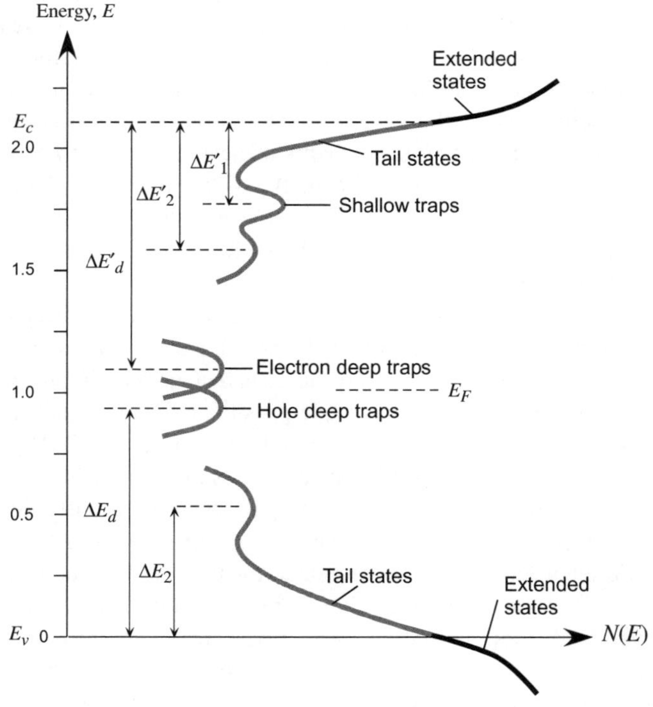

Figure 4.3 A tentative schematic density of states for a-Se

from E_v have not yet been fully resolved. Two different peaks have been shown in Figure 4.3 for electron and hole traps above and below E_F based on the discharge of the saturated cycled-up xerographic potential [29] and cycled time-of-flight experiments [30]. (One 'broad' peak, which straddles the Fermi level within the distribution, cannot be totally ruled out since states above E_F would be deep electron traps, and stated below E_F would be deep hole traps.)

4.5 AVALANCHE MULTIPLICATION AND ULTRA-HIGH-SENSITIVE HARP VIDEO TUBE

The conduction electron in a periodic potential energy function has a traveling Bloch wavefunction and is scattered in the real crystal by its interactions with phonons and impurities. These fundamental ideas are well established. For example, phonons in the crystal spatially deform the potential energy and thereby cause the electron wavefunction to change in a manner that is tantamount to scattering. The mean free path λ of the electron at room temperature is typically of the order of several hundreds of mean interatomic separations, and the drift mobility, taking silicon as an example, is about $1400\,\text{cm}^2\,\text{V}^{-1}\,\text{s}^{-1}$. At sufficiently high fields ($\sim 10^5\,\text{V}\,\text{m}^{-1}$) the electron drift velocity in crystalline Si (c-Si), saturates at about $\sim 10^5\,\text{m}\,\text{s}^{-1}$. As the field increases further, the conduction electron is able to impact ionize the crystal to generate free electron–hole pairs (EHPs) and cause avalanche multiplication; a phenomenon that is commercially utilized in avalanche photodiodes. At high fields, where the electron drift mobility is no longer constant, the electron is said to be a *hot electron* in the sense that its average kinetic energy far exceeds the mean thermal energy, and the electron can be assigned an *electron temperature*. If the original carrier, before multiplication, is part of a signal, as would be the case when the carrier is photogenerated by an absorbed photon in a detector, then the avalanche multiplication provides useful intrinsic (or internal) gain. This is exactly what makes avalanche photodiodes or APDs so indispensable in various optoelectronic applications. This intrinsic gain, however, has a price tag in terms of generating additional noise (called *excess noise*) in the device. Avalanche multiplication is a stochastic process, which has certain statistics associated with it. The

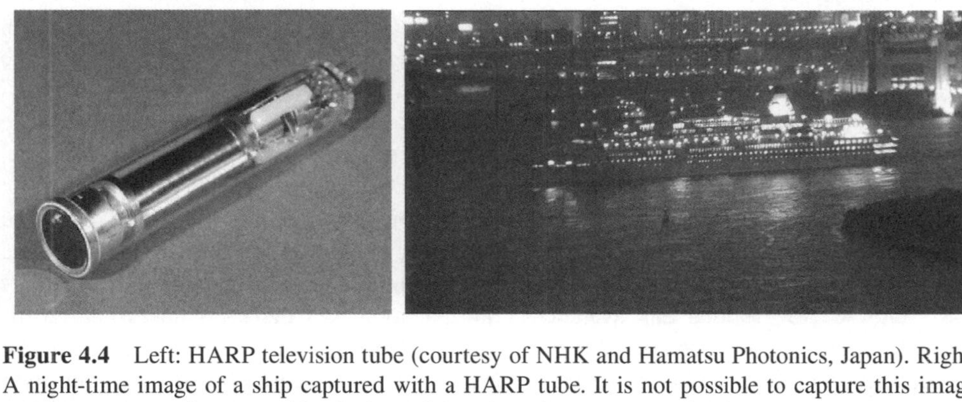

Figure 4.4 Left: HARP television tube (courtesy of NHK and Hamatsu Photonics, Japan). Right: A night-time image of a ship captured with a HARP tube. It is not possible to capture this image with the same quality under identical conditions by using an alternative image sensor (courtesy of NHK, Tokyo, Japan)

randomness of the impact ionization process leads to fluctuations in the gain; an excess noise in the multiplied photocurrent that would not be present if multiplication did not fluctuate. The excess noise is greatest when both carriers avalanche and is minimized when only one type of carrier is allowed to multiply. Another type of excess noise arises when photons are absorbed within the avalanche layer, resulting in a variation in gain due to the depth at which the photon is absorbed. With the development of high-quality multilayer crystalline heterostructures, it has become possible to reduce the excess noise generated by the avalanche processes by separating the photogeneration and the avalanche regions, which has lead to the wide use of avalanche detectors in optical communications [31].

One of the biggest surprises in the amorphous semiconductor field was the experimental observation that it is indeed possible to obtain impact ionization and avalanche multiplication in an amorphous semiconductor [4, 32–37]. Even though the existence of avalanche multiplication was initially controversial and attributed to a probable photoconductive gain (which is distinctly different from impact ionization [38]), it soon became commercialized in the HARP video tube tube—an ultra-high-sensitive TV pick-up tube that uses avalanche multiplication in a-Se to capture images at extremely low light intensities. Soon after the publication of impact ionization in a-Se, others started reporting avalanche type carrier multiplication phenomena in various multilayer or heterojunction-type a-Si:H-based devices [39–46]. There is supporting evidence that an avalanche multiplication M of about 6 can be readily achieved in a-Si:H/a-SiC:H staircase heterostructures [44]. It is believed that the multiplication occurs because, when the electron passes from the wider to the narrower bandgap semiconductor, it is able to use the conduction band mismatch energy ΔE_c between these two semiconductors to cause impact ionization. The electron starts gaining energy, starting from an initial energy ΔE_c and hence can more readily reach the ionization threshold. The evidence for impact ionization within a-Si:H itself however has been limited, and controversial due to experimental difficulties. Nonetheless, avalanche multiplication has been recently reported for reverse-biased a-Si:H *pin*-type photodiode [47] with the onset of avalanche occurring at somewhat higher fields than for a-Se, even though the bandgap of a-Si:H (1.7–1.9 eV) is smaller than that for a-Se (2.0–2.2 eV).

Tanioka, working with coworkers at Japan's Television Broadcast Corporation (NHK) Research Laboratories was able to develop a supersensitive photoconductive target called the HARP (*high-gain avalanche rushing amorphous photoconductor*) [4]. With its avalanche multiplication, the HARP was so sensitive that it was immediately used in a HDTV, high-definition television, camera pick-up tube. The tube incorporating the HARP target has been called the HARP video tube or a-Se *avalanche tube*. The discovery of avalanche in an a-Se camera tube occurred serendipitously and independently of the previous discovery by Juska and Arlauskas [32, 33]. It was generally believed that avalanche would be noisy if it were to occur at all and so was not considered as a worthy project for investigation. However, while measuring what he thought to be the photoconductive gain of an a-Se camera tube, Tanioka noticed that, as he moved the object away from the camera, the image happened to move as well. This contradicted what was known about the relatively long time lag of photoconductive gain, the widely accepted source of gain in many photoconductors [48]. This led to the idea that avalanche multiplication was the mechanism behind his observation. His group was then able to provide evidence for the occurrence of avalanche multiplication in the a-Se layer of the HARP tube for fields above 80 V/μm. Among their observations were the independence of the target capacitance on bias voltage, which excludes the possibility that the charge multiplication is due to electron injection from the scanning beam, the apparently noise-free amplification above 80 V/μm and the variation of the signal current above 80 V/μm with target thickness.

The basic structure of the HARP target and the principle of operation are schematically illustrated in Figure 4.5. The entire target is of the order of 1–35 μm thick. The transparent signal electrode (indium–tin oxide or ITO) is biased positively with respect to the cathode. The CeO_2 and a-Se doped with LiF layers act as a hole blocking contact for hole injection. The Sb_2S_3 layer reduces electron injection from the scanning electron beam and suppresses secondary electron emission. The incident light from the object is absorbed mainly in the a-Se layer (or the photogeneration layer, a-Se:Te). The electron–hole pairs photogenerated in the a-Se layer are then separated by the applied electric field. The electrons are neutralized quickly as they are very close to the positive electrode, whereas holes have to drift across the bulk of the a-Se layer to reach the negative electrode. These drifting holes constitute the signal current. As the photogenerated holes drift through the a-Se layer, as a result of the large applied electric field (greater than 8×10^5 V/cm or 80 V/μm), they experience avalanche multiplication and hence yield an effective quantum gain greater than unity. (The quantum efficiency is the number of EHPs generated per absorbed photon and the quantum gain, or yield, is the number of EHPs collected per absorbed photon.) The effective quantum gain resulting from avalanche multiplication depends on the field as well as the photoconductor thickness. The thicker the a-Se layer, the greater is the multiplication

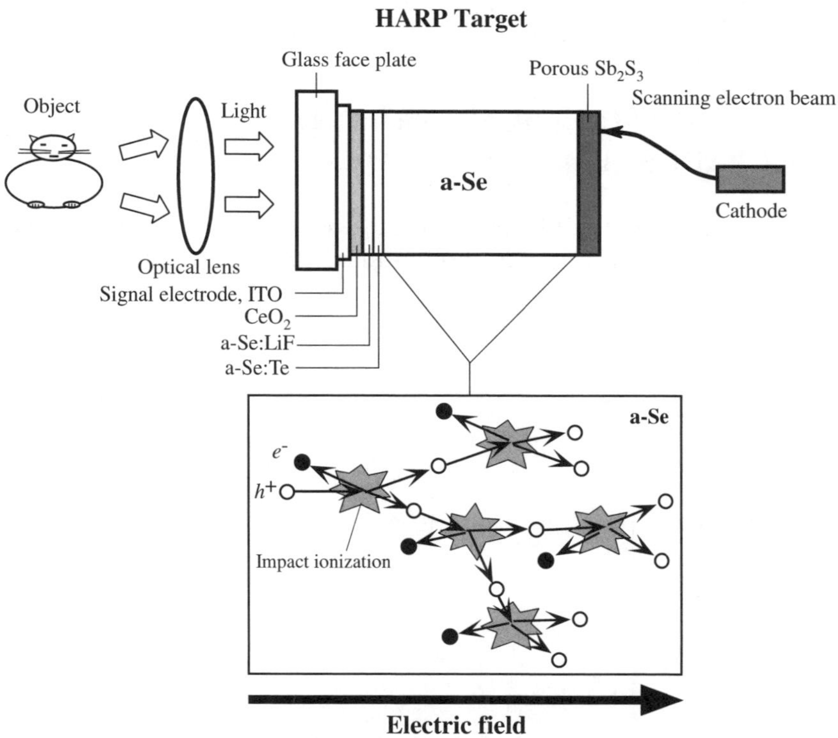

Figure 4.5 A schematic illustration of a HARP video tube and the avalanche that occurs in the photoconductor

at a given field; in fact the multiplication factor increases exponentially with the thickness. For example, in a 2-μm-thick HARP target the quantum gain is about 10 at a field of 120 V/μm, whereas the gain is almost 1000 in an a-Se target of thickness 24.8 μm at a field of 100 V/μm. Avalanche multiplication is normally characterized in terms of an impact ionization coefficient α, which is the probability per unit distance that a carrier, driven by an applied field, will generate an electron–hole pair by impact ionization. The multiplication M therefore depends on the thickness L as

$$M = \exp(\alpha L). \tag{4.1}$$

The ionization coefficient α increases strongly with the field F inasmuch as a carrier can gain more energy from the field and thus has a higher probability of avalanching. The exact dependence of α on the field has been a subject of decades of discussions, and continues to be a topical research area in solid state physics; α depends on the details of the impact ionization process, which will be discussed later. The observed multiplication therefore increases sharply with the field, as illustrated for the HARP in Figure 4.6, which also shows the rise in the dark current with the field. What is interesting is that at $M = 100$ (at $F = 96$ V/μm), the dark current is still below 1 nA—an innocuous level compared with the signal. TV pick-up tubes using such HARP targets have clearly demonstrated a far superior sensitivity than conventional TV pick-up tubes. In fact, these ultra-high-sensitive HARP video tubes have been able to capture even starlight images. The need for high gain vidicons in HDTV is due to the use of small-aperture lenses with much greater depth of field than in conventional TV that allow realistic large-screen effects.

4.6 AVALANCHE MULTIPLICATION IN AMORPHOUS SEMICONDUCTORS

While the experiments have clearly shown the existence of extensive avalanche multiplication in amorphous semiconductors, the theory still lags the experimental work. This is not

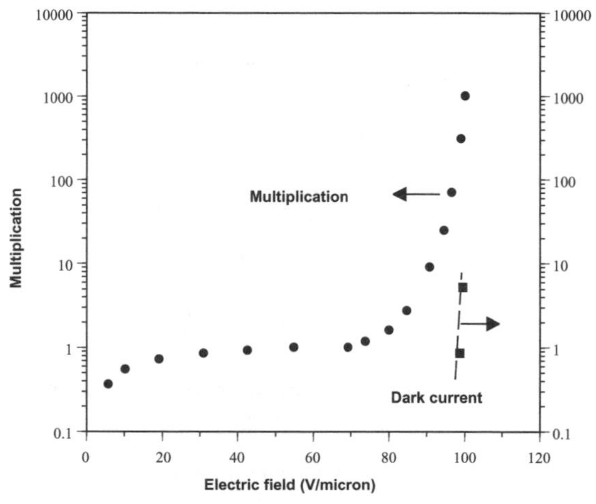

Figure 4.6 Multiplication and dark current vs field in the HARP device

as unusual as it may appear since the theory of amorphous solids lagged behind its counterpart for the crystalline state for many decades. Further, the impact ionization theory in crystalline solids only reached an acceptable level of confidence and understanding in the 1980s and 1990s with the development of the 'lucky-drift' model by Ridley [49, 50]. In both types of semiconductors, the avalanche multiplication within the bulk can be described by an impact ionization coefficient α that has a dependence on the applied field F, at least over the limited fields where avalanche is observed, of the form

$$\alpha \simeq A\exp\left(-\frac{B}{F}\right), \tag{4.2}$$

where A and B are constants; Equation (4.2) is sometimes called *Chynoweth's law.* Equation (4.2) has proven to be indispensable for modeling avalanche photodiodes, and avalanche breakdown in *pn* junction devices. Traditionally, avalanche multiplication data are presented in terms of α vs $1/F$ on a log-linear plot and various straight lines are drawn over limited regions to represent the above behavior (Figure 4.7). The avalanche data for a-Se follows the above expression almost exactly over the avalanche field range, but the avalanche coefficient is significant only at much higher fields than what one would typically expect for crystalline semiconductors of comparable bandgap. There is a known tendency that the α vs $1/F$ curves shift to higher fields for wider bandgap semiconductors so the a-Se behavior stands out as distinctly different from the crystalline cases. The a-Se data for α vs $1/F$ in Figure 4.7 have been taken from Tanioka and co-workers [35, 36], who used the *steady-state* photocurrent measurements and from Juska and Arlauskas [32, 33], who used the TOF *transient*

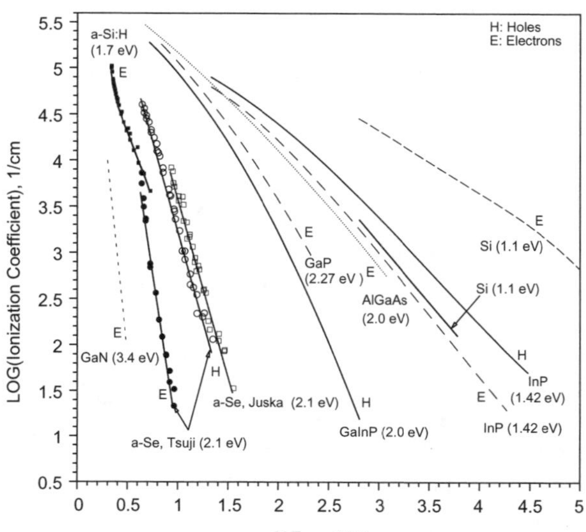

Figure 4.7 Impact ionization coefficient vs reciprocal field for various semiconductors. E for electrons and H for holes (after [59] with permission of the American Institute of Physics)

photoconductivity technique. The two sets of measurements are surprisingly close, given that the samples were different, and the avalanche literature is well known for experimental scatter. The experimental evidence for avalanche in a-Se is impressively convincing.

The origin of Equation (4.2) lies in Shockley's *lucky electron* model that dates back to 1961. When a carrier moves a distance l downstream (along the field) without being scattered, it gains an energy qFl. An unlucky carrier is scattered so frequently that its qFl never reaches the threshold energy E_I for impact ionization. On the other hand, a *lucky electron* is a ballistic electron that avoids scattering for a substantial distance, and hence is able to build its qFl to reach E_I and thereby cause impact ionization as visualized in Figure 4.8. If λ is the mean free path of collisions, then the probability of a ballistic electron gaining the ionization energy E_I is simply given by $\exp(-E_I/qF\lambda)$, which is the exponential factor in Equation (4.2) and the reason for the log-linear plots of α vs $1/F$. (It is assumed that λ is energy independent.). The lucky electron travels a distance $l_I = E_I/qF$ before it impact ionizes, and hence the ionization per unit distance, that is α, must be $(1/l_I)\exp(-E_I/qF\lambda)$, or

$$\alpha = \frac{qF}{E_I}\exp\left(-\frac{E_I}{q\lambda F}\right). \tag{4.3}$$

The main problem with the Shockley model is that there are just not enough ballistic electrons to cause sufficient impact ionizations to explain the experiments. The application of Equation (4.3) to various semiconductors produces nonsensical values for E_I and/or λ. A better model was developed by Baraff [51] in 1962 who numerically solved the Boltzmann transport equation for a simple parabolic band and an energy-independent mean free path

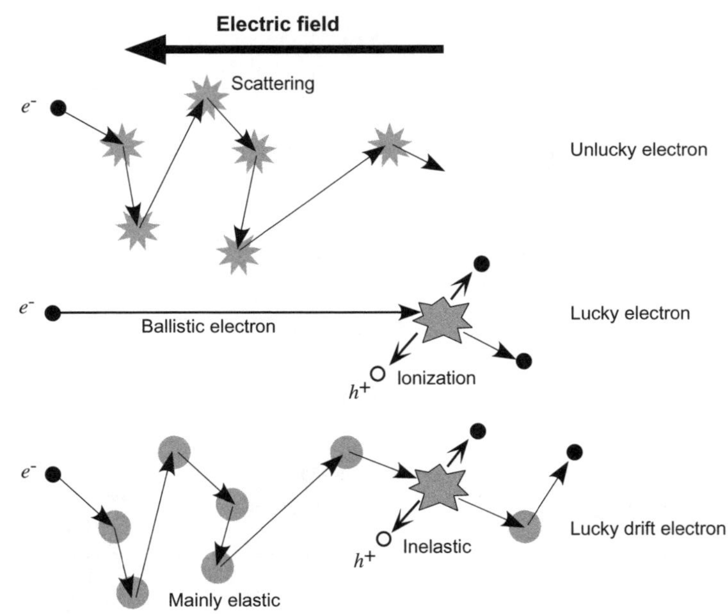

Figure 4.8 Schematic illustration of the Shockley and the Ridley models

to provide a relationship between α and F in terms of four parameters: threshold energy for impact ionization, mean free path associated with ionizations, optical phonon energy, and mean free path for optical phonon scattering. Baraff's theory served experimentalists quite well in terms of comparing their results, even though the model was not intuitive and was limited in terms of its assumptions and applicability to real semiconductors.

Impact ionization theory in crystalline solids only reached an acceptable level of confidence and understanding in the 1980s and 1990 with the development of theoretical models by Ridley [49, 50], Burt [52] and Mackenzie and Burt [53]. (Bringuier has provided a critical overview of the model and comparisons with Monte Carlo simulations [54].) The latter major advancement in the theory appeared as the *lucky-drift* (LD) *model*, and it was based on the realization that at high fields, hot electrons do not relax momentum and energy at the same rates. Momentum relaxation rate is much faster than the energy relaxation rate. An electron can drift, being scattered by phonons, and have its momentum relaxed, which controls the drift velocity, but it can still gain energy during this drift. Stated differently, the mean free path λ_E for energy relaxation is much longer than the mean free path λ for momentum relaxation.

In the Mackenzie and Burt version of the LD model, the probability $P(E)$ that a carrier attains an energy E is given by,

$$P(E) = \exp\left[-\int_0^E \frac{dE'}{qF\lambda(E')}\right] + \int_0^E \frac{dE_1}{qF\lambda(E_1)} \exp\left[-\int_0^{E_1} \frac{dE'}{qF\lambda(E')}\right] \exp\left[-\int_{E_1}^E \frac{dE'}{qF\lambda_E(E')}\right], \tag{4.4}$$

where, as mentioned above, λ is the mean free path associated with momentum relaxing collisions and λ_E is the mean energy relaxation length associated with the energy relaxing collisions. The first term is the Shockley lucky electron probability, i.e., the electrons moves ballistically to energy E. The second term is the lucky-drift probability term which is composed of the following: the electron first moves ballistically to some intermediate energy E_1 ($0 < E_1 < E$) from where it begins its lucky drift to energy E; hence the integration over all possible E_1. The impact ionization coefficient (IIC) can then be readily evaluated from

$$\alpha = \frac{eFP(E_I)}{\int_0^{E_I} P(E)\,dE}. \tag{4.5}$$

As stated in Equation (4.4), there are no specific assumptions that exclude the application of the LD theory to a-semiconductors, provided that one accepts that the same basic concept $\lambda_E > \lambda$ applies to carrier motion in this class of semiconductors. The model above is based on a hard threshold ionization energy E_I; that is, when a carrier attains the energy E_I, ionization ensues. The model has been further refined by the inclusion of soft threshold energies which represent the fact the ionization does not occur immediately when the carrier attains the energy E_I, and the carrier drifts further to gain more energy than E_I before impact ionization [55–57].

Assuming λ and λ_E are energy independent, which would be the case for a single parabolic band in the crystalline state, Equations (4.4) and (4.5) can be solved analytically to obtain

$$\alpha = \frac{1}{\lambda}\frac{(\lambda/\lambda_E)\exp(-E_I/qF\lambda_E) + (\lambda/\lambda_E)^2\exp(-E_I/qF\lambda)}{1-\exp(-E_I/qF\lambda_E) - (\lambda/\lambda_E)^2[1-\exp(-E_I/qF\lambda)]}. \tag{4.6}$$

For $\lambda_E > \lambda$, and in the 'low-field region' where typically $(\alpha\lambda) < 10^{-1}$, or $E_I/qF\lambda > 10$, leads to

$$\alpha = \left(\frac{1}{\lambda_E}\right)\exp\left(-\frac{E_I}{qF\lambda_E}\right). \tag{4.7}$$

Equation (4.4) can also be integrated (at least, numerically) for the case of energy-dependent relaxation lengths. For crystalline semiconductors, one typically also assumes that λ_E depends on the field F, λ and the optical phonon energy $\hbar\omega$ as

$$\lambda_E = \frac{qF\lambda^2}{2\hbar\omega}\coth\left(\frac{\hbar\omega}{2kT}\right). \tag{4.8}$$

As the field increases λ_E eventually exceeds λ, and allows lucky drift to operate and the LD carriers to reach the ionization energy.

Ridley's lucky-drift model has been shown, in principle, to explain the observed avalanche multiplication in a-semiconductors [58, 59]. The LD model with an energy- and field-dependent $\lambda_E = \lambda_E(E, F)$ leads to excellent fits to α vs $1/F$ data for both holes and electrons in a-Se for excitation across the bandgap; $E_I \simeq E_g$. For the a-Se case, $E_I = E_g \simeq 2\,\text{eV}$ leads to the following conclusions: For holes, λ_E has negligibly little field dependence, but increases with energy. At the ionization threshold energy, λ_E is of the order of 4 nm. For electrons, λ_E increases with energy and the field and has the same order of magnitude as λ_E for holes at the ionization threshold and at impact ionization fields. For electron impact ionization in a-Si:H, the data can be readily interpreted in terms of near-bandgap ionization $E_I \simeq E_g$ and a λ_E that decreases with increasing field, and having very little energy dependence; λ_E is about 1–2 nm near the ionization threshold. The energy relaxation length has opposite tendencies in a-Se and a-Si:H, which probably reflects the distinctly different types of behavior of carriers in the transport band in these two amorphous semiconductors. While the overall physics for impact ionization in these two amorphous semiconductors can been identified as that of *lucky-drift*, that is, the momentum relaxation rate is greater than the energy relaxation rate, the details of the theory need to consider how λ_E is related to various carrier scattering mechanisms.

The most significant difference between avalanche in crystalline and amorphous semiconductors is the temperature dependence of the impact ionization rate for the avalanching charge carriers. In crystalline semiconductors, this coefficient *decreases* significantly with increasing temperature due to stronger phonon scattering at higher temperatures. In contrast, the impact ionization rates for both electrons and holes in *a*-Se tend to *increase* with temperature, implying that the mean free path of hot carriers is not controlled by free carrier and phonon interactions, and there must be other underlying causes. The IIC α for a-Se decreases with increasing temperature, and the activation energy E_α for α_h vs T behavior is small, 0.023–0.045 eV, and decreases with increasing field. This can probably be explained by impact excitation to band tails and then thermal excitation into the extended states. The release time must be much shorter than the transit time. As the field increases, the transit time gets shorter, the release time must likewise get shorter, and hence the activation energy, the energy for release from the tail state into the transport band, must get smaller. This would qualitatively explain the observed temperature dependence. We can estimate the release time τ_r. Taking $\upsilon \simeq 10^{12}\,s^{-1}$, the longest release time would correspond to the largest activation energy of 0.045 eV at $F = 1\,MV/cm$. Thus, $\tau_r = \upsilon^{-1}\exp(E_\alpha/kT)$ gives $\tau_r = 6\,ps$, much shorter than the TOF hole transit time observed at this field, and hence consistent with the aforementioned explanation. τ_r gets shorter as E_α decreases with increasing F, and hence remains shorter than the transit time.

4.7 FUTURE IMAGING APPLICATIONS WITH *a*-Se HARP

The HARP video tube is essentially a vacuum-tube-based supersensitive imaging device that uses an electron beam to scan a photoconductive target to capture the image. At present there is much research interest in solid-sate image sensors because they are simply more practical and reliable Recently Tanioka and coworkers have successfully incorporated a HARP layer, operating with avalanche gain, onto a CMOS image sensor to produce a prototype HARP-CMOS high-sensitivity image sensor [5] as schematically illustrated in Figure 4.9. A conventional CMOS imaging array with electrodes at each pixel is coated with a HARP that is biased to operate in the avalanche multiplication mode. When the HARP at a particular pixel receives signal photons, it photogenerates holes that multiply as they drift towards the negative electrode. This multiplied photocurrent charges a capacitor that stores the signal as charge. The signal charge is later read out by appropriately

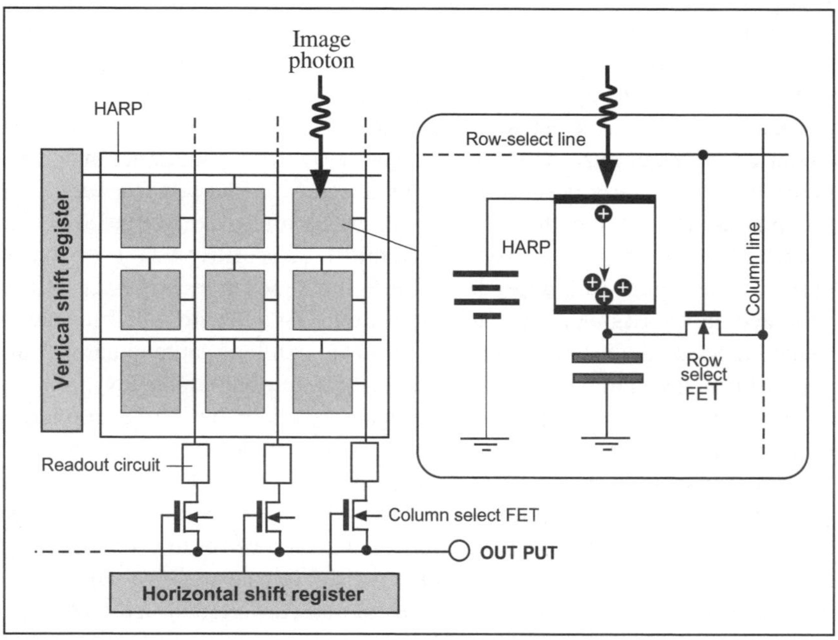

Figure 4.9 The principle of the CMOS-HARP imager

a b

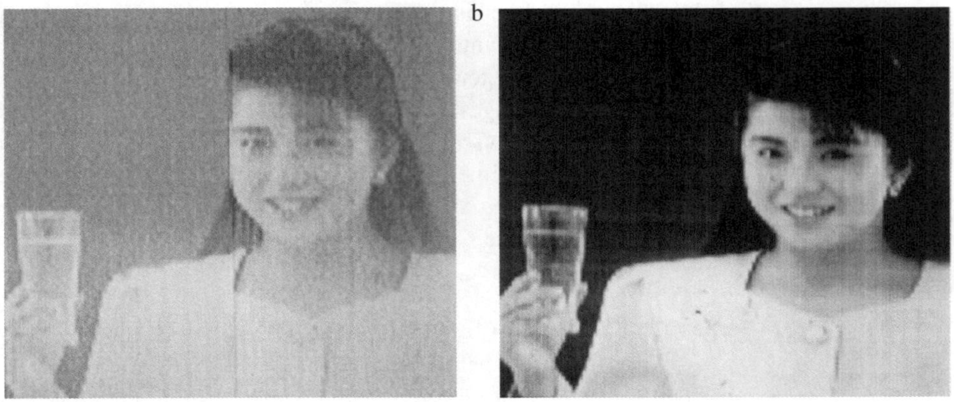

Figure 4.10 Images captured by: (a) a conventional CMOS; (b) the prototype CMOS-HARP image sensors at very low light levels (courtesy of NHK)

addressing that particular pixel. Thus, each pixel of the CMOS sensor receives an avalanche-multiplied photocurrent signal and consequently the overall signal-to-noise ratio becomes substantially improved; the signal is multiplied at each pixel. The images from the HARP-CMOS sensor have been very impressive and superior to those from conventional imagers (Figure 4.10).

One of the problems with a-Se as an X-ray photoconductor is that its ionization energy $W_{\pm}$ is not the best among various competing photoconductors due to the fundamental limitation of X-ray EHP creation in this material. The avalanche multiplication in a-Se at high fields opens the possibility of using avalanche gain to increase its X-ray sensitivity [60]. As recently shown by Hunt, Kirby and Rowlands [60], the increase in signal due to avalanche multiplication would be sufficient to make a-Se-based flat-panel X-ray image detectors quantum noise limited, even at the lowest exposure levels (one photon per image per pixel) needed during the medical procedure called fluoroscopy. In the last few years, NHK has produced an X-ray HARP tube that has supersensitivity, though the target area size so far is only 1 inch (2.5 cm). The X-ray HARP video tube is so sensitive that it can detect single X-ray photons; see Figure 4.11 upper left. The figure shows how an image of a Lincoln penny has been built up using a very low exposure to X-rays, with a high-resolution final image. Such high sensitivities also allow the sensor to exhibit a large dynamic range that, if used in a flat panel, will make the imaging detector usable both in the low X-ray exposure procedure of fluoroscopy as well as the much higher exposure level projection radiography and still obtain images with the theoretical image quality obtainable at each of these exposures. There are essentially two possibilities for using the HARP in low-exposure X-ray imaging. First is the indirect detection system in which the HARP would be coupled to a phosphor screen. The screen converts the X-ray photons it receives to visible photons which are then detected by the HARP. The second is the direct conversion system in which the X-rays are actually absorbed in the photoconductor and converted directly into charge carriers that can be multiplied by avalanche in this layer.

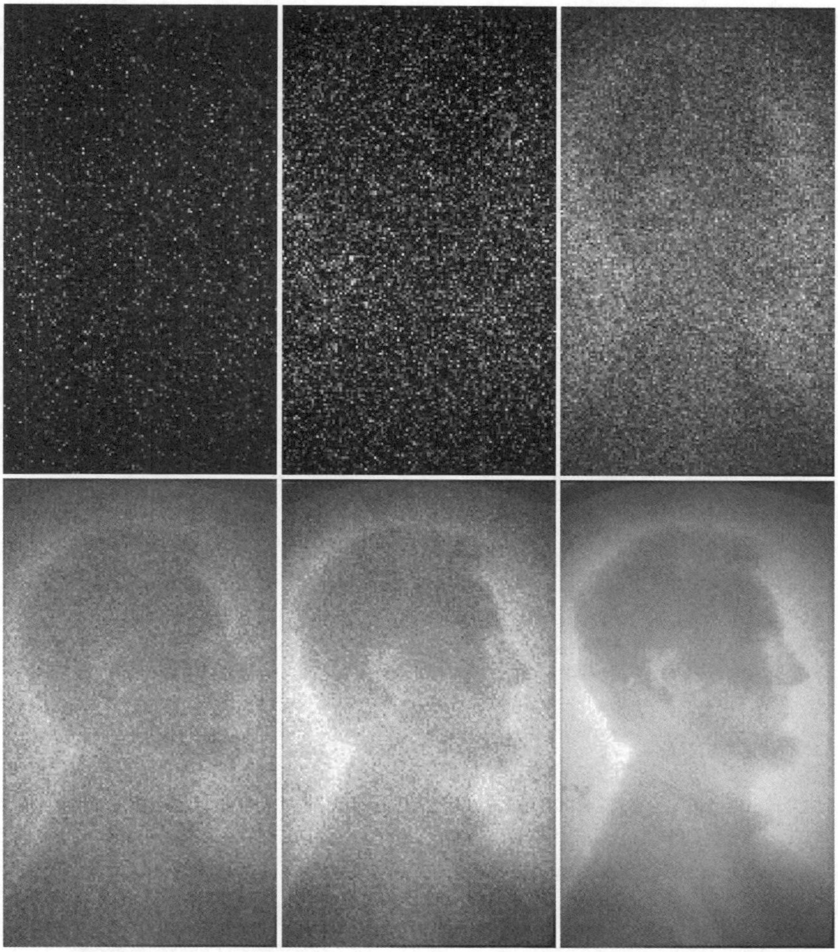

Figure 4.11 X-Ray image of an American one-cent coin captured using the X-ray a-Se HARP camera. The first image at the top left is obtained under extremely low exposure and the subsequent images are obtained with increasing exposure of approximately one order of magnitude between each image. These images demonstrate the huge dynamic range of the sensor that is achieved by changing the avalanche gain, depending on the mean exposure expected (courtesy of Dylan Hunt. © Hunt and Rowlands, 2002)

4.8 HYDROGENATED AMORPHOUS SILICON THIN-FILM TRANSISTORS

The a-Si:H TFT has evolved from its role as a switching device in active matrix arrays [6, 61] to new application areas in large-area electronics where it serves as a linear integrated circuit element. Depending on the application, there are different requirements on device performance. For example in active matrix organic light-emitting diode (AMOLED) displays [62], the forward static characteristics are of great importance, since the organic light-emitting diode requires a high drive current. In active matrix flat-panel imaging (AMFPI) applications [63], the reverse (leakage) regime of operation of the TFT is critical and becomes a key design constraint, which can limit the performance of the array in terms of its signal-to-noise-ratio and dynamic range. In particular, essential for reliable design of the TFT and associated pixel circuitry are physically-based compact models of TFT operation that are not only capable of accurately predicting device and circuit characteristics, but which can also relate TFT performance to its geometrical and fabrication process parameters. Models for the transient behavior of the TFT are of equal importance for accurate prediction of the turn-on and turn-off switching time constants, clock feedthrough, and charge-sharing effects, which strongly influence voltages at floating nodes.

Figure 4.12(a) shows the cross-section of the TFT. To examine the effect of geometric parameters of the TFTs on its characteristics, TFTs with different channel length, channel width, and overlap length values were fabricated. In order to study the leakage characteristics of the TFT, up to 100 TFTs were put in parallel to increase the current to a measurable

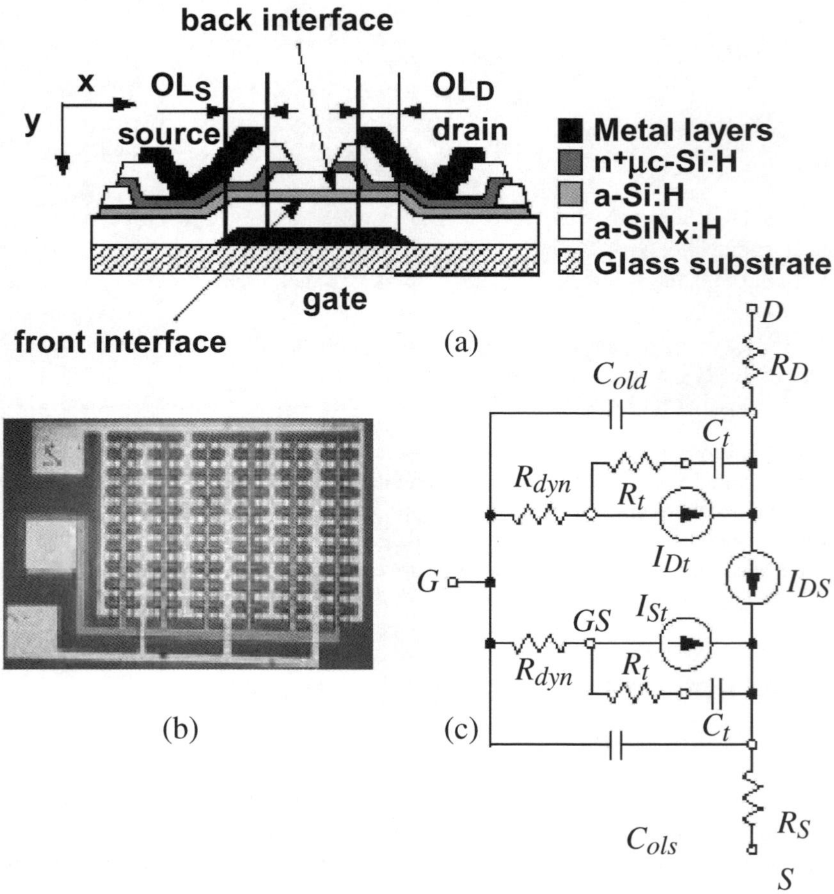

Figure 4.12 (a) Cross-section of the TFT showing its different layers, (b) photomicrograph of an array of TFTs in parallel; (c) the complete equivalent circuit model of the a-Si:H TFT

value [64, 65]. Figure 4.12(b) shows a photograph of the fabricated TFT arrays, including 60 TFTs in parallel ($W/L = 20\,\mu$m / $10\,\mu$m), designed for extraction of meaningful TFT parameters, particularly at very low (<1 pA) DC currents, based on delayed measurements and averaging. Figure 4.12(c) illustrates the complete model of the TFT, for prediction of both static and dynamic characteristics. The current source I_{DS} and contact resistances R_D and R_S constitute the static model. Depending on bias, four regimes of operation can be recognized for I_{DS} (Figure 4.13): above-threshold, forward sub-threshold, reverse sub-threshold (back channel), and front channel (Poole–Frenkel). An elaborate description of the different regimes and associated compact models can be found in [66].

In the above-threshold region ($V_{GS} > V_T$), the quasi-Fermi-level lies close to the conduction band edge, thus the acceptor-like tail states determine the on-current I_{DS} of the TFT [67, 68, 69]. Here, the effect of contact resistances R_S, R_D becomes strongly visible. In particular, R_S plays the role of a feedback resistance that should be considered in the extraction of the DC parameters. Since the value of the contact resistances is not a function of the channel length of the TFT, their values can be retrieved from measurements using different channel length values. In the sub-threshold region ($V_T > V_{GS} > V_{Ts}$), interface charges and deep defects in the a-Si:H bulk constitute the main source of charge. Here, conduction occurs close to the interface where there is extended band bending. The model parameters in general are dependent on the material parameters and thus serve as measures for the quality of the interfaces and bulk a-Si:H. Consequently, they depend on process conditions as well as electrical stress.

In the reverse regime of operation, there are two major mechanisms responsible for the leakage current between drain and source: back-channel (reverse sub-threshold) and front-channel conduction [64]. Exactly which prevails depends on the applied bias and device geometry. The back-channel conduction is responsible for the reverse sub-threshold regime. Its due to a weak electron channel formed at the back a-Si:H/a-SiN$_x$:H interface, as a result of the high defect density at that interface [70], and the presence of a positive drain voltage. Front-channel conduction (via holes) is responsible for the exponential increase in current

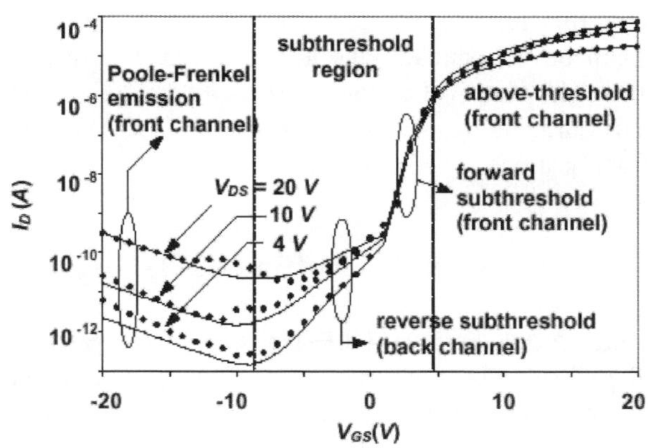

Figure 4.13 Drain-source current as a function of gate-source voltage for V_{DS} = 4, 10 and 20 V, showing the different regimes of operation. Symbols: · · · : experimental; solid line: model

at high negative gate and high positive drain voltages (Figure 4.13). Accumulation of holes at the front a-Si:H/a-SiN$_x$:H interface by virtue of a negative gate voltage provides conduction path for this component of leakage. The holes are generated as a result of the Poole–Frenkel field-enhanced thermoionic emission in the gate–drain overlap vicinity [71, 72].

The transient response of the TFT depends on the distribution of channel charge and the charge accumulated in the gate-source and gate-drain, by virtue of the source/drain overlap capacitances, C_{ols} and C_{old}. The associated equivalent charge share current components at the source and drain are represented as I_{St} and I_{Dt} in Figure 4.12(c). The resistance R_{dyn} is a dynamic resistance that models the delay in TFT response to change in bias voltage [73]. It gives the expected charge and discharge time constants for the channel as predicted in [67] for a-Si:H TFTs and in [74] for polysilicon TFTs. Trapping of carriers at deep energy levels has been reported in the literature and has been found to significantly affect the TFT's transient and static performance [65, 74]. The time constants associated with charge trapping and de-trapping in deep levels can be of the order of milliseconds or more. In the above-threshold operation, trapping of electrons in these states provides a slow decay of current that is observed in experiments at Waterloo and also reported in the literature [65]. The observation of the so-called Fermi level relaxation [74], which is associated to the release time of electrons from the deep energy levels during the turn-off process of the TFT, also signifies the influence of the trapped carriers on the transient characteristics of the TFT. The effect of trapped carriers is modeled by using the capacitance C_t and a resistance R_t [73].

4.9 TFT BACKPLANES FOR ORGANIC LIGHT-EMITTING DIODE DISPLAYS AND FLAT-PANEL X-RAY IMAGERS

Although hydrogenated amorphous silicon (a-Si:H) is inherently disadvantaged in terms of speed and stability compared with polycrystalline or crystalline Si, there is an overwhelming need for this mature technology in several newly emerging and significant application areas. Here, the question arises as to whether circuit techniques can be employed to compensate for its intrinsic material shortcomings so as to meet performance requirements. The following sections review precisely these challenges. Specifically, they address design considerations pertinent to thin-film circuits for on-pixel current drivers and amplifiers, whose integration requires nonconventional design solutions to deal with the high material resistivity and high instability. The family of circuits presented here are critical to two application areas: the active matrix organic light-emitting diode (AMOLED) display and digital fluoroscopy for medical applications.

4.9.1 Active matrix organic light-emitting diode displays

Compared with liquid crystal displays, OLED displays have gained significant interest recently in view of their faster response times, larger viewing angles, higher contrast, lighter weight, lower power, and amenability to flexible substrates [75]. Active matrix addressing, which is needed for high information content formats, involves a layer of backplane electronics based on TFTs to provide the bias voltage and drive current needed in each OLED pixel. The a-Si:H backplane adequately meets many of the drive requirements for small-area displays such as those needed in pagers, cell phones, and other mobile devices. The lower mobility associated with a-Si:H TFTs ($\mu_{FE} \sim 1\,cm^2/Vs$) is not a limiting factor since the drive transistor in the pixel can be scaled up in area to provide the needed drive current, without necessarily compromising the aperture ratio.

In AMOLED displays, it is important to ensure that the aperture ratio or fill factor (defined as the ratio of light-emitting display area to the total pixel area) should be high enough to ensure display quality [76]. Conventional AMOLED displays are based on light emission through an aperture on the glass substrate where the backplane electronics is integrated. Increasing the on-pixel density of TFT integration for stable drive current reduces the size of the aperture. The same happens when pixel sizes are scaled down. The solution to having an aperture ratio that is invariant to scaling or on-pixel integration density is to vertically stack the OLED layer on the backplane electronics, along with a transparent top electrode (Figure 4.14). This implies a continuous back electrode over the OLED pixel. However, this continuous back electrode can give rise to parasitic capacitance, whose effects become significant when the electrode runs over the switching and other TFTs. Here, the presence of the back electrode can induce a parasitic channel in TFTs, giving rise to high leakage current. The leakage current is the current that flows between source and drain of the TFT when the gate of the TFT is in its OFF state. To minimize the conduction induced in all TFTs in the pixel by the back electrode, an alternate TFT structure based on a dual gate structure [77] can be employed, in which the voltage on the top gate can be chosen such so as to minimize the charge induced in the (parasitic) top channel of the TFT. The objective underlying the choice of the voltage on the top gate is to minimize parasitic capacitance in the circuit and leakage currents in the TFTs, so as to enhance circuit performance. In what follows, we describe the theory of operation of the dual gate TFT, which will be central to surface emissive (high aperture ratio) AMOLED displays based on a-Si:H backplane electronics.

The simplest pixel driver circuit possible is the two-TFT voltage-programmed circuit shown in Figure 4.15. Since the current through the OLED depends on the V_{GS} of T2, this circuit is very susceptible to any V_T increase in the drive TFT T2 and to any change in the

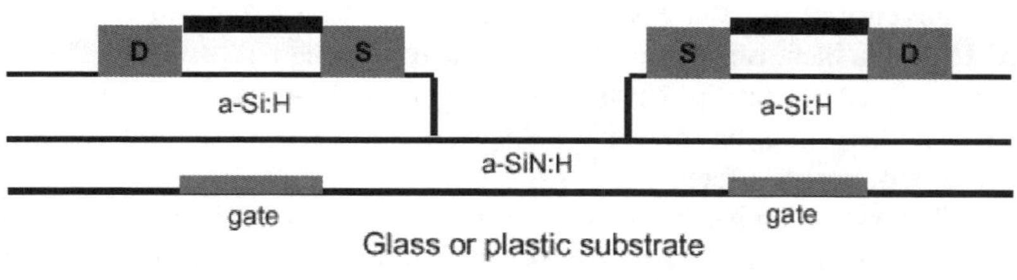

Figure 4.14 Pixel architecture for surface emissive a-Si:H AMOLED displays

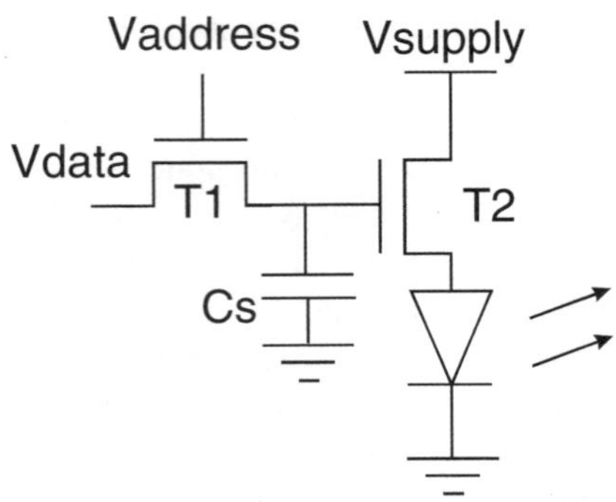

Figure 4.15 Conventional two-TFT pixel circuit

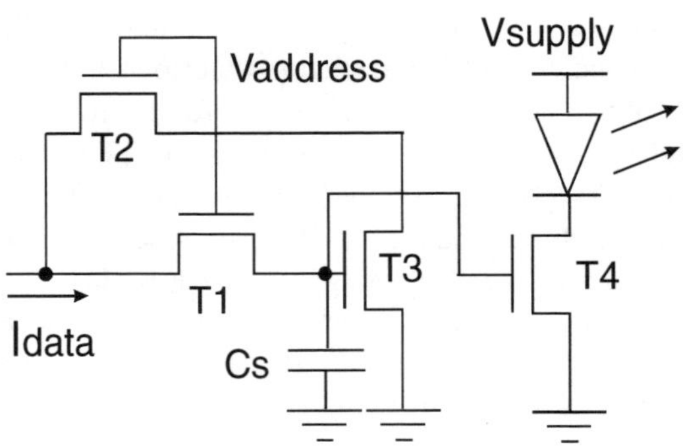

Figure 4.16 Compensated four-TFT pixel circuit

voltage drop across the OLED. This is one of the biggest hurdles in amorphous silicon circuit design since the V_T increase is higher when the TFT gate bias stress is high or applied for long durations [78]. Due to this, the OLED brightness with the two-TFT pixel circuit will gradually decrease and the pixel will eventually turn off.

To overcome this problem, current-programmed pixel circuits have been developed. The example shown here (Figure 4.16) is based on the current mirror circuit family [79, 80]. The OLED current in these circuits is independent of any threshold voltage or mobility variation in the drive TFT as long as it stays in the saturation region of operation. In this circuit, the data current I_{data} charges the capacitor Cs until T3 starts to conduct. The gate voltage of T3 keeps increasing until all of I_{data} passes through T2 and T3. This current is then mirrored to pass through T4 since the gates of T3 and T4 are connected. Since the V_{GS} of both TFTs is the same, the threshold voltages of T3 and T4 will shift equally and the output current will not be affected.

Figure 4.17 shows a comparison of the short-term stability of the OLED current in the two-TFT and four-TFT pixels. After considerable duration of stress, the OLED current in the two-TFT circuit drops by 25% from its initial value, whereas the current in the four-TFT pixel circuit maintains an approximately nominal value.

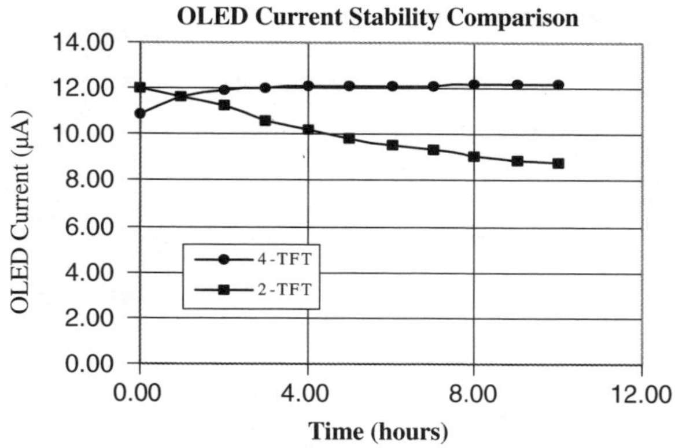

Figure 4.17 Current stability comparison for the two-TFT and four-TFT circuits

4.9.2 Active pixel sensors for digital fluoroscopy

Active matrix flat-panel imagers (AMFPIs) have gained considerable significance in large-area flat-panel digital imaging applications [63], in view of their large-area readout capability. The pixel, forming the fundamental unit of the imager, consists of a detector and readout circuit to efficiently transfer the collected electrons to external readout electronics for data acquisition. The pixel architecture most widely used is based on the passive pixel sensor (PPS). An example is the amorphous selenium (a-Se)-based photoconductor detection scheme where the readout circuit consists of a storage capacitor and a TFT readout switch [81]. The storage capacitor accumulates signal charge during the integration period and transfers the collected charge to an external charge amplifier via the TFT switch during readout. While the PPS architecture has the advantage of being compact and thus amenable to high-resolution imaging, the task of reading the small output signal of the PPS for low input signal, large-area applications (e.g., fluoroscopy) is extremely challenging. More importantly, if external noise sources (e.g., charge amplifier noise and array data line thermal noise) are comparable to the input, there is a significant reduction in pixel dynamic range. These problems can be overcome by integrating an on-pixel amplifier circuit using a-Si TFTs [82], as shown in Figure 4.18, along the lines of the CMOS active pixel sensor (APS) architecture [83]. The APS performs *in situ* signal amplification providing higher immunity to external noise, preserving the dynamic range. In addition, the performance and cost constraints on external charge amplifiers are relaxed.

Unlike a conventional PPS, which has one TFT switch, there are three TFTs in the APS pixel architecture. This could undermine fill factor if conventional methods of placing the sensor and TFTs are used. Therefore, in an effort to optimize fill factor, the TFTs may be embedded under the sensor, along the lines of the architecture shown in Figure 4.14, to provide high-fill-factor imaging systems. Central to the APS illustrated in Figure 4.18 is a source follower circuit, which produces a current output (C-APS) to drive an external charge amplifier. In the C-APS circuit, the characteristic ΔV_T of a-Si TFTs is manageable since the TFTs have a duty cycle of ~0.1% in typical large-area applications. Therefore, appropriate biasing voltages in the TFT ON and OFF states can minimize ΔV_T. Operating the READ and RESET TFTs in the linear region reduces the effect of inter-pixel threshold voltage V_T

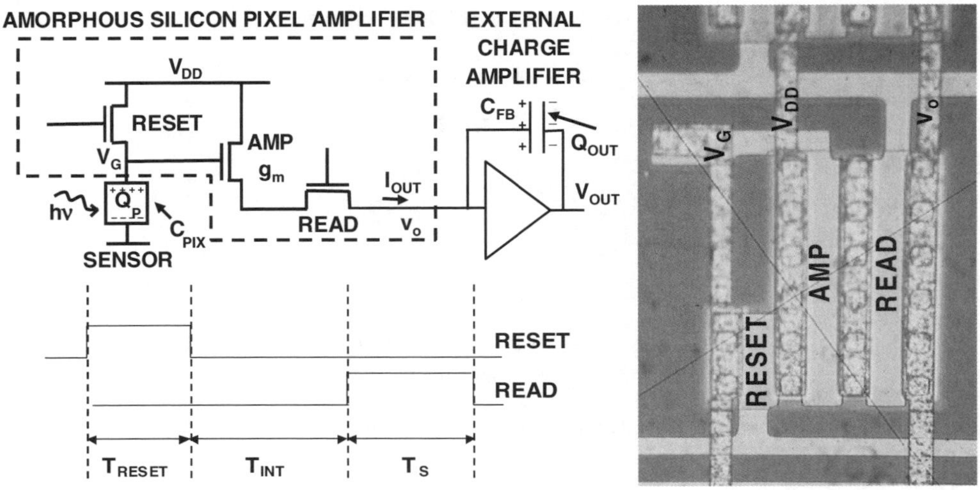

Figure 4.18 Current mode active pixel schematic, readout timing diagram and amorphous silicon circuit micrograph

nonuniformities. However, although the saturated AMP TFT causes the C-APS to suffer from FPN, using CMOS-like off-chip double-sampling techniques [84] can alleviate the problem.

The circuit provides excellent linearity in its transfer characteristics, with nonlinearity less than 5%. High charge gain in these circuits is critical since it provides amplification of the input signal, making it resilient to external noise sources, and can be achieved using a low-capacitance sensor (i.e., small C_{PIX}). However, a tradeoff between pixel gain and amplifier saturation will place an upper limit on the achievable charge gain. From a noise standpoint, it has been found that the intrinsic APS noise is minimized for small C_{PIX}, implying the feasibility of low-capacitance detectors [85]. In addition, minimizing C_{PIX} will also reduce the reset time constant (comprised mainly of the RESET TFT on-resistance and C_{PIX}), hence reducing image lag. For example, assuming column parallel readout, a typical array comprising of 1000×1000 pixels operating in real time at 30 frames/s allows $33\,\mu$s for each pixel's readout and reset. Typical values for a-Si RESET TFT on resistance (~1 MΩ) and C_{PIX} (~1 pF for a-Se) yield an RC time constant of $1\,\mu$s, implying $5\,\mu$s resets would eliminate image lag and still allow sufficient time for readout with double sampling. Like other current mode circuits, the C-APS, operating at 30 kHz, is susceptible to sampling clock jitter. However, off-chip low-jitter clocks using crystal oscillators can alleviate this problem.

The examples given above illustrate how circuitry, coupled with new pixel architectures can compensate for the materials-related issues such as low mobility and threshold voltage shift. As mentioned earlier, one of the key challenges here lies in vertical integration of the on-pixel electronics with the sensor/display layer so as to preserve high fill factor/aperture ratio. Future challenges lie in realization of similar pixilated electronics on mechanically flexible substrates, which require a significant lowering of process temperatures without compromising electrical performance.

ACKNOWLEDGEMENTS

The authors thank NSERC for a collaborative health research project that made this work possible. It is a pleasure to thank Drs Zahangir Kabir, Peyman Servati and K. S. Karim for their contributions and help.

REFERENCES

[1] J. Mort, *The Anatomy of Xerography*, McFarland, London, 1989.
[2] S.O. Kasap and J.A. Rowlands, *Proc. IEEE*, **90**, 591 (2002).
[3] J.A. Rowlands and S.O Kasap, *Physics Today*, **50**, 24–30 (1997).
[4] K. Tanioka, J. Yamazaki, K. Shidara, K. Taketoshi, T. Hirai, and Y. Takasaki, *Adv. Electron. Electron Phys.*, **74**, 379 (1988).
[5] T. Watanabe, M. Goto, H. Ohtake, H. Maruyama, M. Abe, K. Tanioka, and N. Egami, *IEEE Trans. Elec. Dev.*, **50**, 63 (2003).
[6] R. Street, *Technology and Applications of Amorphous Silicon*, Springer, Berlin, 2000.
[7] J.A. Rowlands and J. Yorkston, in: *Handbook of Medical Imaging*, Vol. 1, J Beutel, HL Kundel and RL Van Metter (eds), SPIE Press, Washington, 2000, Chap. 4 and references therein.
[8] S.O. Kasap and J.A. Rowlands, *IEE Proc.-CDS*, **149**, 85 (2002).
[9] S.O. Kasap, M. Zahangir Kabir, and J.A. Rowlands, *Current Applied Physics* (in press).
[10] Data have been extracted and combined from references cited in M.Z. Kabir, S.O. Kasap and J.A. Rowlands, in: *The Springer Handbook of Electronic and Photonic Materials*, S.O. Kasap and P. Capper (eds), Springer, Heidelberg 2006.
[11] M.Z. Kabir and S.O. Kasap, *Appl. Phys. Letts.*, **80** 1664, 2002, and references therein.
[12] S.O. Kasap, *The Handbook of Imaging Materials, Ed.* A Diamond, Marcel Dekker, New York, 1991, Chap. 8 and references therein.
[13] G. Belev and S.O. Kasap, *J Non-Cryst Solids*, **345**, 484 (2004).
[14] S.O. Kasap, K.V. Koughia, B. Fogal, G. Belev, G., and R.E. Johanson, *Semiconductors*, **37**, 816 (2003).
[15] S.O. Kasap, C. Haugen, M. Nesdoly, and J.A. Rowlands, *J. Non-Cryst. Solids*, **266**, 1163 (2000).
[16] I.M. Blevis, D.C. Hunt, and J.A. Rowlands, *J. Appl. Phys.*, **85**, 7958 (1999).
[17] M.F. Stone, W. Zhao, B.V. Jacak, P. O'Conner, B. Yu, B., and P. Rehak, *Med. Phys.*, **29**, 319 (2002).
[18] W. Que and J.A. Rowlands, *Phys. Rev. B*, **51**, 10500 (1995).
[19] E. Fourkal, M. Lachaine, and B.G. Fallone, *Phys. Rev. B*, **63**, 195204 (2001).
[20] D. Mah, J.A. Rowlands, and J.A. Rawlinson, *Med. Phys.*, **25**, 444 (1998).
[21] B. Polischuk, Z. Shukri, A. Legros, and H. Rougeot, *Proc. SPIE*, **3336**, 494 (1998).
[22] M. Abkowitz, *Philos. Mag. Lett.*, **58**, 53 (1988).
[23] H.-Z. Song, G.J. Adriaenssens, E.V. Emelianova, and V.I. Arkhipov, *Phys. Rev. B*, **59**, 10607 (1999).
[24] S.O. Kasap and J.A. Rowlands, *J. Mater. Sci.: Mater in Elect.*, **11**, 179 (2000).
[25] L. Benkhedir, M.S. Aida, and G.J. Adriaenssens, *J. Non-Crystal. Solids*, **344**, 193 (2004).
[26] L. Benkhedir, M. Brinza, and G.J. Adriaenssens, *J. Phys.: Condens. Matter*, **16**, S5253 (2004).
[27] N. Qamhieh, M.L. Benkhedir, M. Brinza, J. Willekens, and G.J. Adriaenssens, *J. Phys.: Condens. Matter*, **16**, 3827 (2004).
[28] K. Koughia, Z. Shakoor, S.O. Kasap, and J.M. Marshall, *J. Appl. Phys.*, **97**, 33706 (2005).
[29] M. Abkowitz and R. Enck, *Phys. Rev. B*, **25**, 2567 (1982).

[30] J. Veres and C. Juhasz, *J. Non-Cryst. Solids*, **164–166**, 407 (1993).
[31] S.O. Kasap, Optoelectronics, in: *Encyclopedia of Optics*, Wiley-VCH, 2003; *Optoelectronics and Photonics: Principles and Practices*, Prentice Hall, 2002.
[32] G. Juska and K. Arlauskas, *Phys. Stat Solidi*, **59**, 389 (1980).
[33] G. Juska and K. Arlauskas, *Phys. Stat Solidi*, **77**, 387 (1983).
[34] K. Tanioka, J. Yamazaki, K, Shidara, K. Taketoshi, T. Kawamura, S. Ishioka, and Y. Takasaki, *IEEE Elec. Dev. Lett.* **EDL-8**: 392 (1987).
[35] K. Tsuji, Y. Takasaki, T. Hirai, and K. Taketoshi, *J. Non-Cryst. Solids*, **14**, 94 (1989).
[36] K. Tsuji, T. Ohshima, T. Hirai, N. Gotoh, K. Tanioka, and K. Shidara, *Mat Res. Symp. Proc.* **219**, 507 (1991).
[37] K. Tsuji, Y. Takasaki, T. Hirai, J. Yamazaki, and K. Tanioka, *Optoelectronics—Devices and Technologies* (Japan), **9**, 367 (1994).
[38] A. Rose, *Concepts in Photoconductivity and Allied Problems*, Robert E. Krieger, New York, 1978.
[39] S.-C. Jwo, M.-T. Wu, Y.-K. Fang, Y.-W. Chen, J.-W. Hong, and C.-Y. Chang, *IEEE Trans. Elec. Dev.*, **35**, 1279 (1988).
[40] S. Sugawa, K. Ohmi, M. Yamada, and Y. Osada, *Sol. Stat. Comm.*, **80**, 373 (1991).
[41] M. Yoshimi, T. Ishiko, K. Hattori, H. Okamoto, and Y. Hamakawa, *J. Appl. Phys*, **72**, 3186 (1992).
[42] A.K. Dutta, N. Morosawa, and Y. Hatanaka, *Sol. State Comm.*, **35**, 1483 (1992).
[43] K.H. Lee, Y.K. Fang, and G.Y. Lee, *IEEE Trans. Elec. Dev.*, **42**, 1929 (1995).
[44] K. Sawada, M. Akiyama, H. Takao, and M. Ishida, *Jpn. J. Appl. Phys.*, **39**, 2364 (2000).
[45] H. Ohtake, T. Nishiguchi, Y. Hirano, F. Sato, M. Abe, N. Saito, K. Sawada, and T. Ando, *Mater. Sci. Eng. B*, **90**, 120 (2002).
[46] K. Sawada, M. Akiyama, H. Takao, and M. Ishida, *Jpn. J. Appl. Phys*, **39**, 2364 (2000).
[47] M. Akiyama, M. Hanada, H. Takao, K. Sawada, and M. Ishida, *Jpn. J. Appl. Phys.*, **41**, 2552 (2002).
[48] A. Rose, *Concepts in Photoconductivity and Allied Problems*, Robert E. Krieger, New York, 1978.
[49] B.K. Ridley, *J. Phys. C*, **16**, 3373 (1983).
[50] B.K. Ridley, *J. Phys. C*, **16**, 4733 (1983).
[51] G.A. Baraff, *Phys. Rev.*, **128**, 2507 (1962).
[52] M.G. Burt, *J. Phys. C*, **18**, L477 (1985).
[53] S. McKenzie and M.G. Burt, *J. Phys. C*, **19**, 1959 (1986).
[54] E. Bringuier, *Phys. Rev. B*, **49**, 7974 (1994).
[55] B.K. Ridley, *Semicond. Sci. Technol.*, **2**, 116 (1987).
[56] J.S. Marsland, *Semicond. Sci. Technol.*, **5**, 177 (1990).
[57] J.S. Marsland, *Solid St. Elec.*, **30**, 125 (1987).
[58] O. Rubel1, S.D. Baranovskii, 1, I.P. Zvyagin, P. Thomas, and S.O. Kasap, *Phys. Stat. Sol.*, (C) **1**, 1186 (2004).
[59] S.O. Kasap, J.A. Rowlands, S. Baranovskii, and K. Tanioka, *J. Appl. Phys.*, **96**, 2037 (2004).
[60] D.C. Hunt, S.S. Kirby, and J.A. Rowlands, *Med. Phys.*, **29**, 2464 (2002).
[61] R. Street, Hydrogenated Amorphous Silicon, Cambridge University Press, Cambridge, 1981.
[62] R.M.A. Dawson and M.G. Kane, Pursuit of active matrix organic light emitting diode displays, *Digest of Technical Papers, SID International Symp.*, San Jose, 5–7 June, 2001, pp. 372–375.
[63] R.A. Street, X.D. Wu, R. Weisfield, S. Ready, R. Apte, M. Nguyen, and P. Nylen, Two dimensional amorphous silicon image sensor arrays, *Mat. Res. Sec. Symp. Proc.*, **377**, 757–766 (1995).
[64] P. Servati and A. Nathan, *IEEE Trans. Elec. Dev.*, **49**, 812 (2002).
[65] F. Lemmi and R.A. Street, *Mat. Res. Soc. Symp. Proc.*, **557**, 671 (1999).
[66] A. Nathan, P. Servati, K.S. Karim, D. Stryahilev, and A. Sazonov, a-Si:H TFTs on Glass and Plastic, in: *Thin Film Transistors*, Vol. 1, Y. Kuo (ed), Kluwer Academic Publishers, Boston, USA, 79–181, 2003.

[67] T. Leroux, *Solid-State Electron.*, **29**, 47 (1986).
[68] K. Khakzar and E.H. Lueder, *IEEE Trans. Elec. Dev.*, **39**, 1428 (1992).
[69] S.-S. Chen and J.B. Kuo, *IEEE Trans. Elec. Dev.*, **41**, 1169 (1994).
[70] R.A. Street, M.J. Thompson, and N.M. Johnson, *Phil. Mag. B*, **51**, 1 (1985).
[71] A. Nathan, M. Austin, and D. Pereira, *Proc., IEEE Workshop on Charge-Coupled Devices and Advanced Image Sensor*, Karuizawa, Japan, 10–12 June 1999, pp. 126–129.
[72] C.T. Angelis, C.A. Dimitriadis, I. Samaras, J. Brini, G. Kamarinos, V.K. Gueorguiev, and Tz.E. Ivanov, *J. Appl. Phys.*, **82**, 4095 (1997).
[73] P. Servati and A. Nathan, *J. Vac. Sci. Technol. A*, **20**, 1038 (2002).
[74] L. Colalongo, M. Valdinoci, A. Pellegrini, and M. Rudan, *IEEE Trans. Elect. Dev.*, **45**, 826 (1998).
[75] S.K. Heeks, J.H. Burroughes, C. Town, S. Cina, N. Baynes, N. Athanassopoulou, and J.C. Carter, Light emitting polymers for full color display applications, *Digest of Technical Papers, SID International Symp.*, San Jose, 5–7 June, 2001, pp. 518–521.
[76] T. Sasaoka *et al.*, A 13.0-inch AM-OLED display with top emitting structure and adaptive current mode programmed pixel circuit (TAC), *Digest of Technical Papers, SID International Symp.*, San Jose, 5–7 June 2001, pp. 384–387.
[77] P. Servati, K.S. Karim, and A. Nathan, *IEEE Trans. Elec. Dev.*, **50**, 926 (2003).
[78] S.M. Jahinuzzaman, A. Sultana, K. Sakariya, P. Servati, and A. Nathan, *Appl. Phys. Lett.*, **87**, 23502 (2005).
[79] A. Nathan, K. Sakariya, A. Kumar, P. Servati, and D. Striakhilev, Amorphous silicon backplane electronics for OLED displays, *Proc. 2002 IEEE/LEOS Annual Meeting*, Glasgow, Scotland, 10–14 November, 2002, pp. 303–304.
[80] K. Sakariya, P. Servati, D. Stryahilev, and A. Nathan, V_t-shift compensated a-Si:H pixel circuits for AMOLED displays, *Proc. Euro Display 2002: The 22nd International Display Research Conference*, Nice, France, 1–4 October, 2002, pp. 609–612.
[81] W. Zhao and J.A. Rowlands, *Med. Phys.*, **22**, 1595 (1995).
[82] K.S. Karim, A. Nathan, and J.A. Rowlands, *IEEE Trans. Elec. Dev.*, **50**, 200 (2003).
[83] S.K. Mendis, S.E. Kemeny, and E.R. Fossum, *IEEE Trans. Elec. Dev.*, **41**, 452 (1994).
[84] N. Matsuura, W. Zhao, Z. Huang, and J.A. Rowlands, Digital radiology using active matrix readout: Amplified pixel detector array for fluoroscopy, *Med. Phys.*, **26**, 672 (1999).
[85] K.S. Karim, A. Nathan, and J.A. Rowlands, Alternate pixel architectures for large area medical imaging, *Proc. SPIE, International Symposium on Medical Imaging* 2001: *Physics of Medical Imaging*, Vol. 4320, 17–22 February, 2001, pp. 35–46.

5 The Investigation of Charge Carrier Recombination and Hopping Transport with Pulsed Electrically Detected Magnetic Resonance Techniques

Christoph Boehme and Klaus Lips

Hahn–Meitner–Institut Berlin, Kekuléstr. 5, D-12489 Berlin, Germany

Charge Transport in Disordered Solids with Applications in Electronics Edited by S. Baranovski.

Hopping transport and recombination in disordered semiconductors are generally determined by a large variety of different electronic transitions whose properties are distributed with regard to their transition energies, transition probabilities as well as the microscopic origin of the electronic states involved. Since all these processes together determine macroscopic observables such as conductivity or photoconductivity, their understanding is the key to the understanding and characterization of electronic material properties. One can gain experimental access to this information with electrically detected magnetic resonance (EDMR) which combines the microscopic selectivity of magnetic resonance with macroscopic conductivity measurements. Here, a review of the theoretical and experimental foundations of the coherent, time-resolved EDMR, the so-called pulsed (p) EDMR and its application for the investigation of recombination and transport in disordered semiconductor materials is presented. The underlying physics of these mechanisms is outlined and it is shown how to take advantage thereof for defect spectroscopy. The technical aspects of pEDMR are discussed and experimental data collected on hopping transport through conduction band tail states of an amorphous silicon thin film in a semiconductor heterostructure consisting of crystalline silicon under amorphous silicon is presented, confirming the theoretically predicted effects and demonstrating the potential of pEDMR for material characterization.

5.1 INTRODUCTION

Disorder of semiconductors is generally reflected by their band structure and therefore by the nature of electronic transitions therein. Consequently, charge carrier transport and recombination in such materials can take place in a variety of different ways. Transport can take place through delocalized states or localized states, in the latter case through hopping transitions or via tunneling processes [1]. The recombination rates can also be dominated by disorder-induced effects such as different kinds of transitions between the conduction band-tail states, deep defects and valance band-tail states. Hence, the microscopic understanding of these different processes is necessary for the prediction of the macroscopic electric properties of disordered semiconductors. Only when the nature of the rate-dominating transport and recombination processes is understood in a given material, is it possible to understand observables such as conductivity, photoconductivity (PC) or photoluminescence (PL) as well as their dependences on temperature, Fermi level or defect and dopant concentrations.

In the past, magnetic resonance spectroscopies such as electron spin resonance (ESR) or nuclear magnetic resonance (NMR) have proven to be viable characterization methods for the microscopic investigation of localized paramagnetic states that influence transport and recombination. The idea of magnetic resonance is to determine the effective Landé factor, also called the g-factor, which depends on local fields and thus on the microscopic vicinity of a given paramagnetic center. Progress in the understanding of semiconductor defects and impurities has always gone hand in hand with progress in the field of defect spin spectroscopy. One of the greatest steps in this regard was taken when the traditional continuous wave (cw) magnetic resonance methods, which are based on adiabatic magnetic field sweep spectroscopy, were replaced by the modern coherent magnetic resonance spectroscopies, the so-called pulsed ESR and NMR [2–4]. The advantage of pulsed excitation schemes is the much better quantitative access to the dynamics of coherence decay processes such as electronic transitions, spin relaxation or spin couplings and interactions from which a detailed microscopic picture can be derived. In spite of this progress, there have also been challenges to the application of ESR and NMR to semiconductor characterization. As low-dimensional semiconductor systems and mesoscopic structures such as quantum wells, quantum dots or quantum wires, or semiconductor thin films incorporated in thin–film devices such as transistors or solar cells, have increasingly become subjects of research, the conventional magnetic resonance methods have reached their sensitivity limits. Since the sensitivity limits of ESR and NMR are usually about the orders of 10^{11} and 10^{15} spins, respectively, small two- or fewer-dimensional samples are hardly characterizable. This limitation holds in particular for the investigation of charge carrier transport and recombination in disordered materials. On one hand, ESR would be the perfect method of choice for these materials since electronic transport therein is usually associated with transitions through highly localized paramagnetic states. On the other hand, since disordered materials are deposited mostly as semiconductor thin films, a sophisticated ESR spectroscopy is difficult because of its sensitivity limitations.

In order to encounter the sensitivity challenge, magnetic resonance methods have been combined in the past with other measurement techniques such as PL or conductivity measurements. The resulting combined techniques are the so-called optically or electrically detected magnetic resonance (ODMR and EDMR, respectively) that take advantage of the spin selection rules of electronic transitions that exist in semiconductors. The spin dependence of recombination processes and its impact on PL rates was discovered as early as 1959 when Geschwind *et al.* [5, 6] carried out the first cwODMR experiments. This discovery actually triggered a series of experiments not only on semiconductors and other solid state systems, but also and in particular on radiative atomic and molecular systems such as recombining radical pair or ion pair species which would eventually establish ODMR as a standard characterization method for chemical analysis [7]. Because of this success, the first transient, pulsed (p) ODMR experiments were soon introduced after the first pulsed ESR spectrometers had been built [8–11]. For the investigation of electronic transitions and their impact on macroscopic electric properties of semiconductors, ODMR has some limitations. Many spin-dependent electronic transitions, in particular charge carrier transport transitions, are either nonradiative or radiative at very long wavelength which makes their detection very difficult or impossible. Moreover, beside these ‘invisible’ transitions, ODMR may detect transitions which do radiate, but do not contribute to the conductivity. An example for this are geminate recombination processes, the luminescent recombination of correlated electrons and holes which do not contribute to PC. Hence, for the investigation of transport phenomena and nongeminate recombination which influence dark conductivity and PC, it is imperative to apply EDMR.

Continuous wave EDMR has been used since Maxwell and Honig [12] investigated spin-dependent scattering of charge carriers at impurities in 1966. In 1972, Lepine and Prejean [13, 14] conducted the first cwEDMR measurements on spin-dependent recombination mechanisms of crystalline silicon. Since then many cwEDMR studies have been performed on many different semiconductor materials [15–21], semiconductor heterostructures [22, 23] and devices [24–26]. Over the course of almost four decades, much experimental and also theoretical work has been performed on the understanding of cwEDMR experiments as well as the underlying processes. In spite of these efforts, transient EDMR experiments and particularly coherent, pEDMR experiments have been demonstrated only recently [27–29]. The challenge for transient EDMR measurements is to detect very small current changes on top of comparatively large constant current offsets at a high time resolution. Unlike pODMR that has been conducted in the past as a simple transient high-speed PL measurement, its is usually impossible to attain a time resolution with electrical measurements that is within the coherence time of the spin systems and that is at the same time sensitive enough to detect the subtle signal currents. This contradiction between sufficient sensitivity and sufficient time resolution was finally overcome by means of an indirect detection scheme of the coherent dynamics during the excitation [28–32]. The idea behind this scheme is to measure the change of the photocurrent after a coherent pESR excitation as a function of the length of the resonant pulse, which reveals the dynamics of the spin systems in resonance during the excitation. As shown in the course of this chapter, this measurement approach makes it possible to observe coherent spin motion effects such as spin-Rabi oscillation, spin-echo effects and coherence decay processes. It is the information obtainable from these phenomena that reveals insights into the nature of the electronic transitions involved and thus about charge carrier transport and recombination.

In the following, the model and the nature of spin-dependent recombination is discussed first. Based thereon, a brief review of the theory behind pEDMR experiments is outlined and the experimental foundations are addressed. This review is then followed by a brief theoretical discussion of spin-dependent hopping which is then applied to the pEDMR experiments on the vertical electronic transport of photo charge carriers through a thin, highly phosphorus-doped hydrogenated amorphous silicon layer (a-Si:H) on top of a crystalline silicon (c-Si) substrate. It will be demonstrated how pEDMR is the method of choice for the measurement of coupling strength between localized conduction band-tail states as well as the determination of different transition times of charge carriers between these states.

5.2 SPIN-DEPENDENT RECOMBINATION

When the first spin-dependent charge carrier recombination channel was discovered through electrical detection by Lepine and Prejean [13, 14], a series of unsuccessful attempts were made to develop models for the explanation of this phenomenon [14, 24, 33–37]. Most of these models failed to predict either correct magnitudes, correct magnetic field dependencies, correct temperature dependencies or combinations thereof, and thus the mechanism behind the ESR-induced recombination changes remained a puzzle until 1978 when Kaplan, Solomon, and Mott [38] (KSM) described spin-dependent recombination in terms of an intermediate pair model where charge carriers first localize in exclusive pairs from which they either recombine or dissociate without recombination. The crucial properties of this model is the pair exclusivity which means that two charge carriers trapped in a pair can recombine only with each other, not with other charge carriers outside the pair. In order to recombine with other charge carriers, they must dissociate from their respective pairs first and then form new pairs with different charge carriers. Figure 5.1 illustrates the electronic transitions associated with the KSM model for the example of spin-dependent recombination through a dangling bond of hydrogenated microcrystalline silicon [39]. The pair is generated by trapping of a conduction electron at the paramagnetic defect center. Note that the double occupation of the defect happens by a two-step process. First, the conduction electron localizes in an energetically high charged excited state of the system before it undergoes a transition into the charged ground state. It is this transition which makes the entire recombination process spin dependent. Once the charged two-electron ground state

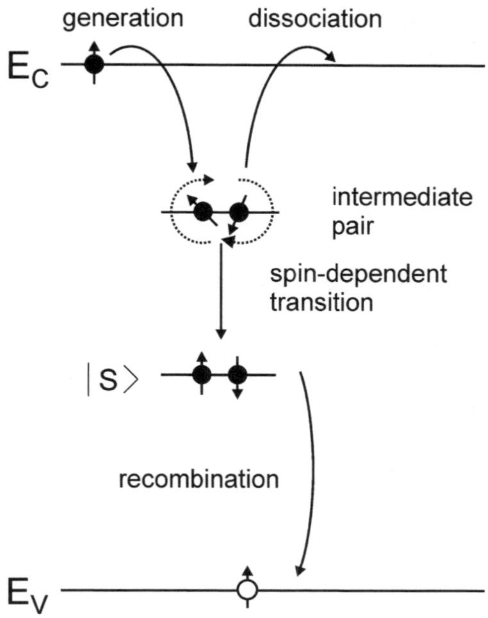

Figure 5.1 Spin-dependent recombination in the picture of Kaplan, Solomon and Mott illustrated for a recombination channel through localized deep level defects which can exist in a charged excited state (e.g., dangling bond defects in hydrogenated microcrystalline silicon). When a conduction electron is trapped into the charged exited state, a KSM pair is formed. This pair can dissociate by emission of one electron back into the conduction band or is trapped through a spin dependent-transition into the charged ground state. Recombination is then completed through the capture of a hole from the valence band, a transition that is not spin dependent

is reached, the conduction electron is unlikely to return to the conduction band and due to fast hole capture, recombination takes place.

The situation illustrated in Figure 5.1 is just one example of spin-dependent recombination that can be described by the KSM model—the nature of the exclusive pairs may differer greatly in terms of the types of electronic states involved (impurity states, excitons, etc.) or the types of electronic transitions by which these pairs dissociate or recombine. Thus, in order to keep the model as general as possible, the physics of the mutual spin–spin interactions within a given pair has to be taken into account. Furthermore, since spin selection rules usually come from the weakness of spin–orbital interaction which may be small, but not necessarily negligible, it is possible that both the recombination probabilities due to the singlet as well as triplet content of a given pair state have to be taken into account, as well as influences by spin–spin relaxation processes. A detailed discussion of the history and the development of a general KSM model taking all these influences into account is given elsewhere [28, 31]. As a result, it leads to a general description of the KSM pairs that can be summarized by seven properties:

1. Spin-dependent recombination takes place by formation of exclusive spin pairs.
2. After pairs are generated, they can be destroyed only by recombination or dissociation.

3. The dynamics of spin-dependent recombination is governed solely by the spin dynamics of the pair ensemble.
4. The intermediate pairs are systems of two $s = \frac{1}{2}$ spins with four spin eigenstates.
5. Spin–exchange and spin–dipole interaction within a pair are determined by the spin pair's nature.
6. Singlet and triplet recombination have to be taken into account.
7. Spin–lattice and spin–spin relaxation may change or destroy the spin pair correlation.

A graphical illustration of the transition rates resulting from these properties is displayed in Figure 5.2. Four eigenstates $|T+\rangle$, $|2\rangle$, $|3\rangle$ and $|T-\rangle$ whose occupation can be represented by their respective density matrix elements ρ_{ii} are populated randomly at rate G, and annihilated either by spin–spin relaxation with probability T_2^{-1}, by dissociation with probability d and recombination with their respective probabilities r_i. Note that in the following, spin–lattice relaxation will be considered negligible for simplicity, an assumption which is reasonable when temperatures are low enough.

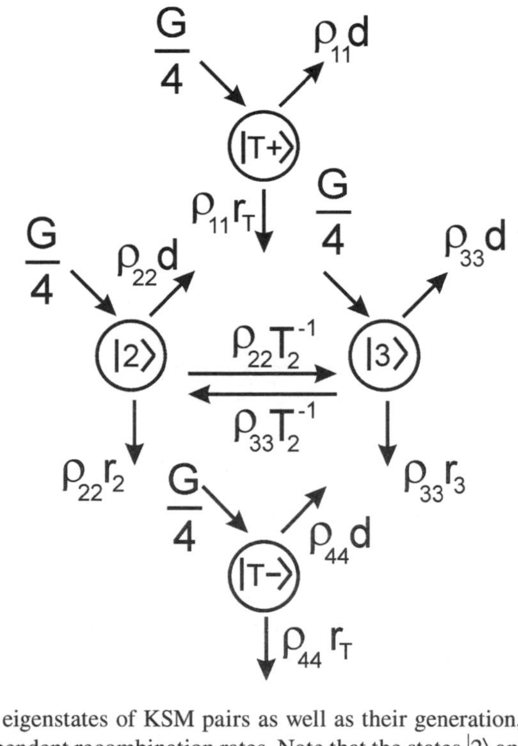

Figure 5.2 The spin eigenstates of KSM pairs as well as their generation, dissociation, spin–spin relaxation and spin-dependent recombination rates. Note that the states $|2\rangle$ and $|3\rangle$ are linear combinations of the $|T_0\rangle$- and $|S\rangle$–states with mixed permutation symmetries that depend on the spin–spin interactions within a given pair and the externally applied magnetic field. For details see text

From the qualitative description given above, quantitative predictions for the transient evolution of PC or dark conductivity can be derived for the case of an immediate ESR excitation (faster than all transition rates) of the steady state [31]. In order to do this, one can describe the physics of an ensemble of charge carrier pairs with time evolution $\hat{\rho} = \hat{\rho}(t)$ by the Hamiltonian

$$\hat{H}_0 = \mu_B g_a \hat{\mathbf{S}}_\mathbf{a} \cdot \mathbf{B} + \mu_B g_b \hat{\mathbf{S}}_\mathbf{b} \cdot \mathbf{B} - J\hat{\mathbf{S}}_\mathbf{a} \cdot \hat{\mathbf{S}}_\mathbf{b} - D^d \left[3S_a^z S_b^z - \hat{\mathbf{S}}_\mathbf{a} \cdot \hat{\mathbf{S}}_\mathbf{b}\right] \tag{5.1}$$

of the individual pairs. $\hat{H}_0$ consists of the Zeeman interaction $g_i\mu_B\hat{\mathbf{S}}_\mathbf{i}\cdot\mathbf{B}$ of the two pair partners a and b with Landé factors g_a and g_b, respectively, the exchange coupling with coupling constant J as well as the dipolar interaction with coupling constant D^d taken into account in the high field approximation ($|D^d| << |g_i\mu_B B|$). The variable μ_B represents the Bohr magneton. Hyperfine coupling effects have been neglected in this description. The evolution of $\hat{\rho}(t)$ can be described by stochastic Liouville equations, as first done for recombining charge carrier pairs by Haberkorn and Dietz [40]. However, since dephasing effects within a coherent ensemble are usually much faster than the time resolution of real-time transient measurements [41], it is in many cases sufficient to limit the description of the coherent propagation to the time during a coherent magnetic resonant excitation whereas the dynamics beyond this short ns time range can be described with transition rates between the four spin eigenstates of this two-spin $\frac{1}{2}$ system. For the calculation of the eigenstates of $\hat{H}_0$ as well as its eigenenergies, $\hat{H}_0$ must be diagonalized, which yields an eigenbase and eigenvalues

$$\begin{aligned}
&|T_+\rangle = |\uparrow\uparrow\rangle && E_{T_+} = \frac{\hbar\omega_0}{2} - J + D^d,\\
&|2\rangle = \cos(\phi)|\uparrow\downarrow\rangle + \sin(\phi)|\downarrow\uparrow\rangle && E_2 = -D^d + \hbar\omega_\Delta,\\
&|3\rangle = -\sin(\phi)|\uparrow\downarrow\rangle + \cos(\phi)|\downarrow\uparrow\rangle && E_3 = -D^d - \hbar\omega_\Delta,\\
&|T_-\rangle = |\downarrow\downarrow\rangle && E_{T_-} = -\frac{\hbar\omega_0}{2} - J + D^d,
\end{aligned} \tag{5.2}$$

respectively. Equation (5.2) shows that both the eigenstates as well as the eigenenergies depend on the external magnetic field B_0 as well as the spin–spin interactions J and D^d since

$$\phi = \frac{1}{2}\arcsin\left(\frac{J + D^d}{\hbar\omega_\Delta}\right) \quad \text{and} \quad \omega_\Delta = \sqrt{\frac{(J + D^d)^2}{\hbar^2} + \frac{\Delta\omega^2}{4}} \tag{5.3}$$

with $\omega_0 = \omega_a + \omega_b$ and $\Delta\omega = \omega_a - \omega_b$ whereas ω_a and ω_b are Larmor frequencies of the spin pair partners a and b, respectively. The model presented above describes transition probabilities from any of the four spin states as solely dependent on their pair permutation symmetry. Thus, for pure singlet and pure triplet pairs, distinct probabilities r_S and r_T can be assigned, respectively. Due to the mixed eigenbase of real pair systems, recombination from $|2\rangle$- and $|3\rangle$-eigenstates will have different recombination probabilities

$$r_i = r_S\,|\langle i|S\rangle|^2 + r_T\,|\langle i|T_0\rangle|^2, \tag{5.4}$$

which, under consideration of Equations (5.2) and (5.3), can be written as

$$r_{T_{+,-}} = r_T \quad r_{2,3} = \frac{r_S}{2}\left[1 \mp \frac{J + D^d}{\hbar\omega_\Delta}\right] + \frac{r_T}{2}\left[1 \pm \frac{J + D^d}{\hbar\omega_\Delta}\right]. \tag{5.5}$$

Note that the two states $|\uparrow\uparrow\rangle = |T_+\rangle$ and $|\downarrow\downarrow\rangle = |T_-\rangle$ are not dependent on the spin–spin interactions and thus, their recombination probabilities are independent from spin–spin interaction changes, too. With Equation (5.5) and the assignment of ρ_{ii} for the eigenstate densities, one can calculate the dynamics of the transition rates indicated in Figure 5.2, which then yield expressions for the time evolution of the charge carrier densities and also for the transient PC $\sigma_{\text{ph}}(t)$. Note that the dependence of the PC on charge carrier densities and mobilities can be different for different materials and thus it seems to be difficult or even impossible to derive a general expression valid for all materials. The approach taken here to overcome this difficulty is to take only first-order contributions of the conductivity functions into account which reduces all these functions to simple proportionalities. Note that this is possible only as long as the conductivity changes are sufficiently small; since, to the knowledge of the authors, there has been no EDMR study over the past four decades that revealed relative conductivity changes of more than 10^{-1} (in fact many inorganic materials usually do not exhibit signals exceeding 10^{-6} to 10^{-4}), the assumption can be considered reasonable.

Under the conditions outlined above, the PC change $\Delta\sigma_{\text{ph}}(t) = \sigma_{\text{ph}}(t) - \sigma^S_{\text{ph}}$ which, like all other variable changes defined in the following, is the difference of its value as a function of time from the steady state value σ^S_{ph}, becomes the sum

$$\Delta\sigma_{\text{ph}}(t) = e[\Delta n_{\text{e}}(t)\mu_{\text{e}} + \Delta n_{\text{h}}(t)\mu_{\text{h}}] \tag{5.6}$$

of the products of the electron and hole density changes $\Delta n_i(t)$ and their respective mobilities μ_i. The variable e denotes an elementary charge. Note that since it is assumed that all changes are small, the pair generation rate G as well as the lifetime τ_{L} can be considered constant because their changes cause only second-order effects, which, as stated above, are neglected here. From Figure 5.2 it becomes clear that when the pair generation rate G and the net charge carrier lifetime τ_{L} are constant, the changes of electron and hole densities

$$\Delta n_{\text{e}}(t) = \tau_{\text{L}} d \sum_{i=1}^{4} \Delta\rho_{ii} \qquad \text{and} \qquad \Delta n_{\text{h}}(t) = -\tau_{\text{L}} \sum_{i=1}^{4} r_i \Delta\rho_{ii} \tag{5.7}$$

are determined by the changes of the dissociation and recombination rates, respectively. Both depend on the spin eigenstate density changes $\Delta\rho_{ii} = \rho_{ii}(t) - \rho^S_{ii}$. Thus, by combining Equations (5.5), (5.6) and (5.7), the transient PC

$$\Delta\sigma_{\text{ph}}(t) = e\tau_{\text{L}} d\mu_{\text{e}} \sum_{i=1}^{4} [\rho_{ii}(t) - \rho^0_{ii}]\left(1 - \frac{r_i}{d}\frac{\mu_{\text{h}}}{\mu_{\text{e}}}\right) \tag{5.8}$$

can be described as a function of the spin eigenstate density changes. Equation (5.8) is a link between the transient PC and the spin state density changes. For long time scales (longer than the coherence times of the transitions investigated), it can be calculated from the simple rate model that is illustrated in Figure 5.2. Connecting the rates that influence the densities ρ_{ii}, with the generation, recombination, spin relaxation and dissociation rates leads to a small system of ordinary, linear, inhomogeneous first-order differential equations

$$\partial_t \rho_{11,44} = \frac{G}{4} - (d + r_T)\rho_{11,44},$$
$$\partial_t \rho_{22,33} = \frac{G}{4} - \left(d + r_{2,3} + \frac{1}{T_2}\right)\rho_{22,33} + \frac{\rho_{33,22}}{T_2} \tag{5.9}$$

from whose solution the steady-state values as well as the transient behavior of $\rho_{ii}(t)$ can be obtained. Experimentally, the dissociation may be much smaller than any of the other rate coefficients [39] while spin–spin coupling is strong and thus, $r_3 \gg \frac{1}{T_2}$, $r_2 > r_T >> d$. Under these conditions, the steady-state solution becomes $\rho_{11}^S = \rho_{44}^s = \frac{G}{4r_T} := \rho^S$ and $\rho_{22} = \rho_{33} = 0$. Obviously, the long-lived pure triplet states are pumped to very high densities while those eigenstates with singlet content are vanishing due to their short lifetimes. The transient solution of the linear ODE type given in Equation (5.9) is a set of exponential decay functions with three different time constants

$$\tau_s^{-1} = d + r_T \quad \text{and} \quad \tau_{m,f}^{-1} = \frac{1}{T_2} + d + \frac{r_2}{2} + \frac{r_3}{2} \mp \sqrt{\frac{1}{T_2^2} + \frac{(r_3 - r_2)^2}{4}}, \tag{5.10}$$

whose indices s, m, and f refer to slow, medium, and fast, respectively.

With Equations (5.8), (5.10) and the expression for ρ_{ii}^S, one can calculate the relaxation of the PC transient when the spin pair ensemble relaxes from an arbitrary initial state $\hat{\rho}$ towards the steady state. If we assume that the ensemble holds in this initial state the same number of spin pairs as in the steady state ($\mathrm{Tr}[\hat{\rho}^S] = \mathrm{Tr}[\hat{\rho}]$), one can introduce a single variable

$$\Delta := -\frac{\rho_{11,44} - \rho_{11,44}^S}{\mathrm{Tr}[\rho^S]} = \frac{\rho_{22,33} - \rho_{22,33}^S}{\mathrm{Tr}[\rho^S]} \frac{\hbar\omega_\Delta}{\hbar\omega_\Delta \pm (J + D^d)}, \tag{5.11}$$

which describes the relative number of spins that are shifted from the occupied states $|T_+\rangle$ and $|T_-\rangle$ into the unoccupied states $|2\rangle$ and $|3\rangle$. A detailed justification of Equation (5.11) is given in [31]. It is introduced since it leads, together with Equations (5.8) and (5.10) and the assumption $r_3 \ll \frac{1}{T_2}$, $r_2 < r_T << d$ made above for the calculation of the steady state, to a sufficiently simple description

$$\Delta\sigma_{\mathrm{ph}}(t) = \frac{e\tau_{\mathrm{L}} d\mu_{\mathrm{e}} G}{2r_T}\left[2\left(\frac{r_{\mathrm{T}}}{d}\frac{\mu_{\mathrm{h}}}{\mu_{\mathrm{e}}} - 1\right)e^{-(r_T+d)t} - \left(\frac{r_2}{d}\frac{\mu_{\mathrm{h}}}{\mu_{\mathrm{e}}} - 1\right)\left(1 + \frac{J + D^d}{\hbar\omega_\Delta}\right)e^{-\left(d + \frac{1}{T_2} + \frac{r_2}{2}\right)t} - \frac{r_3}{d}\frac{\mu_{\mathrm{h}}}{\mu_{\mathrm{e}}}\left(1 - \frac{J + D^d}{\hbar\omega_\Delta}\right)e^{-r_3 t}\right]\Delta \tag{5.12}$$

of the PC relaxation of a coherent excitation. $\Delta\sigma_{\mathrm{ph}}(t)$ is plotted in Figure 5.3 for the example of $r_3 = 10^6\mathrm{s}^{-1}$, $r_2 = 10^4\mathrm{s}^{-1}$, $r_T = 5 \times 10^3\mathrm{s}^{-1}$ and d as well as T_1 and T_2 negligible. One can see that under these conditions, the triple-exponential decay reduces to a double-exponential decay, consisting of a fast relaxing PC decrease with large magnitude and a slow relaxing PC increase with small magnitude. The magnitude of the two components, described by the prefactors of both exponential decays, is proportional to Δ. Thus, by integration of the absolute value of the PC transient, as illustrated in Figure 5.3, a charge $Q \propto \Delta$ is counted which depends on the spin state $\hat{\rho}$ at the moment when the relaxation begins ($t = 0$) and, therefore, the measurement of Q is essentially a measurement of the permutation symmetry within the spin pair ensemble at $t = 0$. It shall be pointed out that the assumption $\mathrm{Tr}[\hat{\rho}^S] = \mathrm{Tr}\,[\hat{\rho}]$ holds when instantaneous changes are imposed on a steady state spin pair ensemble such as by a coherent excitation. The subsequent measurement of Q after a coherent change of spin-dependent recombination, reflects therefore the permutation symmetry of the prepared ensemble state right at the end of the excitation.

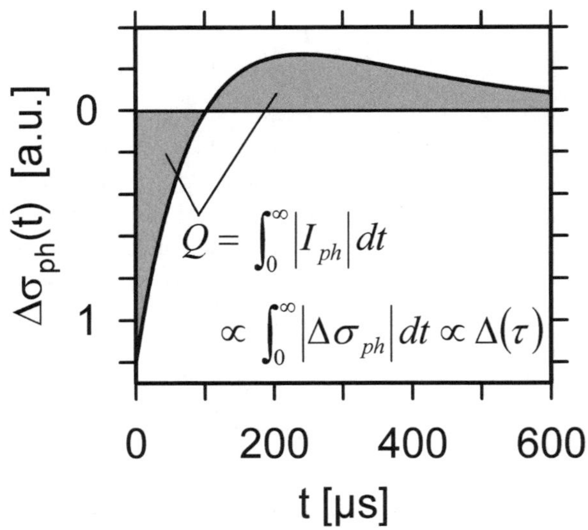

Figure 5.3 A simulated PC transient when the spin ensemble has an initial state $\hat{\rho}(0)$ that is unequal to the steady state $\hat{\rho}_S$ at the time $t = 0$. For the simulation it was assumed that a strong spin–spin coupling was present such that the response is a double exponential decay function. The gray area represents the time integral of the absolute PC change $\Delta\sigma_{\mathrm{ph}}(t)$ which is proportional to the charge Q of an integrated photocurrent and thus the relative density change $\Delta(\tau)$ at the end of the ESR pulse. Thus, the PC measurement is essentially a spin measurement process

5.3 SPIN-DEPENDENT HOPPING TRANSPORT

Similar to spin-dependent recombination, spin selection rules can also influence electronic transport transitions that can be detected by magnetic resonant changes of conductivities, too. There is a variety of experimental work in the literature claiming to observe spin-dependent transport channels by means of cwEDMR [42–49]. In some of these studies it is claimed that the transport can be distinguished from recombination by means of its sign since magnetic resonance enhances the net transition rates such that recombination leads to the observation of a PC quenching while transport causes PC enhancement. It has to be pointed out that this argument does not hold. As one can learn from Equation (5.12), a recombination transient has negative and positive contributions which implies, that the sign of a cwEDMR signal depends solely on the modulation frequency of the magnetic field when lock–in detection is used and thus, a statement about the underlying process of a given EDMR resonance signal is not possible. It will be shown in the following that similar transients with both conductivity enhancement and conductivity quenching can be predicted for the EDMR response of spin-dependent transport channels, too, and thus an identification of spin-dependent transport by means of the signal sign is not possible [50].

Indisputable experimental evidence of spin-dependent transport processes are EDMR signals in bulk semiconductors under dark conditions [26, 51]. In the absence of excess charge carriers, recombination does not exist and spin-dependent recombination processes which enhance transport (as it is the case at *p–n* junctions with forward bias for instance) are not conceivable either. In spite of this clear experimental evidence for the effect of conductivity changes due to magnetic resonant changes of spin states involved in spin-dependent transport, the theoretical understanding has so far remained on a very rudimentary base. Most of the work on spin-dependent transport has been conducted in the past on its influence on disorder-related magnetoresistance effects [52, 53]. While it is understood from this work that spin selection rules may impose different transition probabilities between two different paramagnetic, singly occupied electronic states, it is not obvious where the inequality between transitions of spin pairs with higher singlet and higher triplet symmetry comes from. The latter is crucial for the existence of an EDMR signal. Magnetic resonance can cause a rate change only when an imbalance within the permutation symmetry exists. For spin-dependent recombination, this imbalance comes from the high triplet density that is pumped when the steady state is established (see Equation 5.9). If a similar effect takes place in the course of spin-dependent transport, one has to raise the question of the existence of an intermediate spin pair and its meaning with regard to transitions between states of similar energy levels. Since charge carriers do not annihilate during transport, in contrast to recombination, it is clear that spin-dependent processes have to be described in terms of mobility changes.

In the following, a model for the description of spin-dependent hopping transport through localized states of disordered semiconductors is presented. Therein, the spin selection rules are assumed to be due to the same effect that governs disorder-related magnetoresistance, namely doubly occupied states with correlation energies smaller than the distribution width of the energies of the localized states [53]. Without confinement of generality, only the description of electron transport, but not hole transport will be discussed since the results for electron transport will be applicable to spin-dependent hole transport in a similar way. Note also that only spin-dependent hopping transitions are considered, as they can be found in disordered, organic or highly doped semiconductors. Spin-dependent transport due to delocalized states and spin-dependent charge carrier scattering will be excluded from this discussion. Similarly as for the discussion of spin-dependent recombination, we restrict ourselves to realistic systems where magnetic-resonance-induced spin manipulations introduce only small conductivity changes.

Figure 5.4 depicts the crucial process for spin-dependent hopping between two localized defect states as described by Kamimura *et al.* [54]. One can see a single electron symbolized by its spin-up orientation that occupies site A. For site B, two localized energy levels are indicated which represent the levels of single (0/+) and double (−/0) occupancy, respectively, separated by the two electron correlation energy which may be positive or negative (depicted positive in Figure 5.4 without confinement of the generality). Note that the charge notation chosen here implies that a neutral state is singly occupied which is not the case in general (especially not for band-tail states), but is chosen here as an example which does not confine the generality either. The electron in site A hops into site B, which thereafter is either in its singly or doubly occupied state, depending on the occupation of site B before the transition. If B is empty, the hopping will produce a singly occupied state B and the (+/0) transition probability is not spin dependent. If B is already singly occupied, the hopping will charge site B and the (0/−) transition probability will depend on the permutation symmetry of the two electrons before the hop. Since the electron pair is in a singlet state when both electrons occupy site B, the hopping probability increases with the singlet content of the pair before the transition. Thus, similarly to the spin-dependent recombination mechanism explained above (see Equation 5.5), it is conceivable that three different transition probabilities

$$p_{T_{+,-}} = p_T \qquad p_{2,3} = \frac{p_S}{2}\left[1 \mp \frac{J+D^d}{\hbar\omega_\Delta}\right] + \frac{p_T}{2}\left[1 \pm \frac{J+D^d}{\hbar\omega_\Delta}\right] \tag{5.13}$$

exist for the triplet states $|T_{+,-}\rangle$ and the mixed states $|2\rangle$ and $|3\rangle$, respectively, which depend solely on the spin–spin interactions as well as the probabilities p_S and p_T of transitions between pure triplet and pure singlet states. The latter will depend on the nature of the two defect sites. Note that at temperatures where Mott variable-range hopping is dominant, most of the transitions occur exclusively at energies close to the Fermi or quasi-Fermi level. Thus, when we assume that for the example illustrated in Figure 5.4 this applies to the

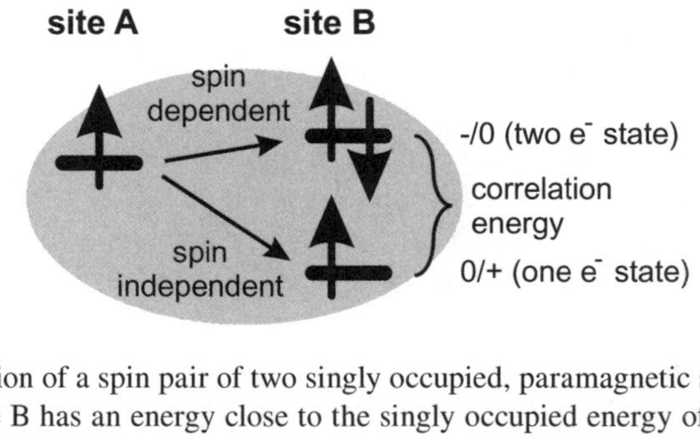

Figure 5.4 Illustration of a spin pair of two singly occupied, paramagnetic sites. The doubly occupied eigenstate of site B has an energy close to the singly occupied energy of site A. Hence, a spin-dependent transition from site A into site B becomes likely. If site B is unoccupied, the transition probability from from site A to site B is spin independent

spin-dependent transition, it is conceivable that site B will always be singly occupied and, therefore, spin-dependent hopping will be very likely.

The spin dependence of hopping transport as described here, requires that: (i) two singly occupied electron sites are in sufficient proximity; (ii) the doubly occupied state of one site is sufficiently close to the singly occupied state of the other site such that a transition is sufficiently likely; and (iii) spin conservation applies for electronic transitions. Applied to transport through band-tail states of disordered semiconductors with large distributions of eigenenergies and geometric properties as well as different densities of the two sites A and B, this means that many and in some cases most transitions will take place without spin selection rules. Therefore, for a description of the spin-dependent processes as well as their influence on macroscopic conductivities, we introduce here the expressions of a 'pair site' and a 'pair'. A pair consists of two singly occupied sites between which a spin-dependent transition can occur. A pair site consists of two electron sites which are not singly occupied, but whose energy levels, correlation energies, coupling strength, etc., would allow spin-dependent transitions if they were singly occupied. Figure 5.5 illustrates this terminology for the two sites A and B as introduced by [53]: Spin-dependent transitions within pairs correspond to the so-called A–B hops. B–A hops also take place at pair sites, yet they do not take place within a pair since they are not spin dependent. The other transitions types

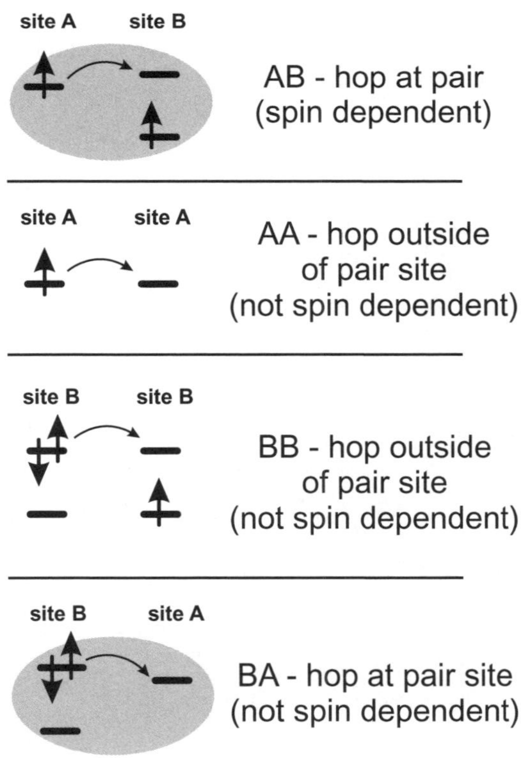

Figure 5.5 Illustration of the four different types of hopping transitions that occur when A-sites and B-sites are present. Note that only AB hops are spin dependent. The shaded area indicates a pair site. For details see text

(A–A and B–B hops) do not occur at pairs sites. They are never spin dependent and therefore, they never contribute to spin-dependent currents.

Figure 5.6 illustrates the different possible transport processes through a pair site containing one electron that is surrounded by two electron sites that do not belong to a pair site. The pair site consists of two electron sites A and B with similar properties to the pair shown in Figure 5.4. The single electron (0/+) state of the defect A has an energy close to the doubly occupied state (−/0) of defect B. Note that transport through empty pair sites will always be spin independent which is why this is neither shown nor discussed here. Figure 5.6 shows the different ways in which an electron outside a singly occupied pair site can propagate through the site. First, the electron undergoes a transition into defect A of the pair site and a pair is formed. Second, the electron can either: (i) undergo a spin-dependent transition into defect B followed by a third transition of one electron out of the pair site; or (ii) the electron of defect B can undergo a spin-independent transition out of the pair followed by a third transition from defect A to defect B. Obviously, there are two different ways for the pair to annihilate: one spin-dependent and one spin-independent path. This represents a similar situation to the recombination model illustrated in Figure 5.2. Here, the pairs can either annihilate by dissociation which is spin independent or by spin-dependent recombination. Consequently, the model for spin-dependent transport can be expressed in terms of an intermediate pair model just as for spin-dependent recombination. The only difference is that the pairs are destroyed by a spin-dependent dissociation which influences transport rates and therefore the charge carrier mobility and not the charge carrier density. With this insight one can formulate the qualitative properties of a spin-dependent transport model:

1. Spin-dependent transport takes place by formation of exclusive spin pairs at pair sites.
2. A pair site consists of two adjacent defect sites whose geometries, distances, mutual interactions and two-electron correlation energies will make spin-dependent transitions likely when both sites are singly occupied.
3. Defect sites outside pair sites cannot contribute to spin-dependent transport.

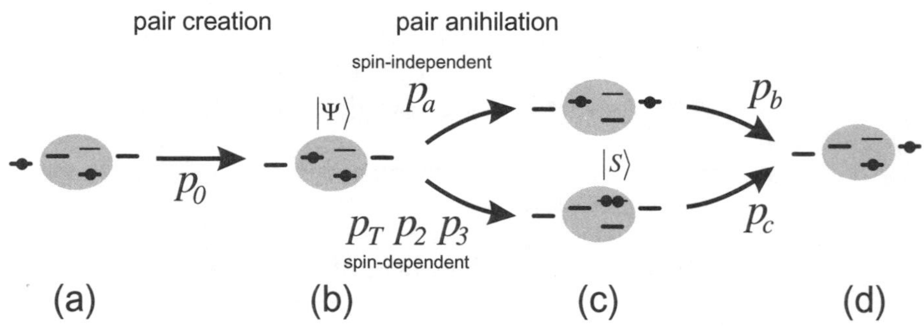

Figure 5.6 The sketch illustrates how an intermediate pair model for spin-dependent hopping through four different sites works. Note that only for the third from the left, the energy level of the double occupied (−/0) state is plotted. All other doubly occupied states are omitted for clarity. For details see text

4. Pairs are destroyed by spin-dependent dissociation. Note that while there may be one or several spin-independent dissociation mechanisms in addition to the spin-dependent dissociation mechanisms for a given pair type, the net dissociation probability describing an averaged value will always be spin dependent.
5. The dynamics of the spin-dependent hopping rate depends on the dynamics of the pair ensemble.
6. The intermediate pairs are systems of two $S = \frac{1}{2}$ spins with four spin eigenstates.
7. Spin–exchange and spin–dipole interaction within a pair are determined by its nature.
8. Both singlet and triplet transitions have to be taken into account.
9. Spin–lattice and spin–spin relaxation may change or destroy the spin–pair correlation.

In the following, quantitative predictions for spin-dependent transitions rates and conductivity changes are derived from the qualitative features listed above. The dependences of the spin-dependent pair dissociation probabilities $d_i (i \in \{T_\pm, 2, 3\})$ can be deduced directly from the rate picture of Figure 5.6. Since both transition paths illustrated therein lead to pair dissociation, one can reduce the different transition probabilities p_a, p_b, p_c and p_i (with p_i corresponding to the only transition that is spin dependent), to three pair dissociation probabilities

$$d_i = \frac{p_a p_b}{p_a + p_b} + \frac{p_i p_c}{p_i + p_c} \qquad \text{and} \qquad i \in \{T_\pm, 2, 3\}, \tag{5.14}$$

which then determine the dynamics of the pair densities and thus the dissociation rates. Equation (5.14) shows nicely why a significance of spin-dependent processes always requires a special combination of energy levels and correlation energies of defect states within pair sites. If the single occupation energy of site B is much closer to the single occupation energy of site A than the double occupation energy, p_a and p_b will become much larger than p_i or p_c and thus, all dissociation probabilities will approach the same values and the transitions through the pair sites will loose their spin dependence.

With the definition of a pair and the understanding of their spin-dependent dissociation probabilities one can address the question of imbalances between permutation symmetry states under steady-state conditions. By analogy to the recombination model, one can assume a constant generation rate $G/4$ for each of the four spin eigenstates of the pairs if thermal polarization effects are neglected (the latter is appropriate due to the small Zeeman splitting that exists under magnetic fields suitable for X-band ESR excitations). G can be considered constant since all changes induced by spin-related hopping would cause only second-order effects that are neglected due to the 'small-signal assumption' made above. Hence, since the dynamics of the pair densities ρ_{ii} leads, under the assumption that T_1 relaxation is slow, to the inhomogeneous ordinary differential equations

$$\begin{aligned} \partial_t \rho_{11,44} &= \frac{G}{4} - (d_T)\rho_{11,44} \\ \partial_t \rho_{22,33} &= \frac{G}{4} - \left(d_{2,3} + \frac{1}{T_2}\right)\rho_{22,33} + \frac{\rho_{33,22}}{T_2} \end{aligned} \tag{5.15}$$

similarly to Equation (5.9), we can find a steady-state solution

$$\rho^S_{11,44} = \frac{G}{4d_T} \quad \text{and} \quad \rho^S_{22,33} = \frac{G}{4}\frac{2T_2^{-1} + d_{3,2}}{d_2 d_2 + (d_2 + d_3)T_2^{-1}}, \tag{5.16}$$

which shows that there is an imbalance between the different spin eigenstates as long as spin selection rules are present. Note that, in contrast to the steady state of ensembles of recombining charge carrier pairs, it is not appropriate to consider the densities of the mixed states $\rho_{22,33}$ to be negligibly small. Even though it is conceivable that $p_T << p_{2,3}$ for realistic systems, it will not generally imply that $d_T << d_{2,3}$ due to Equation (5.14).

For the calculation of the conductivity $\sigma = en_e\mu_e$, we start by analogy to Equation (5.6) except that hole conductivity is ignored (for the reasons stated above). The electron mobility $\mu_e = \mu_e(R_1, \ldots, R_n)$ is a function of all the microscopic transport rates R_n in the material. Since only one particular transport rate R_i is changed and this rate change introduces only slight conductivity changes, one can describe a first-order change of the mobility $\mu_e(\Delta R_i) = \mu_e^S + \frac{\partial \mu_e}{\partial R_i}\Delta R_i$ by a linearity and since $\Delta R_i = \Sigma_i \Delta\rho_{ii} d_i$ when R_i is the transport rate through the pair sites, we obtain a conductivity

$$\Delta\sigma(t) = \lambda \sum_{i=1}^{4} [\rho_{ii}(t) - \rho^S_{ii}] d_i, \tag{5.17}$$

where $\lambda = en_e\partial\mu_e/\partial R_i$ is an arbitrary proportionality factor. Equation (5.15) has the same form as Equation (5.9). Hence, with time constants

$$\tau_s^{-1} = d_T \quad \text{and} \quad \tau_{m,f}^{-1} = \frac{1}{T_2} + \frac{d_2 + d_3}{2} \mp \sqrt{\frac{1}{T_2^2} + \frac{(d_3 - d_2)^2}{4}}, \tag{5.18}$$

and the definition of the relative density change $\Delta(\tau)$ of the spin ensemble during a coherent manipulation as it has been given by Equation (5.11), we attain a similar shape of a conductivity transient

$$\Delta\sigma(t) = \lambda \mathrm{Tr}[\hat{\rho}^S]\left[\left(1 - \frac{J + D^d}{\hbar\omega_\Delta}\right)e^{-d_3 t} + \left(1 + \frac{J + D^d}{\hbar\omega_\Delta}\right)e^{-\left(\frac{1}{T_2} + \frac{d_2}{2}\right)t} - 2e^{-d_T t}\right]\Delta \tag{5.19}$$

between a non-steady state $\hat{\rho}$ at time $t = 0$ and the steady state. Similarly as for Equation (5.12), the definition of Δ as given in Equation (5.11) is valid in Equation (5.19) only if the non-steady state $\hat{\rho}$ consists of the same number of pair systems as the steady state $\mathrm{Tr}(\hat{\rho}^S) = \mathrm{Tr}(\hat{\rho}(\tau))$.

5.4 THE THEORY OF A pEDMR EXPERIMENT

The measurement principle of pEDMR is illustrated in Figure 5.7. When the steady-state ensemble $\rho^S = \frac{G}{4r_T}[|T_+\rangle\langle T_+| + |T_-\rangle\langle T_-|]$, that consists mainly of pure triplet eigenstates, is

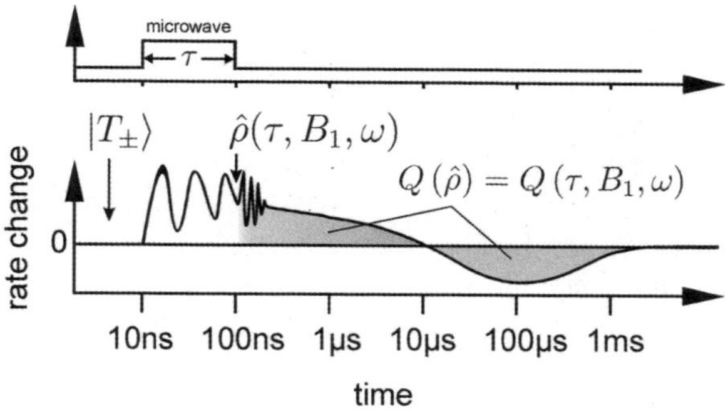

Figure 5.7 Illustration of a recombination or hopping rate transient during a pEDMR experiment on a logarithmic time scale. The experiment is started from the steady state when only $|T_+\rangle$ and $|T_-\rangle$ states exist. During the microwave pulse with strength B_1, frequency ω and length τ, a coherent propagation of the spins (Rabi oscillation) takes place. After the pulse, an ensemble of non-eigenstates $\hat{\rho}$ is prepared which dephases quickly such that a biexponential relaxation towards the steady state takes place. The integral of the transient, the observable Q is proportional to the rate at the end of the pulse and thus, Q reflects the singlet content of $\hat{\rho}$

coherently manipulated, an arbitrary non-steady state, non-eigenstate ensemble $\hat{\rho}(\tau, B_1, \omega)$ is prepared. After the resonant excitation ends, the non-eigenstates will carry out a Larmor precession whose influence on the net transition rate will fade quickly due to the ensemble dephasing [41]. Thus, a short time after the end of the microwave excitation, a non-steady-state transition rate is present that relaxes slowly (on a μs to ms time scale) back to the steady state. Since we know from Equations (5.19) and (5.12) that the integral of the sample current is proportional to the the density change Δ, as introduced in Equation (5.11), and $\Delta = \Delta(\hat{\rho}, \hat{\rho}^S) = \Delta(\hat{\rho}(\tau), \hat{\rho}^S)$ is a function of the ensemble state right at the end of the interaction, it is possible to determine the evolution of $\hat{\rho}(\tau)$ during the excitation by measurement of $Q(\tau)$ as function of the pulse length τ. The time resolution of this measurement scheme is obviously not determined by the current amplifier, but the pulse length generator and thus, a low ns–range time resolution is technically easy to achieve.

The pEDMR experiment described above allows the electrical detection of coherent spin motion of electrons which influences conductivity and photoconductivity through transport and recombination. In the following, the effects on the conductivity transients are explained and it is discussed how these effects can be applied to the investigation and the understanding of electronic transitions.

5.4.1 Rabi oscillation and the discrimination of spin coupling

The simplest pEDMR experiment is the electrical measurement of spin-Rabi oscillation by recording $Q(\tau)$ as a function of τ while the B_1 field of the microwave is strong enough such that Rabi frequencies can attain the higher MHz range and the microwave frequency ω is in ESR with a selected defect or impurity. A quantum mechanical description of this experiment [31] has revealed an expression

$$\Delta(\tau) = \frac{g_i^2 \mu_B^2 B_1^2}{\hbar^2 \Omega_i^2} \sin^2(\kappa \Omega_i \tau). \tag{5.20}$$

under the assumption of homogeneous B_1 fields and narrow ESR resonances (line width < B_1). Therein, $\Omega_i = \sqrt{(g\mu_B B_1)^2 \hbar^{-2} + (w - w_i)^2}$ is the Rabi frequency of the pair partner i with Larmor frequency ω_i and Landé factor g_i whereas κ denotes a factor whose value depends on the spin–spin couplings within a pair. As one can deduce from the prefactor $g_i^2\mu_B^2 B_1^2 \hbar^{-2}\Omega_i^{-2}$ of Equation (5.20), the Lorentzian line shape of a coherently excited narrow pEDMR line is determined by power broadening, it is proportional to the B_1 field strength and not to the relaxation rates as it is the case for cw EDMR lines [20]. This changes for the case of broad, inhomogeneous lines: When B_1 << line width, the Rabi oscillation is due only to those spins which are within the range of the B_1 separation. Hence, the prefactor of the signal is a convolution

$$\Delta(\tau) = \int_{-\infty}^{\infty} \Phi(\omega_i) \frac{g_i^2 \mu_B^2 B_1^2}{g_i^2 \mu_B^2 B_1^2 + \hbar^2 (\omega - \omega_i)^2} \sin^2\left(\kappa \sqrt{\frac{g^2 \mu_B^2 B_1^2}{\hbar^2} + (\omega - \omega_i)^2}\, \tau\right) d\omega_i, \tag{5.21}$$

of the arbitrary inhomogeneous line distribution $\Phi(\omega_i)$ and the power broadened peak. When the distribution $\Phi(\omega_i)$ is smooth on the order of B_1, which means $\partial\omega_i\Phi(\omega_i)g_i\mu_B B_1\hbar^{-1}$ << $\Phi(\omega_i)$, this expression can be simplified into

$$\Delta(\tau) = g_i \mu_B B_1 \hbar^{-1} \Phi(\omega) T(\kappa g_i \mu_B B_1 \hbar^{-1} \tau) \tag{5.22}$$

where $T(\kappa g_i\mu_B B_1\hbar^{-1}\tau) = T(\alpha)$ is a universal function

$$T(\alpha) = \int_{-\infty}^{\infty} \frac{\sin^2\left(\alpha\sqrt{1+x^2}\right)}{1+x^2} \mathrm{d}x, \tag{5.23}$$

which describes the shape of the Rabi oscillation transient in the absence of incoherence and whose argument $\alpha = \kappa g_i\mu_B B_1\hbar^{-1}\tau$ scales this general function to real experimental conditions.

Calculations [31] of the constant κ have been performed in the past for the extremal cases of weak and strong spin–spin interactions in the presence of a sufficiently strong microwave ($g\mu_B B_1$ >> J, D^d). An illustration of these coupling cases is given in Figure 5.8. Strong Larmor ($B_0(g_a - g_b)$ >> gB_1) separation which in this case corresponds to weak coupling ($B_0\mu_B(g_a - g_b)$ >> D^d, J) implies that an ESR excitation can always manipulate either spin a or spin b, depending on the chosen excitation frequency ω. Hence, the Rabi oscillation reflects the transient nutation of a simple $S = 1/2$ electron spin and therefore, $\kappa = 1$. When the Larmor separation is small ($B_0(g_a - g_b)$ << gB_1) which may correspond to strong coupling ($B_0\mu_B(g_a - g_b)$ << D^d, J), the excitation is no longer selective for any pair partner and hence, two $S = 1/2$ electron spins are turned and $\kappa = 2$. Note that the simulated transients plotted in Figure 5.8 are based on the universal oscillation function $T(\alpha)$ for an arbitrary B_1 field and Landé factor. The decay of the apparent oscillation is not due to incoherence (which was neglected), but due to the gradual spectral narrowing of the excitation width due to the increasing pulse length. For the strong coupling case the sketch in

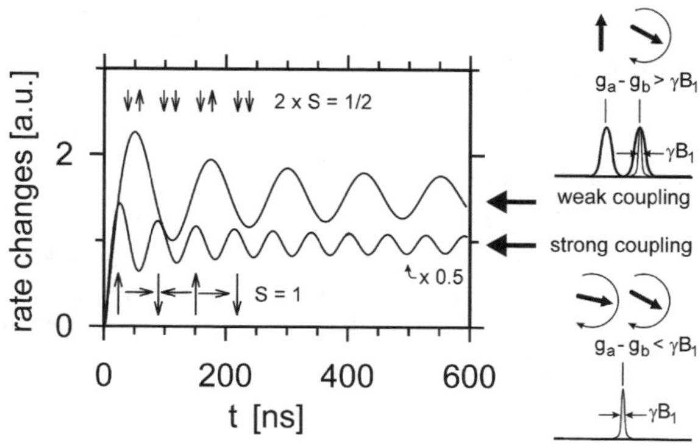

Figure 5.8 The Rabi-oscillation-induced transient nutation of transport and recombination rate changes for the two cases of strong and weak coupling under a strong excitation ($B_1 >> D^d$, J). In the case of weak coupling, the ESR resonance of the two pair partners are well separated and the oscillation reflects the motion of one S = 1/2. For the strong coupling case both spins are always turned and hence, an oscillation with twice the frequency takes place

Figure 5.8 illustrates two moving spins which are not perfectly aligned. This expresses the circumstance that a pEDMR signal can only be detected when there is at least a slight *g*-factor difference between the two pair partners. If both spins moved perfectly aligned (infinite coupling strength) there would be no change within the permutation symmetry of the pair and therefore no rate change.

In addition to the two coupling cases mentioned above and illustrated in Figure 5.8, another case should be mentioned here. When $B_0\mu_B(g_a - g_b) << D^d$, J (strong coupling) but $g\mu_B B_1 << D^d$, J, the factor $\kappa = \sqrt{2}$. This case can become important when a strong distribution of the coupling is present. An example for this case could be dipolar coupling between the pair partners when their distance and their orientation is distributed randomly, as is the case in disordered semiconductors [55]. While this case has so far not been described theoretically for pulsed EDMR experiments, one can deduce it from the description of transient nutation experiments of $S > 1/2$ systems without hyperfine influences, as given by Astashkin and Schweiger [56].

From the description of the electrical detection of Rabi oscillation given above, one can learn that transitions between electronic states can be distinguished by means of their coupling nature. Moreover, from Equation (5.22), one can deduce that, since the resolution of the magnetic field dependence (the line shape) depends on the B_1 strength, one can also determine the *g*-factor accurately that is associated with a certain Rabi oscillation and therefore with a certain coupling situation. Hence, a two-dimensional experiment becomes possible where Q is measured as a function of the pulse length τ and the externally applied magnetic field B_0. This measurement can then be Fourier transformed into a two-dimensional data set where Q is displayed as a function of the Rabi frequency Ω and the Landé factor g. If the *g*-factors of the electronic defects in a given material are known from ESR spectroscopy (as is the case for a-Si : H for instance), the resulting mapping will then display which paramagnetic centers are involved in the detected electronic transitions and what kind of interaction occurs at the different spin pairs.

5.4.2 Recombination and hopping echoes and the determination of transitions times

The electrical detection of Rabi oscillation, a coherent spin effect, suggests that its measurement should allow the measurement of coherence decay times such as the lifetime of the spin pairs. The direct determination of the coherence decay from a detected Rabi oscillation decay is difficult since the latter is not only determined by decoherence, but by the spectral narrowing of the pulse length as well as inhomogeneities of the g-factors and the applied microwave field B_1. As described in Section 5.5, it is especially the B_1 field that is difficult to keep homogeneous throughout a macroscopic sample under real experimental conditions. Hence, for coherence decay measurements, an experiment has to be carried out which allows one to distinguish the influences of dephasing from decoherence effects. There are three pulse sequences which are able to fulfill this task which are all based on so called rotary echoes, the partial rephasing of the spin pair ensemble by means of a sudden (on ps time scale) 180° phase change which is introduced into the resonant microwave pulse after the Rabi oscillation has dephased. In standard pulse ESR literature, rotary echoes have a variety of applications [3]. The theory of pEDMR rotary echoes, here referred to as recombination or hopping echoes is outlined elsewhere [31]. The three different echo sequences which can be used for coherence decay measurements are illustrated in Figure 5.9:

(a) The recombination/ hopping echo–echo decay experiment, consisting of a continuous train of pulses with opposite phase orientations. The decay of the echo amplitudes represents the coherence decay. Note that electrically detected rotary echoes can exhibit a step-like shape due to a second dephasing right after the first phase change. This second dephasing is due to the coupling between the two states and explained in detail in [31]. The echo–echo experiment has the advantage that a single pulse length dependence measurement reveals directly the coherence decay. However, there are two important drawbacks: (i) when the phase difference between the two pulses is not perfectly 180° (which is realistically never the case under experimental conditions), additional dephasing accumulates with the increasing number of phase changes—at some point, the net phase deviation may become relevant and distort the measurement result; (ii) the echo–echo experiment cannot measure long decay times. With increasing pulse length, the noise due to the artefact currents induced by the high power microwave radiation increases. At some point, a signal can no longer be detected by a reasonable number of shot repetitions. Moreover, as the microwave pulses enter the μs time range, the duty cycle of the traveling wave tube amplifier may be exceeded. This can pose another limitation for a maximum echo decay time that can be measured.

(b) The single recombination/ hopping echo decay experiment. In the course of this experiment, there is only one echo after one phase change and the decay is measured by repeating the experiment for different phase change times τ_{180}. The advantage is that there is no accumulation of phase errors, the disadvantages are the higher number of experiments to be carried out and again, similarly to (a), the limitation on the measurable maximum echo decay time due to the noise and the duty cycle limitations of long microwave pulses.

(c) The recombination/ hopping free echo decay experiment. Similarly to the single recombination/ hopping echo decay experiment, only one echo is induced due to two microwave pulses with opposite phase. However, unlike (b), the length of the dephasing and rephasing pulses remain constant and the echo decay is recorded as a function of the off-radiation time τ_{off}, which is the time between the two microwave pulses during which no

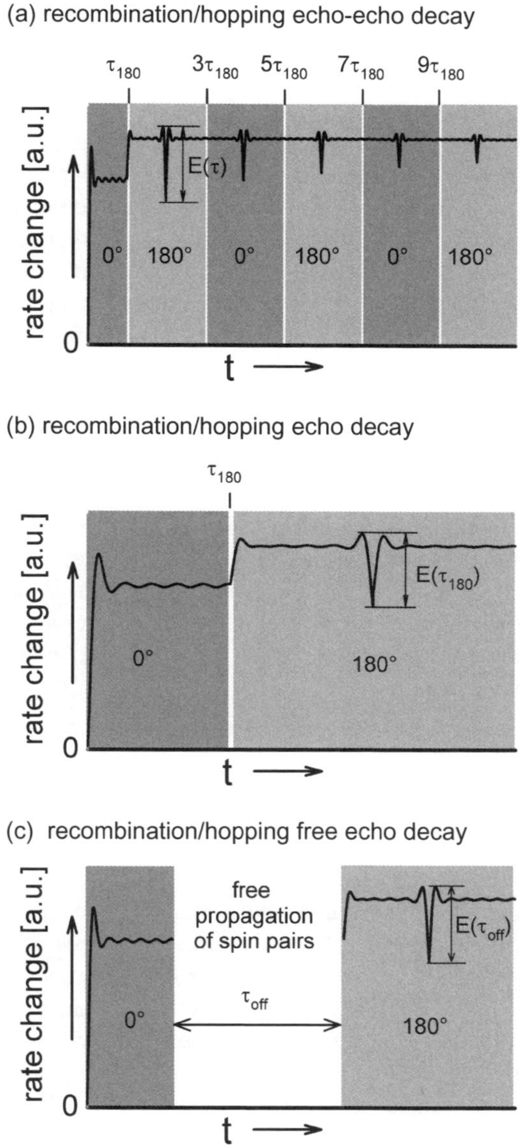

Figure 5.9 The decay of the pair coherence can be determined by echo experiments where the attenuation of the Rabi oscillation due to dephasing is reversed. The evolution of the rate changes are illustrated for three different echo sequences. The grey areas represent times when resonant radiation is applied. (a) Measurement of the recombination/hopping echo–echo decay $E(\tau)$ as a functions of the pulse length τ when the echo occurs. (b) Measurement of a single recombination/hopping echo decay $E(\tau_{180})$ versus the phase change time τ_{180}. (c) Measurement of a single recombination/hopping free echo decay $E(\tau_{off})$ as a function of the 'free propagation time' τ_{off} between the dephasing and the rephasing pulses

radiation is imposed onto the sample at all. The advantage of this pulse scheme is that arbitrarily long decoherence times can be measured since the imposed microwave pulse length during one shot never exceeds $3\tau_{on}$ (= the length of the dephasing pulse) and thus, microwave artifact currents remain constant, independently of τ_{off}, and the duty cycle is never too high. The drawback of this experiment is that during the absence of the resonant microwave radiation, new spin pairs are generated into spin eigenstates. When the second microwave pulse is started, these newly generated pairs start to precess and cause a second oscillation signal. In order to distinguish the contribution of rephased spin pairs from the newly generated ensemble, the pulse length τ_{on} must be sufficiently long such that a complete dephasing of the newly generated spin pairs has taken place before the echo appears. Hence, for slowly dephasing spin ensembles, the recombination/hopping free echo decay experiment can not be used.

The coherence decay measured by recombination or hopping echoes will always reflect the transition rate of only one, namely the fastest coherence time-limiting process. When several processes are present (e.g., singlet and triplet recombination, spin relaxation, dissociation, etc.) one has to determine the origin of the decay time measured by additional experimental means. In this regard it should be pointed out that the measurement of recombination or hopping echo transients can also be combined with the two-dimensional measurement of Rabi frequencies versus the g-factor described in Section 5.4.1. When the two-dimensional mapping of the Rabi oscillation before or after an echo is recorded versus the off-radiation time, the decoherence times of the individual processes that are identified by the mapping can be determined.

5.5 EXPERIMENTAL FOUNDATIONS OF PULSED EDMR

The essential parts of a pEDMR experiment are illustrated in Figure 5.10. The experiment represents a combination of a very sensitive transient photocurrent setup and a pulse ESR spectrometer whose microwave pulse generator should be able to generate pulse lengths τ down to the lower ns time range since τ sets the time resolution of the measurement. Commercially available pulse generators such as a Bruker Elexsys E580 pulse ESR spectrometer that can be equipped with up to eight pulse channels can produce pulse lengths of as little as 1 ns. Naturally, the microwave generator should be equipped with at least two independent pulse channels with opposite phase (180°) orientations such that the echo sequences for the coherence decay measurements are possible. In order to have sufficiently high B_1 fields, the microwave pulses must be amplified. For X-band, commercially available split ring or dielectric resonators provide conversion factors ($B_1 / \sqrt{P}$) which require microwave powers in the lower W to lower kW range in order to achieve reasonably fast Rabi oscillation ($\Omega > 10\,\mathrm{MHz}$).

Typically, such microwave powers are yielded by solid-state or traveling wave tube amplifiers. In addition to a pair of Helmholtz coils for the generation of the constant homogeneous magnetic field and a light source for the generation of photo-charge carriers, the pEDMR setup is equipped with a cryostat and the current detectors. As mentioned above, the measurement is conducted by a fast and sensitive transient detection of subtle current changes on top of strong constant offset currents. This is what makes pEDMR measurements a challenge and therefore, a current detection setup which has proven to be useful for a variety of pEDMR studies in the past is explained in detail in the following.

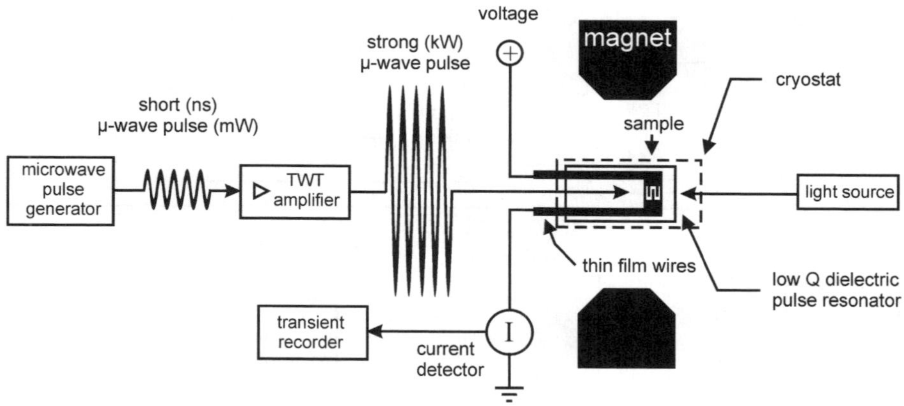

Figure 5.10 Sketch of the pEDMR experiment: The sample is located in a dielectric microwave pulse resonator with low quality factor. The sample is connected electrically by thin-film wiring whose thickness is below the microwave penetration depth in order to preserve the microwave modes of the resonator such that the B_1 field remains sufficiently homogeneous

5.5.1 Current detection

The transient detection of spin-dependent currents requires that the detection setup allows one to measure (i) subtle current changes on top of (ii) comparatively strong constant currents (the so called offset current due to the non-spin-dependent currents in the sample) at (iii) a very high time resolution. Naturally, requirements (i)–(iii) act against each other—if the time resolution is improved, the sensitivity gets worse and vice versa, whereas an increase of the constant offset current leads to a deterioration of both, the absolute sensitivity and the time resolution. This is why current detection for pEDMR experiments will always have to be a compromise between all three requirements to meet the individual conditions of a given material system.

Typically, the detection of a spin-dependent current in inorganic disordered semiconductors requires the fast detection of relative current changes as low as $\Delta I/I = 10^{-7}$ in currents that are in the lower μs range. In order to optimize time resolution and current detection, one should always aim to achieve lowest possible sample resistances so that time constants are low, but the offset current is maximized. Since a low absolute sample resistance can become a problem for semiconductors at low temperatures, the sample geometry and especially the contact design will play an important role for the optimization of the pEDMR experiment. Design rules for sample geometry and interdigited contacts are explained in detail in Section 5.5.2.

For the current detection, a setup as illustrated in Figure 5.11 has proven to reveal optimal results. Therein, the sample current is induced by a constant current source such as provided by the Keithley PCS220, for instance. It shall be pointed out that it is not a contradiction to measure a transient current induced by a constant current source as long as the current dwell time is much longer than the time scale on which the transient is recorded. With typical measurement shot repetition times between $50\,\mu$s and 1 ms, a dwell time of T_D =

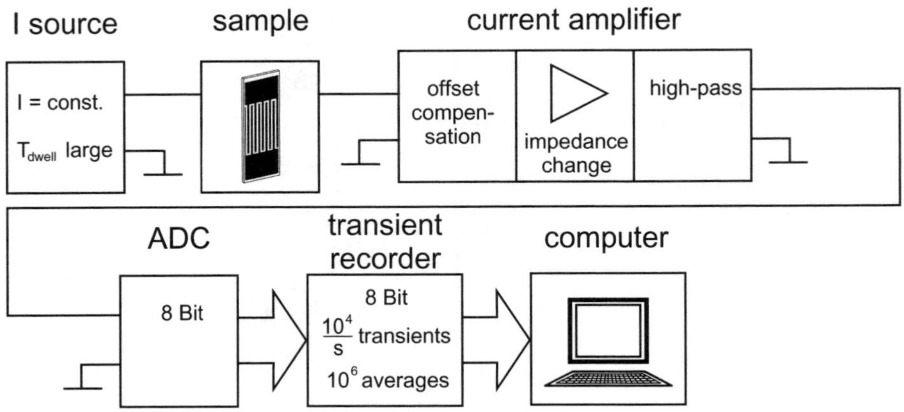

Figure 5.11 The transient detection of photocurrent changes. For details see text

100 ms will not interfere with any signal current, however, effects due to a slow drift of the sample conductivity or the light source will be compensated when the sample current is induced by a constant current source instead of a constant voltage source. For the current measurement, a three-component amplifier has to be used which: (i) is able to subtract the strong offset current first, before the remaining differential changes are then (ii) converted into a voltage signal by an impedance transformer. Due to tolerances and drift of the constant current source and the offset generator, there will never be a complete compensation of the offset current. Consequently, the residual offset is then amplified along with the transient signal current. In order to remove these residual constant offset currents, the output of the impedance transformer is filtered (iii) by a high-pass filter whose transmission frequency is in a similar time range as the dwell time of the current source. For the experiments presented in this study, the offset/amplifier/high-pass combination was provided by a commercial Stanford Research SR570 current amplifier. Its output was connected directly to the input of an 8-bit analog–digital converter which was read by an 8-bit high-speed transient recorder. Since a pulsed ESR spectrometer is necessary for pEDMR experiments, one can use the spectrometer transient recorder such as the Bruker SpecJet which allows real-time averaging of different shots and thus the acquisition of millions of averages within a few minutes. Note that with typical single-shot signal-to-noise (SNR) ratios of less than 1/100, such averaging techniques are a crucial prerequisite of pEDMR measurements.

5.5.2 Sample design

For cw EDMR, a low sample resistance and short sample time constants are sufficient prerequisites for the detectability of spin-dependent signal currents and therefore conditions for a sample design. Microwave field inhomogeneities which can be caused by electrical wiring and sample contacts within the microwave resonator do not (or hardly) play a role and have (to the knowledge of the authors) always been neglected in the literature. For pEDMR, B_1 field inhomogeneities are not negligible since a B_1 distribution will cause a rapid, artificially induced dephasing of the spins in resonance and thus, the observation of Rabi oscillation will become impossible. In order to overcome this problem, the sample and especially the contacts must be designed such that a B_1 distortion is as small as possible. In order to achieve this, sample substrates should preferably be insulators and sample contacts, which naturally cannot be made out of an insulator, must have thicknesses below the microwave penetration depth which depends on the microwave frequency and the contact conductivity. For X-band wavelength and typical contact materials such as Au, Ag or Al, one can anticipate at He temperatures a penetration depth as little as 1–2 μm. This shows that even bonder wires with diameters of 20–40 μm are already much too thick and work for the microwave radiation essentially like short circuits. The solution to this problem presents a complete thin-film contact wiring of the sample within the microwave resonator.

For the experimental demonstration of the pEDMR effects as predicted above, a semiconductor heterostructure consisting of a 20 nm-thin layer of plasma enhanced chemical vapor-deposited, strongly phosphorus-doped, hydrogenated amorphous silicon (n-a-Si : H) on the (100) oriented surface of a floatzone-grown, boron-doped, crystalline silicon (c-Si) wafer will be used. As indicated in Figure 5.12(b), the thin n-a-Si : H layer is covered by a layer of a transparent conducting oxide in order to establish a front contact of the structure and to allow light to reach the heterojunction at the same time. This contact layer consists of 80 nm highly *n*-doped (degenerated) ZnO which itself is connected to a narrow, 100-μm-wide and 100-nm-thin Al grid that connects the sample through a 40-mm-long thin-film wire with the outside of the cavity. As one can see in the device photo shown in Figure 5.12(a), the active device area is 1 mm^2, small. This is enough such that the B_1 field will be sufficiently homogeneous throughout the region where spin-dependent currents occur. Outside the device area, a thick (200 nm) layer of silicon dioxide is grown which serves as

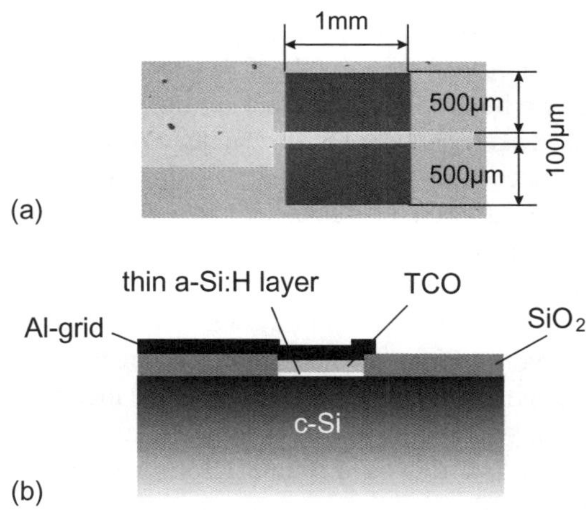

Figure 5.12 (a) Photograph of view on the *n*-a-Si:H/c-Si heterostructures which where used for the measurements presented. One can see the window of the active sample area as well the front contact grid that is connected to the invisible transparent conducting oxide laxer. (b) Cross-sectional sketch of the layer system which forms the heterostructure shown in (a)

an insulating layer between the front contact wires and the underlying substrate. Figure 5.12 shows how a pEDMR sample can be prepared that does not distort B_1 fields. While the actual semiconductor sample with its thin ZnO layer/Al grid contact system is located at the tip of the match-like substrate (the sample area will be positioned in the center of the microwave cavity for the pEDMR measurement), it is connected to the contact pads on the opposite side of the substrate by 40-mm-long and less than 200-nm-thin Al-stripes. Thus, with the sample and contact geometry given, it is conceivable, that (i) low resistances and therefore high currents are achievable and (ii) the actual semiconductor sample will be at the center of the cavity, while (iii) the eigenmodes of the cavity, especially at its center, remain undistorted.

5.5.3 Microwave-induced currents

In contrast to cwEDMR where weak mW-range microwave powers are used for the ESR-induced current changes, the strong radiation used for pEDMR pulses induces non-negligible sample currents that can have magnitudes much higher then the contributions of the spin-dependent currents to be detected. The microwave induced currents are not avoidable since a conducting sample in the center of a microwave resonator will always work to a certain extent as a microwave antenna. It is therefore crucial to learn how to handle the microwave artifacts, how to minimize the contributions and most of all, how to distinguish them from the actual spin-dependent signal transients that may be buried under strong microwave induced artifact transients.

Figure 5.13(a) shows a plot where the photocurrent through an n-a-Si:H/c-Si heterostructure described above is plotted as a function of the magnetic field and the time when a short coherent microwave pulse with length $\tau = 160\,\text{ns}$ and $P = 32\,\text{W}$ was irradiated at a time $t = 0$. The sample temperature is $T = 10\,\text{K}$. One can see that, independently of the magnetic field within the displayed range, there is a transient current response after the short pulse which can obviously not be due to spin-dependent currents since for the microwave frequency used ($\simeq 9.7\,\text{GHz}$), an almost identical response is observed for all of the applied magnetic fields except a short field range at about $B_0 = 345\,\text{mT}$. This illustrates how microwave-induced artifacts can be distinguished from spin-dependent contributions. While the microwave artifact transients reproduce for all magnetic fields as long as the sample conductivity is sufficiently independent of the magnetic field, the magnetic resonant spin contributions are strongly dependent on the magnetic field. Hence, for the extraction of the spin currents from the raw data, one has to subtract the artifact contributions obtained from measurements at magnetic fields well outside all known ESR resonances.

The subtraction of an off-resonant current transient from a measured on-resonant sample current works only as long as the sample resistance is not dependent on the magnetic field B_0 other than due to the spin-resonant effects that are to be measured. When other B_0-field dependencies, so-called magnetoresistances are present, this procedure is not as straightforward. In fact, since the data shown in Figure 5.13(a) have been recorded on a device which contained crystalline silicon, one has to expect magnetoresistance, and thus one can anticipate differing current transients at different magnetic fields even in the absence of any spin-dependent currents. For the c-Si sample and the given magnetic field range, the on-resonant microwave artifacts can be extrapolated sufficiently well with a linear regression. The results of such extrapolations are displayed in Figure 5.13(b): They are based on linear

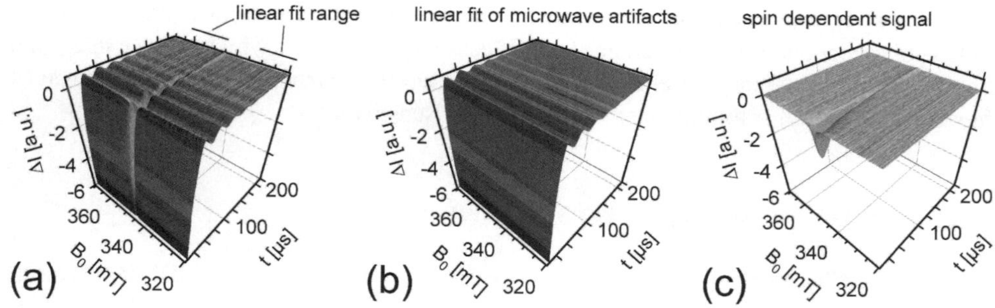

Figure 5.13 (a) The measured PC of an *n*-a-Si:H/c-Si heterostructure as a function of the time and the magnetic field when a short (160 ns) coherent microwave pulse of $P = 32\,\mathrm{W}$ is imposed at $t = 0$. (b) Results of linear fits of the magnetic field dependence of the data in (a) outside the magnetic fields where ESR-induced current changes occur, for all given times t. The plot essentially resembles the magnetic field dependence of the microwave induced artifact currents. (c) The difference of the data in (a) minus the fit results of (b). As a result, only the spin-dependent currents as a function of the magnetic field and the time become visible

fits of the magnetic field dependence of the data outside the field range where the resonance occurs, for each time slice in the range of $0 < t < 200\,\mu\mathrm{s}$. Ideally, the extrapolation of the microwave artifact transients at on-resonance magnetic fields works as long as the expected magnetoresistance can be described by polynomials up to the second order (quadratic terms). Beyond the second order, there may be several extremal points within the fit which will be hard to distinguish from possible magnetic resonance peaks, and thus it will become very difficult or even impossible to distinguish microwave artifact currents from spin-dependent currents. However, for the magnetic field range associated with X-band spectroscopy and most material systems, this situation is highly unusual.

Once the microwave artifact currents within the range of ESR-induced changes are established, they can be subtracted as indicated in Figure 5.13. The result of the subtraction of the extrapolated artifacts in plot (b) from the raw data in plot (a) is displayed in Figure 5.13(c). One can clearly see an immediate current change right after the short resonant pulse which relaxes slowly over the course of about $200\,\mu\mathrm{s}$. Since a current change is visible within a narrow magnetic field range at about $B_0 = 345\,\mathrm{mT}$ only, it is clear that after the subtraction only spin-dependent currents remain, and thus the spin signals have been separated from the microwave artifacts.

Microwave artifacts are an undesirable side effect of the coherent spin excitation during a pEDMR experiment. While the procedure outlined above is able to suppress the artifact currents to a large degree, it is of course not possible to reduce the noise which is associated with these signals. It is the latter though that limits the sensitivity of pEDMR experiments and also the pulse length that can be applied to a given sample because microwave artifact currents, as well as noise amplitudes, increase with the amount of energy that is deposited in a sample during a pulse excitation [28]. Because of this, the reduction of microwave artifacts is an important part for the optimization of a pEDMR experiment. From the comparison of pEDMR measurements on thin-film silicon on insulating substrates [28] with pEDMR experiments on bulk c-Si samples [57] it appears that microwave artifacts are caused to a large extent by shunt currents through non-spin-dependent conduction paths such as photoconductivity due to long-lived excess charge carriers which diffuse deep into the underlying substrate. An elimination of these parallel currents is therefore the most important measure for the increase of the pEDMR sensitivity.

5.5.4 Limitations of pEDMR experiments

In principle, pEDMR should be applicable to any material on which cwEDMR measurements have successfully been applied in the past. Hence, the potential of pEDMR could be particularly promising for the investigation of many disordered semiconductors (organic and inorganic). In spite of this prospect, a variety of limitations exist in addition to the noise limitations posed by microwave artifact and parallel currents mentioned above which shall be discussed briefly in the following.

Important for the applicability of pEDMR is of course the presence of spin-dependent transitions. Since spin selection rules are generally due to spin conservation which itself is due to weak spin-orbit coupling, pEDMR is of course more suitable for group IV materials while some III–V semiconductors are particularly ill suited for pEDMR investigations. Note, that even in absence of weak spin orbit coupling, electronic transitions can be spin dependent only when electron spin systems with $s \neq 0$ are involved. Hence, electronic processes involving doubly occupied or empty states which can be filled with holes or electrons of arbitrary spin orientation will not be spin dependent, and thus they are not detectable with pEDMR.

Another important prerequisite of pEDMR experiments is that the transitions to be investigated are sufficiently slow. At the present, pEDMR has never been conducted at frequencies beyond X-band ($\approx$10 GHz). At these frequencies, one can produce Rabi frequencies of the order of 100 MHz with commercially available resonator and amplifier technology. Hence, coherent spin motion can be made visible with pEDMR as long as the coherence decay is on time scales beyond 10 ns. For transitions faster than this, it is still possible to detect magnetic resonantly changes in spin-dependent transition rates, however, the access to coherent spin-motion effects will not be possible. The time resolution is a particularly strong limitation for the investigation of materials with fast decay processes.

PEDMR can only be performed on spin-dependent transitions where the mutual spin–spin couplings within the pairs are in an intermediate range. When the coupling is weak, the transition probability becomes small, and hence signal currents can get buried under parallel and microwave artifact currents. If the coupling is strong, the eigenbase of the spin pair is tilted towards a pure singlet–triplet base (see Equations (5.2) and (5.3)). ESR-induced spin changes will no longer change the permutation symmetry (pure singlet states are $s = 0$ systems that can not be manipulated by ESR) and therefore, the current imprint of the spin motion vanishes, too.

5.6 PEDMR ON TRANSPORT CHANNELS THROUGH *n*-a-Si:H

In the following, the application of pEDMR to the investigation of charge transport through the *n*-a-Si:H/c-Si test devices described above is outlined and discussed. The band structure of the *n*-a-Si:H/c-Si heterojunction as known from simulation studies [58] is sketched in

Figure 5.14. With the Fermi energy indicated, it is conceivable that at low temperatures, spin-dependent charge carrier hopping through conduction band-tail states and phosphorus donor states can take place. Thus, for the pEDMR experiment, the sample structure was cooled down to $T = 10\,K$ and continuously exposed to a 100 W halogen lamp which, when focused onto the sample surface caused a homogeneous irradiation with visible light (UV and IR components were filtered) of approximately $100\,mW/cm^2$. In order to establish vertical charge transport through the sample, a negative bias was applied such that a negative steady-state current was established. Note that the sign of the current corresponds to the operation point of the device which is in the third quadrant of its IV characteristic. This will be important when the spin-induced currents are discussed since a negative bias of the current changes will then mean that the sample conductivity is enhanced, and thus transport rates are increased.

5.6.1 Detection of transport transitions

In order to identify spin-induced currents, a magnetic field dependence of the current response to a short, coherent microwave pulse was recorded. The result of this measurement has already been shown in Figure 5.13 and has been discussed in Section 5.5.3 with regard to the influence of microwave currents. Once these measurement artifacts are removed and only the transients due to spin-dependent transitions remain, one can deduce the ESR-induced PC changes after the pulse as function of time (the PC transient), as it is displayed for $g = 2.0046$ in Figure 5.15a. The plot indicates the integral of the absolute current which determines the observable $Q(\tau)$ as defined and explained in Section 5.4.

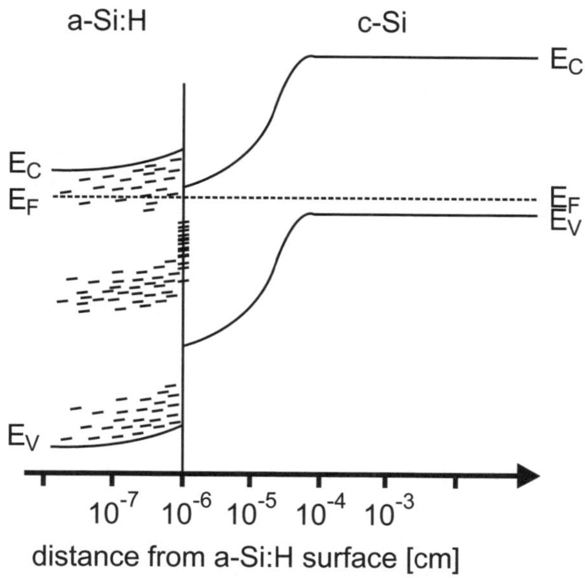

Figure 5.14 Sketch of the band diagram of the *n*-a-Si:H/c-Si heterostructure as well as the different localized defect states involved. The figure is based on a simulation described elsewhere [58]

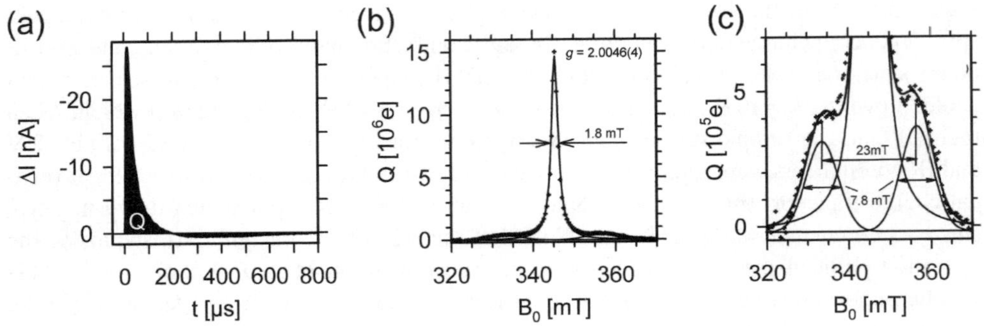

Figure 5.15 (a) Transient of the spin-dependent PC ($\Delta I(t) = I(t) - I^0$ and (I^0 represents the steady state value) through the *n*-a-Si:H/c-Si heterointerface at $T = 10\,\text{K}$ right after a coherent microwave excitation of the $g = 2.0046$ resonance. The plot resembles data obtained from the two-dimensional data set presented in Figure 5.13c. Note the the negative sign of the current units does not indicate a current quenching since the device was operated under reverse bias. (b) and (c). The magnetic field dependence of the integrated current change Q as indicated in (a) plotted on two plots with different ordinate scales. One can see the influence of a strong resonance at $g = 2.0046(4)$ that was fit with a Lorentzian function (solid line) and the two smaller peaks with Gaussian functions that have similar intensities and peak widths

Figures 5.15b and c display Q as a function of the externally applied constant magnetic field B_0 for a fixed pulse length of $\tau = 160\,\text{ns}$. The two plots represent the same data set for two different ordinate scaling factors and prove that the measured current transients are clearly due do spin-resonant PC changes: As displayed in Figure 5.15b, the most pronounced structure can be found at $g = 2.0046(4)$, where the transient displayed in Figure 5.15a was recorded. Within the error margin, the Landé factor of this Lorentzian peak (a fit is displayed in the same plot) is in agreement with literature data of conduction band-tail states of a-Si:H [48, 51, 59, 60]. Its full width at half maximum is $w = 1.8(2)\,\text{mT}$. Under consideration of the power broadening due to $B_1 = 0.178\,\text{mT}$ and an excitation broadening of $\delta = 0.2\,\text{mT}$ due to the τ = 160-ns-long excitation, one can deduce a peak to peak width for the Lorentzian function of $H_{\text{pp}} = \sqrt{(w^2 - 4B_1^2 - 4\delta^2)/3} = 1(1)\,\text{mT}$ [4]. This is close to the literature values of $H_{\text{pp}} = 0.6$–$0.8\,\text{mT}$. Beside this peak, two hyperfine lines with a comparatively small intensity, very broad (7.8 mT) line widths and a wide separation ($\approx$23 mT) become visible as displayed in Figure 5.15c. These properties are indicative of hyperfine separated transitions out of neutral phosphorus donor states [51, 59, 60]. In consequence, the measurement of $Q(\tau)$ shows that conduction band-tail states and phosphorus donor states of highly *n*-doped a-Si:H influence the PC at low temperature where hopping determines the macroscopic conductivity (see Chapter 2 of this book).

In Section 5.3 it was shown that the rate relaxation after a short, coherent excitation can reveal information on whether an observed signal may be due to excess charge carrier recombination or transport. Applied to the PC transient displayed in Figure 5.15a this means that the signal observed must be due to spin-dependent transport and not recombination. Right after $t = 0$ when the pulse is imposed, the current change is negative which means that, because of the negative sign of the PC, the magnitude of the current is enhanced. This positive signal relaxes within approximately 200 μs and changes into a negative signal which relaxes on a much longer time scale and has a much smaller amplitude. Qualitatively, this transient resembles exactly the double-exponential decay behavior described by Equations (5.12) and (5.19) in Sections 5.2 and 5.3, respectively and illustrated by Figures 5.3 and 5.7. Since the measured data represents a PC transient where the current increases first before it then changes into a quenching signal, we can attribute the observed conduction band-tail signal to a transport path. Note that this conclusion is possible only as long as it can be excluded that spin-dependent recombination currents are not responsible for the transport in the sample as it is conceivable for interface recombination at the c-Si/a-Si:H heterojunction. However, the observed integrated signal intensity correlates monotonically to the sample thickness (not shown here) and, therefore, it can be associated with a bulk effect and it is not due to interface recombination.

5.6.2 Observation of Rabi oscillation

According to the theoretical predictions of Section 5.4, the observable Q should represent a spin measurement of the conduction band-tail states right after the pulse. For the experimental proof of this, a second two-dimensional pEDMR experiment was conducted on the same sample and under the same experimental conditions as described above where the observable Q was measured as a function of the pulse length τ and the magnetic field B_0 for 64 field positions in a range of 12.8 mT width around the resonance of $g = 2.0046$. The irradiated microwave pulse power was 250 W. The result of this measurement is plotted in Figure 5.16. While the magnetic field dependence confirms again the electron spin resonant behavior of the detected conductivity responses, the pulse length dependence shows a clearly recognizable oscillatory behavior, the spin–Rabi oscillation of localized electrons in a-Si:H conduction band-tail states.

The oscillation observed in Figure 5.16 exhibits all the qualitative features of an electrically detected electron spin–Rabi oscillation as predicted by Equation (5.21). The oscillation dephases gradually, it never passes the steady-state level (zero line on the charge axis) and it approaches gradually an on-resonant steady state that is unequal to the steady-state level. In spite of these qualitatively compelling arguments, an indisputable proof can be given only by the B_1 dependences of the oscillation frequency and amplitude. As one can see from the definition of the Rabi frequency Ω given in Section 5.4.1, when the resonance condition $\omega - \omega_i = 0$ is fulfilled, Ω becomes proportional to B_1. Hence, a measurement of the oscillation frequency at the peak maximum ($g = 2.0046$) as a function of B_1 can confirm the underlying nature of the observed oscillation. For this experiment, the pulse length dependence of Q was measured at the same sample and under identical conditions to the data displayed in Figure 5.16, whereas the magnetic field B_0 was kept constant at a value corresponding to $g = 2.0046$ and the microwave power was tuned to values of 250, 125, 64, 32, 16, 8 and 4 W, corresponding to B_1 values indicated in the figure. The results of these measurements are plotted in Figure 5.17a. One can clearly recognize that the frequency of the oscillation increases with the applied microwave field. A plot depicting the measured oscillation frequency as a function of the applied B_1 field as shown in Figure 5.17b demonstrates clearly the proportionality between B_1 and Ω. Note the excellent result (solid line) that is achieved when the measured data is fit with a linear function through the origin.

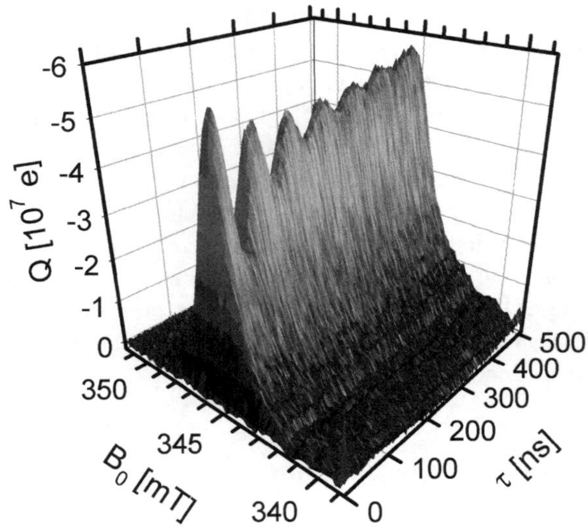

Figure 5.16 Plot of Q as a function of the pulse length τ and the magnetic field at $T = 10\,\mathrm{K}$. One can recognize the Rabi oscillation that takes place during the resonant pulse excitation. For details see text

The theory outlined in Sections 5.2–5.4 can be tested even further. One can learn from Equation (5.21) that when the resonance has a Lorentzian line shape

$$\Phi(\omega_i) = \frac{A\xi}{\xi^2 + (\omega_i - \omega_c)^2} \tag{5.24}$$

with center frequency ω_c and line width ξ (this is the case for the conduction band-tail state transport discussed above as one can deduce from the fit result displayed in Figure 5.15, the intensity of the first maximum $Q_{\max}$ as defined in Figure 5.17a of the pEDMR transient becomes

$$\Delta(\tau_{\max}) = \int_{-\infty}^{\infty} \frac{A\xi}{\xi^2 + (\omega_i - \omega_c)^2} \frac{g_i^2 \mu_B^2 B_1^2}{g_i^2 \mu_B^2 B_1^2 + (\omega - \omega_i)^2} \mathrm{d}\omega_i . \tag{5.25}$$

Note that $Q_{\max}$ is an arbitrary measure of the current response which is chosen for convenience to be the value $\Delta(\tau_{\max})$ at the pulse length $\tau_{\max}$ where the first maximum of the Rabi oscillation is reached since then

$$\sin^2\left(\kappa\sqrt{g^2\mu_B^2 B_1^2 + (\omega - \omega_i)^2}\,\tau_{\max}\right) \simeq 1.$$

Obviously, $\tau_{\max}$ will have a different value for different microwave field strengths, so $\Delta(\tau_{\max}) = \Delta(\tau_{\max}(B_1)) := \Delta^{\max}(B_1)$ with

$$\Delta^{\max}(B_1) \propto \frac{(\xi + g_i \mu_B B_1) B_1}{g_i^2 \mu_B^2 B_1^2 + \xi^2} \tag{5.26}$$

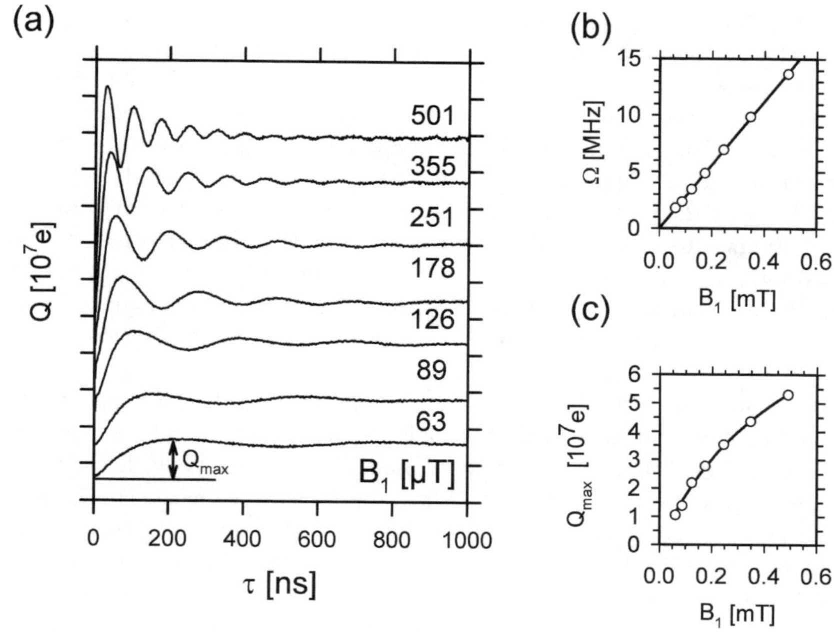

Figure 5.17 (a) Electrically detected Rabi oscillation during a coherent microwave excitation measured for different microwave fields B_1 at the $g = 2.0046$ resonance. (b) The frequency of the Rabi oscillation displayed in (a) as a function of the microwave field strength B_1. A fit with a linear function (solid line through the origin) reveals an excellent agreement. (c) The intensities of the first maxima of the pEDMR signals Q_{max} that is defined in (a) as a function of the microwave field strength B_1. A fit with a hyperbolic function (solid line) reveals an excellent agreement

at the peak of the resonance when $\omega = \omega_c$. Equation (5.26) is the consequence of the circumstance that Equation (5.25) is the convolution of two Lorentzian functions. This results again in a Lorentzian with a width that is the geometrical sum of the widths of the two convoluted functions. In order to verify this experimentally, the intensities of the first maxima measured at $g = 2.0046$ are displayed in Figure 5.17c along with a fit of these data points with Equation (5.26). The good agreement between the fit and the experimental data is obvious and the fit result, the half-width at half-maximum of $\xi = 0.48(3)\,\text{mT}$, corresponding to a peak-to-peak width $H_{pp} = 2\xi/\sqrt{3} = 0.56(4)\,\text{mT}$ of the derivative spectrum of the Lorentzian function [4] matches the literature values of the $g = 2.0044$ conduction band-tail state resonance of about $H_{pp} \simeq 0.6\text{–}0.8\,\text{mT}$ [48, 51, 59, 60]. This result also proves that the discrepancy between the line width of the measurement displayed in Figure 5.15b and discussed in Section 5.6.1 and the literature values is caused by power broadening of the intensive ($P = 32\,\text{W}$) radiation that was applied.

5.6.3 Coherence decay and hopping times

From the data displayed in Figure 5.17a, one can deduce the decoherence and dephasing times of the Rabi oscillation in a range between 30 and 1000 ns. The exact values characterizing this decay requires the fit of the measured transients with appropriate decay functions. However, for the determination of coherence times, this procedure would be of little benefit since, as explained in Section 5.4, decay times of the Rabi oscillation can depend on both coherence-limiting processes or coherent-dephasing processes. The Rabi oscillation as displayed in Figure 5.17 is a good example for this problem. Figure 5.18 shows the decay of the Rabi oscillation under irradiation with $B_1 = 501\,\mu$T as already depicted in Figure 5.17a. Here, the local maxima and the local minima of the experimental data were fit with two single-exponential decay functions, respectively, whose fit results are displayed by the two solid lines. As one can see visually and also deduce from the printed time constants, the fit results differ greatly. This shows that the only way to distinguish decoherence and dephasing times is to refer to hopping echo experiments, as explained in Section 5.4.2.

In order to investigate the coherence decay within the first $2\,\mu$s, a hopping echo–echo decay experiment was conducted whose result is displayed in Figure 5.19. The measurement took place under equal conditions as the Rabi oscillation measurements displayed in Figure 5.17. The microwave power was $P = 128$ W and the 180° phase changes of the microwave radiation were introduced, as indicated in the data plot. Beyond $\tau = 1900$ ns, the measurement became impossible due to the microwave artifact induced increase of the noise level. The data displayed in Figure 5.17 show that: (i) photocurrent detected transport echoes are detectable; and (ii) these transport echoes reflect the coherence decay within the measured time scale which is clearly beyond the decay times of the Rabi oscillation, as shown in Figure 5.18. Note that the hopping echoes in the data plot are the charge spikes that occur exactly between two phase changes (100 ns after the last which is 100 ns before the next pulse). The smaller spikes about the time of the phase changes are not hopping echoes. They are Rabi wiggles before and after the echoes which are visible in the data plot since the dephasing times of the spin ensembles are smaller than the time between a hopping echo and a subsequent phase change. The fit with the exponential decay was possible only with a strong constant offset. This is indicative of different processes with coherence times: (i) in the time range of the decay with constant $\tau_{decay} = 0.6(4)\,\mu$s; and (ii) with a much longer coherence time beyond the measured time range.

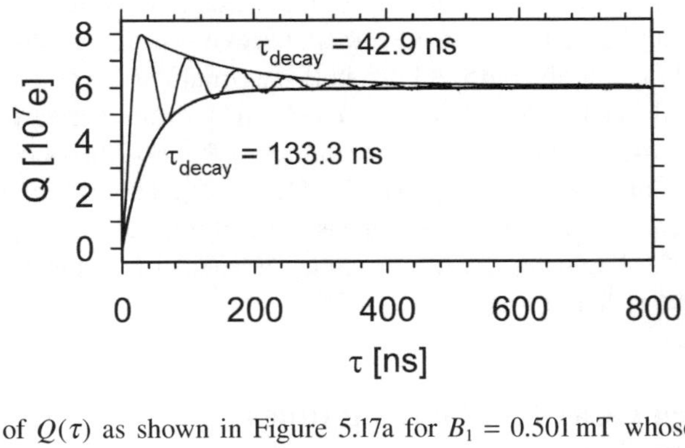

Figure 5.18 Plot of $Q(\tau)$ as shown in Figure 5.17a for $B_1 = 0.501$ mT whose local maxima and minima where fit with two single-exponential decay functions (solid lines). The different fit results show that single exponential decay functions do not accurately describe the behavior of the oscillation

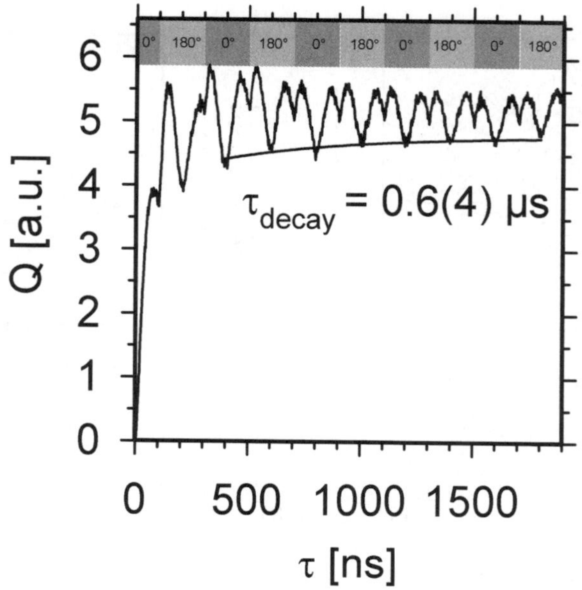

Figure 5.19 Hopping echo-echo decay measurement for the resonance observed at $g = 2.0046$. The plot displays $Q(\tau)$ when the phase of the microwave radiation is suddenly shifted by 180° after every 200 ns between $\tau = 100$ ns and $\tau = 1700$ ns. The fit of the decay with an exponential decay function (solid line) with constant offset reveals a time constant that differs significantly from the Rabi oscillation decay displayed in Figure 5.18

In order to confirm that the different ways to determine decoherence times lead to identical results, a hopping free-echo decay measurement as illustrated in Figure 5.9 was conducted in addition to the hopping echo–echo measurement described above. For this, two pulses with opposite phase were irradiated under same sample conditions as described above, whereas the delay time τ_{off} between the two pulses was changed gradually between 0 and 6 μs. In order to minimize the acquisition time, $Q(\tau)$ was recorded only around the pulse lengths of the transport echoes ($|\tau - \tau_{\text{echo}}| < 100$ ns when τ_{echo} is the pulse length where an echo reaches its maximum. The results of these measurements are displayed in Figure 5.20a. The data clearly show how the echo intensity Q_{echo} as defined in the plot declines with increasing τ_{off}. Figure 5.20b displays a plot of Q_{echo} as a function of τ_{off}. Similar to the decay of the echo–echo amplitude shown in Figure 5.19 one can fit the data with an exponential decay function with significant offset. The decay constants of both, the hopping free-echo decay measurement as well as the hopping echo–echo agree within the margin of error and thus, the measured coherence decay reproduce independently of the method of its determination.

5.7 DISCUSSION OF THE EXPERIMENTAL RESULTS

From the measurements displayed in Figures 5.18–5.20 it became apparent that the resonantly excited electronic transition at $g = 2.0046$, namely the electron hopping through

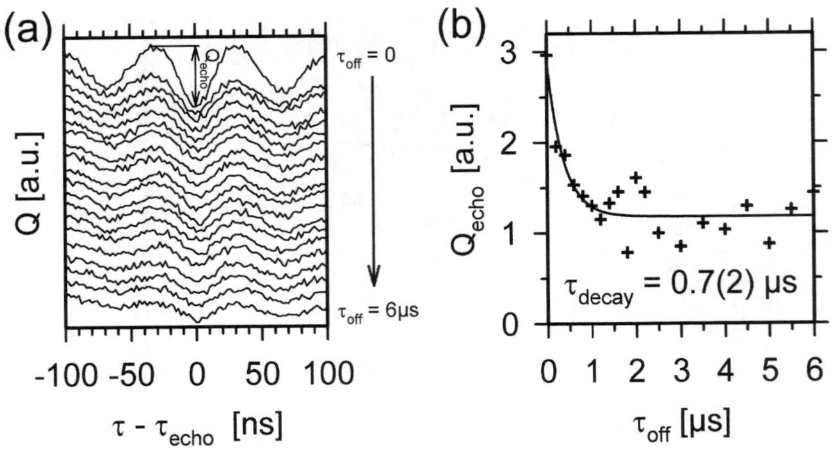

Figure 5.20 (a) Hopping free-echo decay measurement for the resonance observed at $g = 2.0046$. The plot displays $Q(\tau)$ for $|\tau - \tau_{echo}| < 100\,\text{ns}$ for different τ_{off} between 0 and $6\,\mu\text{s}$. (b) Plot of the echo intensity Q_{echo} as defined in (a) as a function of τ_{off}. The fit of the decay with an exponential decay function (solid line) with constant offset reveals a time constant that is within the margin of error in agreement with the coherence decay determined by hopping echo echoes displayed in Figure 5.19

conduction band-tail states in a-Si:H, is not one sharply defined process which is homogeneous throughout the material, but a rather complex group of qualitatively different transitions. Further information about these inhomogeneities can be gained from the spectrum of Rabi frequencies because of their dependence on the spin–spin couplings between the states where transitions occur (see Section 5.4.1). Figure 5.21a displays the result of a fast Fourier transform (FFT) of the data set displayed in Figure 5.18. The data could be fit with three Lorentzian lines whose intensities, widths and center frequencies are shown in the plot. Note that the intensities decrease with increasing center frequencies whereas at the same time, the peak widths increase. Within the margin of the fit errors, the center frequency $\Omega_1 = 13.7(1)\,\text{MHz}$ of peak 1 is just half the center frequency of peak 3 with $\Omega_3 = 29(2)\,\text{MHz}$. Hence, since both oscillation frequencies are observed on resonance with $g = 2.0046$, it becomes obvious with the theory described in Section 5.4 that pairs of conduction band-tail states can exist with different coupling strength. The three different widths of the three peaks can be associated with the inverse of the decay constants of their respective Rabi oscillation components in the time domain. Obviously, w_2^{-1} and w_3^{-1} reflect the decay constants determined from the plot in Figure 5.18 within the margin of the fit errors whereas w_1^{-1} is about half the coherence decay τ_{decay} as obtained from the plots in Figures 5.19 and 5.20. Thus, all three time constants w_1^{-1}, w_2^{-1}, and w_3^{-1} are lower than the measured coherence time. This is in agreement with the hypothesis (but not a proof) that the peak widths are determined by dephasing and not decoherence. A proof of this hypothesis can be given with additional FFTs of the Rabi oscillation after hopping echoes. These FFTs ought to contain the same components as the FFT of Rabi oscillation right after the pulse begins except those which are dissipated due to incoherence. Two FFTs of the echo dephasing obtained from the hopping free-echo decay measurements displayed in Figure 5.20 are plotted in Figure 5.21b and c for the echo times $\tau_{echo} = 400\,\text{ns}$ (corresponding to an off-time $\tau_{off} = 0$)

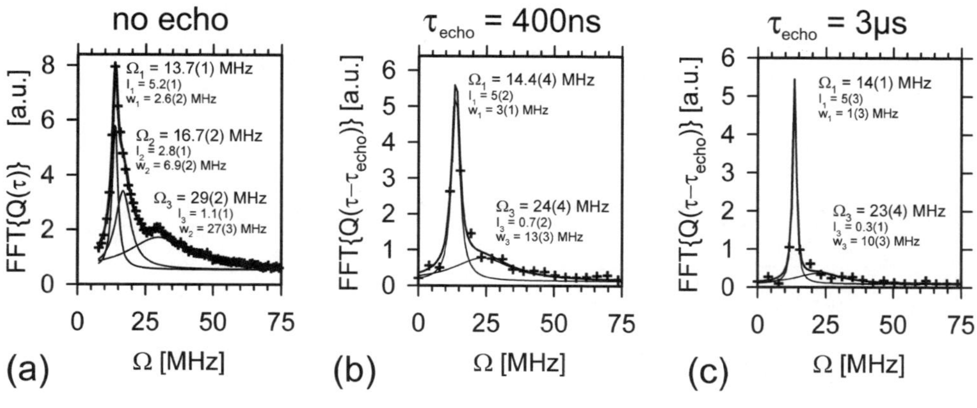

Figure 5.21 (a) Fourier transform of the Rabi oscillation data displayed in Figure 5.18 for $B_1 = 501\,\mu$T. The fit (solid line) consists of three Lorentzian functions whose parameters are displayed. Within the error margins of the fit results, the ratios of the peak center frequencies are $1:1.3:2$ and the inverse values of the peak widths correspond to the decay times determined by exponential decay functions in Figures 5.18, 5.19 and 5.20, respectively. (b) and (c) FFT of the Rabi oscillation recorded after the echo maxima obtained from the free-echo decay data displayed in Figure 5.20 for $\tau_{\text{off}} = 0$ and $\tau_{\text{off}} = 2600\,$ns, respectively

and $\tau_{\text{echo}} = 3\,\mu$s (corresponding to an off-time $\tau_{\text{off}} = 2.6\,\mu$s), respectively. Note that due to the limited number of Rabi wiggles recorded after the echoes, the frequency resolution of the FFT plots in Figure 5.21b and c is worse than in Figure 5.21a which makes the fit of narrow peaks particularly difficult and therefore the error margins larger. Nevertheless, one can deduce from the FFT plots that contributions from peak 1 and 3 can be found after 400 ns and $3\,\mu$s whereas peak 2 is no longer recognizable. This shows that the widths of peaks 1 and 3 are determined by dephasing and therefore by inhomogeneities whereas their coherence time is on a μs time scale or beyond. Since in the plots from Figure 5.21a–c the decline of peak 3 is much faster than the decline of peak 1, peak 1 can be associated with the long coherence time which causes the decay offset in the plots of Figures 5.19 and 5.20 whereas the visible decay can be associated with $\tau_{\text{decay}} = 0.7\,\mu$s. For peak 2 whose center frequency is $\Omega_2 \simeq \sqrt{2}\Omega_1$, the peak width $w_2^{-1} \simeq 350\,$ns must be determined by decoherence since its peak has already vanished after 400 ns.

5.8 CONCLUSIONS

The measurements presented above showed that spin–Rabi oscillation of electron spins localized in conduction band-tail states of a-Si:H can be observed electrically in the integrated PC transient by means of a pEDMR experiment. There are different Rabi frequency domains which we conclude to be associated with three different groups of hopping transitions that are depicted in Figure 5.22a–c. The first (in reference to peak 1 in Figure 5.21a and process (a) in Figure 5.22) can be associated with transitions between weakly coupled-pairs with $2\times\frac{1}{2}$ spin systems and therefore, the Larmor separation $\Delta\omega$ must be larger than

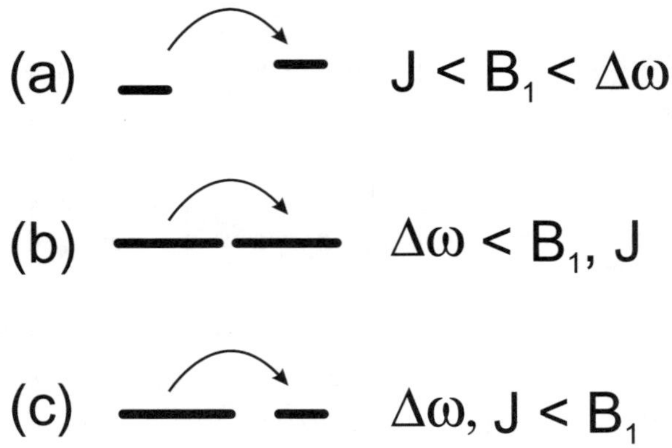

Figure 5.22 Three qualitatively different coupling regimes between conduction band-tail states in a-Si:H. (a) weak exchange and large Larmor separation; (b) strong exchange coupling; (c) weak exchange and small Larmor separation

the exchange coupling J and the microwave field B_1. From the FFT plots, we know that these transitions have little inhomogeneities and very long coherence times, and thus we conclude that they are due to transitions between band-tail states with large spatial separation, high localization, different local environments which may be due to slightly different electronic energies. The second Rabi frequency domain (with reference to peak 2 in Figure 5.21a and process (b) in Figure 5.22) is due to strong exchange coupled pairs ($J > B_1$, $\Delta\omega$) with little spatial separation, and the highest transition times (less than 400 ns) and little inhomogeneity (at least less than the transition frequencies). We conclude that this channel is representative of transitions between states with stronger delocalization and therefore larger wavefunction overlap. The third Rabi frequency domain (with reference to peak 3 in Figure 5.21a and process (c) in Figure 5.22) is due to pairs with narrow Larmor separation which can be larger or weaker than the exchange coupling, whereas both are smaller than the microwave field ($\Delta\omega$, $J < B_1$). Under the given conditions, the transition times are shorter than those of the weakly coupled pairs, but longer than the strongly coupled pairs. It is conceivable that there must be a continuous range of spatial localizations and spatial separations between states which fall into this category. We conclude that this range is reflected by the strongest inhomogeneity of peak 3.

The observed pEDMR resonances and their respective Rabi frequency components give an insight into the electronic processes of n-doped a-Si:H since the properties of the interface system used for the experiments appear to be determined by the a-Si:H bulk processes. Thus, the devices used here allow a low-temperature EDMR access to spin-dependent processes in a-Si:H which is difficult to achieve with simple a-Si:H thin films due to the low photoconductivity of a-Si:H below $T = 10\,\text{K}$—in fact most of the cwEDMR work on band-tail states on a-Si:H presented in the literature has been conducted at $T > 50\,\text{K}$. Hence, the pEDMR measurements presented here are complementary to pODMR measurements on a-Si:H films which have to be carried out at low temperatures in order to have sufficiently strong luminescence signals. There has been a report of pODMR on intrinsic a-Si:H [55]. Due to the different doping, the material used in this study in not comparable with the n-a-Si:H used here. There, different optically active electronic transitions (mostly recombination) through conduction and valence tail states and silicon dangling bonds could be observed. Similarly to the pEDMR measurements presented here, pODMR was able to resolve different coupling regimes such as the weak $2 \times s = \frac{1}{2}$ system as well as strongly diplolar and strongly exchange coupled systems. Thus, from the pEDMR measurements presented here, one can conclude that the test structures used could be appropriate for combined pEDMR/pODMR measurements which are conducted at the same time on the same sample. Such measurements, which to the knowledge of the authors have not been conducted so far on a-Si:H, could demonstrate how different electronic processes in a single sample can influence electric and optical properties in a different way at the same time, and thus experimental access to the direct observation and the separation of geminate and distant pair processes could be possible.

5.9 SUMMARY

A theoretical review of spin-dependent recombination and hopping in disordered solids has been given and based thereon, the investigation of these phenomena by means of the pulsed electrically detected magnetic resonance has been discussed. This theoretical description explains how coherent spin motion effects can be observed by simple photoconductivity measurements and how these effects have to be interpreted in order to gain information about the material system and its electronic transitions that are to be investigated.

In addition to the theoretical description, the experimental foundations of pEDMR are outlined and technical aspects are discussed to a degree which will allow the reader to reproduce a pEDMR setup and pEDMR experiments. The limitations of pEDMR are discussed with regard to sensitivity and the applicability to different material and device systems.

In order to demonstrate the potential of pEDMR for the investigation of hopping transport in disordered materials, experimental data are presented for the example of spin-dependent hopping through phosphorus donor states and conduction band-tail states of an ultrathin hydrogenated amorphous silicon layer on top of a crystalline silicon substrate. For the hopping transitions through conduction band-tail states, the electrical detection of spin–Rabi oscillation, its decay and dephasing as well as rephasing (echo) effects was demonstrated. It was shown how information about coherence times and spin coupling can be obtained from these observations.

ACKNOWLEDGEMENTS

We would like to acknowledge the work of Karsten von Maydell and Manfred Schmidt on the deposition of the amorphous to crystalline silicon samples which served as such a beautiful model system for the experimental part of this study.

We would also like to thank our colleagues Jan Behrends, Kai Petter, Robert Mueller and Peter Kanschat who have inspired and enriched our work with helpful discussions and Sergei Baranovski for the enriching and exciting collaboration on the topic.

Finally, we want to express our gratitude to Walther Fuhs from the Hahn-Meitner–Institut and Gerald Lucovsky from North Carolina State University. Both have worked as student advisors for the authors (Walther Fuhs for both, Gerald Lucovsky for C.B.) and have had a strong impact on the developments within the research fields of disordered semiconductors in general and on the work presented here in particular.

REFERENCES

[1] R.A. Street, *Hydrogenated Amorphous Silicon*, Cambridge University Press, Cambridge, UK, 1991.

[2] N.M. Atherton, *Priciples of Electron Spin Resonance*, Ellis Horwood PTR Prentice Hall, Chichester, England, 1993.

[3] A. Schweiger and G. Jeschke, *Principles of Pulse Electron Paramagnetic Resonance*, Oxford University press, Oxford, 2001.

[4] C.P. Poole, Jr., *Electron Spin Resonance*, 2nd edn, Dover Publications, Inc., Mineola, New York, 1996.

[5] S. Geschwind, R.J. Collins, and A.L. Schawlow, *Phys. Rev. Lett.*, **3**, 545 (1959).

[6] J. Brossel, S. Geschwind, and A.L. Schawlow, *Phys. Rev. Lett.*, **3**, 548 (1959).

[7] U. Steiner and T. Ulrich, *Chemical Rev.*, **89**, 51 (1989).

[8] A.J. Hoff, E.J. Lous, and R. Vreeken, in: *Pulsed EPR a New Field of Application*, C.P. Keijzers, E.J. Reijerse, and J. Schmidt (eds), North Holland, Amsterdam, 1989, Chap. 19, pp. 219–226.

[9] V. Weis, K. Möbius, and T. Prisner, *J. Magn. Res.*, **131**, 17 (1998).

[10] K.L. Purvis, S.P. Wiemelt, T. Maras, M. Blue, V. Melkonian, P.D. Ashby, S.A. Riley, L.S. Fifield, K.A. Martin, and A.M. Nishimura, *J. Lumin.*, **71**, 199 (1997).

[11] E. van Oort and M. Glasbeek, in: *Pulsed EPR a New Field of Application*, C.P. Keijzers, E.J. Reijerse, and J. Schmidt (eds), North Holland, Amsterdam, 1989, Chap. 19, pp. 227–231.

[12] R. Maxwell and A. Honig, *Phys. Rev. Lett.*, **17**, 188 (1966).

[13] D.J. Lepine and J.J. Prejean, in: *Proceedings of the 10th International Conference on the Physics of Semiconductors*, S.P. Keller, J.C. Hensel, and J. Stern (eds), US Atomic Energy Commission, USA, 1970, p. 805.

[14] D.J. Lepine, *Phys. Rev. B*, **6**, 436 (1972).

[15] M. Stutzmann, M.S. Brandt, and M.W. Bayerl, *J. Non-Cryst. Solids*, **266–269**, 1 (2000).

[16] P. Kanschat, K. Lips, and W. Fuhs, *J. Non–Cryst. Solids*, **266–269**, 524 (2000).

[17] B. Stich, S. Greulich–Weber, and J.-M. Spaeth, *J. Appl. Phys.*, **77**, 1546 (1995).

[18] H. Dersch, L. Schweitzer, and J. Stuke, *Phys. Rev. B*, **28**, 4678 (1983).

[19] I. Solomon, *J. Non–Cryst. Solid*, **35–36**, 625 (1980).

[20] T. Eickelkamp, S. Roth, and M. Mehring, *Mol. Phys.*, **95**, 967 (1998).

[21] I. Hiromitsu, Y. Kaimori, and T. Ito, *Solid State Commun.*, **104**, 511 (1997).

[22] M. Dobers, K.V. Klitzing, J. Schneider, G. Weimann, and K. Ploog, *Phys. Rev. Lett.*, **61**, 1650 (1988).

[23] C.F.O. Graeff, M.S. Brandt, M. Stutzmann, M. Holzmann, G. Abstreiter, and F. Schäffler, *Phys. Rev. B*, **59**, 13242 (1999).

[24] I. Solomon, *Solid Stat. Commun.*, **20**, 215 (1976).

[25] R. Müller, P. Kanschat, S. von Aichberger, K. Lips, and W. Fuhs, *J. Non-Cryst. Sol.*, **266–269**, 1124 (2000).

[26] K. Lips and W. Fuhs, *J. Appl. Phys.*, **74**, 3993 (1993).

[27] I. Hiromitsu, Y. Kaimori, M. Kitano, and T. Ito, *Phys. Rev. B*, **59**, 2151 (1999).

[28] Christoph Böhme, *Dynamics of Spin-dependent Charge Carrier Recombination*, Cuvillier Verlag, Göttingen, 2003.

[29] C. Boehme and K. Lips, *Phys. Rev. Lett.*, **91**, 246603 (2003).

[30] C. Boehme and K. Lips, *Appl. Phys. Lett.*, **79**, 4363 (2001).

[31] C. Boehme and K. Lips, *Phys. Rev. B*, **68**, 245105 (2003).
[32] C. Boehme and K. Lips, *Appl. Magnetic Resonance*, **27**, 109 (2004).
[33] I. Solomon, D. Biegelsen, and J.C. Knights, *Solid Stat. Commun.*, **22**, 505 (1977).
[34] D.J. Lepine, V.A. Grazhulis, and D. Kaplan, *Proc. 13th Int. Conf. on the Physics of Semiconductors*, Rome, North Holland, Amsterdam, 1976), Vol. 2440, p. 1081.
[35] V.S. L'vov, D.V. Tretyak, and I.A. Kolomiets, *Sov.-Phys.-Semicond.*, **11**, 661 (1977).
[36] G. Mendz, D.J. Miller, and D. Haneman, *Phys. Stat. Sol. B*, **20**, 5246 (1979).
[37] G. Mendz and D. Haneman, *J. Phys. C:Solid St. Phys.* **13**, 6737 (1980).
[38] D. Kaplan, I. Solomon, and N.F. Mott, *J. Phys. (Paris)—Lettres*, **39**, L51 (1978).
[39] C. Boehme and K. Lips, *Phys. Stat. Sol. C*, **1**, 1255 (2004).
[40] R. Haberkorn and W. Dietz, *Solid State Commun.*, **35**, 505 (1980).
[41] C. Boehme, P. Kanschat, and K. Lips, *Europhys. Lett.*, **56**, 716 (2001).
[42] D.D. Thornton and A. Honig, *Phys. Rev. Lett.*, **30**, 909 (1973).
[43] K. Morigaki and M. Onda, *J. Phys. Soc. Japan*, **33**, 1031 (1972).
[44] K. Morigaki and M. Onda, *J. Phys. Soc. Japan*, **36**, 1049 (1974).
[45] N. Kishimoto and K. Morigaki, *J. Phys. Soc. Japan*, **42**, 137 (1977).
[46] N. Kishimoto, K. Morigaki, and K. Murakami, *J. Phys. Soc. Japan*, **50**, 1970 (1981).
[47] H. Dersch and L. Schweitzer, *Phil. Mag. B*, **50**, 397 (1984).
[48] G. Kawachi, C.F.O. Graeff, M.S. Brandt, and M. Stutzmann, *Phys. Rev. B*, **54**, 7957 (1996).
[49] T. Sommer, R.C. Barklie, A. Davey, and W. Blau, *Synthetic Metals*, **76**, 259 (1996).
[50] B. Movaghar, B. Ries, and L. Schweitzer, *Phil. Mag. B*, **41**, 159 (1980).
[51] M.S. Brandt and M. Stutzmann, *Phys. Rev. B*, **43**, 5184 (1991).
[52] H. Mell, B. Movaghar, and L. Schweitzer, *Phys. Stat. Sol.*, **88**, 531 (1978).
[53] K.A. Matveev, L.I. Glazman, Penny Clarke, D. Ephron, and M.R. Beasley, *Phys. Rev. B*, **52**, 5289 (1995).
[54] H. Kamimura, T. Takemori, A. Kurobe, D. Ephron, and M.R. Beasley, in: *Springer Series in Solid State Science*, Y. Nagaoka and H. Fukuyama (eds), Springer, Berlin, 1982, Vol. 39, p. 156.
[55] K. Lips, C. Boehme, and T. Ehara, *J. of Optoelectronics and Advanced Materials*, **7**, 13 (2004).
[56] A.V. Astashkin and A. Schweiger, *Chem. Phy. Lett.*, **174**, 595 (1990).
[57] F. Friedrich, C. Boehme, and K. Lips, *J. Appl. Phys.*, **97**, 056101 (2005).
[58] W.R. Stangl, A. Froitzheim, *Proc. of the 14th European PV Conf.*, Rome, Italy, 2002, pp. 123–126.
[59] M.S. Brandt and M. Stutzmann, *Appl. Phys. Lett.*, **58**, 1620 (1991).
[60] K. Lips, S. Schütte, and W. Fuhs, *Phil. Mag. B*, **65**, 945 (1992).

6 Description of Charge Transport in Disordered Organic Materials

Sergei Baranovski and Oleg Rubel

Faculty of Physics and Material Sciences Center, Philipps Universität Marburg, Germany

Charge Transport in Disordered Solids with Applications in Electronics Edited by S. Baranovski.

6.1 INTRODUCTION

In this chapter we focus on the theoretical description of charge transport in disordered organic materials. This topic is enormously broad since the variety of organic materials with different conducting properties is very rich. In some organic substances superconductivity has been found with transition temperatures as high as 10 K [1]. Some polymeric conductors based on polyacetylene evidence the values of the electrical conductivities which are comparable to those of the best conducting metals. For example, in iodine-doped polyacetylene prepared by the Naarmann method a value of the conductivity above $10^5\,\Omega^{-1}\,\text{cm}^{-1}$ was obtained [2]. Many organic materials such as polyethylene with conductivities below $10^{-8}\,\Omega^{-1}\,\text{cm}^{-1}$ can be considered as good insulators. It is not possible, of course, to cover the properties of all these materials in one chapter. In [3] some device applications of organic materials are described. Until now electrophotographic image recording has been the main technique that exploited the electrical conducting properties of organic solids on a broad industrial scale [4]. The main efforts of researchers are focused presently on the development of organic materials for applications in light-emitting diodes (LEDs) as well as in photovoltaics. Materials used for these purposes are mostly random organic, notably conjugated, or/and molecularly doped polymers with semiconducting properties [4, 5]. Therefore we restrict our description in this chapter to considering electrical conduction in such organic semiconducting materials.

Several handbooks have been published recently, which are devoted to the description of physical properties and device applications of organic materials. For example, numerous details of the chemistry, physics and engineering of semiconducting polymers can be found in recent books [5, 6]. To the interested reader we can also recommend the comprehensive monograph of Pope and Swenberg [7] where various electronic processes in organic crystals and polymers are beautifully described. We will discuss below the charge carrier transport problems raised in Chapter XIV of that monograph.

Discussing charge transport properties of organic materials, one should clearly distinguish between the properties of ordered systems such as molecular crystals on one hand and those of essentially disordered systems like molecular doped polymers on the other. Transport models used to describe electrical conduction in these two distinct classes of materials are essentially different.

In order to describe good conducting properties of such organic materials as molecular crystals, one usually employs rather standard methods of solid-state physics developed for charge transport in crystalline inorganic solids. For applications to organic materials one slightly modifies the standard theory by taking into account the strong electron–phonon interaction leading to polaron effects and nonlinear excitations such as solitons [1, 7]. A completely different set of ideas is exploited in order to describe charge carrier transport in essentially disordered organic materials, such as molecularly doped polymers [8], low-molecular-weight glasses [9, 10] and conjugated polymers [11]. In these materials charge transport is assumed to be due to the variable-range hopping (VRH) of electrons or holes via randomly distributed in space and energy localized states [1, 5, 7, 8, 12, 13]. It is this transport mode that is described in this chapter. The question may arise as to why one needs to describe this transport mode once more if the VRH in application to disordered organic semiconductors has already been described in numerous review articles [1, 8, 12, 13] and also in recent book chapters (for instance, Chapter XIV in [7] and the chapter of H. Bässler in [4]). We do it because we find the treatment of the VRH in application to organic disordered materials given in the above literature to be incomplete. Some examples of the incompleteness are given below.

(i) It has already become a tradition to claim that the Gaussian shape of the energy distribution of localized states (DOS) assumed for random organic systems prevents analytical solution of the hopping transport problem [7, 8, 12, 13]. We show, to the contrary, that such a solution can be easily obtained within the standard set of ideas for the VRH introduced in [14].

(ii) It has already become a tradition to claim that the best theoretical approach to describe the VRH in random organic systems is based on Monte Carlo computer simulations [1, 4, 5, 7, 8, 12, 13]. We show below that although computer simulations often provide valuable information on transport properties, the simulation results should be taken with caution. For instance, some dependences of transport coefficients on material parameters have been considered as universal simply because other material parameters were not changed in the course of the simulations. Furthermore finite size effects in computer simulations have not always been treated appropriately.

(iii) In most review articles and book chapters published so far the effect of carrier mobility decreasing with increasing electric field was specified as one of the most challenging for theoretical explanation [4, 5, 7, 8, 12, 13]. We will show below, for exactly solvable models [15, 16] that this effect might be an artifact that has been caused by misinterpretation of experimental results.

(iv) There is no consensus between researchers on the very basic question on whether the conventional Einstein relation between the charge carrier mobility and diffusivity is valid for hopping transport in disordered systems. We clarify this problem below.

(v) Recently many contradictory papers have been published on the theoretical treatment of hoping transport in a random system with Gaussian DOS. In all disordered organic and inorganic materials a very strong nonlinear dependence of the carrier mobility μ on the concentration of localized states N is observed experimentally in the hopping regime. This strong dependence is caused by the very strong exponential dependence of the transition rates on the distances between localized states. While for inorganic materials the dependence $\mu(N)$ has been well described theoretically within the VRH approach, and has already become a subject of textbooks [17], the very same dependence $\mu(N)$ still looks puzzling for many theoreticians working with organic materials. Sometimes it is claimed that the dependence $\mu(N)$ in the hopping regime should be linear [18–21], in marked contradiction to experimental data. Sometimes a dependence of the form $\ln(\mu) \propto -\gamma(N\alpha^3)^{-1/3}$ has been suggested [22, 23], where α is the localization length and γ is a numerical coefficient. The spread of values for γ in the literature is enormous. For example, in [22] γ was estimated as $1.056 < \gamma < 1.076$ while in [23] it was claimed to be in the range $1.54 < \gamma < 1.59$, depending on temperature. This difference in γ leads to the difference in μ values at low concentrations $(N\alpha^3)^{1/3} < 0.02$ of more than 10 orders of magnitude! Since there is no cross-citation between [22] and [23], published in the same year, it is difficult to guess which of these contradictory results the authors consider as correct. We show below that neither of these results seem correct. Furthermore, while in some recent publications it is correctly claimed that averaging of hopping rates leads to the omission of the concentration dependence of the carrier mobility $\mu(N)$ [23] in other publications by the same authors it is claimed that averaging of hopping rates is capable of describing the concentration dependence of the mobility [24, 25]. Remarkably, what the authors call averaging of hoping rates in [24, 25] is not what they call averaging of hopping rates in [18–21, 23]. Even more remarkable is the statement of a recent publication [26] that the results of [18–21], claiming the linear dependence $\mu(N)$, are correct. There are other recent papers that lead to this conclusion [27, 28]. The belief of the researchers working with disordered organic semiconductors in the averaging of hopping rates is so strong that sometimes the agreement between experimental results and the correct formulas with exponential dependence $\mu(N)$ are called occasional, because the latter dependence cannot be obtained by averaging of hopping rates [29].

This unsatisfactory situation with theoretical description of hopping conductivity in organic disordered solids shows that many researchers working with hopping transport in organic semiconductors are not familiar with the basic fundamental ideas, well approved and known in the parallel field of hopping transport in inorganic disordered systems. Our aim in this chapter is to provide the description of these elementary ideas in application to random organic systems. For example, using the routine VRH approach [30, 31] we calculate the dependence $\mu(N)$ in organic disordered solids. It appears to be in agreement with experimental data. Only in the limit of very dilute systems is the dependence $\ln(\mu) \propto -\gamma(N\alpha^3)^{-1/3}$ valid, but with $\gamma \simeq 1.73$, in agreement with the classical percolation result [17] and at variance with the results of [18–21, 23–25, 27, 28].

In the next section we briefly describe the model for charge carrier transport in disordered organic semiconductors formulated and justified in previous reviews [1, 5, 7, 8, 12, 13]. In subsequent sections we describe theoretical methods which provide transparent solutions of various transport problems in the framework of this model.

6.2 CHARACTERISTIC EXPERIMENTAL OBSERVATIONS AND THE MODEL FOR CHARGE CARRIER TRANSPORT IN RANDOM ORGANIC SEMICONDUCTORS

Although the variety of disordered organic solids is very rich, the details of charge transport in most of such materials are common. The canonical examples of disordered organic materials with the hopping transport mechanism are the binary systems that consist of doped polymeric matrixes. Examples include polyvinylcarbazole (PVK) or bis-polycarbonate (Lexan) doped with either strong electron acceptors such as, for example, trinitrofluorenone acting as an electron transporting agent, or strong electron donors such as, for example, derivatives of triphenylamine or triphenylmethane for hole transport. To avoid the necessity of specifying each time whether charge transport is carried by electrons or holes, in the general discussion below we use the notation 'charge carrier'. The results are valid for each type of carriers—electrons or holes. Charge carriers in random organic materials are believed to be highly localized. Localization centers are molecules or molecular subunits, henceforth called sites.

In order to understand the charge transport mechanism in random organic solids, let us briefly recall some experimental observations that are decisive for formulating the transport model. Among these are the dependences of the carrier drift mobility, diffusivity and conductivity on temperature T, on the strength of applied electric field F, and on such important material parameter as the spatial concentration of localized states N.

(i) *Dependence of transport coefficients on the concentration of localized states.* Already in the early stages of the study of molecularly doped polymers it was established that the dependence of the carrier kinetic coefficients on the concentration of localized states is very strong and that it is essentially nonlinear [32]. In Figure 6.1 the dependence of the logarithm of the carrier drift mobility, $\ln(\mu)$, on $R = N^{-1/3}$ obtained in time-of-flight experiments is shown. This extremely nonlinear dependence at low concentrations can be fitted by the expression

$$\mu \propto \exp\left(-\frac{2N^{-1/3}}{\alpha}\right), \tag{6.1}$$

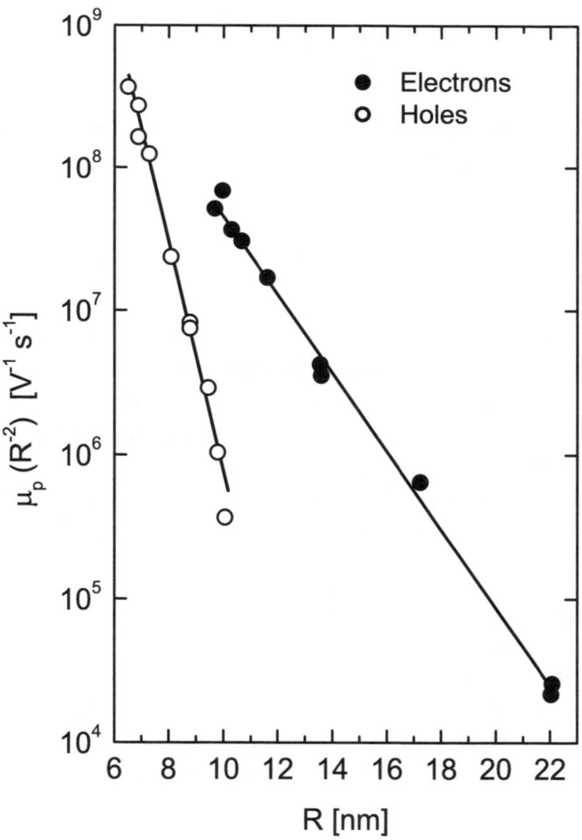

Figure 6.1 Electron and hole mobility data plotted as a function of the average separation between uncomplexed molecules treated as localized states (reproduced with permission from [32]. Copyright 1972 by the American Institute of Physics)

which Gill [32] suggested in this form, assuming that α is the localization length of charge carriers in the localized states. As shown in [14], such dependence of transport coefficients on the concentration of localized states N is characteristic for incoherent hopping transport mechanism in which the charge carrier hops between the nearest available localized states. Indeed, such a transport mode is expected for the case of low concentrations N, at which the main limiting factor for carrier transitions between localized states is provided by the exponential distance dependence of hopping probabilities (see [14]). At high concentrations, the deviations from the dependence described by Equation (6.1) were observed experimentally [33], as expected for hopping transport in the VRH mode that succeeds the nearest-neighbor transport mode with rising N. This result should be considered as the key observation leading to the conclusion that transport mechanism in disordered organic solids is the incoherent hopping of charge carriers between spatially localized states.

(ii) *Dependence of transport coefficients on temperature.* The temperature dependence of the drift mobility obtained in time-of-flight experiments in random organic solids usually takes the form

$$\mu \propto \exp\left(-\left(\frac{T_0}{T}\right)^2\right), \tag{6.2}$$

where T_0 is a parameter [1, 4, 5, 7, 8, 12, 13]. However, in several studies an Arrhenius dependence has also been reported

$$\mu \propto \exp\left(-\frac{\Delta}{kT}\right), \tag{6.3}$$

where Δ is the activation energy. Usually the latter dependence in organic materials is observed by measurements of the electrical conductivity in field-effect transistors (see, for instance, [34]). We will show below that both temperature dependences described by Equations (6.2) and (6.3) are predicted theoretically in the frame of the same transport model, depending on the total concentration of charge carriers in the system. At low carrier densities the dependence described by Equation (6.2) is predicted, while at high carrier densities the dependence described by Equation (6.3) should be valid.

(iii) *Dependence of transport coefficients on the strength of the applied electric field.* Numerous experimental studies have shown that at high electric fields the charge carrier transport coefficients increase with the field strength F, being approximately proportional to exp $[(F/F_0)^{1/2}]$ in a broad range of field strengths, with some parameter F_0 [5, 7, 8, 12, 13]. More curiously, it has been reported that at high temperatures the carrier drift mobility measured in time-of-flight experiments increases with decreasing field at rather low field strengthes [5, 7, 8, 12, 13].

When formulating the appropriate transport mechanism for charge carriers in random organic materials such as molecular glasses and molecular doped polymers, one should keep in mind these experimental observations.

The corresponding model established to treat the charge carrier transport in random organic solids is as follows [1, 5, 7, 8, 12, 13]. It is assumed that charge carriers move via incoherent hopping transitions between localized states randomly distributed in space with some concentration N. All states are presumed to be localized. The energies of charge carriers on these states are assumed to have a Gaussian distribution so that the density of states (DOS) takes the form

$$g(\varepsilon) = \frac{N}{\sigma\sqrt{2\pi}} \exp\left(-\frac{\varepsilon^2}{2\sigma^2}\right), \tag{6.4}$$

where σ is the energy scale of the distribution and the energy ε is measured relative to the center of the DOS. The origin of the energetic disorder is the fluctuation in the lattice polarization energies and the distribution of segment length in the π- or σ-bonded main-chain polymer [8]. The Gaussian shape of the DOS was assumed on the basis of the Gaussian profile of the excitonic absorption band and by recognition that the polarization energy is determined by a large number of internal coordinates, each varying randomly by small amounts [8]. Later we will see that the Gaussian shape of the DOS accounts for the observed temperature dependence of the kinetic coefficients such as carrier drift mobility, diffusivity, and conductivity. The energy scale σ of the DOS in most random organic materials is of the order of ~0.1 eV [8]. In the initial model, no correlations between spatial positions of localized states and their energies were included [8]. For the sake of simplicity we will assume below that this assumption is valid unless the contrary is specified.

A tunneling transition rate of a charge carrier from a localized state i to a lower in energy localized state j depends on the spatial separation r_{ij} between the sites i and j as $\nu(r_{ij}) = \nu_0 \exp(-2r_{ij}/\alpha)$, where α is the localization length which we assume equal for sites i and j. This length determines the exponential decay of the carrier wavefunction in the localized states. The decay length on single sites has been evaluated in numerous studies of the concentration-dependent drift mobility. For example, for trinitrofluorenone in PVK the estimates $\alpha \simeq 1.1 \times 10^{-8}$ cm and $\alpha \simeq 1.8 \times 10^{-8}$ cm were obtained for holes and electrons, respectively [32]. For dispersions of N-isopropylcarbazole in polycarbonate the estimate $\alpha \simeq 0.62 \times 10^{-8}$ cm for holes has been obtained [35]. The preexponential factor ν_0 depends on the electron interaction mechanism that causes the transition. Usually it is assumed that carrier transitions contributing to charge transport in disordered materials are caused by interactions with phonons. Often the coefficient ν_0 is simply assumed to be of the order of the phonon frequency ~$10^{13}\,\mathrm{s}^{-1}$, although a more rigorous consideration is in fact necessary to determine ν_0. Such a consideration should take into account the particular structure of the electron localized states and also the details of the interaction mechanism [36, 37].

When a charge carrier performs a transition upward in energy from a localized state i to a higher in energy localized state j, the transition rate also depends on the energy difference between the states. This difference should be compensated, for example, by absorption of a phonon with the corresponding energy [38]. Generally, the transition rate from the occupied site i to an empty site j can be expressed as

$$\nu(r_{ij}, \varepsilon_i, \varepsilon_j) = \nu_0 \exp\left(-\frac{2r_{ij}}{\alpha}\right) \exp\left(-\frac{\varepsilon_j - \varepsilon_i + |\varepsilon_j - \varepsilon_i|}{2kT}\right). \tag{6.5}$$

If the system is in thermal equilibrium, the occupation probabilities of sites with different energies are determined by the Fermi statistics. This effect can be taken into account by modifying Equation (6.5) and adding terms which account for the relative energy positions of sites i and j with respect to the Fermi energy ε_F. Taking into account these occupation probabilities one should write the transition rate between sites i and j in the form [38]

$$\nu_{ij} = \nu_0 \exp\left(-\frac{2r_{ij}}{\alpha}\right)\exp\left(-\frac{|\varepsilon_i - \varepsilon_F| + |\varepsilon_j - \varepsilon_F| + |\varepsilon_j - \varepsilon_i|}{2kT}\right). \tag{6.6}$$

With the help of these formulas the problem of the theoretical description of hopping conduction can be easily formulated. One has to calculate the conductivity which is provided by transition events with the rates described by Equations (6.5) or (6.6) in the manifold of localized states with the DOS described by Equation (6.4).

Unfortunately, it has become a common belief that the Gaussian form of the DOS prevents closed-form analytical solutions of the hopping transport problems in random organic systems and therefore the best way to study these problems is a computer simulation [7, 8, 12, 13]. We will do our best in this report to prove the opposite. However, before doing so we would like to present several important results obtained previously by computer simulations.

6.3 ENERGY RELAXATION OF CHARGE CARRIERS IN A GAUSSIAN DOS. TRANSITION FROM DISPERSIVE TO NONDISPERSIVE TRANSPORT

One of the most remarkable results known for energy relaxation of charge carriers in a Gaussian DOS is the existence of the so-called equilibration energy $\langle \varepsilon_\infty \rangle$ [7, 8, 12, 13]. This situation is in contrast to the case of the exponential DOS, where in the empty system at low temperatures the charge carrier always relaxes downward in energy, as discussed in [14]. The energy relaxation in the exponential DOS leads to the dispersive character of the charge transport, in which kinetic coefficients are time dependent and the carrier mobility slows down in the course of the energy relaxation, as described in [14]. In the Gaussian DOS the charge carrier on average relaxes from high-energy states downward in energy only until it arrives at the equilibration energy $\langle \varepsilon_\infty \rangle$, even in the empty system without any interactions between the relaxing carriers [7, 8, 12, 13]. In computer simulations [8, 12, 13], noninteracting carriers were initially distributed uniformly in energy and their further relaxation via hopping processes with the rates described by Equation (6.5) was traced. The temporal evolution of a packet of noninteracting carriers relaxing within a Gaussian DOS is schematically shown in Figure 6.2. Initially the energy distribution of carriers coincides with that of the DOS. In the course of time the carrier energy distribution moves downward until its maximum $\langle \varepsilon(t) \rangle$ arrives at the energy $\langle \varepsilon_\infty \rangle = \lim_{t\to\infty} \langle \varepsilon(t) \rangle = -\sigma^2/kT$ [8, 12, 13]. This result can be easily obtained analytically [39]. In thermal equilibrium

$$\langle \varepsilon_\infty \rangle = \frac{\int_{-\infty}^{\infty} d\varepsilon\, \varepsilon g(\varepsilon) \exp(-\varepsilon/kT)}{\int_{-\infty}^{\infty} d\varepsilon\, \varepsilon g(\varepsilon) \exp(-\varepsilon/kT)} = -\frac{\sigma^2}{kT}. \tag{6.7}$$

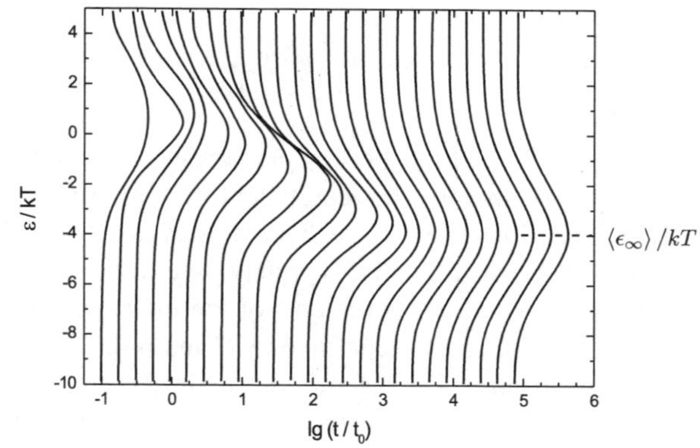

Figure 6.2 Temporal evolution of the distribution of carrier energies in a Gaussian DOS ($\sigma/kT = 2.0$). $\langle\varepsilon_\infty\rangle$ denotes the theoretical equilibrium energy determined by Equation (6.7) (reproduced with permission from [8]. Copyright 1993, Wiley-VCH)

The time required to reach this equilibrium distribution (called the relaxation time) τ_{rel} is of key importance for the analysis of experimental results [40]. Indeed at time scales shorter than τ_{rel}, charge carriers initially randomly distributed over localized states perform a downward energy relaxation during which transport coefficients, such as the carrier drift mobility, essentially depend on time, and charge transport is dispersive, as described in [14]. At time scales longer than τ_{rel}, the energy distribution of charge carriers stabilizes around the equilibration energy $\langle\varepsilon_\infty\rangle$, even in a very dilute system with noninteracting carriers. In such a regime, transport coefficients are time independent. In other words, at $t \simeq \tau_{rel}$ dispersive transport is succeeded by the nondispersive (Gaussian) transport behavior. This is one of the most important results for charge carrier transport in disordered organic media [8]. While at short times dispersive current transients were observed in such materials, at long times transport characteristics are nondispersive, time independent and hence they can be well characterized and described in contrast to analogous quantities in inorganic disordered materials with exponential DOS. In the latter materials, as shown in [8], transport coefficients in dilute systems are always dispersive (time dependent). Hence, they depend on such experimental conditions as the length of a sample and the strength of the electric field. Therefore the transport coefficients in disordered materials with exponential DOS can hardly be characterized. On the contrary, in organic disordered materials with Gaussian DOS, transport coefficients do not depend on the experimental conditions at $t > \tau_{rel}$ and hence they can be well characterized. It has been established by computer simulations that τ_{rel} strongly depends on temperature [8, 40],

$$\tau_{rel} \propto \exp\left[\left(\frac{B\sigma}{kT}\right)^2\right], \tag{6.8}$$

where numerical coefficient B is close to unity: $B \simeq 1.1$. It has also been found by computer simulations that the carrier drift mobility μ at long times corresponding to the nondispersive transport regime has the following peculiar temperature dependence [7, 8, 12, 13, 40]

$$\mu \propto \exp\left[-\left(\frac{C\sigma}{kT}\right)^2\right]. \tag{6.9}$$

For coefficient C the value $C = 2/3$ has become a conventional one [7, 8, 12, 13, 40]. In fact, computer simulations [41] give for this coefficient the value $C \simeq 0.69$, while rather sophisticated analytical calculations [39, 42] predict a close value $C \simeq 0.64$. Equation (6.9) with $C = 2/3$ is believed to be universal, and it is widely used to determine the energy scale σ of the DOS from the experimental measurements of the $\ln(\mu)$ versus T^{-2} [43, 44].

However, it looks rather strange that the coefficients C and B in Equations (6.8) and (6.9) are considered as the universal ones. In fact, it is the main feature of the VRH transport mode that the spatial and energy parameters are interconnected in the final expressions for transport coefficients [17]. Equations (6.5) and (6.6) show that both the energy difference between localized states and the spatial distance between them determine the hopping probability. Hopping transport in a system with spatially and energetically distributed localized states is essentially a VRH process as described in [14]. In such a process, the transport path used by charge carriers is determined by both energy and spatial variables in transition probabilities. It means that the temperature dependence of transport coefficients should include the spatial parameter $N\alpha^3$, while the dependence of transport coefficients on the concentration N of localized states should depend on the temperature normalized by the energy scale of the DOS, kT/σ. In computer simulations [41], a particular value of the spatial parameter was taken: $N\alpha^3 = 0.001$ and the magnitude $C \simeq 0.69$ was obtained for this particular value of $N\alpha^3$. The question then arises on whether this value of C is stable against variations of $N\alpha^3$.

There is another important question, already raised in the scientific literature [40], that is related to the apparent difference in the temperature dependences of τ_{rel} and μ expressed by the difference between coefficients C and B in Equations (6.8) and (6.9). The relaxation time τ_{rel} at which a transition from dispersive to nondispersive transport should take place depends on the ratio σ/kT more strongly than the carrier drift mobility μ. This has an important consequence that a time-of-flight signal produced by a packet of charge carriers drifting across a sample of some given length must become dispersive above a certain degree of disorder, i.e., below a certain temperature at otherwise constant system parameters, because eventually τ_{rel} will exceed the carrier transient time [40]. The time evolution of the time-of-flight signal is shown in Figure 6.3 as a function of the ratio σ/kT. The early analytical theories [39, 42] already indicated the difference between coefficients C and B, although they predicted $C > B$ in contrast to the simulation results. Therefore a better analytical theory for description of the VRH transport in a system of random sites with a Gaussian distribution is definitely desirable.

6.4 THEORETICAL TREATMENT OF CHARGE CARRIER TRANSPORT IN RANDOM ORGANIC SEMICONDUCTORS

6.4.1 Averaging of hopping rates

The attempts to develop a more transparent and usable theory than that in [39, 42] were performed by Arkhipov and Bässler [18–21]. The result obtained for the dependences of

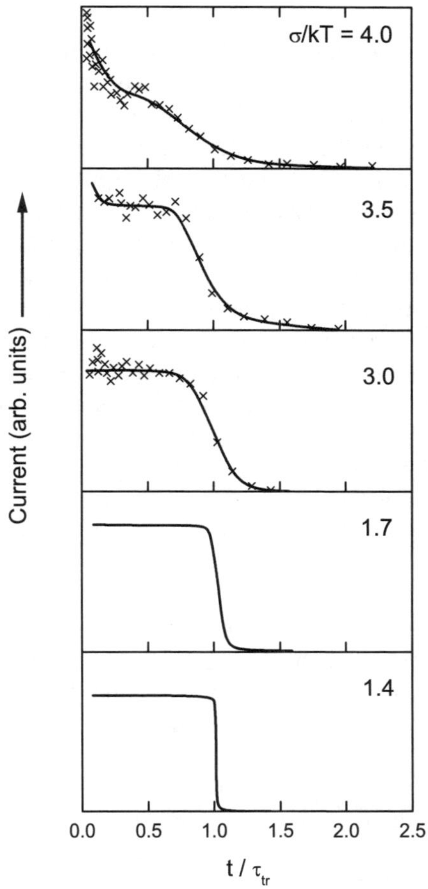

Figure 6.3 Time-of-flight signal, parametric in σ/kT (sample length 8000 lattice planes, $F = 6 \times 10^5\,\mathrm{V\,cm^{-1}}$) (reproduced with permission from [8]. Copyright 1993, Wiley-VCH)

the carrier mobility on temperature, T, and on the concentration of localized states, N, reads [18–21]

$$\mu = \frac{\pi^{1/2} e v_0 N \alpha^5}{2\sigma} \exp\left[-\frac{1}{4}\left(\frac{\sigma}{kT}\right)^2\right]. \tag{6.10}$$

A similar expression is provided by the model of Roichman and co-workers [27, 28]. This result gives for the temperature dependence a numerical coefficient 1/4 in front of $(\sigma/kT)^2$ in the exponent, which is twice as small as the one ($\simeq 1/2$) predicted by computer simulations [8, 40, 41] and by previous analytical theories [39, 42]. More important, this expression predicts a linear dependence of the carrier mobility μ on N in contradiction to experimental results [32].

The reason for these contradictions is rather transparent. Equation (6.10) is obtained by the configurational averaging of hopping rates [18–21]. Although this method has been already analyzed in textbooks (see, for instance, [17]) and qualified as inappropriate for treatment of hopping transport in disordered materials, it is repeatedly used by numerous researchers. This occurs not only in the studies of charge transport, but also in the studies of the hopping carrier energy relaxation. For example, it is often claimed in the literature that exponential dependence of the local transition rates on the distances between localized states is 'averaged out' in the final results for the relaxation rate leading to the linear dependence of the relaxation rate on the concentration of states N (see, for instance [45]). However, it had been already emphasized by Ambegaokar *et al.* in 1971 [46] at a very early stage of the study of hopping transport that an essential ingredient of a successful theory for hopping transport in disordered materials is that it takes into account that hopping is not determined by the rate of 'average' hops, but rather by the rate of those hops that are 'most difficult but still relevant': hopping conduction is in fact a percolation problem. Unfortunately, the belief of the researchers working with disordered organic semiconductors in the averaging of hopping rates is so strong that sometimes the agreement between experimental results on the exponential dependence $\mu(N)$ with Equation (6.1) are called occasional, because the latter equation cannot be obtained by averaging of hopping rates [29]. We therefore find it necessary to clarify once more the invalidity of the ensemble averaging of hopping rates for the description of hopping transport in disordered materials, although this invalidity has been already explained in textbooks (see, for instance, [17]).

In the approach based on the averaging of hopping rates, one assumes that carrier mobility is proportional to the average hopping rate $\langle \nu \rangle$ multiplied by the squared typical displacement r^2 of a charge carrier in single hopping events: $\mu \propto r^2\langle \nu \rangle$. The shortcoming of such a treatment is mostly transparent in the case of high temperatures, $kT > \sigma$, at which charge transport takes place via nearest-neighbor hopping. The latter is true under the condition $N\alpha^3 << 1$ necessary for electron states to be localized, as assumed in the model [7, 8, 12, 13, 40]. At high T, the energy-dependent terms in transition probabilities do not play an essential role and the hopping rates are determined mostly by spatial separations between localized states: $\nu(r) \simeq \nu_0 \exp(-2r/\alpha)$. Multiplying this transition probability by the probability density to find the nearest neighbor at a given distance r, provided the total concentration of sites is N: $\varphi(r)\mathrm{d}r = 4\pi r^2 N \exp(-4\pi r^3 N/3)\mathrm{d}r$ and integrating over distances, one obtains for the average hopping rate:

$$\langle \nu \rangle = \int_0^\infty \mathrm{d}r\, \nu_0 \exp\left(-\frac{2r}{\alpha}\right) 4\pi r^2 N \exp\left(-\frac{4\pi}{3} r^3 N\right) \simeq \pi \nu_0 N \alpha^3. \tag{6.11}$$

It is easy to understand this result. Due to the very strong decrease of the function $\nu(r) = \nu_0 \exp(-2r/\alpha)$ with increasing r at the scale $r \simeq \alpha$ and due to the weak dependence of the function $\varphi(r) = 4\pi r^2 N \exp(-4\pi r^3 N/3)$ on r at $r \simeq \alpha$, the average hopping rate in Equation (6.11) is determined by transitions with $r \simeq \alpha$, since the main contribution to the integral comes from such distances $r \simeq \alpha$. Assuming $\mu \propto r^2\langle \nu \rangle$ with $r \simeq \alpha$ and $\langle \nu \rangle$ described by Equation (6.11) one obtains [18–21, 27, 28] $\mu \propto \nu_0 N\alpha^5$, as expressed in Equation (6.10). This result is however invalid for a dilute system of localized states, for which the condition $N\alpha^3 << 1$ should be fulfilled to justify the localization. The average hopping rate is determined by transitions over the distances $r \simeq \alpha$ with the rate $\nu \simeq \nu_0$. The probability of finding such a close pair of sites with $r \simeq \alpha$ at the total concentration N of sites is proportional to $N\alpha^3$. This is the reason why the result for the average hopping rate $\langle \nu \rangle$ is linearly proportional to N. However, the charge carrier cannot move over considerable distances using only transitions with the length $r \simeq \alpha$ in a system with low concentration of sites

$N\alpha^3 << 1$. Therefore, the averaging of hopping rates cannot describe the charge carrier kinetic coefficients in a random organic materials and hence other theoretical methods should be used.

6.4.2 Percolation approach

One of the most powerful theoretical tools to describe charge carrier transport in disordered systems is provided by the percolation theory as described in numerous handbooks (see, for instance, [17]). We briefly described this method in Chapter 2 [14]. According to the percolation theory, one has to connect sites with fastest transition rates in order to fulfill the condition that the average number Z of bonds per site on the transport path is equal to the so-called percolation threshold [17, 47, 48]:

$$Z \equiv B_c = 2.7 \pm 0.1. \tag{6.12}$$

This method has been very successfully applied to the theoretical description of hopping transport in doped crystalline semiconductors [17] and also in disordered materials with exponential DOS [49]. The treatment of charge transport in disordered systems with a Gaussian DOS in the framework of the percolation theory can be found in [50, 51]. However, this theory is not easy for calculations. Therefore it is desirable to have a more transparent theoretical description of transport phenomena in disordered systems with a Gaussian DOS. In the next section we present such an approach based on the well-approved concept of the transport energy, that was successfully applied earlier to describe transport phenomena in inorganic disordered systems with exponential DOS.

6.4.3 Transport energy for a Gaussian DOS

The routine and so far widely accepted interpretation of the temperature dependence of the carrier mobility described by Equation (6.9) claims that activation of carriers from the energy level $\langle\varepsilon_\infty\rangle$ determined by Equation (6.7) to some transport level is responsible for the drift mobility μ [4, 8, 12, 13, 40]. Furthermore, it has been claimed [52] that Equation (6.9) along with (6.7) would ascertain the transport level to be at $\langle\varepsilon_\infty\rangle + (C\sigma)^2/kT \approx -5\sigma^2/(9kT)$. However, in such interpretation one treats the hopping transport as a pure activation of charge carriers from the level $\langle\varepsilon_\infty\rangle$ over the activation barrier with the hight $(C\sigma)^2/kT$ necessary to obtain Equation (6.9). The transport mechanism, in which transition probabilities are limited solely by the energy activation, might be valid in ordered crystalline semiconductors. Transport in random disordered materials is, however, a hopping process limited not only by the energy activation, but also by the necessity for charge carriers to tunnel in space between the localized states. It is the interplay between spatial and energy-dependent terms in the transition probabilities that determines the hopping transport process in the VRH regime. Remarkably it is sometimes claimed [52] that in disordered systems with a Gaussian DOS there is no transport energy in the VRH sense. We disagree with such statements and show below that the transport energy for the Gaussian DOS can be calculated in full analogy with the corresponding derivation for the exponential DOS described in [14], where it was shown, following [53–57] that a particular energy level ε_t, called by Monroe [53] the transport energy (TE), determines all hopping transport phenomena.

The concept of transport energy can be easily extended to describe hopping transport phenomena in a disordered system with the Gaussian DOS [58]. Since the derivation of the TE for the Gaussian DOS is absolutely analogous to that for the exponential DOS described in [14], we present here this concept in a compacted form. The introduction of the TE makes sense only at low temperatures, $kT < \sigma$, in the VRH regime, since otherwise the trivial nearest-neighbor approximation is valid (see [14]). Qualitatively, the energy relaxation of carriers in a Gaussian DOS is schematically shown in Figure 6.4. At low temperatures, $kT < \sigma$, carriers placed in the high-energy part of the Gaussian DOS perform downward energy transitions until they reach a particular energy level ε_t called the transport energy. At ε_t the character of the relaxation changes. After a hop to a state below ε_t, the carrier hops upward in energy to a state in the vicinity of ε_t. This hopping process near and below ε_t resembles a multiple-trapping process where ε_t plays the role of the mobility edge. In order to calculate ε_t, one should find the maximum of the hopping transition probability with respect to the energy of the final state ε_x for a carrier placed initially at a site with deep energy ε_i:

$$\frac{\mathrm{d}\nu_{\uparrow}(\varepsilon_i, \varepsilon_x)}{\mathrm{d}\varepsilon_x} = 0. \tag{6.13}$$

It is easy to show [58] that the final energy ε_t determined by Equation (6.13) does not depend on the initial energy ε_i and hence it is universal for given parameters σ/kT and $N\alpha^3$. The TE in a random system with the Gaussian DOS is determined as

$$\varepsilon_t = \sigma X\left(N\alpha^3, kT/\sigma\right), \tag{6.14}$$

where $X(N\alpha^3; kT/\sigma)$ is the solution of the equation [58]

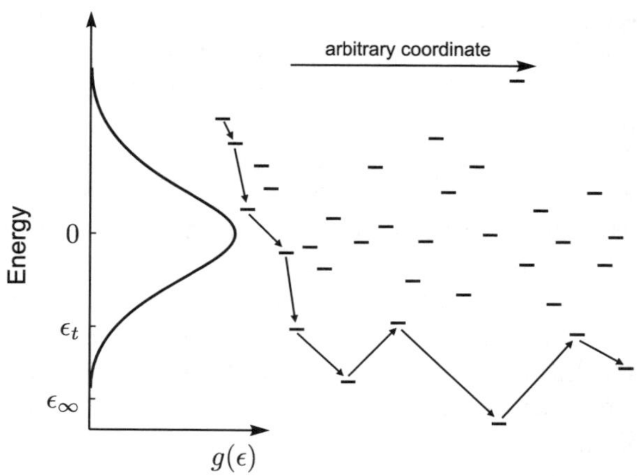

Figure 6.4 Schematic picture of carrier energy relaxation in the Gaussian DOS via the transport energy ε_t (reproduced with permission from [31]. Copyright 2000 by the American Physical Society)

$$\exp\left(\frac{X^2}{2}\right)\left[\int_{-\infty}^{X/\sqrt{2}} dt \exp(-t^2)\right]^{4/3} = \left[9(2\pi)^{1/2} N\alpha^3\right]^{-1/3} kT/\sigma. \quad (6.15)$$

It has been shown [58] that at low temperatures, $kT < \sigma$, the maximum determined by Equation (6.13) is sharp and therefore the introduction of the TE for a Gaussian DOS makes sense. An equation for the TE in a Gaussian DOS literally coinciding with Equation (6.15) was later published by Arkhipov *et al.* [59], who also claimed that the concept of transport energy is applicable to practically all realistic DOS distributions. We cannot agree with the latter statement. It has been shown in [58] that, for example, in a system with the DOS in the form $g(\varepsilon) \propto \exp\left(-\sqrt{\varepsilon/\varepsilon_0}\right)$ the maximum determined by Equation (6.13) is so broad and the position of ε_t is so deep in the tail of the DOS that introduction of the TE makes no sense. The DOS described by the latter formula is known to be valid for such a broad class of disordered systems as the mixed crystals with compositional disorder [60]. Hence it should be considered as a realistic one. Furthermore, sometimes the TE approach is applied to a system with a constant, energy-independent DOS [61]. We do not think that such a procedure is meaningful. If the energy dependence of the DOS is weak, the transport path in the equilibrium conditions corresponds to the vicinity of the Fermi level. Concomitantly, the temperature dependence of the conductivity obeys the classical Mott formula: $\sigma(T) \propto (T_0/T)^{1/4}$, as described in [14]. In the next two subsections we describe how one can use the concept of transport energy in order to explain experimentally observed dependences of the transport coefficients on temperature and on the concentration of localized states, N, in a system with Gaussian DOS.

6.4.4 Calculations of τ_{rel} and μ

After having understood the relaxation kinetics, it is easy to calculate τ_{rel} and μ. Let us start with temperature dependences of these quantities. We consider the case $\langle\varepsilon_\infty\rangle < \varepsilon_t < 0$, which corresponds to any reasonable choice of materials parameters $N\alpha^3$ and kT/σ [58]. After generation of carriers in a nonequilibrium situation, the carrier energy distribution moves downward in energy with its maximum $\langle\varepsilon(t)\rangle$ logarithmically dependent on time t [8, 40, 41]. States above $\langle\varepsilon(t)\rangle$ achieve thermal equilibrium with ε_t at time t, while states below $\langle\varepsilon(t)\rangle$ have no chance at time t to exchange carriers with states in the vicinity of ε_t, and hence the occupation of these deep energy states does not correspond to the equilibrium occupation. The system of noninteracting carriers comes into thermal equilibrium when the time-dependent energy $\langle\varepsilon(t)\rangle$ achieves the equilibrium level $\langle\varepsilon_\infty\rangle$ determined by Equation (6.7). The corresponding time is the relaxation time τ_{rel} [8, 40, 41]. At this time, dispersive conduction is replaced by the Gaussian transport [8, 40, 41]. As long as the relaxation of carriers occurs via thermal activation to the level ε_t, the relaxation time τ_{rel} is determined by the activated transitions from the equilibrium level $\langle\varepsilon_\infty\rangle$ to the transport energy ε_t [31]. Hence according to Equation (6.5), τ_{rel} is determined by the expression [31]

$$\tau_{rel} = \nu_0^{-1} \exp\left[\frac{2r(\varepsilon_t)}{\alpha} + \frac{\varepsilon_t - \langle\varepsilon_\infty\rangle}{kT}\right], \quad (6.16)$$

where

$$r(\varepsilon) = \left[\frac{4\pi}{3} \int_{-\infty}^{\varepsilon} \mathrm{d}\varepsilon\, g(\varepsilon) \right]^{-1/3}. \tag{6.17}$$

From Equations (6.14–6.17) it is apparent that the activation energy of the relaxation time in Equation (6.16) depends on both parameters σ/kT and $N\alpha^3$, and therefore, generally speaking, this quantity cannot be represented by Equation (6.8) and, if at all, the coefficient B should depend on the magnitude of the parameter $N\alpha^3$.

Let us calculate the exponent in the temperature dependence of τ_{rel} in Equation (6.16), $\ln(\tau_{rel}\nu_0)$, and plot it as a function of $(\sigma/kT)^2$ for various values of $N\alpha^3$. Solving Equation (6.15) numerically and using Equations (6.7), (6.14)–(6.17), one obtains results given in Figure 6.5. For $N\alpha^3 = 0.001$, the dependence described by Equation (6.8) with $B \simeq 1.0$ is in good agreement with computer simulations that give $B \simeq 1.1$ [8, 40]. Hence the result of the analytic calculations confirms Equation (6.8). However, Equation (6.8) can be considered only as an approximation because it originates from rather different temperature

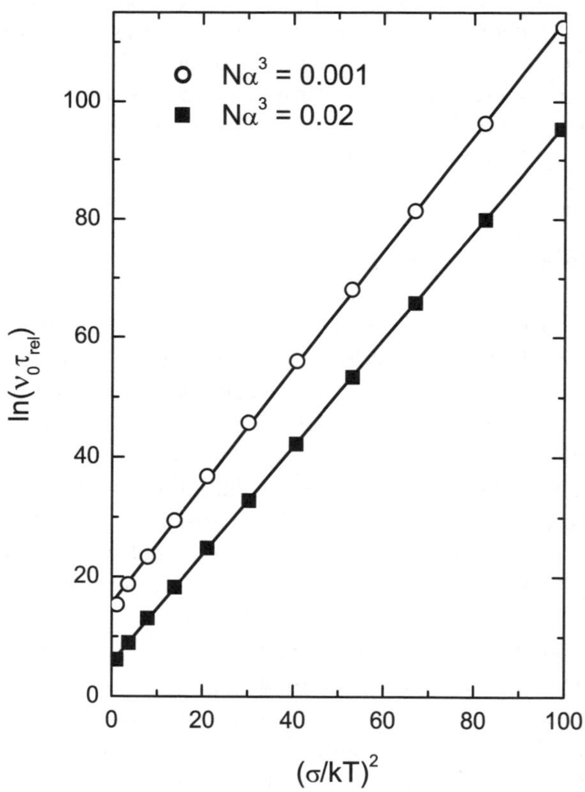

Figure 6.5 Temperature dependences of the relaxation time for different values of $N\alpha^3$. Solid lines represent the best fits in the form of Equation (6.8) for dependences obtained by numerical solution of Equations (6.15) and (6.16). The values of the parameter B obtained from such a fit are specified in the text (reproduced with permission from [31]. Copyright 2000 by the American Physical Society)

dependences of different terms in the exponent of Equation (6.16). This possibility of effective description via Equation (6.8) is provided by the strong temperature dependence of $\langle\varepsilon_\infty\rangle$ given by Equation (6.7) while the temperature dependences of the quantities ε_t and $r(\varepsilon_t)$ are weaker and they almost cancel each other in Equation (6.16). However, at another value of the parameter $N\alpha^3 = 0.02$, the relaxation time τ_{rel} is described by Equation (6.8) with another value of the coefficient $B \simeq 0.9$. This clearly shows that the temperature dependence in Equation (6.8) cannot be universal with respect to the concentration of localized states N and the decay parameter of the carrier wave function α.

Now we turn to the calculation of the carrier drift mobility μ. We assume that the transient time t_{tr}, necessary for a carrier to travel through a sample is longer than τ_{rel}, and hence the charge transport takes place in the equilibrium conditions. As described above, every second jump brings the carrier upward in energy to the vicinity of ε_t, being succeeded by a jump to the spatially nearest site with deeper energy determined solely by the DOS. Therefore, in order to calculate the drift mobility, μ, it is correct to average the times of hopping transitions over energy states below ε_t, since only these states are essential for charge transport in thermal equilibrium [30, 31]. Hops downward in energy from the level ε_t occur exponentially faster than upward hops towards ε_t. Therefore, one can neglect the former in the calculation of the average time $\langle t\rangle$. The carrier drift mobility can be evaluated as

$$\mu \simeq \frac{e}{kT}\frac{r^2(\varepsilon_t)}{\langle t\rangle}, \tag{6.18}$$

where $r(\varepsilon_t)$ is determined by Equations (6.17), (6.14), (6.15) and (6.4). The average hopping time has the form [30]

$$\langle t\rangle = \left[\int_{-\infty}^{\varepsilon_t} d\varepsilon\, g(\varepsilon)\right]^{-1} \int_{-\infty}^{\varepsilon_t} d\varepsilon\, v_0^{-1} g(\varepsilon) \exp\left[\frac{2r(\varepsilon_t)B_c^{1/3}}{\alpha} + \frac{\varepsilon_t - \varepsilon}{kT}\right], \tag{6.19}$$

where $B_c \simeq 2.7$ is the percolation parameter taken from Equation (6.12). This numerical coefficient is introduced into Equation (6.19) in order to warrant the existence of an infinite percolation path over the states with energies below ε_t. Using Equations (6.4), (6.14), (6.15), (6.18) and (6.19), one obtains for the exponential terms in the expression for the carrier drift mobility the relation

$$\ln\left[\mu\left(\frac{er^2(\varepsilon_t)v_0}{kT}\right)^{-1}\right] = -2\left[\frac{4\sqrt{\pi}}{3B_c}N\alpha^3 \int_{-\infty}^{X_t/\sqrt{\pi}} dt \exp(-t^2)\right]^{-1/3} - \frac{X_t\sigma}{kT} - \frac{1}{2}\left(\frac{\sigma}{kT}\right)^2. \tag{6.20}$$

It is Equation (6.20) that determines the dependence of the carrier drift mobility on the parameters $N\alpha^3$ and σ/kT.

In Figure 6.6 the dependence of the drift mobility on temperature at $N\alpha^3 = 0.01$ is shown for several values of σ. Equation (6.20) confirms the validity of Equation (6.9), though with the coefficient C slightly dependent on the value of the parameter $N\alpha^3$. In computer

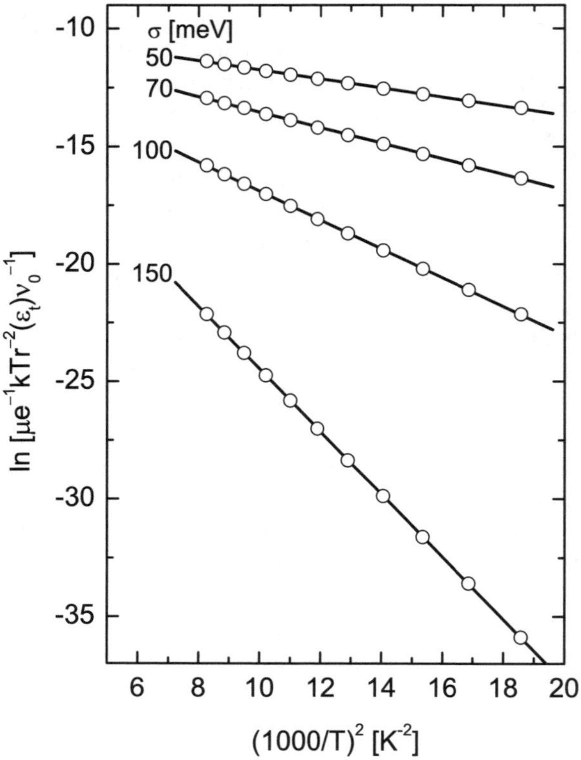

Figure 6.6 Temperature dependence of zero-field mobility in organic semiconductors obtained via Equation (6.20) for different disorder energy scales, σ (reproduced with permission from [109]. Copyright 2006 by Springer)

simulations the value $N\alpha^3 = 0.001$ was chosen and $C \simeq 0.69$ was obtained [41]. Equation (6.20) predicts a very close value $C \simeq 0.68$ for $N\alpha^3 = 0.001$. However, at $N\alpha^3 = 0.02$ we obtain from Equation (6.20) a different value $C \simeq 0.62$, which shows that this coefficient in Equation (6.9) indeed depends on the value of parameter $N\alpha^3$ as expected for the VRH transport process. Very similar data for C were obtained in several other papers [61–63].

It is clear from this picture how sensitive is the mobility to the value of temperature. Comparison of these dependences with experimental measurements of the $\ln(\mu)$ versus T^{-2} (some are shown in Figure 6.7) provides information on the energy scale σ of the DOS (see, for example, [8, 44]). We would like to emphasize here once more that the choice of the numerical coefficient C in Equation (6.9) is important for the estimation of σ from the comparison with experimental data. This parameter is not universal, being dependent on the value of $N\alpha^3$.

In Figure 6.8, the dependence of the drift mobility on $N\alpha^3$ is shown for $kT/\sigma = 0.3$. Also experimental data of Gill [33] are shown in the figure. It is well illustrated that the slope of the mobility exponent as a function of $(N\alpha^3)^{-1/3}$ given by the theory described above agrees with the experimental data. Comparison between the theory [30] and experimental results [32, 33, 35, 64] provides an estimate for the decay parameter α of the carrier wavefunction in localized states. The values between $\alpha \simeq 0.1\,\text{nm}$ and $\alpha \simeq 0.3\,\text{nm}$ are obtained

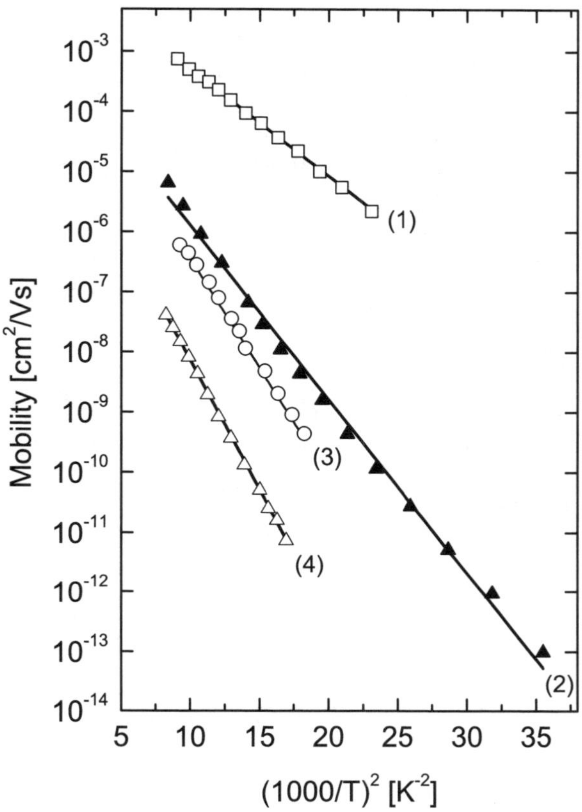

Figure 6.7 Experimental data for temperature dependence of zero-field mobility in organic semiconductors: (1) di-p-tolylphenylamine containing (DEASP)-traps [110]; (2) (BD)-doped polycabonate [43]; (3) (NTDI)-doped poly(styrene) [111]; (4) (BD)-doped TTA/polycarbonate [112] (reproduced with permission from [109]. Copyright 2006 by Springer)

depending on the chosen polymer and the doping species. At very low concentration of the localized states, N, the probability of carrier tunneling in space dominates the transition rate in Equation (6.5). In this regime of the nearest-neighbor hopping, the concentration dependence of the drift mobility is described by the expression [17]

$$\mu \propto \exp\left(-\frac{\gamma N^{-1/3}}{\alpha}\right), \tag{6.21}$$

where $\gamma = 1.73 \pm 0.03$ [17] as explained in [14]. Equation (6.21) is illustrated by the dashed line in Figure 6.8.

In several recent publications, the dependence described by Equation (6.21) for random organic solids was claimed, but with different numerical coefficients γ [22–25]. We suppose that these deviations in the obtained values of the coefficient γ from the well-known text-book result $\gamma = 1.73 \pm 0.03$ are caused by the unjustified assumption that carriers always hop to the neighboring site with the optimal exponent in the transition rate described by

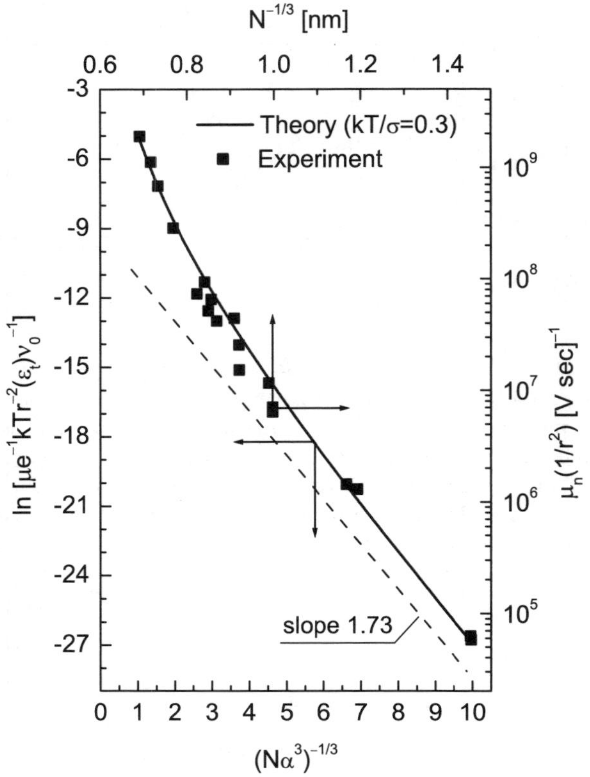

Figure 6.8 Concentration dependence of the drift mobility evaluated from Equation (6.20) (solid line) and the dependence observed experimentally [33] (squares) for TNF/PE and TNF/PVK. The dashed line references the result of Equation (6.21) with $\gamma = 1.73$ for the nearest-neighbor hopping mode (reproduced with permission from [15]. Copyright 2004 by the American Physical Society)

Equation (6.5) [22–25]. Such an assumption is often made in theoretical studies of hopping transport in organic materials (see, for instance [65]). Expression (6.12) shows however that, when calculating the DC hopping transport, one should take into account on average not a single neighbor, but at least $B_c \simeq 2.7$ neighbors per site. This average number $B_c \simeq 2.7$ is responsible for the correct value of the coefficient $\gamma = 1.73 \pm 0.03$ in Equation (6.21) as described in [14] following the textbooks [17].

We would also like to comment here on the contradictory statements in the literature about the invalidity/validity of the averaging procedure for hopping rates to describe the hopping conductivity in random systems. In Section 6.4.1, it is shown that this procedure in invalid (see also [17]). This has been confirmed in several recent publications (see, for instance, [23]). However, in other publications the averaging procedure for hopping rates was claimed capable of describing the hopping conductivity [22, 24, 25, 66]. It is worth noting that the procedure, which the authors call the averaging of hopping rates in [22, 24, 25, 66] is not the procedure, which they call the averaging of hopping rates in [23] and which was used in [18–21] and described in Section 6.4.1. In [22, 24, 25, 66], the average hopping rate is defined via the following chain of arguments. On one hand, the general expression for the DC conductivity in the following form is used [67]

$$\sigma_{\mathrm{DC}} = e\int \mathrm{d}\varepsilon \tilde{\mu}(\varepsilon)\tilde{n}(\varepsilon), \tag{6.22}$$

where e is the elementary charge, $\tilde{n}(\varepsilon)\mathrm{d}\varepsilon$ is the concentration of electrons in the states with energies between ε and $\varepsilon + \mathrm{d}\varepsilon$ and $\tilde{\mu}(\varepsilon)$ is the mobility of these electrons.

On the other hand, one uses the general expression for the DC conductivity in the form $\sigma_{\mathrm{DC}} = e\mu n$, where n is the total concentration of the charge carriers

$$n = \int_{-\infty}^{\infty} \mathrm{d}\varepsilon \tilde{n}(\varepsilon), \tag{6.23}$$

and $\tilde{n}(\varepsilon) = g(\varepsilon)f(\varepsilon)$ is the product of the density of states $g(\varepsilon)$ and the Fermi function $f(\varepsilon)$ dependent on the position of the Fermi energy ε_{F}.

The quantity μ obtained as $\mu = \sigma_{\mathrm{DC}}/en$ is called to be proportional to the 'average hopping rate' $\langle v\rangle : \mu \propto (e/kT)\langle r^2\rangle\langle v\rangle$, where $\langle r\rangle$ is the average length for local hopping events. In other words the 'average hopping rate' $\langle v\rangle$ is determined via this relation. After calculating σ_{DC} via Equation (6.22) and n via Equation (6.23) and using the expression $\mu = \sigma_{\mathrm{DC}}/en$ it is claimed that μ was calculated via averaging of hopping rates. We would just like to remark that this definition of the average rate looks unjustified. It is not the definition of the average hopping rate used in [23] and it has nothing to do with the averaging procedure used in [18–21] leading to Equation (6.10), which is still often claimed as correct (see, for instance, [26–28]).

6.4.5 Saturation effects

So far we have discussed the drift mobility of charge carriers under the assumption that their concentration is much less than that of the localized states in the energy range essential for hopping transport. In such a case, one can assume that carriers perform hopping motion independently from each other and calculate the conductivity as a product

$$\sigma_{\mathrm{DC}} = e\mu n, \tag{6.24}$$

where n is the concentration of charge carriers in the material and μ is their drift mobility calculated via Equation (6.20) under the assumption that the system is empty and a charge carrier is not affected by the possibility that localized states can be occupied by other carriers. In such a regime, Equations (6.20) and (6.24) describe the dependences of the conductivity on temperature and on the value of the parameter $N\alpha^3$, while the dependence of the conductivity on the concentration of charge carriers n is linear in accord with Equation (6.24). If, however, the concentration n of charge carriers is increased so that the Fermi energy ε_{F} in thermal equilibrium or the quasi-Fermi energy under stationary excitation is located energetically higher than the equilibrium energy $\langle\varepsilon_\infty\rangle$ determined by Equation (6.7), one should use a more sophisticated theory in order to calculate σ_{DC}.

Let us first estimate the critical concentration of charge carriers n_c below which the consideration based on Equations (6.20) and (6.24) is valid. In order to estimate n_c, the value of the Fermi energy ε_{F} should first be calculated. The position of the Fermi level is determined by the equation

$$\int_{-\infty}^{\infty} \frac{g(\varepsilon)\mathrm{d}\varepsilon}{1+\exp[(\varepsilon-\varepsilon_{\mathrm{F}})/kT]} = n. \tag{6.25}$$

At low n, for all energies in the domain giving the main contribution to the integral in Equation (6.25), the exponential function in the denominator of the integral is large compared with unity (the nondegenerate case). Then a straightforward calculation [50] gives for the DOS determined by Equation (6.4) at low temperatures

$$\varepsilon_{\mathrm{F}} \simeq -\frac{1}{2}\left(\frac{\sigma^2}{kT}\right) - kT \ln\left(\frac{N}{n}\right). \tag{6.26}$$

From the equation $\varepsilon_{\mathrm{F}}(n_c) = \langle\varepsilon_\infty\rangle$, where $\langle\varepsilon_\infty\rangle$ is determined by Equation (6.7), one obtains for n_c the estimate

$$n_{\mathrm{c}} \simeq N \exp\left[-\frac{1}{2}\left(\frac{\sigma}{kT}\right)^2\right]. \tag{6.27}$$

At $n < n_c$, the Fermi level is situated below the equilibration energy, $\langle\varepsilon_\infty\rangle$, and the charge transport can be described for independent carriers via Equations (6.20) and (6.24). At $n > n_{\mathrm{c}}$, the theory for charge transport should be essentially modified. These arguments can be easily converted in order to consider the case of constant concentration of charge carriers and changing temperature. Then at some particular temperature T_{c} equation $\varepsilon_{\mathrm{F}}(T_{\mathrm{c}}) = \langle\varepsilon_\infty\rangle$ will be fulfilled. At $T > T_{\mathrm{c}}$, the carrier mobility and conductivity can be described by Equations (6.20) and (6.24), while at $T < T_{\mathrm{c}}$, an essential modification of the theory is needed. Using Equations (6.7) and (6.26) and solving equation $\varepsilon_{\mathrm{F}}(T_{\mathrm{c}}) = \langle\varepsilon_\infty\rangle$ for T_{c}, one obtains [50]

$$T_{\mathrm{c}} \simeq \frac{\sigma}{2^{1/2} k \ln^{1/2}(N/n)}. \tag{6.28}$$

In order to develop a theory for hopping transport at $n > n_{\mathrm{c}}$ (or at $T < T_{\mathrm{c}}$) one should take into account explicitly the filling probabilities of the localized states by charge carriers. One possibility is to solve the percolation problem with transition rates between localized states described by Equation (6.6) that includes the value of the Fermi energy related via Equation (6.25) to the concentration of carriers, n [50, 51]. An alternative theoretical approach to describe hopping conductivity in the Gaussian DOS, taking into account the occupation of the essential fraction of localized states by charge carriers, was recently suggested by Schmechel [62, 63]. Schmechel extended the concept of transport energy described in Section 6.4.3 taking into account the possibility that the localized states can be essentially filled by charge carriers. Another kind of percolation approach to the problem was suggested by Martens *et al.* [61]. We will not describe these theories in detail; the interested reader can find a comprehensive analysis of some of them in the recent paper of Coehoorn *et al.* [68]. We would like, however, to emphasize one very pronounced result of those theories. As soon as the Fermi energy determined by Equation (6.25) lies essentially above the equilibration energy determined by Equation (6.7), the temperature dependence of the electrical conductivity is no longer proportional to $\exp[-(C\sigma/kT)^2]$ as in the case of low carrier concentrations at which Equations (6.9), (6.20) and (6.24) are valid. The conductivity instead closely follows the routine Arrhenius behavior [50, 51]. For example, the percolation approach predicts at $n > n_{\mathrm{c}}$ the temperature dependence [50, 51]

$$\sigma_{\mathrm{DC}} = \sigma_0 \exp\left(-\frac{\varepsilon_{\mathrm{t}} - \varepsilon_{\mathrm{F}}}{kT}\right), \tag{6.29}$$

where σ_0 is the preexponential factor, only slightly dependent on temperature and on the concentration of sites N; the transport energy ε_{t} is determined via

$$\varepsilon_{\mathrm{t}} = -2^{1/2}\sigma \ln^{1/2}\left[\frac{\beta N\alpha^3}{s^4 kT}\left(\frac{\sigma}{kT}\right)^3\right], \tag{6.30}$$

where s is the solution of transcendental equation

$$s = -\ln^{1/2}\left[\frac{\beta N\alpha^3}{s^4 kT}\left(\frac{\sigma}{kT}\right)^3\right] \tag{6.31}$$

with $\beta \simeq 0.045$. The Fermi energy ε_{F} is determined via the equation [50, 51]

$$\varepsilon_{\mathrm{F}} = -2^{1/2}\sigma \ln^{1/2}\left[\frac{\sigma N}{(2\pi)^{1/2}\, n\varepsilon_{\mathrm{F}}}\right], \tag{6.32}$$

where σ is the energy scale of the DOS in Equation (6.4).

It is worth noting that there are numerous observations of a simple activated temperature dependence of the conductivity in disordered organic materials [69–72]. Particularly the Arrhenius temperature dependence of the conductivity is often observed in field-effect transistors, where charge carrier concentration is usually high (see, for instance, [34]).

Often the dependence $\ln(\mu) \propto T^{-2}$ or $\ln(\sigma_{\mathrm{DC}}) \propto T^{-2}$ is considered as evidence for a Gaussian DOS, while the Arrhenius temperature dependence $\ln(\mu) \propto T^{-1}$ or $\ln(\sigma_{\mathrm{DC}}) \propto T^{-1}$ is claimed to indicate a pure exponential DOS, which was described in [14]. The important conclusion from the above consideration is the possibility to account successfully for both kinds of temperature dependence of hopping conductivity described by Equations (6.2) and (6.3) in the framework of the universal theoretical model based on the Gaussian DOS. The temperature dependence of the conductivity is sensitive to the concentration of charge carriers n.

6.5 THEORETICAL TREATMENT OF CHARGE CARRIER TRANSPORT IN ONE-DIMENSIONAL DISORDERED ORGANIC SYSTEMS

In this section we show how the hopping charge transport can be described theoretically in one-dimensional (1D) disordered organic solids. Particular interest to this topic is twofold. First, 1D disordered organic systems have been intensively studied experimentally, aiming at their applications in various electronic devices. Furthermore, in some vitally important processes such as electron transport along DNA molecules or charge transport in ionic channels through cell membranes, charge transport can be effectively treated as one-dimensional. Second, in many cases hopping transport in 1D systems can be described theoretically with much higher precision than hopping transport in three-dimensional systems. Theoretical study of three-dimensional transport is complicated because it is sometimes difficult to find the spatial structure of the transport path [17]. In 1D systems, this problem does not exist. A charge carrier in such systems performs a series of successive transition events between localized states placed on a 1D chain. Provided the rates for hopping transitions between localized states and the DOS function are known, the problem of calculating the transport coefficients can be in many cases solved exactly. Using this advantage of the precise theoretical description, one can clarify various transport phenomena discovered experimentally. Among such phenomena we will discuss the dependence of the carrier kinetic coefficients on the applied electric field and in particular the rather puzzling observation of the drift mobility increasing with decreasing field strength at low fields [4, 7, 8, 12, 13, 40].

In recent years particular attention of researchers has been devoted to the study of charge transport processes in columnar discotic liquid-crystalline glasses due to their potential technical applications for electrophotography and as transport materials in light-emitting diodes [73, 74]. Dielectric measurements have clarified that charge transport in most of such materials is extremely anisotropic [75] so that it is reasonable to describe the transport of charge carriers as a 1D process [76]. Indeed, the values of conductivity along and perpendicular to the discotic columns differ in such materials by three orders of magnitude [75]. Experimentally observed dependences of the conductivity on the frequency, on the strength of the applied electric field, and on temperature show that an incoherent hopping process is the dominant transport mechanism in such materials [44, 75, 76]. It has been found reasonable to describe the transport of charge carriers as a hopping process via molecules arranged in spatially ordered 1D chains with a Gaussian distribution of site energies described by Equation (6.4) [44, 76, 77].

Let us first consider, following [15, 76], a simplest case allowing only hops to nearest neighbors and assuming that 1D chains of localized states are spatially regular. In such a case, one can omit the r-dependence of transition probabilities using the effective preexponential factor ν_0 that contains the term $\exp(-2b/\alpha)$, where b is the distance between the neighboring sites on the conducting chain. The Miller–Abrahams transition rates have in such a case the form [15, 76]

$$\nu_{ij}(\varepsilon_{i,}\varepsilon_j) = \nu_0 \exp\left(-\frac{\varepsilon_j - \varepsilon_i + |\varepsilon_j - \varepsilon_i|}{2kT}\right). \tag{6.33}$$

Other forms of transition probabilities have also been considered in the literature. Seki and Tachiya [78] studied the 1D conduction with transition rates for charged carriers described by the Markus law that takes into account the polaron effect [79]. Another, symmetric form was also used in order to analyze analytically the field and temperature dependences of the hopping conductivity in 1D systems [77]:

$$\nu_{i\pm1,i} = \nu_0 \exp\left(-\frac{\varepsilon_{i\pm1} - \varepsilon_i}{2kT}\right). \tag{6.34}$$

Two distinct models have been discussed in the literature with respect to the space–energy correlation of localized states responsible for charge transport. In the so-called Gaussian disordered model (GDM), which we considered in previous sections of this chapter, such correlations are neglected. In the alternative correlated disorder model (CDM), it is assumed that energy distributions of spatially close sites are correlated, for example, due to dipole or quadrupole interactions [77, 80–82].

Much attention has been paid in recent years to the dependence of the carrier mobility on the electric field. This dependence in disordered organic materials in a broad range of high field strengthes can be fitted by the function $\mu \propto \exp\left(\sqrt{F/F_0}\right)$, where F_0 is a parameter [8]. It has been shown that in three-dimensional systems such field dependence can be explained only in the framework of the CDM [77, 80–82]. It is challenging to check this result by the exactly solvable 1D models.

The study of the temperature dependence of the drift mobility in 1D systems is also of great interest. Researchers agree that this dependence in the empty system can be described by Equation (6.9), although there is no agreement on the magnitude of the coefficient C in this formula. Ochse *et al.* [44] used the three-dimensional value $C \simeq 2/3$ for columnar systems with 1D transport, while Bleyl *et al.* [76] obtained $C \simeq 0.9$ in their computer simulation for the 1D case with asymmetric transition probabilities. The analytic calculations of Dunlap *et al.* [77] for symmetric probabilities predict $C = 1$ in the 1D case. The value of C is decisive for characterization of the disorder parameter σ from comparison with experimental data in 1D systems. We show in the next subsection analytical formulas that allow one to calculate exactly transport coefficients in 1D systems for hopping transitions between the nearest sites.

6.5.1 General analytical formulas

The steady-state drift velocity of charge carriers in 1D periodic systems with the number of sites in the period N and the distance between the nearest sites on the chain b can be written exactly using the general solution found by Derrida [83]

$$\upsilon = \frac{Nb\left(1-\prod_{k=0}^{N-1}\frac{\nu_{k+1,k}}{\nu_{k,k+1}}\right)}{\sum_{k=0}^{N-1}\nu_{k,k+1}^{-1}\left(1+\sum_{i=1}^{N-1}\prod_{j=1}^{i}\frac{\nu_{k+j,k+j-1}}{\nu_{k+j,k+j+1}}\right)}. \tag{6.35}$$

This formula was used by Dunlap *et al.* [77] to study the drift mobility in a 1D system with symmetric transition rates. The drift mobility μ is related to the steady-state velocity υ as

$$\mu = \frac{\upsilon}{F}. \tag{6.36}$$

However in the experiments, the drift mobility is usually obtained by the time-of-flight technique [8]. Here charge carriers pass only once through a sample of finite thickness and drift mobility is calculated by dividing the sample length Nb by the mean transit time $\langle t_{0N}\rangle$ and electric field F:

$$\mu = \frac{Nb}{\langle t_{ON}\rangle F}. \tag{6.37}$$

Murthy and Kehr [84, 85] have derived the analytical expression for the mean transit time of carriers through a given chain of $N + 1$ sites:

$$\langle t_{0N} \rangle = \sum_{k=0}^{N-1} v_{k,k+1}^{-1} + \sum_{k=0}^{N-2} v_{k,k+1}^{-1} \sum_{i=k+1}^{N-1} \prod_{j=k+1}^{i} \frac{v_{j,j-1}}{v_{j,j+1}}. \tag{6.38}$$

This formula gives the transit time for a charge carrier that starts on site 0 and finishes on site N. The time is averaged over many transits through the same chain with given values of transition probabilities v_{ij}.

One should be cautious with the application of Equations (6.37) and (6.38) at low electric fields. Even without electric field, carriers starting at site 0 will pass through the finite system solely due to the diffusion process. At low fields, diffusion dominates the motion of carriers and it would be wrong to use Equation (6.37). More appropriate in such a case is the estimate of the mobility via the diffusion formula

$$\mu = \frac{e}{kT} \frac{N^2 b^2}{2\langle t_{0N} \rangle}, \tag{6.39}$$

which presumes the validity of the Einstein relationship.

Note that Equations (6.35) and (6.38) are valid for any type of nearest-neighbor hopping of noninteracting carriers in 1D systems. We show in the next section the results of the exact theory applied to a model system represented by a 1D regular chain with equal site energies separated by barriers with random heights, a so-called random-barrier model (RBM). In the subsequent sections we consider more realistic models, GDM and CDM, with distribution of site energies.

6.5.2 Drift mobility in the random-barrier model

Transition probabilities between the neighboring sites under the influence of electric field F in the RBM model are given by expressions

$$v_{k,k+1} = v_0 \exp\left(-\frac{\Delta_k - ebF/2 + |\Delta_k - ebF/2|}{2kT} \right), \tag{6.40}$$

and

$$v_{k+1,k} = v_0 \exp\left(-\frac{\Delta_k + ebF/2 + |\Delta_k + ebF/2|}{2kT} \right), \tag{6.41}$$

where Δ_k is the energy barrier between sites k and $k + 1$.

In the limit of long chains, $N >> 1$, calculation of the drift mobility via Equations (6.35) and (6.36) or Equations (6.37) and (6.38) for any chain with a given distribution of transition rates is equivalent to the averaging of Equations (6.35) or (6.38) over all possible distributions of transition rates determined by disorder. We assume the Gaussian distribution of barrier heights with the typical scale σ. Performing the averaging, one obtains the following expression for the drift mobility in the infinite chain [15]

$$\mu = \frac{\nu_0 b}{F}\left\{1 - \exp\left[\frac{1}{2}\left(\frac{\sigma}{kT}\right)^2 - \frac{ebF}{2kT}\right]\left[\operatorname{erf}\left(\frac{ebF}{\sqrt{8}\sigma} + \frac{\sigma}{\sqrt{2}kT}\right)\right.\right.$$
$$\left.\left. - \operatorname{erf}\left(\frac{\sigma}{\sqrt{2}kT}\right)\right] - \exp\left(-\frac{ebF}{kT}\right)\left[1 - \operatorname{erf}\left(\frac{ebF}{\sqrt{8}\sigma}\right)\right]\right\} \times \left\{\operatorname{erf}\left(\frac{ebF}{\sqrt{8}\sigma}\right)\right.$$
$$\left. + \exp\left[\frac{1}{2}\left(\frac{\sigma}{kT}\right)^2 - \frac{ebF}{2kT}\right]\left[1 - \operatorname{erf}\left(\frac{ebF}{\sqrt{8}\sigma} - \frac{\sigma}{\sqrt{2}kT}\right)\right]\right\}^{-1}, \tag{6.42}$$

where $\operatorname{erf}(x) = (2/\pi) \int_0^x dy \exp(-y^2)$ is the error function. In Figure 6.9 the calculated field dependences of the drift mobility are shown for parameters $b = 3.6$ Å, $\sigma = 50$ meV, $kT = 25$ meV [15]. The solid line shows the exact solution for an infinite chain given by Equation (6.42). All points in the figure correspond to mobilities for finite chains of $N = 500$ sites averaged over 1000 different chains. Circles were obtained via Equations (6.37) and (6.38) while squares came from Equations (6.38) and (6.39). Results of the Monte Carlo computer simulations [16] are shown by crosses to demonstrate the excellent agreement of the simulation results with the analytic theory. At low fields, the drift approach [Equations (6.37) and (6.38)] for finite systems leads to an increase of the calculated mobility with decreasing field strength. Similar results were obtained for all considered models of disorder and various forms of the transition probabilities between neighboring sites [15]. This result is apparently artificial, reflecting the fact that charge carriers can penetrate through a finite system via diffusive motion, even in the absence of electric field. By using Equation (6.37) one overestimates the mobility at low fields. The same happens when using Equation (6.39) at higher fields. Comparison of Equations (6.37) and (6.39) reveals the strength of the electric field F_c at which a transition from the diffusion-controlled to the drift-controlled transit takes place: $F_c \simeq 2kT/(eNb)$ [15].

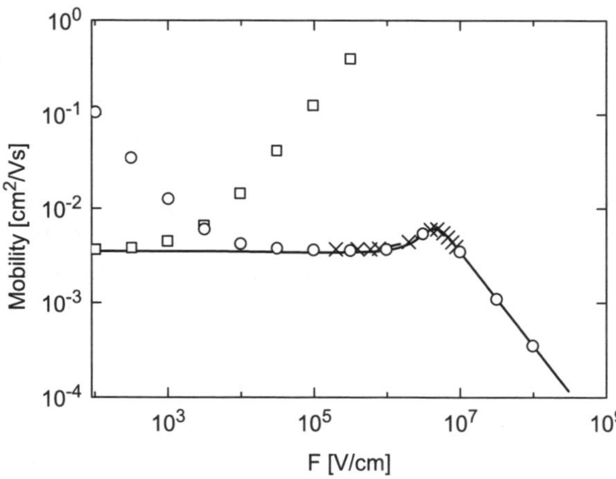

Figure 6.9 Field dependence of the carrier mobility in the RBM with Gaussian distribution of barriers. The solid line represents exact solution for the infinite chain. Data shown by circles and squares were calculated via Equation (6.38) using drift (Equation 6.37) and diffusion relations (Equation 6.39), respectively. Data shown by crosses were obtained by Monte Carlo simulations [16] (reproduced with permission from [15]. Copyright 2001 by the American Physical Society)

At low fields, the transit time does not depend on the strength of the electric field, being determined mostly by diffusion. Using Equation (6.37) one artificially obtains an increasing drift mobility with decreasing field F. In numerous publications, experimental results were reported that show an increase of the drift mobility with decreasing field at low fields [11, 86, 87]. This was always observed at high temperatures, at which diffusion can dominate the motion of charge carriers. Experimental evidence for decreasing mobility with increasing field is usually obtained using equations similar to Equation (6.37), where the drift mobility is determined via the measured transit time by dividing the sample length by the value of this time and by the value of the field strength. We believe that a diffusion equation similar to Equation (6.39) should be used at low fields and high temperatures. In the three-dimensional case, this equation should be slightly modified, though the conclusion is the same: decreasing drift mobility with increasing field strength at low fields can be an artifact caused by using the drift equation instead of the diffusion equation for the evaluation of the mobility on the basis of the measured transit time. This conclusion was also suggested by Hirao *et al.* [88]. In the 1D calculations, this idea is illustrated by using Equation (6.39) instead of Equation (6.37) for finite systems at low electric fields. The result is shown by squares in Figure 6.9. The excellent agreement of the diffusion equation with the exact calculation for the infinite system at low fields confirms our conclusion. For the chosen parameters, the linear transport regime with mobility independent of the electric field is valid up to a field strength of approximately 10^6 V/cm, at which a narrow nonlinear regime starts with the mobility increasing with electric field. At higher electric fields, this regime is replaced by the trivial decrease of the mobility as F^{-1}. At such high electric fields, all energy barriers between sites on the chain are eliminated by the field and the transit time becomes field independent. Equation (6.37) predicts a F^{-1} dependence in this regime.

6.5.3 Drift mobility in the Gaussian disorder model

In this section we consider a random-energy model with distribution of site energies described by Equation (6.4), presuming that there are no correlations between spatial positions of chain sites and their energies. This model is called in the literature a Gaussian disorder model (GDM). In Equation (6.33), the site energies ε_k depend on the strength of the electric field F. They are related to the zero-field site energies, which we here denote as φ_k, by the expression $\varepsilon_k = \varphi_k - ekbF$. In accordance with Equation (6.33), the ratio of the forward $\nu_{k,k+1}$ and the backward hopping rates $\nu_{k+1,k}$ for any pair of neighboring sites is

$$\frac{\nu_{k+1,k}}{\nu_{k,k+1}} = \exp\left(\frac{\varepsilon_{k+1} - \varepsilon_k}{kT}\right) \tag{6.43}$$

equivalent to the condition of the detailed balance.

Inserting Equation (6.43) into (6.35), one obtains

$$\upsilon = \frac{bN[1 - \exp(-ebNF/kT)]}{\sum_{k=0}^{N-1} \nu_{k,k+1}^{-1} + \sum_{k=0}^{N-1} \exp(-\varphi_k/kT) \sum_{i=0}^{N-1} \frac{\exp(-iebF/kT)}{\nu_{k+i,k+i+1} \exp(-\varphi_{k+i}/kT)}}. \tag{6.44}$$

The only random variables in Equation (6.44) are the zero-field site energies φ_k and the forward transition rates $\nu_{k+i,k+i+1}$, which in turn depend only on the neighboring site energies φ_{k+i} and φ_{k+i+1}. Since in all products appearing in the second term of the denominator, φ_k is not correlated with φ_{k+i} or $\nu_{k+i,k+i+1}$, averaging over disorder can be carried out for $\exp(-\varphi_k/kT)$ and $\exp(\varphi_{k+i}/kT)\nu^{-1}_{k+i,k+i+1}$ separately. Straightforward calculations of υ and $\langle t_{0N}\rangle$ then lead via Equation (6.37) to the expression for the drift mobility [15]

$$\mu = \frac{2\nu_0 b}{F}\left(1+\mathrm{erf}\left(\frac{ebF}{2\sigma}\right)+\exp\left[\left(\frac{\sigma}{kT}\right)^2-\frac{ebF}{kT}\right]\left[1+\mathrm{erf}\left(\frac{\sigma}{kT}-\frac{ebF}{2\sigma}\right)\right]\right.$$
$$+\exp\left[\left(\frac{\sigma}{kT}\right)^2-\frac{ebF}{kT}\right]\left[1-\exp\left(\frac{ebF}{kT}\right)\right]^{-1}\left\{\exp\left(-\frac{ebF}{kT}\right)\right.$$
$$\left.\left.\times\left[1+\mathrm{erf}\left(\frac{\sigma}{2kT}-\frac{ebF}{2\sigma}\right)\right]+1+\mathrm{erf}\left(\frac{\sigma}{2kT}+\frac{ebF}{2\sigma}\right)\right\}\right)^{-1}. \qquad (6.45)$$

At low electric fields this exact expression can be approximated by

$$\mu \simeq \frac{\nu_0 b}{2F}\exp\left[-\left(\frac{\sigma}{kT}\right)^2\right]\left[\exp\left(\frac{ebF}{kT}\right)-1\right], \qquad (6.46)$$

while at high fields $\mu \simeq \nu_0 b/F$. Interpolation between both approximations gives [15]

$$\mu \simeq \frac{\nu_0 b}{F}\left\{1+\frac{2\exp\left[\frac{\sigma^2}{(kT)^2}\right]}{\exp\left(\frac{ebF}{kT}\right)-1}\right\}^{-1}. \qquad (6.47)$$

Field dependences of the drift mobility in the GDM are shown in Figure 6.10 for $b = 0.36\,\mathrm{nm}$, $\sigma = 50\,\mathrm{meV}$, and $kT = 25\,\mathrm{meV}$. The solid line represents the exact solution for the infinite chain given by Equation (6.45), while dashed lines show the low- and the high-field approximations. One can easily check that for these parameters the interpolation formula (6.47) is in excellent agreement with the exact solution. Circles and triangles in the figure show the calculated results for finite systems of 2000 sites averaged over 1000 different chains, once using the averaging of inverse transit times (circles) and once using the averaging of inverse velocities (triangles). Qualitatively these results in Figure 6.10 resemble those for the random-barrier model described in the previous section. In particular, increase of the mobility with decreasing electric field, shown in Figure 6.10 by circles, is caused by the invalidity of the drift approximation in finite systems at low fields.

In Figure 6.11 the field dependences of the drift mobility are shown in the Poole–Frenkel representation (ln μ versus $\sqrt{F}$) for two different temperatures and two different averaging procedures. The figure clearly shows that in the chosen model the carrier mobility can hardly be described by the Poole–Frenkel law, ln $\mu \propto \sqrt{F}$. This conclusion is in agreement with the results of Garstein and Conwell [80], Dunlap *et al.* [77] and Novikov *et al.* [81, 82]. Following these authors we consider in Section 6.5.5 the drift mobility in the model with correlated disorder (CDM). Before doing so, we focus in the next section on the temperature dependence of the drift mobility at low fields in the GDM.

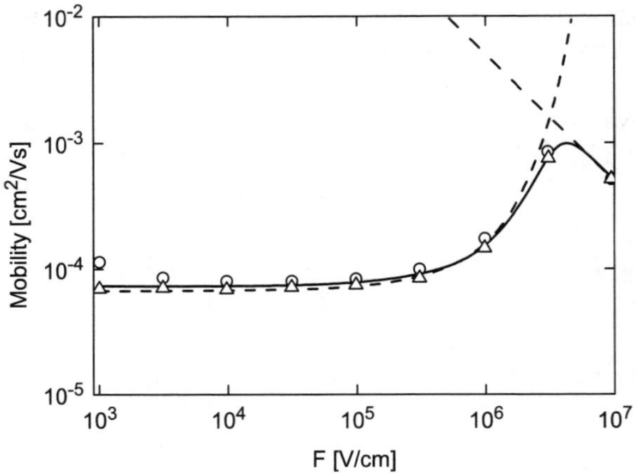

Figure 6.10 Field dependence of the carrier mobility for nearest-neighbour hopping with Miller–Abrahams rates between sites with Gaussian energy distribution. Solid line shows the exact solution for an infinite chain. Dashed lines correspond to the low-field and high-field approximations. Circles show the averaged mobilities calculated according to Equations (6.37) and (6.38). Triangles represent the results obtained by averaging of the inverse mobilities calculated by Equations (6.36) and (6.44) for the same chains (reproduced with permission from [15]. Copyright 2001 by the American Physical Society)

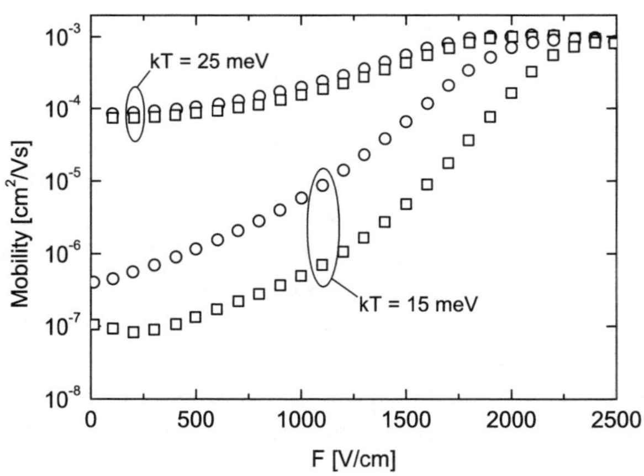

Figure 6.11 Poole–Frenkel plots of the carrier mobilities. Circles show the averaged mobilities calculated by Equations (6.36) and (6.44); squares show the corresponding results obtained by averaging of the inverse mobilities given by Equation (6.45). The number of chains was $c = 10^4$ with $N = 500$ sites each (reproduced with permission from [15]. Copyright 2001 by the American Physical Society)

6.5.4 Mesoscopic effects for the drift mobility

The low-field mobility for a finite chain described by Equations (6.44) and (6.36) in the limit of $F \to 0$ is given by the equation

$$\mu = \frac{eb^2}{kT}\left[N^{-2}\sum_{k=0}^{N-1}\exp(-\varphi_k/kT)\sum_{k=0}^{N-1}\exp(\varphi_k/kT)\nu_{k,k+1}^{-1}\right]^{-1}. \tag{6.48}$$

For the infinite chain with transition rates described by Equation (6.33) the low-field mobility in the GDM is exactly given by [15]

$$\mu = \frac{\nu_0 eb^2}{kT}\exp\left[-\left(\frac{\sigma}{kT}\right)^2\right]\left[1+\mathrm{erf}\left(\frac{\sigma}{2kT}\right)\right]^{-1}. \tag{6.49}$$

The temperature dependence of the drift mobility given by Equation (6.49) is shown by the solid line in Figure 6.12. The slope of ln μ versus $(\sigma/kT)^2$ in this curve differs slightly from −1 due to the temperature dependence of the preexponential factor in Eqution (6.49). This result is in good agreement with the computer simulations of Bleyl *et al.* [76] and with the analytic calculations of Dunlap *et al.* [77], although the latter analytic calculations were carried out for correlated systems with symmetric transition rates described by Equation (6.34). This shows that in 1D systems Equation (6.9) with $C \simeq 1$ is correct and even stable against the choice of the form of the transition rates.

The temperature dependences of the low-field drift mobility in finite systems calculated according to Equation (6.48) are also shown in Figure 6.12. These results are really remark-

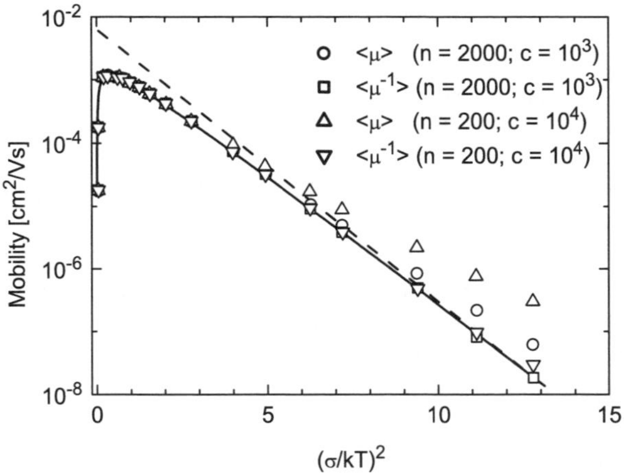

Figure 6.12 Temperature dependence of the low-field mobility for $\sigma = 50$ meV. Solid line represents the solution for the infinite chain given in Equation (6.49). Circles and squares show the results obtained by the avereging of mobilities and averaging of inverse mobilities calculated by Equation (6.48), respectively, with averaging over $c = 10^3$ chains of $N = 2000$ sites. Upward and downward triangles are the corresponding values for $c = 10^4$ and $N = 200$. Dashed line illustrates the temperature dependence of the drift mobility in the form of Equation (6.9) with $C = 1$ (reproduced with permission from [15]. Copyright 2001 by the American Physical Society)

able. They clearly show that the temperature dependence of the mobility is related to the size of the system. This is a manifestation of the mesoscopic character of the hopping transport problem in finite systems, which is most pronounced in 1D case. The averaging of transit times or inverse mobilities over various finite chains corresponds to the calculation of the mobility in a long system consisting of all those chains connected sequentially one after another. On the contrary, the averaging of the mobilities over various finite chains corresponds to the calculation of the carrier mobility in the system in which these chains are arranged in parallel to each other. In the latter case, the 'fast' chains with untypically short transit times strongly contribute to the average value of the mobility and they dominate the transport of charge carriers. This is the very essence of the mesoscopic effects [89].

Having in mind an application to discotic organic disordered systems where many quasi-1D current channels are connected in parallel, one should conclude that the temperature dependence of the electrical current in such systems does change with the thickness of the sample. For example, comparison between the data obtained by the averaging of the mobility values for chains with $N = 2000$ sites and chains with $N = 200$ suggests that, for shorter chains and hence for the thinner samples, the temperature dependence of the drift mobility should be weaker than that for thicker samples. Indeed 'fast' channels dominating electrical conduction in finite systems arise due to the narrower variation of site energies than in the infinite systems. This effect is caused by statistical fluctuations in the energy distribution [89]. Closer site energies lead to a weaker temperature dependence.

The mesoscopic character of the temperature dependence of the carrier drift mobility can also be illustrated by the following consideration. In the infinite 1D chain with hopping transitions only to the nearest-neighboring sites, there exists a characteristic value of the local resistance that determines the overall chain resistivity [90]. The corresponding energy ε_{opt} of one of the involved sites is, in some respect, similar to the transport energy introduced in two- and three-dimensional systems. Indeed, let $p(\varepsilon)$ be the probability for a site to have an energy above some value ε in a Gaussian DOS. Then the product

$$2[1-p(\varepsilon)]\left[-\frac{\mathrm{d}p(\varepsilon)}{\mathrm{d}\varepsilon}\right] \simeq \frac{2}{\sigma\sqrt{2\pi}}\exp\left(-\frac{\varepsilon^2}{2\sigma^2}\right) \tag{6.50}$$

is the probability distribution that one of the energies of a neighboring pair of sites is below ε, whereas the upper energy lies in the interval between ε and $\varepsilon + \mathrm{d}\varepsilon$. This equation gives therefore the distribution of the nearest-neighbor resistances. Averaging the resistances, one easily finds that the exponent of the average resistivity is determined by the energy $\varepsilon_{opt} \simeq \sigma^2/kT$ and the exponent is equal to $2^{-1}(\sigma/kT)^2 + \varepsilon_{opt}/kT - \varepsilon_{opt}^2/2\sigma^2 = (\sigma/kT)^2$. This leads to asymptotic temperature dependence described by Equation (6.9) with $C \simeq 1$. We see that this asymptotic temperature dependence of the drift mobility is determined by the sites with energies in the vicinity of $\varepsilon_{opt} \simeq \sigma^2/kT$. This energy increases with decreasing temperature and it might happen that in a finite system there are no sites with such high energies as ε_{opt}. If so, the temperature dependence of the resistivity and that of the carrier mobility should be weaker than the dependence described by Equation (6.9) with $C \simeq 1$. This effect has recently been observed in a straightforward computer simulation of the nearest-neighbor carrier hopping in a 1D chain [91]. Studying a finite system, Pasveer *et al.* [91] obtained a weakening of the temperature dependence of the carrier drift mobility with decreasing temperature in accord with the above arguments [90].

This mesoscopic effect should be taken into account when studying material properties at different samples. For time-of-flight studies, researchers prefer to deal with thick samples in order to obtain longer transit times and to determine the drift mobility with higher precision. However, for device applications thin samples are mostly used. One should be well aware that some properties of the thin samples used for device applications may essentially differ from those of thick samples used in the research. It is, for instance, true for the temperature dependences of the conductivity, as shown above.

So far we have considered systems without correlations between spatial positions of sites and their energies. In the next section we abandon this assumption and consider the effects of such correlations on the carrier drift mobility.

6.5.5 Drift mobility in the random-energy model with correlated disorder (CDM)

In order to construct a model with correlated disorder, we first generate provisional site energies ϕ_k with a Gaussian distribution

$$g(\phi) = \frac{1}{\sigma\sqrt{2\pi\lambda}} \exp\left(-\frac{\phi^2}{2\lambda\sigma^2}\right), \tag{6.51}$$

where λ is an integer number. The zero-field energy φ of a charge carrier on site k is then determined by the arithmetic average of provisional energies ϕ of neighboring sites:

$$\varphi_k = \sum_{j=-m}^{m} \phi_{k+j}. \tag{6.52}$$

Here $\lambda = 2m + 1$ is the correlation length in units of the intersite separation b. This recipe to create a system with correlated site energies is similar to that suggested by Garstein and Conwell [80]. The result is the set of site energies at zero field that have a Gaussian distribution described by Equation (6.4). Due to the correlation, two site energies φ_k and φ_{k+i} are not independent as long as $i \leq \lambda$. This makes the analytical calculation of the drift mobility slightly more elaborate than in the model without correlations described in Sections 6.5.3 and 6.5.4. Nevertheless, one can perform the averaging over disorder straightforwardly and obtain the following exact result for the carrier drift mobility in the infinite 1D chain in the form [15]

$$\begin{aligned}\mu = \frac{2\nu_0 b}{F}\Biggl(&\frac{1-\exp\left[\left(\frac{\sigma}{kT}\right)^2 - \frac{\lambda ebF}{kT}\right]}{1-\exp\left[\lambda^{-1}\left(\frac{\sigma}{kT}\right)^2 - \frac{ebF}{kT}\right]}\Biggl\{1+\operatorname{erf}\left(\frac{\lambda^{1/2}ebF}{2\sigma}\right)\\ &+\exp\left[\lambda^{-1}\left(\frac{\sigma}{kT}\right)^2 - \frac{ebF}{kT}\right]\left[1+\operatorname{erf}\left(\frac{\sigma}{\lambda^{1/2}kT} - \frac{\lambda^{1/2}ebF}{2\sigma}\right)\right]\Biggr\}\\ &+\frac{\exp\left[\left(\frac{\sigma}{kT}\right)^2 - \frac{\lambda ebF}{kT}\right]}{1-\exp\left(\frac{ebF}{kT}\right)}\Biggl\{1+\operatorname{erf}\left(\frac{\sigma}{\lambda^{1/2}kT} + \frac{\lambda^{1/2}ebF}{2\sigma}\right)\\ &+\exp\left(-\frac{ebF}{kT}\right)\left[1+\operatorname{erf}\left(\frac{\sigma}{\lambda^{1/2}kT} - \frac{\lambda^{1/2}ebF}{2\sigma}\right)\right]\Biggr\}\Biggr)^{-1}.\end{aligned} \tag{6.53}$$

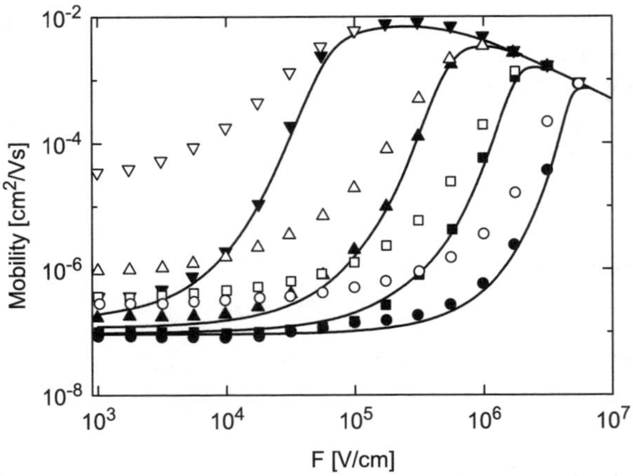

Figure 6.13 Influence of the space-energy correlations on the field dependence of the carrier mobility at $kT = 15\,\text{meV}$ and $\sigma = 50\,\text{meV}$. The different correlation lengths are $\lambda = 1$ (circles), $\lambda = 3$ (squares), $\lambda = 11$ (upward triangles), and $\lambda = 101$ (downward triangles). Open symbols show the averaged mobilities calculated by Equation (6.44). Solid symbols show the corresponding results obtained by averaging of the inverse mobilities. Solid lines show the solutions for the infinite chain given by Equation (6.53) (reproduced with permission from [15]. Copyright 2001 by the American Physical Society)

In Figure 6.13 the corresponding field dependences are plotted for different correlation lengths λ for finite chains (symbols) calculated via Equation (6.35) and (6.36) and for the infinite chain calculated via Equation (6.53). Results for finite chains of $N = 10^3$ sites were averaged over $c = 10^3$ realizations of the chain. The particular shape of the field dependence of the drift mobility in the CDM for the infinite system is considered in detail in the papers of Garstein and Conwell [80], Dunlap *et al.* [77] and Novikov *et al.* [81, 82] and we do not discuss it here.

Instead we would like to focus our attention on the other aspect of the phenomenon, namely, on the increasing difference between the results obtained by averaging of mobilities (open symbols in Figure 6.13) and those obtained by averaging the inverse mobilities or transit times (filled symbols in Figure 6.13) with increasing correlation length. This trend is clearly related to the smaller number of high barriers in systems with longer correlation lengths, which favors the mesoscopic effects. The increase of the correlation length in the CDM is analogous to the decrease of the total number of sites in the finite chains in the GDM.

6.5.6 Hopping in 1D systems: beyond the nearest-neighbor approximation

So far we have considered hopping in various 1D systems, taking into account carrier transitions only between the nearest sites. The question then arises of how transitions to further neighbors on the chain can modify the results obtained. The VRH philosophy tells us that the temperature dependence of the conductivity and that of the carrier drift mobility should be determined at low temperatures by transitions of charge carriers to further neighbors than the nearest ones. The effect of such long-distance tunneling transitions on the temperature dependences of the kinetic coeffcients in 1D systems was studied analytically by Zvyagin *et al.* [90] and by computer simulations by Kohary *et al.* [16] and by Pasveer *et al.* [91]. Zvyagin *et al.* [90] considered only the GDM model and showed that transitions to m neighbors beyond the nearest ones lead to the asymptotic temperature dependence described by Equation (6.9) with $C = \sqrt{(m+1)/2m}$. For instance, hopping to the second nearest neighbors leads to Equation (6.9) with $C = \sqrt{3/4}$. However, this asymptotic behavior with $C = \sqrt{(m+1)/2m}$ can hardly be achieved at realistic conditions. Zvyagin *et al.* [90] have shown that even at $(\sigma/kT)^2 \simeq 100$, the coeffcient C in Equation (6.9) is close to 0.8.

An analogous result on the weak effect of distant transitions has been obtained by computer simulations [16]. Kohary *et al.* [16] studied by computer simulations hopping transport in 1D GDM and CDM systems using transition rates in the form of Equation (6.5) instead of Equation (6.33) that was used for the nearest-neighbor hopping. In these simulations the lattice constant $b = 0.36\,\text{nm}$ was chosen as known for discotic liquid-crystalline glasses, and the values of parameter α in Equation (6.5) were varied between 0.1 and 0.3 nm. This parameter determines the decay length of the carrier wavefunction in the localized states. Transitions to six neighbors in each direction were allowed in the simulation procedure. The main result of the simulations is the following. For the CDM, i.e., for a system with correlated disorder, tunneling to further sites than the nearest neighbors does not play any essential role, while for the GDM, i.e., for uncorrelated systems, these distant hopping transitions slightly affect the transport coeffcients. The unimportance of distant transitions for systems with correlated disorder is not surprising. In such systems, energies of neighboring sites are close to each other due to correlation effects. Thus the variable-range hopping is not significant for such systems, because the closest energy and space states are the nearest neighbors. The quantitative confirmation of this qualitatively clear statement [16] implies that for systems with correlated disorder one can use analytic theories described above and one can be sure that the restriction of the analytic theory which considers only the nearest-neighbor hopping is not essential for the results obtained.

For systems with uncorrelated disorder the result of the simulation is also rather optimistic with respect to the validity of the exact analytic solutions described above. Even for a rather high value of the decay constant $\alpha = 0.2\,\text{nm}$, which makes the distant hops favorable, the magnitude of the coeffcient $C \simeq 0.9$ in Equation (6.9) [16] is not much different from its value for the nearest-neighbor hopping. Thus, even for systems with uncorrelated disorder, exact analytic solutions obtained in previous sections, taking into account only the nearest-neighbor hopping, give reasonable values for transport coeffcients [16].

6.6 ON THE RELATION BETWEEN CARRIER MOBILITY AND DIFFUSIVITY IN DISORDERED ORGANIC SYSTEMS

High interest of researchers in charge transport in organic disordered systems motivated numerous studies of the relation between such kinetic coeffcients as the mobility μ and the diffusion coeffcient D of charge carriers in such systems. However, one can find rather contradictory statements on this subject in the scientific literature and therefore we would like to clarify this problem.

Borsenbereger *et al.* [92] studied experimentally the behavior of the ratio eD/μ in 1,1-bis(di-4-tolylamionophenyl)cyclohexane (TAPC) and found that this ratio increases approximately linearly with applied electric field at high fields. In the low-field region, the ratio eD/μ becomes independent of the field strength F. Nevertheless, the limiting value of this ratio does not agree with that given by the classical Einstein relationship. The latter reads [93]

$$\mu = \frac{e}{kT} D. \tag{6.54}$$

Also Monte Carlo computer simulations were carried out that claimed invalidity of Equation (6.54) for the relation between the mobility and diffusivity of hopping carriers in a random system with Gaussian energy distribution of site energies [94–96]. The amount of disorder is characterized by the energy scale σ of the DOS described by Equation (6.4). With increasing disorder and field, significant deviations from Einstein's law were obtained in the simulations. These results contributed to the general belief of many researchers that the Einstein law expressed by Equation (6.54) is violated in random media.

Discussing the validity or invalidity of the Einstein relationship for hopping electrons one should clearly distinguish between the equilibrium and nonequilibrium conditions on one hand and between the degenerate and nondegenerate systems of charge carriers on the other. Furthermore, one should distinguish between the regime of low electric fields with field-independent transport coeffcients and the regime of high fields, in which the nonlinear effects caused by electric fields become significant. Computer simulations [94–96] have shown that at high fields in a nonlinear transport regime the diffusion coeffcient D depends more strongly on the electric field than the carrier drift mobility μ. Therefore the ratio eD/μ increases with increasing field and the relation described by Equation (6.54) is violated. We will not consider this nonlinear regime here due to the rather complicated definition of the diffusion constant in the case of high electric fields which cause strong anisotropy in the diffusion process. Interested readers can find the necessary information in the literature [94, 95]. Instead we discuss below the regime of low electric fields with field-independent transport coeffcients D and μ.

As shown in Section 2.5 of Chapter 2 [14] Einstein's law cannot be valid in the strong nonequilibrium conditions at low temperatures, when transport processes are governed by the downward energy hopping relaxation of charge carriers. For the energy-loss hopping in the exponential DOS, a relation between D and μ similar to that in Equation (6.54) was found, although thermal energy kT in this relation is replaced by the energy scale of the DOS (see Equation (2.53) in Chapter 2). It is, unfortunately, not possible to obtain such a relation for the energy-loss hopping in a Gaussian DOS. The exponential DOS considered in Chapter 2 represents the only exception among possible DOS functions, for which the relation between $\mu(\varepsilon)$ and $D(\varepsilon)$ is independent of the localization energy ε and therefore the relation between the effective D and μ for the whole system of charge carriers performing the energy-loss hopping can be formulated in the universal form (see Equation (2.53) in Chapter 2). Therefore we will not consider nonequilibrium conditions for hopping in the Gaussian DOS and restrict our consideration here to studying the validity of the Eisntein law in the equilibrium conditions.

The problem of the relation between D and μ in thermal equilibrium for a disordered system with Gaussian DOS has recently been solved by Roichman and Tessler [97]. The authors used the formula derived for a general energy distribution of charge carriers and a general DOS function [98, 99]

$$\mu = eD\frac{1}{n}\frac{\partial n}{\partial \varepsilon_{\mathrm{F}}}, \tag{6.55}$$

where n is the total concentration of charge carriers and ε_{F} is the Fermi energy. Using Equation (6.25), one can rewrite this expression for a given density of states $g(\varepsilon)$ in the form [97]

$$\mu = \frac{e}{kT}D\frac{\int_{-\infty}^{\infty} \mathrm{d}\varepsilon\, g(\varepsilon)\dfrac{\exp[(\varepsilon-\varepsilon_F)/kT]}{\{1+\exp[(\varepsilon-\varepsilon_F)/kT]\}^2}}{\int_{-\infty}^{\infty} \mathrm{d}\varepsilon\, g(\varepsilon)\dfrac{1}{1+\exp[(\varepsilon-\varepsilon_F)/kT]}}. \tag{6.56}$$

In the case when the Fermi energy is very low and the major part of the carrier energy distribution is far above ε_{F} , the Fermi distribution can be replaced by the Boltzmann function and the ratio of integrals in Equation (6.56) becomes unity. This corresponds to the nondegenerate energy distribution of charge carriers. In such a case, the generalized Einstein relation described by Equation (6.56) approaches the classical form represented by Equation (6.54). For the Gaussian DOS described by Equation (6.4) this condition is valid at small disorder parameter σ for deep Fermi energy. In Figure 6.14 the ratio $\mu kT/eD$ is plotted for Gaussian DOS as a function of the ratio $\varepsilon_{\mathrm{F}}/kT$ for different values of σ/kT (following Roichman and Tessler [97]). Using for the disorder parameter realistic values $\sigma \simeq 0.1\,\mathrm{eV}$, one comes to the conclusion that at room temperature the classical Einstein relation can hold only at rather small concentrations of charge carriers. Computer simulations for the ratio μ/D in such conditions in Gaussian DOS are described in [100]. For a

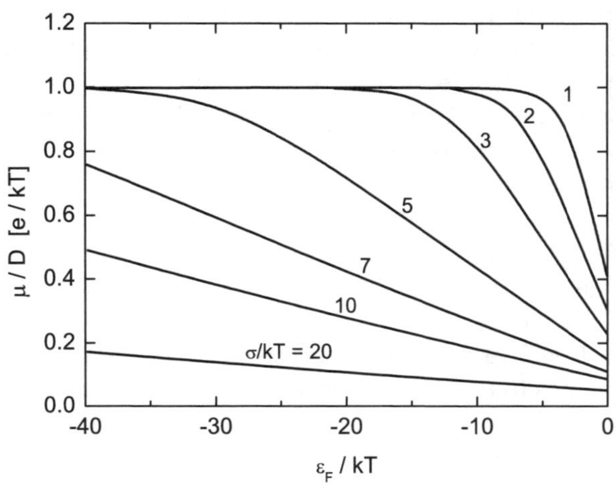

Figure 6.14 The ratio μ/D as a function of the Fermi level $\varepsilon_{\mathrm{F}}/kT$ for different σ/kT (reproduced with permission from [97], Copyright 2002, American Institute of Physics)

degenerate system the generalized Einstein relation has to be calculated in its full form using Equation (6.56) with $g(\varepsilon)$ described by Equation (6.4) [97].

A question could arise as to why some computer simulations provide results that look contradictory to the above conclusions. For instance, it has been claimed on the basis of straightforward Monte Carlo computer simulations that, in disordered systems with Gaussian DOS, hopping mobility and diffusivity do not obey the classical Einstein relation given by Equation (6.54) even in the case of noninteracting carriers when the latter are considered as independent entities [96]. On the other hand, Equation (6.56) for a charge carrier in an empty system should coincide with Equation (6.54). This problem has been solved in [100], where it was shown that in computer simulations that demonstrate apparent deviations from Equation (6.54) the system of charge carriers was not in thermal equilibrium because of the particular choice of the simulation parameters. As soon as the equilibrium conditions were established, the relation between D and μ for independent carriers became in agreement with Equation (6.54) [100].

6.7 ON THE DESCRIPTION OF COULOMB EFFECTS CAUSED BY DOPING IN DISORDERED ORGANIC SEMICONDUCTORS

One of the interesting topics in research on disordered materials is the effect of Coulomb potentials of charged species on transport properties. The decisive role of such effects for various charge transport phenomena has been clarified for hopping transport in doped crystalline materials and in inorganic disordered materials such as amorphous semiconductors. This topic has already become a subject of textbooks [17, 101]. With respect to disordered organic materials the situation is not as favorable, although Coulomb effects in such materials can play even a more pronounced role. The dielectric constant of the organic matrix is usually several times smaller than that in inorganic materials. This should make Coulomb effects more significant in organic materials. Not much has been done yet in the study of Coulomb potentials in organic disordered solids and only several initial treatments have been attempted so far (see, for instance [102]). Unfortunately, the results of these treatments are in contradiction to the results on Coulomb effects from textbooks devoted to inorganic materials. This situation resembles that of the initial study of hopping transport in organics. Researchers began this study not trying to use the experience already gathered in the field of inorganic systems. Therefore we would like to analyze briefly the shortcomings of the suggested treatments of Coulomb effects in organic materials in order to warn researchers with respect to possible drawbacks in such treatments.

It has been established experimentally that the doping effciency of disordered materials is much lower than that of crystalline semiconductors. For example, in amorphous inorganic semiconductors, such as hydrogenated amorphous silicon, a-Si:H, the concentration of impurities as deduced from experiments involving electronic states is considerably lower than that determined from the study of local bonding configurations by extended X-ray fine structure or nuclear magnetic resonance [103]. Also in disordered organic materials it has been claimed that at low doping levels, electrochemical doping is much less effcient than the field-effect doping in which the same amount of charge carriers is injected into the system without inducing chemically foreign impurities [104, 105]. Moreover, at low dopant concentrations, chemical doping can even lead to decreasing carrier mobility [106]. At high doping levels the mobility steeply increases with dopant concentration [104–106]. Qualitatively this result was interpreted by assuming that in chemically doped materials charge carriers can be trapped by Coulomb potentials of ionized dopant species at low doping levels [105]. Concomitantly, the carrier mobility could be much smaller than the field-effect mobility measured without introducing charged dopants into the sample. At high doping levels in the electrochemical process the energy landscape might become more uniform, leading to the increase of the carrier mobility [105].

Two identical attempts [102, 107] were recently performed in order to put this argument onto a quantitative theoretical basis. We briefly describe these attempts and show that the set of equations suggested in such an approach is irrelevant for the description of the problem under study since neither charge neutrality nor screening effects were properly taken into account.

The authors considered an array of localized states of the host system with some high concentration $N_t \sim 10^{21}\,\text{cm}^{-3}$ and an array of dopant atoms with much lower concentration $N_d \sim 10^{18}\,\text{cm}^{-3}$. To be specific we will consider the case of donors as dopant species. The case of acceptors can be treated in an analogous way. Donors are supposed to give their electrons into localized states of the host system and to become positively charged. Arkhipov *et al.* [102, 107] considered a localized state and estimated the Coulomb energy shift of this state due to the presence of the charged donors with concentration N_d. They considered the contribution to the Coulomb potential from only the nearest donor. The probability density, $w(r)$, of having a nearest donor at a distance r to a chosen localized state is determined by the Poisson distribution

$$w(r) = 4\pi r^2 N_d \exp\left(-\frac{4\pi}{3} N_d r^3\right). \tag{6.57}$$

The energy of the localized state under consideration is shifted downward by the Coulomb potential of the nearest donor. Let the energy of this localized state in the absence of donors be ε. In [102, 107] it is argued that in the doped system the energy of this state becomes equal to $E = \varepsilon + E_c$, where

$$E_c = -\frac{e^2}{\kappa r} \tag{6.58}$$

is the energy shift due to the Coulomb potential of the nearest donor. In Equation (6.58) e is the electron charge and κ is the dielectric constant of the host material. A weak doping was considered when the concentration of dopant ions N_d remains much smaller than the total density of intrinsic hopping sites N_t. The authors claim that, under these conditions, the energy of almost every localized state will be essentially affected only by the nearest dopant ion. In such a case one can easily find the distribution $G(E)$ of site energies, $E = \varepsilon + E_c$ from the given density of states (DOS) for intrinsic energies $g(\varepsilon)$ and from Equations (6.57) and (6.58) for the distribution of Coulomb energies E_c. Due to the effect of the positively charged donors, the distribution $G(E)$ appears shifted to lower energies as compared with the distribution of intrinsic energies $g(\varepsilon)$. Hence the extra electrons supplied into the system by chemical doping are not forced to fill the intrinsic density of states (thus pushing the Fermi level upwards), but rather they fill the density of states already shifted towards lower energies by Coulomb potentials of the positively charge donors [102, 107]. Therefore the increase of the Fermi level with doping occurs much more slowly than would be the case in the intrinsic DOS $g(\varepsilon)$ filled by the same amount of electrons. After having calculated $G(E)$, Arkhipov *et al.* [102] studied the hopping transport of electrons, treating the system as spatially homogeneous with DOS $G(E)$. We have to point out the deficiency of such treatment.

In the picture described above, one assumes that only a single donor, namely, the nearest one, causes the Coulomb shift of the energy of a localized state in the host material [102, 107]. In [102, 107] it is claimed that this assumption is justified by the inequality $N_d \ll N_t$. Moreover the authors claim that this assumption under condition $N_d \ll N_t$ is obvious [102]. We argue however that this assumption is neither obvious, nor correct. Of course, the contribution of donors to the Coulomb potential on a localized site decreases with the distance from the site as r^{-1}. However, the number of donors in a sphere of radius r around the chosen site increases as r^3. Therefore, the contribution to the Coulomb potential of a chosen hopping site from more distant donors than the nearest one increases proportional to r^2. In the absence of screening, as considered in [107], distant donors contribute more to the Coulomb energy shifts on localized states than the nearest ones. The condition expressed via inequality $N_d \ll N_t$ is irrelevant for this conclusion. Furthermore, one should realize that, in the picture suggested in [102, 107], the energy diverges if one considers only donors as charged centers and takes into account the Coulomb contributions to the energy of an intrinsic site from more and more distant donors. This is a trivial result discussed in several textbooks (see, for example, [17, 101]). In order to avoid the divergence of the Coulomb energy, one should not restrict the consideration to charges of only a given polarity as done in [102, 107]. Instead, one should consider both kinds of charges—positive and negative—keeping the system electrically neutral. Herewith we come to the important though trivial question: are electrons electrically charged? The answer to this question is definitely 'yes'. A donor becomes charged positively only because it can get rid of a valence electron. In [102, 107], electrons brought into the system by donors were given just a passive role to fill the density of states shifted to lower energies by positively charged donors. One should, however, take into account that electrons are also charged with opposite polarity to that of the donors. The concentration of the negatively charged extra electrons introduced into the system by donors is equal to that of charged donors N_d. These electrons are the cause for the effect on the energies of intrinsic sites, which is exactly opposite to that of positively charged donors exclusively considered in [102, 107]. Being negatively charged, electrons shift the energies of intrinsic hopping sites upward. The authors of [102, 107] deliberately took into account only positive charges. If one would do the same, taking into account only negative charges of extra electrons, one would come to the conclusion exactly opposite to that in [102, 107], namely to the conclusion that DOS function would be shifted upward in energy with respect to that in an undoped sample. Of course, none of such deliberate considerations can be correct. One should consider an electrically neutral system, taking into account both positively charged donors and negatively charged electrons as described in textbooks [17, 101]. The crucial point in such treatments is the question of the spatial distribution of charges. Donors are assumed to be distributed randomly in space. What about electrons?

Let us, following [102, 107], consider a lightly doped sample when condition $N_d \ll N_t$ is fulfilled. This assumption is plausible since estimates give $N_d \sim 10^{18}\,\mathrm{cm}^{-3}$ and $N_t \sim 10^{21}\,\mathrm{cm}^{-3}$

[102, 107]. Condition $N_d << N_t$ cannot justify that only the nearest donor to a localized state should be taken into account, though this strong inequality will help us to answer the question of where electrons are situated. A Coulomb energy shift of an intrinsic localized state from the nearest donor is described by Equations (6.57) and (6.58). It is determined by the typical distance between donors $r_d \simeq N_d^{-1/3}$. Inserting such estimate in Equation (6.57), and taking $\kappa = 3$ as known for organic semiconductors, one obtains the typical Coulomb contribution to the energy of a localized state from the nearest donor of the order $E_c \sim 0.08\,\text{eV}$ for $N_d \sim 10^{18}\,\text{cm}^{-3}$ [102, 107]. Since the width of the energy distribution $g(\varepsilon)$ of intrinsic hopping sites is of the order of ~0.1 eV, the effect of Coulomb centres on intrinsic localized states seems essential. We claim however that in the above picture one should consider not a localized state and the nearest donor to it, but rather the localized state which is the nearest to a donor. While the typical distance between a localized state and the nearest donor is determined by the distance of the order $r_d \simeq N_d^{-1/3}$, the distance between a donor and the nearest localized state is determined by the distance $r_t \simeq N_t^{-1/3}$, which is much smaller than r_d. Inserting such a value for r_t into Equation (6.58) and taking for the concentration of localized states the magnitude $N_t \sim 10^{21}\,\text{cm}^{-3}$, as suggested in [102, 107], we find that the Coulomb energy shift of the localized state nearest to a donor E_c is about 0.8 eV. This value is by an order of magnitude larger than the shift $E_c \sim 0.08\,\text{eV}$ caused on a localized state by the donor nearest to it. In the ground state of the system, electrons would tend to occupy deeper energy levels and therefore they will be situated on the intrinsic sites, which are closest to dopant ions (donors). Therefore one should conclude that most charges brought into the system by donors are gathered into dipoles formed by positively charged donors and the negatively charged intrinsic centers nearest to them. One could suppose that in [102, 107] the effect with typical scale of about 0.1 eV was considered, neglecting the effect with typical scale of about 1 eV. The situation is, however, worse. We have just seen that due to the effect of the typical scale of ~1 eV, the effect of the scale ~0.1 eV considered in [102, 107] does not exist, since not the point charges, but rather very short dipoles affect localized states in the intrinsic material. The extra electrons brought by donors are trapped into states with very deep energies ($\simeq -1\,\text{eV}$ at $N_t = 10^{21}\,\text{cm}^{-3}$) which are the nearest to donors. This might be the reason why no effect of increasing conductivity has been observed experimentally at low dopant concentrations when condition $N_d << N_t$ is fulfilled. The effect of heavy doping when the strong inequality $N_d << N_t$ breaks down needs a special treatment, which is beyond our scope. Let us instead estimate the effect of Coulomb potentials on a given intrinsic localized state in a lightly doped material.

Following [102, 107] we consider a given localized site in the intrinsic material. The distance $r_d \simeq N_d^{-1/3}$ from this localized site to the nearest dipole is determined by the concentration of dipoles, which is equal to the concentration of donors N_d. The length of the dipole is determined by the concentration of localized states $r_t \simeq N_t^{-1/3}$. The contribution of such a dipole to the energy of a given localized state is of the order

$$E_c \simeq -\frac{e^2}{\kappa r_d}\frac{r_t}{r_d} \tag{6.59}$$

This equation replaces the estimate given by Equations (6.57) and (6.58). The result of Equation (6.59) for the energy shift E_c on 'a localized state' is by an order of magnitude smaller than that obtained in [102, 107], since in Equation (6.59) the Coulomb contribution of a point charge at a distance r_d is multiplied by a small factor r_t/r_d, which is equal to 0.1

for the ratio $N_d/N_t = 10^{-3}$ chosen in [102, 107]. This contribution is less than 0.001 eV and it can be neglected.

One should emphasize that the study of the effects of Coulomb potentials on charge transport in disordered organic materials is still in its initial phase and more study is needed to clarify the role of these effects. In particular, the calculation of transport properties affected by charge distribution at high doping levels is a challenging theoretical problem, which still awaits its solution.

6.8 CONCLUDING REMARKS

In this chapter we have presented several basic concepts developed for description of charge carrier transport in organic disordered semiconductors, such as molecularly doped, conjugated polymers, and organic glasses. These concepts are, to a great extent, analogous to the theoretical concepts developed earlier for description of charge transport in inorganic disordered materials such as amorphous and microcrystalline semiconductors. Therefore, we have tried to keep the presentation of these ideas parallel to that in Chapter 2 of this book [14]. However, contrary to Chapter 2, two important topics were not considered here—the thermally stimulated currents and the nonlinear transport effects in high electric fields. The description of thermally stimulated currents in organic disordered materials can be found in the work of Schmechel and von Seggern [108]. The description of the nonlinear field effects in such materials can be found in [77, 80–82]. Furthermore, we focused in this chapter only on the description of the motion of charge carriers through the disordered material. Such important topics as the injection of charge carriers from the contacts into the system as well as the description of the space-charge-limited currents remained beyond our scope. Readers interested in these topics can find comprehensive descriptions, for example, in the recent review article of H. Bäassler [4] and in Chapter 7 of this book [3].

A comparison between the results of this chapter and those described in Chapter 2 of this book show the role of the DOS function on the transport phenomena. Although the transport concepts used in these two chapters are very similar to each other, some results for Gaussian DOS differ essentially from those for the exponential DOS. Therefore we have considered the charge transport effects for the Gaussian DOS and for the exponential DOS in two separate chapters.

ACKNOWLEDGEMENTS

The authors are indebted to numerous colleagues for stimulating and enlightening discussions. In particular, we would like to express our gratitude to Igor Zvyagin (Moscow State University) for clarifying to us the role of distant hops in 1D systems and to Peter Thomas (Philipps University Marburg) for close collaboration on topics discussed in this chapter. Financial support of the Deutsche Forschungsgemeinschaft, of the Fonds der Chemischen Industrie, of the Optodynamic Centre at the Philipps University Marburg, of the European Community [IP 'FULLSPECTRUM' (Ref. N: SES6-CT-2003-502620)] and that of the European Graduate College 'Electron-Electron Interactions in Solids' Marburg–Budapest is gratefully acknowledged.

REFERENCES

[1] S. Roth, Hopping conduction in electrically conducting polymers. in: *Hopping Transport in Solids*. M. Pollak and B.I. Shklovskii, (eds), Elsevier, 1991, 377.

[2] N. Naarmann, in: *Electronic Properties of Conjugated Polymers*. H. Kuzmany, M. Mehring, and S. Roth, (eds), Springer, Heidelberg, 1987, 12.

[3] E. von Hauff, C. Deibel, and V. Dyakonov, Device applications of organic materials. in: *Charge transport in disordered solids with applications in electronics*. S.D. Baranovski, (ed.), John Wiley & Sons, Ltd, Chichester, 2006.

[4] H. Bässler, Charge transport in random organic semiconductors. in: *Semiconducting Polymers*. G. Hadziioannou and P.F. van Hutten, (eds), John Wiley & Sons, Inc., New York, 2000, 365.

[5] G. Hadziioannou and P.F. van Hutten, *Semiconducting Polymers*. John Wiley & Sons, Inc., New York, 2000.

[6] C. Brabec, V. Dyakonov, J. Parisi, and N.S. Sariciftci, *Organic Photovoltaics: Concepts and Realization*. Springer, Berlin, 2003.

[7] M. Pope and C.E. Swenberg, *Electronic Processes in Organic Crystals and Polymers*. Oxford University Press, Oxford, 1999.

[8] H. Bässler, *Phys. Stat. Sol. (b)*, **175**, 15 (1993).

[9] P.M. Borsenberger, E.H. Magin, and J. Shi, *J. Phys. B*, **217**, 212 (1996).

[10] P.M. Borsenberger, W.T. Greenbaum, and E.H. Magin, *Physica B*, **228**, 226 (1996).

[11] M. Abkowitz, *Phil. Mag. B*, **65**, 817 (1992).

[12] P.M. Borsenberger, E.H. Magin, M. van der Auweraer, and F.C. de Schryver, *Phys. Stat. Sol. (a)*, **140**, 9 (1993).

[13] M. van der Auweraer, F.C. de Schryver, P.M. Borsenberger, and H. Bässler, *Adv. Matter.*, **6**, 199 (1994).

[14] S.D. Baranovski and O. Rubel, Description of charge transport in amorphous semiconductors. in: *Charge transport in disordered solids with applications in electronics*. S.D. Baranovski, (ed.), John Wiley & Sons, Ltd, Chichester, 2006.

[15] H. Cordes, S.D. Baranovski, K. Kohary, P. Thomas, S. Yamasaki, F. Hensel, and J.H. Wendorff, *Phys. Rev. B*, **63**, 094201 (2001).

[16] K. Kohary, H. Cordes, S.D. Baranovski, P. Thomas, S. Yamasaki, F. Hensel, and J.H. Wendorff, *Phys. Rev. B*, **63**, 094202 (2001).

[17] B.I. Shklovskii and A.L. Efros, *Electronic Properties of Doped Semiconductors*. Springer, Heidelberg, 1984.

[18] V.I. Arkhipov and H. Bässler, *Phil. Mag. Lett.*, **67**, 343 (1993).

[19] V.I. Arkhipov and H. Bässler, *Phil. Mag. Lett.*, **69**, 241 (1994).

[20] V.I. Arkhipov and H. Bässler, *Phil. Mag. B*, **68**, 425 (1993).

[21] V.I. Arkhipov and H. Bässler, *Phil. Mag. B*, **70**, 59 (1994).

[22] V.I. Arkhipov, E.V. Emelianova, and H. Bässler, *Phil. Mag. B*, **81**, 985 (2001).

[23] V.I. Arkhipov, E.V. Emelianova, and G.J. Adriaenssens, *Phys. Rev. B*, **64**, 125125 (2001).

[24] V.I. Arkhipov, E.V. Emelianova, and H. Bässler, *J. Non-Cryst. Solids*, **299–302**, 1047 (2002).

[25] V.I. Arkhipov, P. Heremans, E.V. Emelianova, G.J. Adriaenssens, and H. Bässler, *J. Phys.: Condens. Matter*, **14**, 9899 (2002).

[26] I.I. Fishchuk, A. Kadashchuk, H. Bässler, and S. Nespurek, *Phys. Rev. B*, **67**, 224303 (2003).

[27] Y. Roichman and N. Tessler, *Synth. Metals*, **135–136**, 443 (2003).

[28] Y. Roichman, Y. Preezant, and N. Tessler, *Phys. Stat. Sol. (a)*, **201**, 1246 (2004).

[29] B. Ries, H. Bässler, and M. Silver, *Phil. Mag. B*, **54**, 141 (1986).

[30] O. Rubel, S.D. Baranovskii, P. Thomas, and S. Yamasaki, *Phys. Rev. B*, **69**, 014206 (2004).

[31] S.D. Baranovski, H. Cordes, F. Hensel, and G. Leising, *Phys. Rev. B*, **62**, 7934 (2000).

[32] W.D. Gill, *J. Appl. Phys.*, **43**, 5033 (1972).
[33] W.D. Gill, in J. Stuke and W. Brenig, (eds), *Proceedings of the Fifth International Conference of Amorphous and Liquid Semiconductors*, Taylor Francis, London, 1974, 901.
[34] G. Horowitz, Physics of organic field-effect transistors. in: *Semiconducting polymers*. G. Hadziioannou and P.F. van Hutten, (eds), John Wiley & Sons, Inc., New York, 2000, 463.
[35] S.J. Santos Lemus and J. Hirsch, *Phil. Mag. B*, **53**, 25 (1986).
[36] T. Holstein, *Phil. Mag. B*, **37**, 49 (1978).
[37] H. Scher and T. Holstein, *Phil. Mag. B*, **44**, 343 (1981).
[38] A. Miller and E. Abrahams, *Phys. Rev.*, **120**, 745 (1960).
[39] B. Movaghar, M. Grünewald, B. Ries, H. Bässler, and D. Würtz, *Phys. Rev. B*, **33**, 5545 (1986).
[40] H. Bässler, Transport and relaxation of excitations in random organic solids: Monte Carlo simulation and experiment. in: *Advances in Disordered Semiconductors*. M. Pollak, and H. Fritzsche, (eds), Volume 2. World Scientific, Singapore, 1990, 491.
[41] G. Schönherr, H. Bässler, and M. Silver, *Phil. Mag. B*, **44**, 369 (1981).
[42] M. Grünewald, B. Pohlmann, B. Movaghar, and D. Würtz, *Phil. Mag. B*, **49**, 341 (1984).
[43] A. Nemeth-Buhin and C. Juhasz, in O. Millo and Z. Ovadyahu (eds), *Hopping and Related Phenomena*, Jerusalem, Israel, 27–30 August, 1995.
[44] A. Ochse, A. Kettner, J. Kopitzke, J.H. Wendorff, and H. Bässler, *Phys. Chem. Chem. Phys.*, **1**, 1757 (1999).
[45] C. Gourdon and P. Lavallard, *Phys. Stat. Sol. (b)*, **153**, 641 (1989).
[46] V. Ambegaokar, B.I. Halperin, and J.S. Langer, *Phys. Rev. B*, **4**, 2612 (1971).
[47] N.W. Dalton, C. Domb and M.F. Sykes, *Proc. Phys. Soc.*, **83**, 496 (1964).
[48] C. Domb and N.W. Dalton, *Proc. Phys. Soc.*, **89**, 859 (1966).
[49] M.C.J.M. Vissenberg and M. Matters, *Phys. Rev. B*, **57**, 12964 (1998).
[50] S.D. Baranovski, I.P. Zvyagin, H. Cordes, S. Yamasaki, and P. Thomas, *Phys. Stat. Sol. (b)*, **230**, 281 (2002).
[51] S.D. Baranovski, I.P. Zvyagin, H. Cordes, S. Yamasaki, and P. Thomas, *J. Non-Cryst. Solids*, **299–302**, 416 (2002).
[52] B. Hartenstein and H. Bässler, *J. Non-Cryst. Solids*, **190**, 112 (1995).
[53] D. Monroe, *Phys. Rev. Lett.*, **54**, 146 (1985).
[54] M. Grünewald, and P. Thomas, *Phys. Stat. Sol. (b)*, **94**, 125 (1979).
[55] F.R. Shapiro and D. Adler, *J. Non-Cryst. Solids*, **74**, 189 (1985).
[56] B.I. Shklovskii, E.I. Levin, H. Fritzsche, and S.D. Baranovski, Hopping photoconductivity in amorphous semiconductors: dependence on temperature, electric field and frequency. in: *Advances in Disordered Semiconductors*. H. Fritzsche, (ed.), Volume 3. World Scientific, Singapore, 1990, 161.
[57] S.D. Baranovski, P. Thomas, and G.J. Adriaenssens, *J. Non-Cryst. Solids*, **190**, 283 (1995).
[58] S.D. Baranovski, T. Faber, F. Hensel, and P. Thomas, *J. Phys. C*, **9**, 2699 (1997).
[59] V.I. Arkhipov, U. Wolf, and H. Bässler, *Phys. Rev. B*, **59**, 7514 (1999).
[60] S.D. Baranovski and A.L. Efros, *Sov. Phys. Semicond.*, **12**, 1328 (1978).
[61] H.C.F. Martens, I.N. Hulea, I. Romijn, H.B. Brom, W.F. Pasveer, and M.A.J. Michels, *Phys. Rev. B*, **67**, 121203 (2003).
[62] R. Schmechel, *Phys. Rev. B*, **66**, 235206 (2002).
[63] R. Schmechel, *J. Appl. Phys.*, **93**, 4653 (2003).
[64] D.M. Pai, J.F. Yanus, and M. Stolka, *J. Phys. Chem.*, **88**, 4714 (1984).
[65] A. Kadashscuk, Y. Skryshevskii, A. Vakhnin, N. Ostapenko, V.I. Arkhipov, E.V. Emelianova, and H. Bässler, *Phys. Rev. B*, **63**, 115205 (2003).
[66] V.I. Arkhipov, P. Heremans, E.V. Emelianova, G.J. Adriaenssens, and H. Bässler, *Appl. Phys. Lett.*, **82**, 3245 (2003).
[67] N.F. Mott, *Phil. Mag.*, **19**, 835 (1969).

[68] R. Coehoorn, W.F. Pasveer, P.A. Bobbert, and M.A.J. Michels, *Phys. Rev. B*, **72**, 155206 (2005).
[69] P.M. Borsenberger and D.S. Weiss, *Organic Photoreceptors for Xerography*. Marcel Dekker, New York, 1998.
[70] D.M. Pai, *J. Chem. Phys.*, **52**, 2285 (1970).
[71] L.B. Schein, D. Glatz, and J.C. Scott, *Phys. Rev. Lett.*, **65**, 472 (1990).
[72] L.T. Pautmeier, J.C. Scott, and L.B. Schein, *Chem. Phys. Lett.*, **197**, 568 (1992).
[73] A. Bacher, I. Bleyl, C.H. Erdelen, D. Haarer, W. Paulus, and H.W. Schmidt, *Adv. Mater.*, **9**, 1031 (1997).
[74] T. Christ, B. Glüsen, A. Greiner, A. Kettner, R. Sander, V. Stümpflen, V. Tsukruk, and J.H. Wendorf, *Adv. Mater.*, **9**, 48 (1997).
[75] N. Boden, R.J. Bushby, J. Clements, B. Movaghar, K.J. Donovan, and T. Kreozis, *Phys. Rev. B*, **52**, 13274 (1995).
[76] I. Bleyl, C. Erdelen, H.W. Schmidt, and D. Haarer, *Phil. Mag. B*, **79**, 463 (1999).
[77] D.H. Dunlap, P.E. Parris, and V.M. Kenkre, *Phys. Rev. Lett.*, **77**, 542 (1996).
[78] K. Seki and M. Tachiya, *Phys. Rev. B*, **65**, 014305 (2002).
[79] R.A. Marcus, *J. Chem. Phys.*, **24**, 966 (1956).
[80] Y.N. Garstein and E.M. Conwell, *Chem. Phys. Lett.*, **245**, 351 (1995).
[81] S.V. Novikov, D.H. Dunlap, V.M. Kenkre, P.E. Parris, and V. Vannikov, *Phys. Rev. Lett.*, **81**, 4472 (1998).
[82] S.V. Novikov, D.H. Dunlap, and V.M. Kenkre, *Proc. SPIE*, **3471**, 181 (1998).
[83] B. Derrida, *J. Stat. Phys.*, **31**, 433 (1983).
[84] K.P.N. Murthy and K.W. Kehr, *Phys. Rev. A*, **40**, 2082 (1989).
[85] K.P.N. Murthy and K.W. Kehr, *Phys. Rev. A*, **41**, 1160 (1990).
[86] A. Peled and L.B. Schein, *Chem. Phys. Lett.*, **153**, 422 (1988).
[87] L.B. Schein, *Phil. Mag. B*, **65**, 795 (1992).
[88] A. Hirao, H. Nishiwawa, and M. Sugiuchi, *Phys. Rev. Lett.*, **75**, 1787 (1995).
[89] M.E. Raikh and I. Ruzin, Transmittancy fluctuations in randomly nonuniform barriers and incoherent mesoscopics. in: *Mesoscopic Phenomena in Solids*. B. Altshuler, P.A. Lee, and R.A. Webb, (eds), North-Holland, Amsterdam, 1991, 315.
[90] I.P. Zvyagin, S.D. Baranovski, K. Kohary, H. Cordes, and P. Thomas, *Phys. Stat. Sol. (b)*, **230**, 227 (2002).
[91] W.F. Pasveer, P.A. Bobbert, and M.A.J. Michels, *Phys. Stat. Sol. (c)*, **1**, 164 (2004).
[92] P.M. Borsenberger, L. Pautmeier, R. Richert, and H. Bässler, *J. Chem. Phys.*, **94**, 8276 (1991).
[93] R.P. Feynman, R.B. Leighton, and M. Sands, *The Feynman lectures on physics*. Volume 1. Addison-Wesley Publishing, Palo Alto, 1963.
[94] R. Richert, L. Pautmeier, and H. Bässler, *Phys. Rev. Lett.*, **63**, 547 (1989).
[95] L. Pautmeier, R. Richert, and H. Bässler, *Phil. Mag. B*, **63**, 587 (1991).
[96] J.M. Casado and J.J. Mejias, *Phil. Mag. B*, **70**, 1111 (1994).
[97] Y. Roichman and N. Tessler, *Appl. Phys. Lett.*, **80**, 1948 (2002).
[98] N.W. Aschcroft and D. Mermin, *Solid State Physics*. Holt, Rinehart and Winston, New York, 1988.
[99] R.A. Smith, *Semiconductors*. Cambrudge University Press, Cambridge, 1978.
[100] S.D. Baranovski, T. Faber, F. Hensel, and P. Thomas, *J. Non-Cryst. Solids*, **227–230**, 158 (1998).
[101] H. Overhof and P. Thomas, *Electronic Transport in Hydrogenated Amorphous Semiconductors*. Springer, Heidelberg, 1989.
[102] V.I. Arkhipov, P. Heremans, E.V. Emelianova, and H. Bässler, *Phys. Rev. B*, **71**, 045214 (2005).
[103] R.A. Street, *Hydrogenated Amorphous Silicon*. Cambridge Solid State Science Series, Cambridge University Press, 1991.

[104] C. Tanase, E.J. Meijer, P.W.M. Blom, and D.M. de Leeuw, *Phys. Rev. Lett.*, **91**, 216601 (2003).

[105] H. Shimotani, G. Diguet, and Y. Iwasa, *Appl. Phys. Lett.*, **86**, 022104 (2005).

[106] X. Jiang, Y. Harima, K. Yamashita, Y. Tada, J. Ohshita, and A. Kunai, *Chem. Phys. Lett.*, **364**, 616 (2002).

[107] V.I. Arkhipov, E.V. Emelianova, and G.J. Adriaenssens, *Phys. Rev. B*, **63**, 081202 (2001).

[108] R. Schmechel and H. von Seggern, *Phys. Stat. Sol. (a)*, **201**, 1215 (2004).

[109] S.D. Baranovski and O. Rubel, Description of charge transport in disordered materials. in: *Handbook of Electronic and Photonic Materials*. S. Kasap and P. Capper, (eds), Springer, 2006.

[110] P.M. Borsenberger, W.T. Gruenbaum, E.H. Magin, S.A. Visser, and D.E. Schildkraut, *Journal of Polymer Science Part B: Polymer Physics*, **37**, 349 (1999).

[111] P.M. Borsenberger, W.T. Gruenbaum, E.H. Magin, and S.A. Visser, *Phys. Stat. Sol. (a)*, **166**, 835 (1998).

[112] J. Veres and C. Juhasz, *Phil. Mag. B*, **75**, 377 (1997).

7 Device Applications of Organic Materials

Elizabeth von Hauff[1], Carsten Deibel[2] and Vladimir Dyakonov[2]

[1]*Energy and Semiconductor Research Laboratory, Institute of Physics, Carl von Ossietzky University of Oldenburg, 26111 Oldenburg, Germany*
[2]*Experimental Physics VI, Faculty of Physics and Astronomy, University of Würzburg, 97074 Würzburg, Germany*

7.1 INTRODUCTION

In recent years there has been a lot of research done on organic semiconductors. The first report of conductivity in organic solids was in 1953 when dark conductivity was observed

Charge Transport in Disordered Solids with Applications in Electronics Edited by S. Baranovski.

in aromatic hydrocarbon compounds doped with halogens [1]. Small molecule organic semiconductors were being investigated shortly thereafter [2]. These materials could be vacuum deposited on substrates to form thin films. The discovery of conducting polymers and the work done to increase their conductivity in the late 1970s [3], however, offered to revolutionize electronics as these materials could be processed in solution form. The discovery and continued work on conducting polymers led to the Nobel Prize in Chemistry in 2000 for Heeger, MacDiarmid and Shirakawa. The initial work was done on the conjugated polymer polyacetylene, and since then, many new materials have been investigated for various applications. Electronic devices made from solution-processed organic semiconductors can be fabricated at minimal cost and energy expenditure, and make good candidates for large-scale and flexible applications for which conventional semiconductors are too expensive or brittle.

It is not appropriate to simply transfer general understanding from the well–established field of inorganic semiconductors to the field of organic semiconductors, as the organic materials are inherently different from their inorganic counterparts. These differences lead to unique properties and the potential for novel applications. Studies performed on organic semiconductors have focused on explaining the charge transport within these materials, and much work has been done toward improving device performance. Many device architectures have been investigated, including organic light emitting diodes (OLEDs), organic field-effect transistors (OFETs), solar cells, lasers, batteries, and supercapacitors, to name a few. Generally, we will focus on applications made from disordered organic semiconductors such as polymers, but also state the progress made with vacuum-deposited small molecules, which are either amorphous or polycrystalline.

In the scope of this chapter some common electrical and optical characterization techniques used by researchers in the field, and the current state of the performance and the production of organic devices is outlined. We offer by no means a complete picture, but rather, attempt to offer an overview of this growing area of research. The first section briefly reviews some basics of charge transport in disordered organic materials, relevant as background for the following section which deals with experimental techniques to characterize charge transport in organic devices. The focus is on the determination of the charge carrier mobility, an important material parameter. The results of experimental work that has been performed toward studying and improving the charge carrier injection in organic devices are also presented. The final section of this chapter is dedicated to organic devices, and some advances and research issues concerning this field of work are discussed.

7.2 CHARGE TRANSPORT IN DISORDERED ORGANIC SEMICONDUCTORS

In this section, the foundation for the electrical conduction in disordered organic semiconductors is laid. The scope comprises the description of the origin of conduction on the molecular level as well as a brief introduction to charge transport where it is relevant as background for the section about device characterization. In particular, charge transport in the space-charge-limited current regime, relevant for many low mobility materials, and injection processes into organic semiconductors are discussed. Also, theoretical models describing the charge carrier mobility in disordered organic semiconductors are briefly reviewed. A more comprehensive theoretical description can be found in Chapter 6 'Description of Charge Transport in Disordered Organic Materials' by S.D. Baranovskii and O. Rubel in this book.

7.2.1 Electrical conduction in carbon-based materials

Organic semiconductors are carbon-based materials. The carbon atoms are joined by alternating single and double bonds, and the conjugated nature of the system leads to the semiconducting properties in these materials. In the carbon atom there are four electrons in the outer shell, which in the ground state are found in the $1s^2 2s^2 2p^2$ configuration. In organic semiconductors, there is a hybridization of the s and p orbitals to form three sp^2 orbitals. These orbitals are positioned 120° apart, forming a triangle coplanar to the carbon atoms. These orbitals form the stronger σ bonds. The fourth orbital is the p_z orbital, which is perpendicular to the plane of the atoms. The overlap of the p_z electrons from the carbon atoms form a delocalized band, the π orbitals, which are responsible for the conductive nature of the material [2].

The system may be expected to behave as a one-dimensional metal, however, symmetry breaking reduces the energy of the system, resulting in two delocalized energy bands, the bonding and antibonding π and π^* orbitals, respectively (Peierls instability). These orbitals are also called highest occupied molecular orbital (HOMO) and lowest unoccupied molecular orbital (LUMO). They are separated by a forbidden gap of the order of one to a few electron volts, thus with optical transitions usually in the visible range of the spectrum. The value of the energy gap depends on the structure of the material, and decreases for an increasing number of repeat units comprising the molecule [2, 4, 5]. A single charge on a molecule causes a structural relaxation by electron–phonon coupling, and is then called polaron. This relaxation leads to two additional localized energy levels within the gap.

The rather large energy gap makes organic materials insulators by nature, consequently the intrinsic carrier concentration is very low. In order to increase the carrier density, as required for many device applications, different mechanisms such as carrier injection or field effect, photocarrier generation, or doping can be applied.

The molecules are bound by strong covalent forces, the bonds between the molecules, however, are bound by weaker van der Waals forces. The properties in organic semiconductors are then determined by the more prominent molecular characteristics rather than those of the whole solid [2].

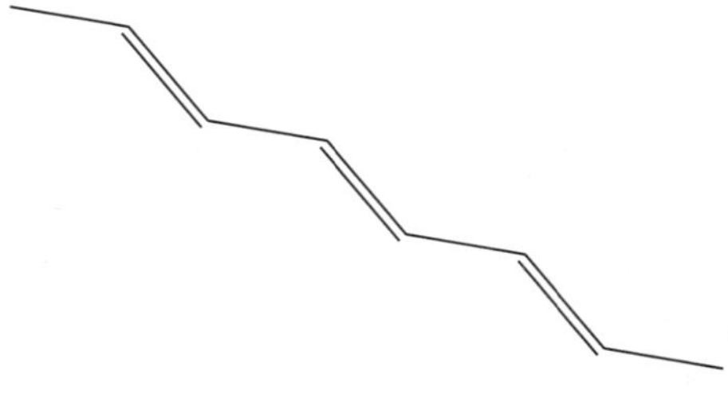

Figure 7.1 Molecular structure of polyacetylene. The alternating single and double bonds make it a conjugated carbon chain. The hydrogen atoms are not shown for clarity

There is a large variety of organic semiconducting materials, and the possibilities for new materials are next to endless, as semiconductors can potentially be designed and synthesized according to the application. Organic semiconductors can be very roughly devided into two groups. The first group consists of the small molecule organic semiconductors which are deposited by means of vacuum deposition to form ordered films. Organic crystals of this sort were already being investigated in the 1950s [2]. The discovery of conductive polymers in the late 1970s [3], however, signified a breakthrough in electronics. A new group of materials was being investigated that could be deposited from solution. These materials can be processed at low temperatures, and offer the potential for economical device production, as well as large-scale and flexible applications, for which crystalline semiconductors are unsuitable.

7.2.2 Hopping transport

Disordered organic semiconductors show a smaller density of states in comparison with their inorganic counterparts. The charge carriers are strongly localized on molecules or molecular segments. Due to the corresponding lack of extended (delocalized) states in amorphous organic materials, charge transport is governed by carrier hopping between the localized sites. Different environments of the electronic polarization on every site lead to a disorder of site energies [6]. An additional positional disorder follows from the amorphous nature of the disordered organic materials. Also, considering that the absorption bands of amorphous organic semiconductors are typically found to exhibit a Gaussian shape, it is generally accepted that charge transport in these materials is appropriately described with a Gaussian density of states (DOS) [6]

$$g(E) = \frac{N}{\sqrt{2\pi}\sigma} \exp\left(-\frac{E^2}{2\sigma^2}\right). \tag{7.1}$$

The distribution is centered around zero energy, with a distribution width of σ, which is a measure of the energetic disorder within the materia; N is the concentration of localized states.

Hopping transport in organic materials is commonly described by Miller–Abrahams formalism [6, 7]. The modern description of hopping transport in a disordered system is be simplified by employing the concept of the effective transport energy [9–11]. This concept was originally used for transport in an exponential DOS as described in Chapter 2 ‘Description of Charge Transport in Amorphous Semiconductors’ in this book. Later this concept has been extended to account for a Gaussian DOS [12–15]. Such theoretical considerations concerning hopping transport form the basis of models for the charge carrier mobility. Some relevant experimental and theoretical work describing the mobility dependence on temperature, electric field, and carrier concentration will be discussed in Section 7.2.5. We touch this topic only briefly as it is described in detail in Chapter 6.

7.2.3 Injection into organic semiconductors

The injection from a metal into an organic semiconductor is still poorly understood, and a deeper knowledge of the dynamics at the metal–semiconductor interface is one of the keys

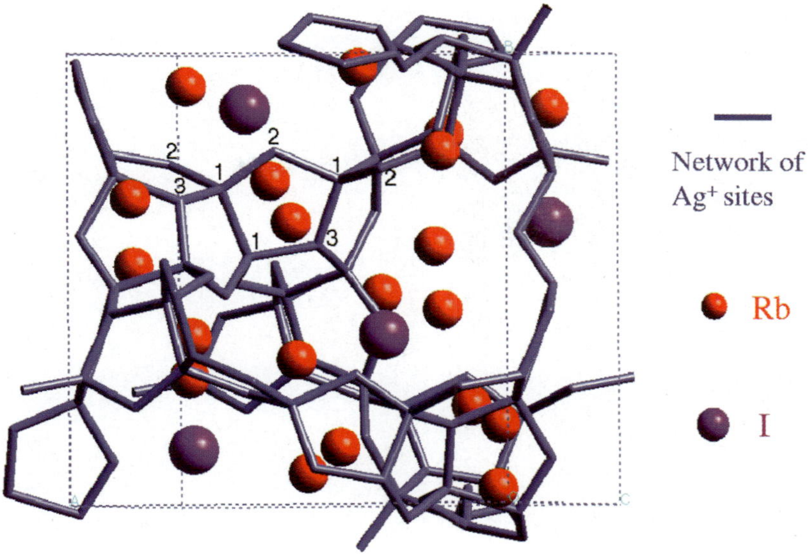

Figure 10.1 Structure of the silver fast ion conductor α-$RbAg_4I_5$. The average occupation number of the silver sites forming a percolative network is 1/3. Together with low energy barriers between the sites this leads to fast ion conduction (reproduced with permission from Stefan Adams, Habilitation Thesis, Goettingen 2000)

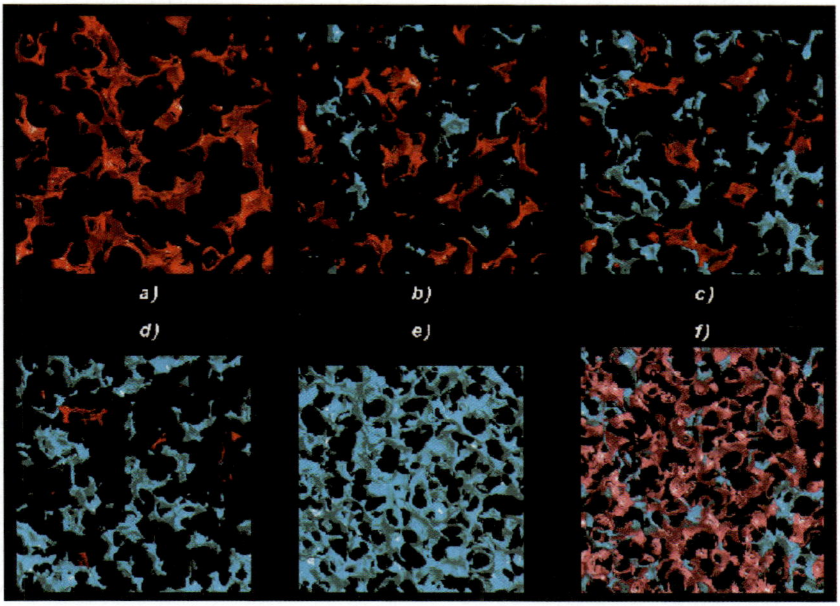

Figure 10.4 Slices through conduction pathways for Li^+ and Rb^+ ions in $Li_xRb1_{1-x}PO_3$ glasses. The picture are from bond valence calculations applied to reverse Monte Carlo models of the glasses. (a) $x = 0$; (b) $x = 0.25$; (c) $x = 0.5$; (d) $x = 0.75$; (e) $x = 1$. The pathways for the Li^+ and the Rb^+ ions are shown as blue and red isosurfaces, respectively. (f) Li^+ pathways (blue) in a $Li_{0.5}Rb_{0.5}PO_3$ glass and those regions that are blocked by Rb^+ ions (pink), but otherwise would have a matching bond valence and would therefore be conduction pathways for the Li^+ ions (reproduced with permission from [22], Copyright 2003 by the American Physical Society)

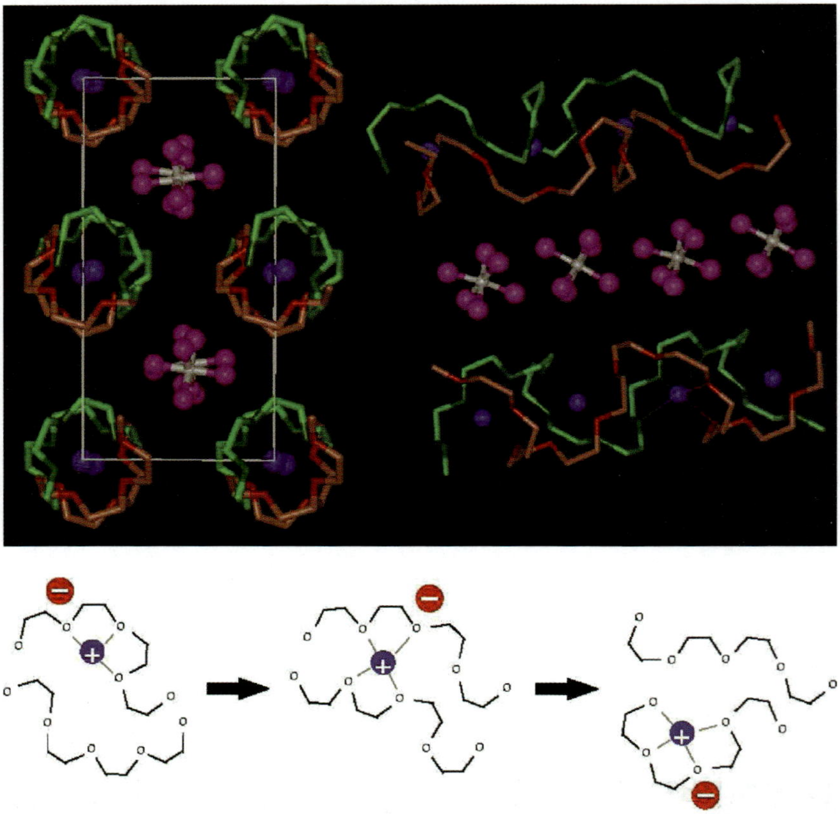

Figure 10.6 Upper: structure of the crystalline polymer electrolyte $(PEO)_6 \cdot LiAsF_6$. Magenta spheres: Li; white spheres: As; light magenta spheres: F; light green: C in chain 1; dark green: O in chain 1; light red: C in chain 2; dark red: O in chain 2 (reproduced with permission from [38], Copyright 2001 by Nature Publishing Group)

Lower: schematic illustration of the coupling between ion dynamics and polymer chain dynamics in an amorphous polymer electrolyte based on PEO

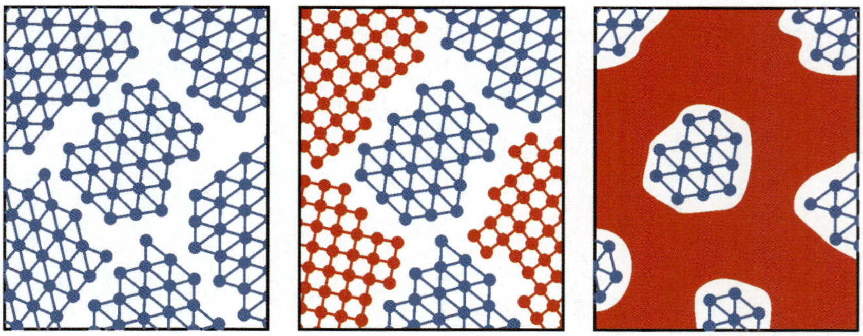

Figure 10.7 Schematic illustration of the structure of a conductor–insulator nanocomposite.

Left-hand: Material composed of ion conducting nanocrystallites. Middle: Composite prepared by mixing conductor (blue) and insulator (red) crystallites. The interfaces between dissimilar crystallites provide fast conduction pathways for the ions. A percolation of these pathways leads to a high ionic conductivity. Right-hand: At high concentrations of insulator crystallites (red area), the pathways are interrupted, and the ionic conductivity is low (reproduced with permission from [57], Copyright 2003 by Institute of Physics Publishing Ltd)

Figure 10.8 Schematic illustration of space charge regions with enhanced ionic conductivity in BaF_2/CaF_2 heterostructures.

Left-hand: When the thickness of the layers is large, individual space charge regions close to the interfaces are formed, and the ionic conductivity is proportional to the interface density.

Right-hand: At small layer thicknesses, the space charge regions overlap, the individual layers loose their identity, and an artifical ionic conductors with anomalous transport properties is formed (reproduced with permission from [62], Copyright 2000 by the Nature Publishing Group)

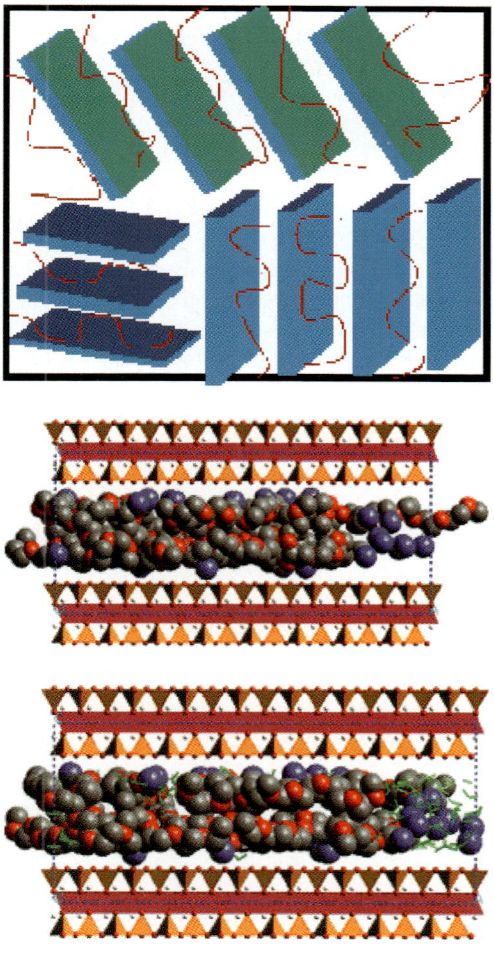

Figure 10.10 Upper: Schematic illustration of a nanogel electrolyte with ordered layers of nanoparticles.

Middle: PEO chains (gray spheres: C; red spheres: O) between nanocrystals of layered silicates. The nanocrystals are negatively charged, and the Li^+ ions (blue spheres) are located close the surfaces of the crystals (results from Monte Carlo and molecular dynamics simulations).

Lower: After addition of a solvent (green molecules), such as propylene carbonate, most Li^+ ions are coordinated by solvent molecules and are highly mobile (reproduced with permission from [73], Copyright 2003 by the Royal Society of Chemistry)

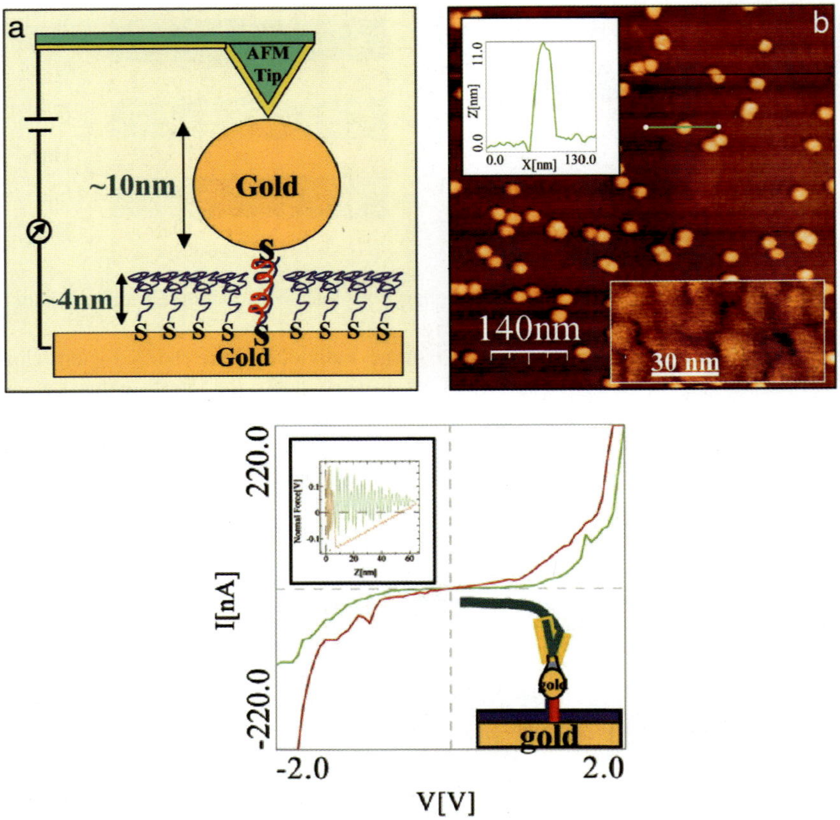

Figure 12.9 Upper panel (left): Schematic presentation of the gold particle-connected dsDNA that is embedded in an ssDNA monolayer and the measurement configuration; (right) AFM topography image showing a top view of the GNPs connected to dsDNA molecules on the background of the ssDNA monolayer. Lower panel: Collection of current–voltage curves measured through various dsDNA molecules connected on opposite ends to a gold substrate and a GNP using a metallized AFM tip. This measurement was performed on a metal particle without pressing on it, as seen in the inset. The current measured is larger than 220 nA at 2 V. The hysteresis in the backward curves in the insets is caused by adhesion and indicates that good contact was established with the surface or the particle underneath. The green curves are the 'forward' and were measured first; the red curves are the 'backward' and were measured after the green ones (reproduced from [82] with permission, Copyright 2005 National Academy of Sciences)

to improving device performance. As organic semiconductors intrinsically have virtually no free charge carriers, charge carrier injection is one of the major steps in charge transport through an organic device. Inefficient injection or extraction of charge will hamper the device performance.

As a rule of thumb, the currents in organic devices with injection barriers greater than 0.25–0.3 eV [2] at zero field are found to be 'injection limited', i.e., the maximum current is determined by the injection process of the charge carriers into the device, as opposed to 'bulk limited' (or space-charge limited) devices.

Injection barriers can be difficult to estimate based on the work function of the metal and the energy levels of the semiconductor alone. Actual injection barrier heights can differ quite strongly from the expected values. In the literature, the deviation between experimentally determined and expected values for the injection barrier height are attributed to chemical reactions between the metal and semiconductor leading to interface dipoles [16, 17], band bending [18, 19] or Fermi level pinning [20].

Until the mid 1990s injection models from inorganic semiconductor physics, such as Richardson–Schottky (RS) thermionic emission [21, 22] and Fowler–Nordheim (FN) field emission [23], were often employed by experimentalists to describe the charge carrier injection process into organic devices. At high electric fields or high injection barrier heights, the FN model describes tunneling currents through a triangular barrier into a delocalized conduction band. At high temperatures or low injection barrier heights, RS thermionic emission predicts the injection of a charge carrier from a metal contact into a semiconductor if the thermal energy of the carrier is greater than the Schottky barrier height.

When applied to organic semiconductors the FN model was successful to some extent in describing the shape of current–voltage curves of some organic diodes [24, 25]. The predicted injection currents were found to differ from the experimental values by several orders of magnitude [26], however, without any physical explanation.

Similarly, it was demonstrated that the RS equation is not sufficient for describing thermionic emission currents into low-mobility materials [27, 28]. Injection currents in organic devices are experimentally found to be many orders of magnitude lower than those predicted by the RS model. Also, deviations from the expected temperature dependence [29] and mobility dependent injection currents [30] were observed in organic devices.

In the mid 1990s, new injection models began to emerge that took the disordered energetic structure of organic semiconductors into account. Abkowitz *et al.* [31] proposed an injection model in which charge carriers undergo a thermally assisted tunneling from the Fermi level of the metal contact to localized sites within the organic semiconductor. The results from the model were found to successfully describe the temperature and injecting contact-dependent current–voltage characteristics in a polytetraphenylbenzidine polymer. Later, Conwell and Wu [32] proposed an injection model for charge carrier tunneling into polaron levels, which depend on the conjugation length of the system.

Arkhipov *et al.* [33] proposed an analytical model for injection into a Gaussian DOS. The model describes injection as a two-step process. First, carriers hop from the electrode into states close to the metal–semiconductor interface. Coulomb binding of the carriers to their image charges in the metal result in a potential well, which the carriers have to either overcome or they will recombine. Based on this model, a Monte Carlo simulation was carried out by Wolf *et al.* [34] and compared with the analytical model [35] and experimental data [36]. Charge carrier injection from a metal into a Gaussian DOS via hopping was simulated, where the significant simulation parameters were the width of the Gaussian DOS and the injection barrier height.

An alternative approach to injection was proposed by Scott and Malliaras [37], who rederived the results for the diffusion-limited thermionic emission model of Simmons and Emtage *et al.* for organic semiconductors.

Based on electrical measurements alone it can be challenging to distinguish between the results predicted by individual injection models, or even between contact and bulk-limited devices. Only more experimental work in combination with refined theoretical models can shed more light on this topic. It is accepted, however, that a successful injection model must be able to describe the temperature, electric field, mobility, and charge density dependence observed in injection currents, as well as interface effects such as the image potential and trapping [38].

7.2.4 Space-charge-limited currents

The charge carrier mobilities in organic semiconductors are typically low. In the case that the injection barriers between the metal and semiconductor are small, and charges can be efficiently injected into the device, the device limiting factor is the ability of the material to transport the charge through the bulk. Charge transport in such 'bulk-limited' devices can be described using the theory of space-charge-limited currents (SCLC) [2, 39–41].

The general approach for deriving the SCLC current–voltage characteristics of insulators exploits the Poisson equation, and is thus independent of the microscopic transport mechanism. The Poisson equation describes the relationship between the electric field F and the local charge density,

$$\frac{\mathrm{d}F}{\mathrm{d}x} = \frac{q}{\varepsilon_0 \varepsilon_\mathrm{r}} (p_\mathrm{c}(x) + p_\mathrm{t}(x)), \tag{7.2}$$

where q is the elementary charge, ε_0 is the permittivity of free space, and ε_r is the dielectric constant of the material. The total density of carriers is given by $p(x) = p_\mathrm{c}(x) + p_\mathrm{t}(x)$, where $p_\mathrm{c}(x)$ is the density of carriers in conductive states, and $p_\mathrm{t}(x)$ is the density of carriers in trapped states.

The current density is given by the current-flow equation for the drift current,

$$j = q\mu(F)F(x)p_\mathrm{c}(x), \tag{7.3}$$

where $\mu(F)$ is the field-dependent mobility.

Equations (7.2) and (7.3) can be combined to give the current–voltage relation

$$\frac{\mathrm{d}F(x)}{\mathrm{d}x} = \frac{j}{\varepsilon_0 \varepsilon_\mathrm{r} \mu(F) F(x)} + \frac{q}{\varepsilon_0 \varepsilon_\mathrm{r}} p_\mathrm{t}(x). \tag{7.4}$$

To obtain the current–voltage characteristics from Equation (7.4), the differential equation has to be solved numerically according to the boundary equation $F(0) = 0$, i.e., an ohmic contact. The voltage is given by $V = \int_0^L F(x)\mathrm{d}x$, where L is the thickness of the semiconducting layer.

In the limiting case that the mobility is assumed to be not strongly field dependent, and trapping is ignored so that only the free charge carriers are considered, Equation (7.4) is rewritten as [41, 42]

$$j = qp_c\varepsilon_0\varepsilon_r\mu F(x)\frac{dF(x)}{dx}. \tag{7.5}$$

Integrating Equation (7.5) from $x = 0$ to L results in the Mott–Gurney law [39]

$$j = \frac{9}{8}qp_c\varepsilon_0\varepsilon_R\mu\frac{V^2}{L^3}. \tag{7.6}$$

Generally, at lower voltages the current–voltage characteristics demonstrate Ohmic behavior. At higher voltages, the SCLC behavior becomes apparent, and the current–voltage characteristics follow Equation (7.5) for the trap-free case. The transition is marked by a change from slope 1 to slope 2 of the current–voltage characteristics, with the transition voltage being proportional to the carrier concentration. The influence of single-level traps and traps distributed in energy is very well analyzed by Lampert and Mark [41].

Experimental work to date on SCLC in disordered organic semiconductors in the literature [43–46] mainly applies the general SCLC theory for insulators without or with traps in order to extract the charge carrier mobility. For disordered organic semiconductors, the SCLC theory has been adapted to account for hopping transport in a Gaussian density of states distribution [47, 48]. At low voltages and current densities, the SCLC indeed shows a $j \propto V^2$-like behavior similar to SCLC in the presence of shallow traps. Almost all states below the Fermi level are filled by carriers, but the total carrier concentration is controlled by states above the Fermi level. At higher voltages, the space charge is formed by carriers occupying states below the Fermi level. In this regime, the slope of the current–voltage characteristics in the logarithmic representation exceeds the value 2, rendering it more problematic to recognize SCLC behavior in disordered organic semiconductors.

7.2.5 Charge carrier mobility

Charge carrier drift mobility is an important parameter in organic semiconductors. The hopping transport in disordered organic semiconductors results in rather low charge carrier mobilities that are electric field dependent and thermally activated. Investigations of these dependences can offer a lot of information about the charge transport processes in these disordered materials. Some models describing the charge carrier mobility in disordered organic semiconductors are reviewed here, whereas experimental methods for their determination are looked at in Section 7.3.

The solution processing of organic semiconductors can lead to higher disorder in the film and therefore lower mobilities compared with other deposition techniques such as vacuum deposition [49]. The difference in intramolecular and intermolecular charge carrier mobilities in the solution-processed films illustrates this point. Charge carrier mobilities on the individual molecules can be orders of magnitude larger than charge carrier mobilities through the films [50, 51]. A higher degree of order in the semiconducting film can be attained through improved processing conditions and can lead to higher mobilities. For example, field-effect mobilities in regioregular polythiophene of the order of $0.1\,cm^2V^{-1}s^{-1}$ have been reported for films in which tempering or self-organization was used to improve polymer chain ordering [52–54]. One major focus point of the research in the area of organic electronics is the improvement of charge carrier transport through the semiconducting films for better device performance.

In the 1970s, the electric field dependence of the mobility of a polymer was empirically described by [55, 56]

$$\mu(F) = \mu(0)\exp\left(\gamma\sqrt{F}\right), \tag{7.7}$$

F is the electric field, $\mu(0)$ is the thermally activated zero field mobility [57], and γ is a coefficient related to the field dependence. The dependence $\ln\mu \propto \sqrt{F}$ is commonly observed experimentally by TOF [58, 59] and other methods [57, 60], and resembles the Poole–Frenkel effect [61], a phenomenon that arises due to traps in the material. The magnitude of the proportionality factor and the temperature dependence, however, deviate from the theory's predictions. Rather, the spatial and energetic disorder of the hopping sites in the presence of a material containing permanent electric dipole moments seems to be the origin of this dependence in amorphous organic materials [62]. The field dependence of the mobility then reflects a field-induced barrier lowering for hopping transport [63].

A major step forward in the description of charge transport in amorphous organic films was the introduction of an uncorrelated Gaussian disorder model by Bässler in 1993 [6]. This model was already able to capture many experimental features properly, e.g., the temperature dependence of the time-of-flight mobility, but fitted the field dependence only in a limited range. Another experimental feature now commonly observed, however, was not addressed. Mobilities in polymers and other disordered materials can depend strongly on the charge carrier concentration and on trapping effects. The highest mobilities are usually determined by field effect transistor measurements. For regioregular P3HT, FET mobilities of up to $0.1\,cm^2V^{-1}s^{-1}$ were presented [52], whereas TOF measurements at room temperature yielded mobilities between 10^{-4} and $10^{-3}\,cm^2V^{-1}s^{-1}$ [64]. These two measurement techniques have two fundamental differences. First, in TOF photogenerated carriers are studied, compared with dark transport in FET structures. Second, OFETs operate with charge carrier densities typically of the order of 10^{18}–$10^{22}/cm^3$, depending on the magnitude of the gate voltage [65]. These carrier concentrations are relatively high compared with those in organic diodes or solar cells.

The influence of the charge carrier concentration on the mobility is impressively documented by Tanase *et al.* [66]. Charge carrier mobilities from SCLC and FET measurements are shown to differ in magnitude due to the different charge carrier concentrations in the respective devices. In Figure 7.2, the mobilities determined in the two polymers P3HT and PPV are plotted versus the carrier concentration. In order to explain their data, the authors proposed a unified transport model. The theoretical description of the FET mobility is based on an analytical expression derived by Vissenberg and Matters [67], considering variable-range hopping in an exponential DOS within the framework of percolation theory. The authors claim that such an approximation of a Gaussian DOS is valid for the energy range relevant for FET measurements. Maennig *et al.* used the same model to fit the experimentally observed dependence of the mobility on the charge carrier concentration measured for amorphous and polycrystalline *p*-doped zinc phthalocyanine [68]. However, a proper expla-

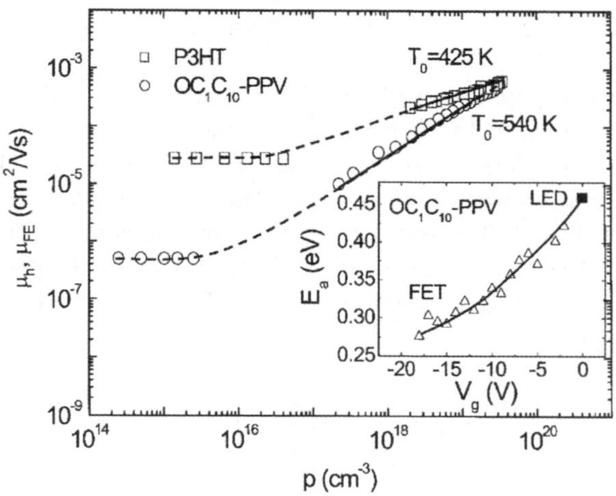

Figure 7.2 The dependence of the mobility on the hole concentration. The mobilities were determined in PPV and P3HT samples using field-effect transistor measurements and current–voltage measurements in the SCLC regime. The dashed lines are a guide to the eye (reproduced with permission from [66], Copyright (2003) by the American Physical Society)

nation of the temperature and concentration dependence could not be provided within framework of the model by Vissenberg and Matters. This limitation was recently overcome by Schmechel [14, 15] and Pasveer *et al.* [69], who independently presented advanced models based on hopping transport in a system with a Gaussian DOS. Both models account for a strong dependence of the mobility on the charge carrier concentration. The latter model also accounts for a (weaker) electric field dependence, and offers a parametrization scheme in order to simplify the fits of experimental data.

7.3 EXPERIMENTAL CHARACTERIZATION OF CHARGE TRANSPORT PROPERTIES

Here we give a brief summary of some common methods to experimentally obtain information on the charge transport in organic semiconductors. The focus is on the determination of the charge carrier mobility in complement to Section 7.2.5 which covered different mobility models for disordered organic materials. The experimental methods discussed here comprise time-of-flight transient photoconductivity, charge extraction by linearly increasing voltage, current–voltage measurements in the space-charge-limited current regime, and field-effect transistor measurements. In the former two methods, the extraction of charges is studied, i.e., these methods are not affected by injection barriers. The latter two methods can also be applied to learn something about the injection from an electrode into the disordered semiconductor. Of course, other methods such as transient electroluminescence [70–72] or admittance spectroscopy in the space-charge-limited regime [73] have also been used for charge transport investigations, but will not be discussed in this section.

7.3.1 Time-of-flight transient photoconductivity

Time-of-flight transient photoconductivity (TOF) is a widely used method to study the charge carrier drift mobility in disordered semiconductors [58, 74]. The measurement principle is sketched in Figure 7.3. A thin 'sheet' of charge carriers is photogenerated by a laser pulse at one (semi)transparent electrode of the device, and drift to the other electrode is due to an external field. The transient of the corresponding displacement current is recorded. The type of extracted charge carrier is determined by the polarity of the applied electric field. From the measured photocarrier transit time τ, the charge carrier mobility can be determined as

$$\mu = \frac{L^2}{V_\tau}, \tag{7.8}$$

where L is the device thickness and V the applied external voltage.

Generally, one should bear in mind that the transport of photogenerated charges is studied using the TOF technique which might result in different characteristics from those achieved under equilibrium conditions (dark case).

The sample needs to be thick compared with the absorption length, otherwise the identification of the transit time is obstructed, or a thickness-dependent mobility is extracted. Thus, the TOF method is usually not directly applied to organic thin films (~100 nm thickness) as used in most devices, but in films with thicknesses in the order of micrometers. This might be problematic as solution-processed organic materials can show a thickness-dependent morphology. Another necessary condition for the application of the TOF technique is that the dielectric relaxation needs to be slower than the charge carrier transit time. This is a prerequisite for a uniform field distribution within the sample. Also, the time constant of the resistance–capacitance of the measurement setup, including the sample, has to be considered.

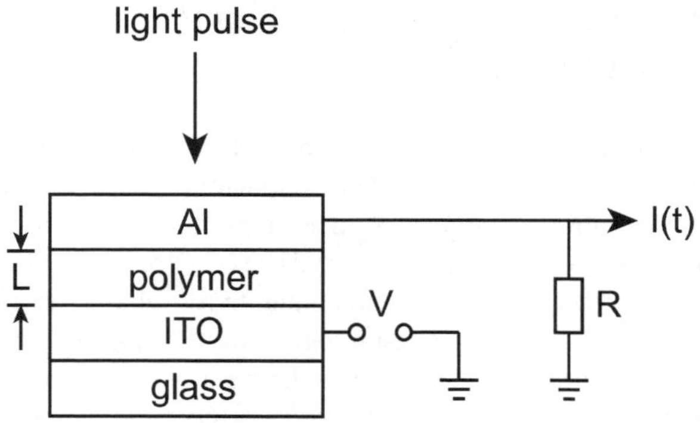

Figure 7.3 Experimental setup of the time-of-flight transient photoconductivity measurement. Charge carriers are generated by a short laser pulse in the polymer close to the semitransparent Al electrode. They drift across the polymer layer to the other electrode due to an external field V/L. The photocurrent transient $I(t)$ is recorded by an oscilloscope

TOF allows for the possibility of determining mobilities in materials showing nondispersive and dispersive transport. In the former, the photogenerated charge carriers propagate through the material as Gaussian carrier packets with similar velocity. In dispersive transport, a considerable part of the carriers remains at the point of their generation for a longer time. This leads to an exponential distribution of charge carriers across the lateral extent of the sample, with the maximum close to the semitransparent electrode [75]. The energetic and positional disorder in amorphous materials lead to a wide distribution of hopping times, which can cause dispersive transport. Dispersive and nondispersive (Gaussian) transport modes are described in detail in Chapter 2 in this book. Organic disordered semiconductors can show a transition from dispersive to nondispersive transport depending on temperature [58, 74], as shown in Figure 7.4 for photocurrent transients for samples made from regioregular P3HT.

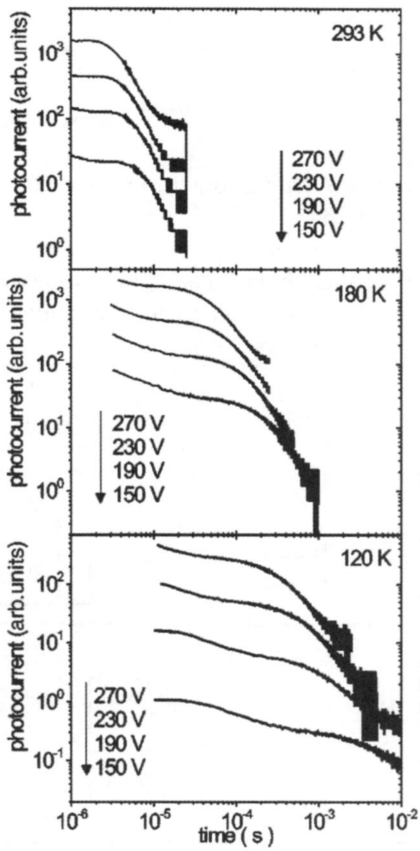

Figure 7.4 The photocurrent transients of a regioregular P3HT sample show dispersive transport at low temperatures, but gradually change to a nondispersive propagation of charge carriers at room temperature (from [74] reproduced by permission of the American Physical Society)

7.3.2 Charge extraction by linearly increasing voltage

The charge extraction by linearly increasing voltage (CELIV) method [76] is closely related to the transient photoconductivity measurement. Indeed, it was shown by Mozer *et al.* [74] on P3HT diodes that the mobilities determined by TOF and CELIV agree very well. The experimental setup is almost identical, but CELIV has a wider applicability as it can also be used in cases where the dielectric relaxation time τ_σ is shorter than the transit time τ. On the other hand, the data analysis is not well suited for measurements taken in the dispersive transport regime as it yields somewhat distorted mobility values.

The extraction of charge carriers is initiated by triangular voltage pulses, $V(t) = At$, applied to samples with one blocking contact, e.g., a diode. Here, A is the rise speed of the voltage. A schematic representation of the principle of the CELIV method is shown in Figure 7.5; $t_{\max}$ is the time for which the charge extraction is maximum, $j(0) = A\varepsilon_0\varepsilon_r/L$ is the initial current proportional to the geometric capacitance of the sample, and Δj the maximum displacement current due to the charge extraction. The mobility can be determined using

$$\mu = \frac{2L^2}{3At_{\max}^2\left(1-0.36\dfrac{\Delta j}{j(0)}\right)}, \tag{7.9}$$

which is based on a numerical estimate in order to account for both, high ($\Delta j > j(0)$) and low ($\Delta j < j(0)$) conductivity materials [76].

In cases where the intrinsic charge carrier concentration is very low, charge carriers can additionally be generated by a short laser pulse; after a time delay, they are extracted by the triangular voltage pulse [77]. It also allows for the comparison of equilibrium (dark case) and photogenerated charge carriers.

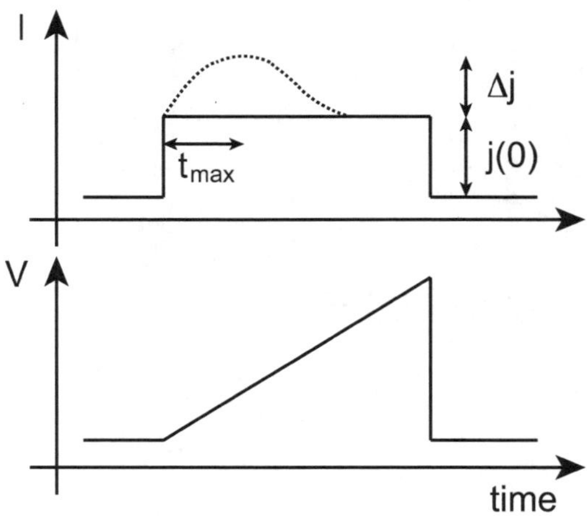

Figure 7.5 Principle of the CELIV measurement. A triangular voltage pulse extracts charge carriers from a diode. The mobility is proportional to the time $t_{\max}$ at which the extraction current is maximum

7.3.3 Current–voltage measurements

The maximum current density flowing through an organic semiconductor device is limited either by the contacts or the bulk material properties. Therefore, current–voltage measurements are frequently used in order to determine the dominant limiting mechanism. In the case that charge carrier transport is limited by the space charge and not by injection, the charge carrier mobility can be determined using the analytical background covered in Section 7.2.4. Otherwise, information about the charge carrier injection process can be extracted, given that an appropriate model (see Section 7.2.3) is used for the interpretation of the experimental data.

The unambiguous identification of the space-charge-limited current (SCLC) regime, as compared with an injection limit, is not trivial in disordered organic materials [47]. de Boer *et al.* compared TOF with SCLC measurements [78]. For more than 100 tetracene single crystals, the former method yielded comparable mobilities. SCLC, however, yielded a very broad distribution of different apparent mobilities, covering a range of at least six orders of magnitude, with maximum mobilities similar to the TOF results. The difference between these two methods lies in the susceptibility of current–voltage measurements to injection, and thus the applicability of the SCLC analysis is subject to the injection barrier. The wide range of apparent SCLC mobilities observed by de Boer *et al.* was indeed due to strongly differing contact qualities. Other studies show well-agreeing field-dependent mobilities measured by the TOF and SCLC methods on small molecules [79].

The dependence of the current–voltage characteristics on the sample thickness as well as on temperature can aid the identification of the correct current-limiting case (if Joule heating of the sample is avoided). As an example, a comparison of the SCLC regime with injection-limited currents of a poly(paraphenylene vinylene) (PPV) diode at different temperatures is shown in Figure 7.6.

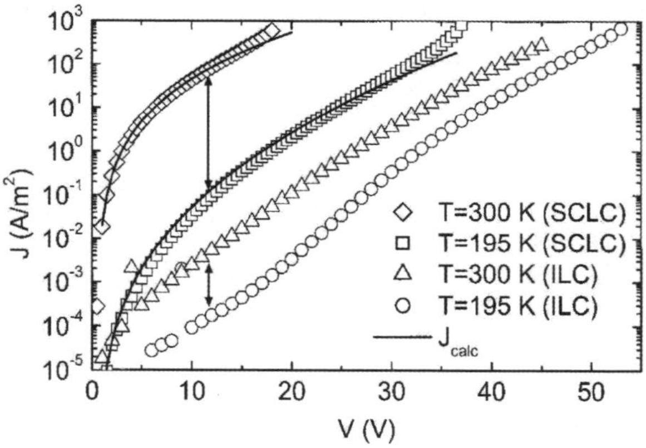

Figure 7.6 Injection-limited current (ILC) versus space-charge-limited current for a ITO/PPV/Ag device. For hole injection from an ITO electrode, SCLC is observed. For hole injection from Ag, the current is injection limited (after [80] reproduced by permission of the American Institute of Physics)

It is possible to discriminate between the electron and hole SCLC mobilities by choosing appropriate (and usually symmetric) injecting contacts, yielding devices with electron-only or hole-only currents. Doing so, the influence of recombination within the bulk of the organic material can be neglected.

In the metal–insulator–metal picture [81], when a semiconductor is brought into contact with a metal, band bending will occur. In the case of an organic semiconductor, the material is depleted of charges when the device is under equilibrium conditions, and the band bending will result in a tilting of the HOMO–LUMO levels between the metal contacts, as shown in Figure 7.7. The result is a built-in potential across the diode [82, 83]. The built-in potential has to be overcome by the applied voltage before the device will operate, and thus needs to be considered when evaluating current–voltage characteristics according to the SCLC model [84, 85].

Bohnenbuck *et al.* [85] studied diodes based on poly[2-methoxy,5 ethyl(2′hexyloxy) paraphenylenevinylene] (MEH-PPV). The anode is a sputtered layer of indium tin oxide (ITO), with a layer of poly-ethylene dioxythiophene:poly-styrene sulphonic acid (PEDOT: PSS) spun over the top. The cathodes were varied between samples: Cu, Au, Al. Figure 7.8a shows the experimental current–voltage characteristics for the diodes. The dependence on the cathode material is unexpected, as in the single-carrier devices, the cathode is assumed to function as a collecting contact, with little effect on device behavior. This dependence cannot be explained by the SCLC model alone. The SCLC model was then investigated in combination with other effects, including accounting for deep traps in the material, recombination effects, and the built-in field across the diode due to the asymmetric workfunctions of the metal contacts. It was found that the current–voltage characteristics of the diodes could best be explained by an SCLC model with a field-dependent mobility and accounting for the built-in field [85], as illustrated in Figure 7.8b.

7.3.4 Field-effect transistor measurements

Organic field-effect transistors (OFETs) offer a straightforward method to investigate the charge transport properties of organic semiconductors. The electric field and temperature

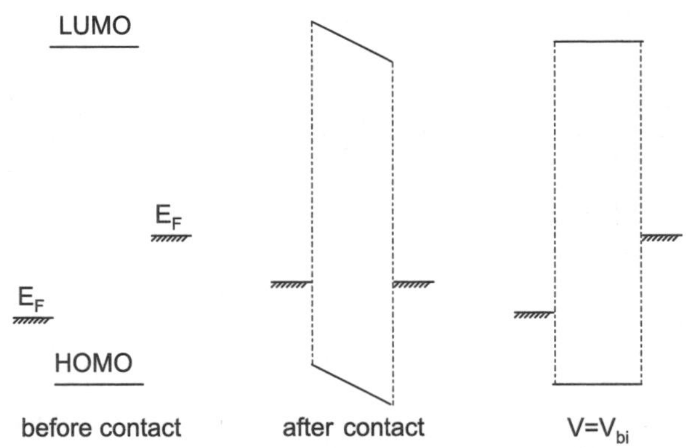

Figure 7.7 Left: energy levels of the metals and the semiconductor before contact. Middle: energy diagram of the diode in equilibrium with band tilting effects. Right: energy diagram of the diode when a voltage equal to the built-in field is applied across the device

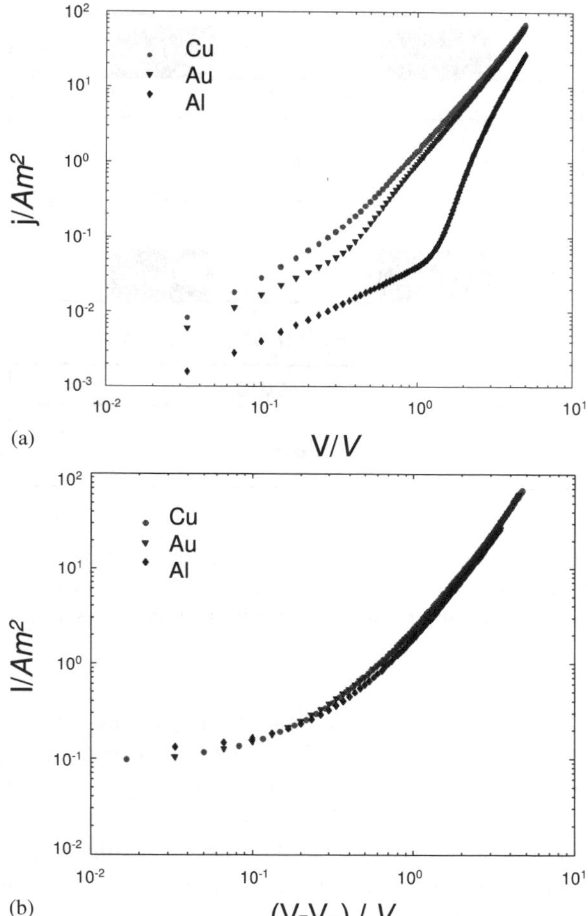

Figure 7.8 (a) Experimental current–voltage characteristics for MEH-PPV diodes with Au, Cu, and Al electrodes; (b) the same data, corrected for the built-in field. The apparent cathode dependence of the current–voltage characteristics is no longer present (after [85] with permission of the American Institute of Physics)

dependence of the mobility can be investigated for high charge carrier concentrations. However, it has to be considered that impurities in the chemical structure or defects in the semiconducting film [86], contact effects, and general artifacts from processing can seriously affect the output characteristics of these devices [87, 88].

An OFET is a three-terminal device. Two variations, the top contact and bottom contact structure, are shown in Figure 7.9. A voltage applied to the gate electrode is used to form a conductive channel in the semiconducting layer at the interface to the dielectric. The channel is contacted by the source and drain contacts, the source is connected to ground. OFETs make use of the thin-film transistor (TFT) structure which is suitable for low-conductivity materials. The devices operate in accumulation mode, and the current is made up of majority-charge carriers. For a *p*-type OFET, applying a gate–source voltage $V_{gs} < 0$ will cause an accumulation of holes near the semiconductor–insulator interface. The charge

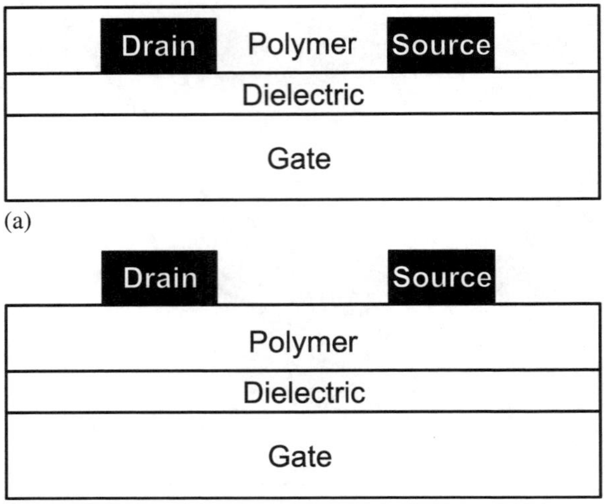

Figure 7.9 Schematic structures of OFETs. (a) Bottom contact structure: the source and drain contact are deposited onto the gate oxide and the semiconductor is applied as the final step; (b) top contact structure: the source and drain contacts are deposited onto the semiconducting layer

carrier density is proportional to the gate voltage. Then applying a drain–source voltage $V_{ds} < 0$ will cause a current to flow across the channel, which is dominated by the charges closest to the semiconductor–dielectric interface. Figure 7.10 [89] shows the current–voltage characteristics of a bottom contact OFET made on a glass substrate with a sputtered gate contact, and a soluble fullerene derivative as the semiconducting layer. The gate insulator is a spin-coated organic resin. Field-effect mobilities of the order of $10^{-3}\,cm^2V^{-1}s^{-1}$ were reported for this device structure.

The device is then in principal turned on for any $V_{gs} < 0$, but the actual on/off behavior is controlled by the low conductivity of the organic material, i.e., usually no current flow is observed at lower values of V_{gs}. This means that the concept of the threshold voltage, which generally is defined as the gate voltage at which the conducting channel starts to form, does not strictly apply to OFETs [49]. In contrast, some organic semiconductors have been observed to always exhibit on behavior, even when no gate voltage is applied, something that is attributed to doping effects in the organic layer [90, 91].

For material characterization, the bottom contact OFET structure (see Figure 7.9a) is more practical from the point of view of device processing than the top contact OFET structure (shown in Figure 7.9b), as the contacts can be deposited or patterned directly onto the substrate, and the semiconducting layer is applied as the final step. The risk of damaging the sensitive material during contact deposition is reduced. However, for integrated circuits the bottom contact structure is unsuitable. Additionally, this geometry results in higher electrical losses at the contacts than the top contact configuration [92, 93].

The current–voltage characteristics at lower fields can be approximated by the gradual channel, or Shockley approximation, for a field-independent mobility. The Shockley approximation is based on the assumption that the variation in the electric field due to source–drain voltage is much smaller than the variation in the field due to the gate voltage. This assump-

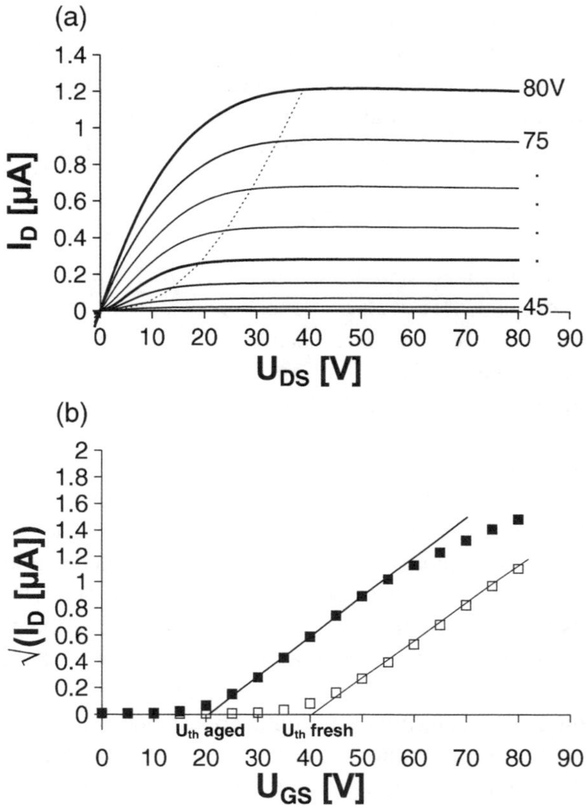

Figure 7.10 Current–voltage characteristics from a fullerene OFET. (a) Current versus drain–source voltage for various gate voltages; (b) current vs gate voltage for a low, constant drain–source voltage (from [89] reproduced by permission of Wiley-VCH)

tion is justified for OFETs, as the thickness of the gate insulator and semiconducting layer is of the order of a few hundred nanometers, while the channel length is in the order of micrometers. The field effect mobility μ_{FE} can then be determined from the linear regime of the transfer characteristics (I_{ds} vs V_{gs}), i.e., for a low source–drain voltage, as

$$\mu_{FE} = \frac{L}{WC_i V_{ds}} \frac{\partial I_{ds}}{\partial V_{gs}}\bigg|_{V_{ds}\to 0}, \tag{7.10}$$

where I_{ds} is the source–drain current, W is the channel width, L is the channel length, and C_i the the capacitance of the insulating layer [94]. The source–drain voltage is chosen as low as possible in order to minimize the variation of the electric field along the conducting channel. A high uniformity of the electric field is desirable as the field effect mobility depends directly on this field, which is due to either filling of traps [65, 94] or a charge carrier concentration, depending on the gate voltage [95].

In addition to mobility measurements, OFET measurements have been found to be useful in gaining information about charge injection into organic semiconductors via analysis of the parasitic contact effects in these devices. The total device resistance R_{tot} is a sum of the resistances between the source and drain contacts and the semiconductor R_{con} and the resistance of the semiconducting film across the channel R_{ch}, and is given by

$$R_{tot} = R_{ch} + R_{con}. \tag{7.11}$$

In the case that the OFET is contact limited ($R_{con} > R_{ch}$), the contact resistance has an impact on the device output. The effects of contact resistances on the performance of silicon TFTs are well known [96]. Recently, there has been a lot of focus on the influence of contact resistances on the performance of OFETs. Contact resistances in OFETs at room temperature are often found to be in the kΩ to MΩ range [93, 97–99], and can affect determined mobility values by up to several magnitudes [98, 100], and as previously mentioned, can be influenced by the geometry of the device [92, 93].

The contact resistance manifests itself as a voltage drop in the drain and source contact regions. This results in an effective voltage drop across the channel, resulting in a lower current. There are several methods that can be employed to determine the magnitude of contact resistances in a device. Some groups have performed studies where the voltage drop across the channel is directly measured by means of conducting probe potentiometry [101], noncontact scanning probe potentiometry [38, 97, 102, 103], or by the four-probe method [99, 104].

Another direct method to determine the contact resistance is to vary the channel length between samples. R_{tot} can be determined from the linear region of the I_{ds}–V_{ds} characteristics for a constant V_{gs}. The total device resistance will increase with increasing channel length, but the contact resistance remains constant between the samples. Plotting the channel length versus device resistance will yield the channel resistance per unit length as the slope, and the contact resistance as the ordinate [98–100].

After isolation of the contact resistance the field effect mobility can be determined from [98]

$$\frac{\partial\left[(\partial R_{tot}/\partial L)^{-1}\right]}{\partial V_{gs}} = \mu_{FE} W C_i. \tag{7.12}$$

The contact effects in organic field-effect transistors (OFETs) have also been used to investigate charge injection into organic semiconductors. It has been shown that the contact resistances in OFETs are temperature and field activated [97–99], like the field-effect mobility, and in fact that the contact resistance varies inversely with the mobility with comparable activation energies [97, 98]. The injection characteristics from the source contact into the channel is the commonality. An injection current described by diffusion-limited thermionic emission (DLTE) [27, 28, 37] is suggested by these findings [97, 98].

Microscopic studies [38, 97] on OFETs by means of noncontact scanning probe potentiometry, however, revealed some discrepancies between the DLTE model and the experimental results when the theory was strictly applied. It was found that the effective injection barrier height predicted by the model was too large to explain the experimental data. This indicates that either other injection mechanisms are at hand, such as tunneling, or that the disorder in the system must be considered here, as a disorder term results in a reduced effective barrier height [33, 80, 97].

7.4 ADVANCES IN ORGANIC ELECTRONICS

Organic semiconductors were already being investigated decades ago. However, the solution-processable organic semiconductors which emerged in the late 1970s promised low-temperature device production as well as the possibility for applications that were not possible with inorganic semiconductors. Since then, much headway has been made in the area of organic electronics. These advances reflect increasing knowledge about charge transport in organic semiconductors, leading to new materials being synthesized, improved processing conditions and device structures.

Organic electronics are interesting because of potential economical device production and possible new applications. Although organic light-emitting diodes are already on the market, there are some issues that must be resolved before the potential offered by organic electronic devices can be fully realized on a larger scale. The feasibility of organic electronics depends on parameters such as the stability and suitability of a material for a specific application, the charge carrier mobilities of both carrier types in a device, the efficiency of device interfaces such as the metal–semiconductor interface, and the cost of device production, to name just a few.

In this section we briefly touch device-processing techniques before going over to discuss several important types of organic applications. Organic light-emitting diodes are the most studied of the organic devices, and research and developments on materials and contacts used for these devices have paved the way for advances in other areas of organic electronics. Here, we outline the principles and state of the art of the organic devices commonly investigated in the literature. Of course, this is only a sample of the work that has been done and of the device architectures that are possible.

7.4.1 Device fabrication

Due to the unique properties of these materials, organic semiconductors have some advantages over conventional semiconductor technologies. One of these advantages is cost-effective device production. Organic semiconducting materials are cheap, but the major cost for device production lies in the processing steps. The flexible nature of organic semiconductors allows for large-scale roll-to-roll processing with various low-cost patterning techniques [105]. Due to the low cost, and the flexible nature of the materials, organic semiconductors can also be used for applications that are not possible with conventional semiconductor technologies, e.g., lightweight portable electronics, and disposable products such as electronic barcodes [106].

Patterning devices made from solution-processed organic semiconductors is problematic when the films are applied via the conventional means of spin coating, doctor blading, etc., as small structures are difficult to achieve and applying multiple layers can result in the interdissolving of the layers. Several means of device patterning have been suggested and tested: photochemical patterning to produce all-polymer integrated circuits [106]; inkjet printing techniques to print organic light-emitting diode displays [107] and organic transistor circuits [108, 109]; dye-diffusion [110] and dye-diffusion thermal transfer processes applied to OLEDs [111], and thermal imaging techniques which allow for the dry printing of large-area organic electronics [112].

A technique called organic vapor-phase deposition (OVPD) [113] for the patterning of electronic devices fabricated with vacuum-deposited organic semiconductors was developed which is analogous to the inkjet printing of solution-processed organic semiconductors.

7.4.2 Organic light-emitting diodes

Organic semiconductors have been recognized as candidates for light-emitting diodes (OLEDs) and multicolor display technologies for decades as many of these materials have high fluorescence quantum efficiencies in the visible spectrum. Electroluminescence was first observed in organic compounds in the 1950s [114], and much work was done in the next decade toward developing electroluminescent devices from organic semiconductors [115, 116]. The early work resulted in device characteristics dominated by SCLC behavior or limited by impurities in the semiconductor, leading to extremely high driving voltages, of the order of hundreds of volts, and therefore low power conversion efficiencies. Improvements to the device structure, such as modified cathode materials for better electron injection, and the use of thin organic films [117], resulted in driving voltages in the tens of volts range. A breakthrough in the field of organic light-emitting diodes (OLEDs) was reported by Tang *et al.* [118] in 1987, who used a double thin-film layer of p-type diamine and n-type 8-hydroxyquinoline aluminum (Alq_3). The organic films were purified and sandwiched between a semitransparent ITO anode and a magnesium alloy electrode for efficient charge injection. This was the first report of OLEDs that could be operated with voltages below 10 V. Sometimes dubbed the 'Kodak breakthrough' [119], this device structure became the blueprint for future work on OLEDs. Since then, OLEDs have become the most studied organic device structure, and are now commercially available in display applications. The work done in the field of OLEDs has been beneficial for researchers in other areas of organic electronics, as many materials have already been synthesized and screened for developments in OLEDs.

The energy diagram of a single-layer conjugated polymer poly(paraphenylene vinylene) (PPV) diode is shown in Figure 7.11a. The contacts are chosen so that the anode is a high-workfunction metal that will form an ohmic contact with the HOMO of the semiconductor for efficient hole injection, and the electrode is a low-workfunction metal that will form an ohmic contact with the LUMO for electron injection. Electroluminescence was observed in PPV in 1989 [120]. A layer of PEDOT:PSS is used to improve the hole injection conditions [121, 122]. The chemical structure of PPV is shown in Figure 7.11b. Light, with a wavelength corresponding to the HOMO–LUMO gap, is emitted from the diode when the charge carriers recombine within the device.

The characteristically low mobilities and disordered nature of the materials are issues for charge carrier transport through the material. Most organic semiconductors behave as p-type materials and if ambipolar transport is observed, the electron currents are usually orders of magnitude lower than the hole currents. An OLED operates on the principle that currents due to both charge carrier types are present; one of the limiting factors for OLEDs is the unbalanced hole and electron currents in the device, a fact which led to the introduction of hole-blocking layers in PPV-based OLEDs for more efficient devices [123] and the

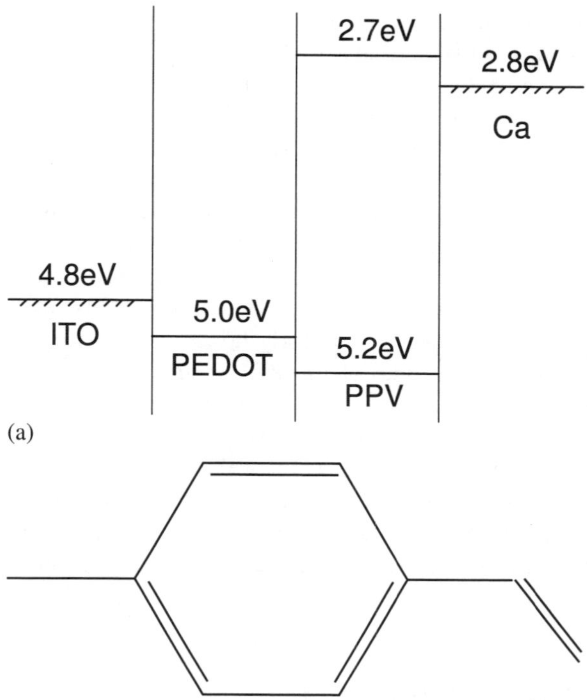

Figure 7.11 (a) Energy diagram of an organic light emitting diode; (b) chemical structure of PPV

development of electroluminescent multilayer devices which consist of a hole transport layer, an emitting layer, and an electron transport layer [124].

The efficient injection of electrons into organic semiconductors remains an issue. In order to inject electrons into the LUMO level of an organic semiconductor, lower-workfunction metals (Ca, Mg, Al) must be used as contacts. These metals are unstable, however, and oxidize quickly and react with the organic layer [125], which can lead to unexpectedly high injection barrier heights for charge carrier injection. It was found that using a very thin interlayer (~1 nm) of LiF or MgO between the organic material and metal contact could improve the electron injection efficiency considerably in some devices [126]. In the case of LiF, this has contributed to the lowering of the metal workfunction by the insulating interlayer [127], or a doping of the organic material by Li^+ cations [128] or F^- anions [129]. Similarly improved device behavior due to interlayer effects has been observed for electrodes deposited in the presence of some oxygen, resulting in a thin insulating layer between the contact and semiconductor, which may serve to prevent a direct reaction between the cathode and the organic semiconductor [130, 131]. An enhancement in device performance was also observed for interlayers between the anode and semiconducting layer, resulting in improved hole injection [132].

OLEDs are extremely good candidate materials for display technologies. OLED displays are potentially cheap, energy efficient, lightweight and thin; only a driving voltage is needed across the panel instead of back lighting. The displays have a large viewing angle and good resolution. OLED displays appeared on the market in 1999 when Pioneer introduced a multicolor small molecule OLED car stereo display. In 2002, Phillips brought out a polymer LED (PLED) monochrome display for an electric razor. Since then OLED displays have made their way into cell phones, and car components, and larger prototype displays exist.

Limiting factors for the performance of OLED and PLED displays remain the efficiencies of the devices and the stability of the organic materials, however considerable progress has been made on these fronts [133]. The commercialization of OLEDs and PLEDs has also been slowed by the short lifetimes of the blue emitters [134] which are necessary for the development of white light displays. Recently, however, Cambridge Display Technology announced blue PLEDs with lifetimes of 100 000 h (Cambridge Display Technology Press Release from May 2005) and reports of high efficiency white light emission from single molecular emitter layers [135] and from multilayer PLEDS [136] have emerged.

An interesting alternative approach to increase the performance of LEDs might be the concept of triplet emitters [137, 138]. These phosphorescent organo–transition-metal complexes can circumvent the limitation of standard OLEDs that only singlet excitons, constituting one quarter of all excitons, are available for radiative recombination due to spin statistics.

7.4.3 Organic field-effect transistors

The transistor is one of the most important building blocks in the modern world. The advantages of organic semiconductors have already been mentioned, however in the field of transistor technology the potential to produce smaller, cheaper, faster electronics that can be processed at low temperatures and on flexible substrates is especially significant. The device structure and basic principles behind the OFET were already introduced in Section 7.3.4. Here we discuss some of the issues and advances made toward developing OFETs, improving device performance, and some applications for OFETs.

The transistor was invented in 1947 by Shockley, Brattain and Bardeen. It can operate as a switch or an amplifier, and is a major component in all digital electronics. The first silicon-based metal–oxide–semiconductor field-effect transistor (MOSFET) [139] appeared in 1960. The field effect was being investigated in organic semiconductors just a decade later, in 1970 [140, 141]. The first actual organic field effect transistor (OFET) device was reported in 1987 [142], and since then, there has been a lot of progress made toward the technological development of the organic-based devices as well the understanding of the physics behind them.

OFETs are candidates for, among other things, inverters [143], complementary circuits [144], rectifiers [145] and ring oscillators [146], low-end display driving circuits [147, 148], chemical sensors [149], and integrated circuits [106, 150]. The first light-emitting OFETs (O-LEFTs) have recently been reported [151]. Many good reviews have appeared about the state of the art of OFET technology [152, 153] and single-crystal OFETs [88].

Much more work has been done on p-type OFETs than on n-type OFETs. This is due to several factors: until recently, reports of ambipolar transport were absent from the literature and most organic semiconductors were observed to behave as p-type materials. The lack of n-type behavior was attributed to traps in the material that impede electron transport

[43]. Creating ambipolar devices from the combination of *p*-type and *n*-type materials was limited: the few *n*-type materials available were found to be unsuitable for many applications due to instability [154]. Recently, however, time-of-flight photoconductivity measurements on purified regioregular poly(3-hexylthiophene) films have revealed electron and hole mobilities of the same order of magnitude [155]. These results indicate that improved processing conditions allow for ambipolar transport in materials previously observed to display only unipolar characteristics. Reports of ambipolar transport in OFETs have also emerged. Field-effect measurements on a soluble fullerene derivative, typically an *n*-type material, revealed hole mobilities that were only slightly lower than the electron mobilities [156], and it was reported [157] that ambipolar transport could be achieved in polymer OFETs using an appropriate hydroxyl-free gate oxide. The lack of *n*-type conduction reported in the literature for conjugated polymer OFETs was attributed to the trapping of electrons at the semiconductor–insulator interface by the hydroxyl groups present in commonly used gate oxides. Ambipolar transport is particularly interesting in OFETs for organic complementary metal–oxide semiconductor (CMOS) circuits. CMOS circuits are low-energy-consumption devices comprised of both *n*-type and *p*-type channels.

There are a few factors that influence the performance of OFETs. The material parameters field effect mobility and conductivity influence the on/off behavior of the device: a high mobility and a low conductivity will result in higher on/off ratios [153]. Unintentional doping due to impurities leads to an increase in the conductivity, and a reduction in the on/off behavior. Generally, the mobilities in organic semiconducting films are lower than those in inorganic semiconductors. Single-crystal OFETs, however, are quicker, smaller and offer higher mobilities than the thin-film OFETs. Mobilities in rubrene single-crystal transistors have been reported to be as high as $15.4\,cm^2/Vs$ [158]. These are higher than the mobilities found in amorphous silicon TFTs. As already discussed, contact resistances can also play a major role in the OFET performance.

The field-effect mobilities in thin films are also limited by morphological factors. Impurities and grain boundaries behave as traps, and lead to reduced effective mobilities. For the thin-film OFETs, the highest mobilities reported to date are for pentacene, one of the more widely studied organic semiconductors. Mobilities in the order of $3\,cm^2\,V^{-1}s^{-1}$ and on/off ratios of 10^5 have been reported for pentacene OFETs using a polymer gate dielectric [159]. Pentacene has a poor solubility and is therefore typically vacuum deposited. Recent efforts have focused on soluble precursor forms of pentacene which are applied to the substrate as a solution; the resulting film is then treated in order to convert the precursor to pentacene. Some studies report precursor conversion by heat treatment [160–162] and recent work reports on a photosensitive precursor that can be used to pattern the resulting pentacene layer with UV light [163]. The mobility values reported for the solution-prepared pentacene layers are lower than those reported for vacuum-deposited pentacene, of the order of $0.1–0.2\,cm^2\,V^{-1}s^{-1}$ [161, 162].

Also important are good preparation techniques for the insulator–semiconductor interface. The conductive channel forms in the vicinity of this interface, and it is known that the insulating material used influences the performance of the device. The preparation and deposition techniques for the semiconductor and a surface treatment for the insulator can be decisive for reducing trapping effects. Traditionally, SiO_2 was used as the gate insulator for samples prepared on silicon substrates. The movement toward all-organic devices, however, has led to the use of insulating polymers for the gate insulator material. These materials typically have low dielectric constants compared with inorganic materials which result in high operating voltages, and thicker insulating layers are needed to avoid leakage currents which occur at the higher fields. Many advances have been made in this direction, though, including a molecular monolayer gate dielectric [164] and a polymer electrolyte gate dielectric [165] which resulted in lower operating voltages for pentacene OFETs.

7.4.4 Organic memory

Electrical bistability has been demonstrated in organic materials. Bistability means that the material demonstrates two different states of conductivity at the same applied voltage, typically a highly conductive on state and a poorly conductive off state. These states are stable, and transition can occur between them when a defined criteria such as a high voltage bias is met. This property is ideal for switching and rewritable memory applications.

Several methods have been used to achieve bistable switching in organic materials over the last three decades. In 1971, reproducible bistable switching, attributed to filament formation, was observed in metal–polymer–metal structures [166]. In the later 1970s, switching using charge transfer complexes was observed in copper–tetracyanoquinodimethane (Cu-TCNQ) polycrystalline films sandwiched between Cu and Al contacts [167]. Much later, in 2003, it was demonstrated that using a thin layer of Al_2O_3 between the Al anode and the semiconductor in ITO:Al/Cu/TCNQ/Al devices improved the reproducibility of the switching of these devices [168].

The first reports of stable and high-performance memory devices came from Yang's group in 2002. The original device was a simple three-layer structure consisting of a metal layer embedded between two organic layers, with metal contacts sandwiching the device. On/off ratios of 10^4 were observed, the response time was found to be under 10 ns, and more than 1 million write–erase cycles could be conducted [169]. The memory effect was attributed to the diffusion of metal ions within the organic layer due to an applied voltage. Since then Yang *et al.* have reported several variations of memory device structures with large on/off ratios (10^6) and low operating voltages [169–171].

Other stable and high-performance devices have also recently been reported which rely on charge trapping levels which are controlled by the applied voltage: in the on state the transport sites and traps are not filled and charges can readily enter the transport sites, at higher fields charges tunnel into the trap sites and the resulting space-charge field impedes the flow of current, and reducing the voltage rapidly to zero preserves the state [172, 173]. Making use of this principle, Alq_3 diodes with an additional thin trapping layer made up of either silver islands or an organic dye have recently been reported [174] which behave as OLEDs when a forward bias is applied, and as memory devices when a reverse bias is applied. Such a technology opens up the possibility of organic displays capable of storing information.

Bistable switching has also been reported using molecules; the different conductive states are achieved by conjugation modification via electroreduction of the molecules [175, 176]. Further work in this direction using Rose Bengal for the molecular layer led to a large on/off ratio, of the order of 10^5, longer memory effects due to the acceptor groups on the molecule, and a low operating voltage, between –3 and 3 V [177]. These results were expanded upon by another group [178] who used the Rose Bengal memory cells as the memory elements in a cross-point matrix, resulting in a 32-fold increase in memory and an on/off ratio of 200 for the entire device.

The ferroelectric properties of many organic materials [179] have also been used to develop organic memory devices. Ferroelectric FETs are attractive memory elements due to the fast nonvolatile switching and the low power consumption. All-organic permanent memory transistors (FerrOFET), which make use of the ferroelectric properties of the gate insulator, have been reported [180, 181] with on/off ratios of the order of 10^4, response times of 0.1–1 ms, and operating voltages below 10 V [181].

7.4.5 Organic photovoltaics

As in other areas of electronics, in the field of photovoltaics organic semiconductors have the potential to become a very cost effective alternative to the currently used materials. This, in combination with advances in the understanding and processing of organic semiconductors, has led to much focus recently in the area of organic solar cells [182]. The production of photovoltaics from monocrystalline and polycrystalline silicon is very expensive, making the price of conventional silicon photovoltaics too high for wide use or large-scale applications. The cost for silicon photovoltaics is around $4 per Watt and is estimated to level off within the next decade at best in the range of $1–1.50 per Watt, which is still far above the target price for photovoltaics (for a good review of the economic standpoint of photovoltaics see [183]). Other materials have been investigated in order to avoid the use of the expensive silicon substrates and lower the cost of solar cell production [184].

Organic solar cells promise lower processing costs and thus low prices per Watt, even though the absolute power conversion efficiencies will remain lower than for the inorganic devices. In addition, organic solar cells can potentially be prepared on any type of substrate, allowing for flexible applications. The goal of reel-to-reel processing and printing techniques [185] with no high-energy steps involved makes organic materials a serious candidate for photovoltaic applications.

The color of the absorber layer and thin-film structure allows for interesting possibilities that can be incorporated into the design and architecture of buildings, as incident light will pass through the device. An absorber layer used as window tint could serve as shading as well as an additional energy source for buildings. Doping, and the synthesis of new materials with different bandgaps can lead to a wide color variation in the cells. This is a fact that, together with the light weight, makes organic photovoltaic materials good candidates for portable electronics [186] or e-textiles [187].

Polymer photovoltaics became a focus of research after the discovery of conducting polymers and the continued developments made in this field. Generally, the success of OLEDs has contributed to the advances in organic photovoltaics, as many materials have already been screened and investigated. Although organic semiconductors have relatively low charge carrier mobilities, they have fairly strong absorption coefficients, exceeding $10^5\,cm^{-1}$ [188], which leads to high absorption even for the ~100-nm-thin absorber layers used.

Organic solar cells must meet certain criteria before they can be introduced onto the market. Brabec *et al.* [189] recently examined the feasibility of commercial organic solar cells. Based on low-cost production, the qualifications for market entry would be an efficiency of 5% for modules, and a lifetime of 3–5 years (correspondingly an operational lifetime of 3000–5000 h). The introduction of printing techniques have the potential to lower the cost of production to considerably under $1 per Watt. The stability of organic photovoltaic devices, however, remains an important issue. Organic materials are to some degree sensitive to oxygen (in particular in combination with light), water, and temperature [190, 191]. Much research is needed in exploring means to protect and prolong the endurance of these devices.

Power conversion efficiencies reported for organic solar cells have been steadily increasing, as new suitable semiconducting materials are synthesized and investigated and device architectures improved [182, 192–194]. The first organic solar cells consisted of single layers of evaporated organic semiconductors sandwiched between contacts, and efficiencies reached 0.7% [195, 196] for vacuum-deposited merocyanine dyes in these simple structures. Solar cells with two semiconducting components to form a heterojunction bilayer [197] or tandem structures [198] led to even higher efficiencies. The single polymer layer device architectures yielded very low efficiencies and high recombination effects which led to the motivation to introduce an electron acceptor into the system [199]. The use of an acceptor material can lead to a long-lived charge separation state so that charge carriers can be collected at the device contacts before they recombine. A breakthrough came with the discovery of the ultra-fast charge transfer, which was reversible and metastable, between a conjugated polymer and a fullerene [199, 200]. This led to the development of the polymer–fullerene solar cell. The original cells consisted of the polymer donor and fullerene acceptor separated by a single interface [201, 202]. Combining the polymer and fullerene components in a bulk-heterojunction structure, maximizing the interface between the polymer and fullerene in order to minimize the influence of the rather low exciton diffusion length, led to much higher efficiencies [203]. Recently, bulk-heterojunction solar cells made from the polymer poly(3-hexylthiophene) (P3HT), have exhibited power conversion efficiencies of almost 5% [204]. Studies on donor–acceptor systems on the same molecule, dyads [205, 206], however, led to low efficiencies, attributed to the increased chance for recombination. Optimizing the morphology is clearly a tradeoff between reducing the charge carrier recombination with some phase separation between the donor and acceptor, and keeping the phase separation on a scale suitable for allowing for the dissociation of excitons in the blend. For a good overview, see [188].

The basic structure of the bulk heterojunction polymer–fullerene solar cell is shown in Figure 7.12. A semitransparent contact, such as ITO, sputtered onto a glass substrate serves as the anode. The blended layer of conjugated polymer–fullerene is applied to the substrate, typically by spin coating, screen printing or doctor blading [185]. The electrode is then applied on the semiconducting layer by means of vapor deposition or printing [207].

To improve the efficiencies of organic solar cells, more knowledge about the principles behind these relatively complex devices is needed. The efficiency of a solar cell is dependent on the open-circuit voltage and the short-circuit current, which in turn are dependent up the energetics of the materials, the absorption and the morphology of the active layer, respectively. Unfortunately the current–voltage characteristics of an organic solar cell cannot be modeled completely using the models conventionally used for inorganic photovoltaics, and no model which sufficiently describes the behavior of organic solar cells has been suggested to date. In the next part, the current–voltage characteristics of organic solar cells, the factors which influence the open circuit voltage, and the effect of morphology on the solar cell performance are discussed.

The transport properties of organic semiconductors are much different under illumination than in the dark. In an illuminated solar cell, the optically driven charge transfer between the polymer and fullerene, and the nature of the generated exciton within the bulk

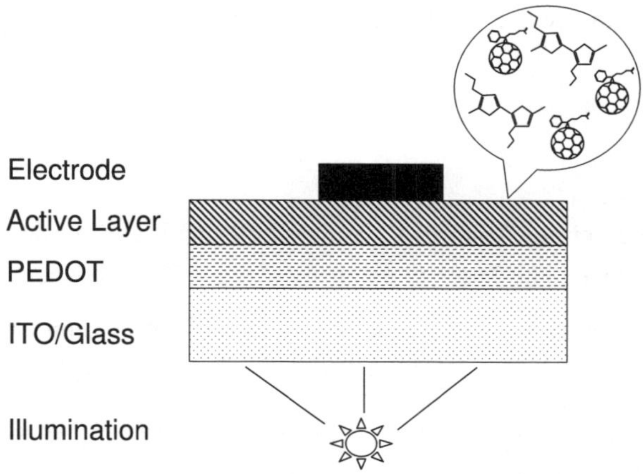

Figure 7.12 Structure of a polymer–fullerene bulk heterojunction solar cell

are specific to the solar cell. The primary excitations within the bulk of organic materials upon illumination are Frenkel-type excitons, Coulombically bound electron–hole pairs with a rather high binding energy [208]. These neutral and mobile excitons can be dissociated into charge carriers only by strong fields, defect sites, or—the dominant process in organic solar cells—they diffuse to an interface of the absorbing material and an electron-accepting material and are separated by an ultra-fast charge transfer of the electron to this acceptor [199, 200]. The separated carriers are then extracted by an applied field and selective contacts before they can recombine.

Ideally, the hole and electron mobilities within a polymer solar cell should be balanced. Differing mobilities result in differing mean free paths of the individual charge carriers, which can lead to an imbalanced distribution of charges within the semiconducting layer, whereby the slowest charges will determine the device performance [209]. The build-up of one type of charge carrier within the bulk will lead to an electric field in the device opposing the flow of charge [210].

Generally the current in inorganic solar cells is described by the Shockley equation,

$$I = I_0\left[\exp\left(\frac{q}{nkT}(V - IR_s)\right)\right] + \frac{V - IR_s}{R_{sh}} - I_{ph}, \tag{7.13}$$

where I_0 is the dark current, n is the ideality factor of the diode, R_s is a series resistance, R_{sh} is a shunt resistance, and I_{ph} is the photocurrent. Unfortunately, organic solar cells are not adequately described by this equation, although it is commonly used in the literature to either fit the current–voltage characteristics in a limited voltage range [211] or to describe the temperature dependence of the open-circuit voltage and short-circuit current [212, 213]. The reverse-bias region would have to be described by a field and light dependent resistor in order to fit the data properly, but no physical model has been proposed to explain this behavior. In forward bias, the influence of space-charge-limited currents commonly observed is not accounted for. An analytic expression describing the current–voltage characteristics of organic solar cells does not yet exist.

The efficiency of the device can be determined by

$$\eta = \frac{\mathrm{FF} j_{sc} V_{oc}}{P_{illum}}, \tag{7.14}$$

where V_{oc} is the open-circuit voltage, j_{sc} is the short-circuit current density, and FF is the fill factor of the device. The latter is given by $\mathrm{FF} = \frac{j_{MPP} V_{MPP}}{j_{sc} V_{oc}}$ where V_{MPP}, j_{MPP} are the voltage and current density values taken at the maximum power point of the device, respectively. P_{illum} is the illumination power density of the incoming light. Thus, in order to improve the efficiencies of these solar cells, the device parameters, such as the open-circuit voltage, the short-circuit current, and the fill factor [188], have to be enhanced. More specifically, the understanding of the electrical losses at the metal contacts and semiconductor interface [203, 214] due to recombination or large extraction barriers, or within the bulk of the device [209, 213, 215], resulting in charge carrier trapping or recombination, are necessary to improve device efficiencies. The current is determined by the width of the absorption spectrum of the organic absorber material, and by the morphology of the active layer. The former is intrinsically limited by the rather narrow density of states distribution in disordered organic materials. Poor morphology limits the charge carrier transport through the film due to increased recombination.

The open-circuit voltage is a material parameter which is determined by the energetics of the components in the solar cell. In the metal–insulator–metal picture, a built-in field results when an insulator is sandwiched between two contacts with asymmetric workfunctions. For a single-component organic solar cell, the V_{oc} is then determined by the difference in the workfunctions of the two contacts or the Schottky barrier formation between the metal and doped organic layer [81]. In the case of single-component organic photovoltaic devices, the built-in potential in a device can then be easily determined as it is equal to the V_{oc} when the device is under (sufficiently strong) illumination [216].

In the case of the more complicated donor–acceptor systems, Brabec *et al.* have shown that the open-circuit voltage is dominated by the energetics of the donor–acceptor system, i.e., V_{oc} depends strongly (almost linearly) on the difference between the HOMO of the donor and the LUMO of the acceptor, and only weakly on the metal work functions [217]. Other researchers claim a stronger scaling of the V_{oc} with the difference between the metal workfunctions [218, 219]. The use of LiF as an interlayer for electron injection has been observed to increase the open-circuit voltage [193, 220].

It was proposed that for ohmic contacts, the open-circuit voltage is determined by the energetics of the donor–acceptor system, while for nonohmic contacts, the V_{oc} is limited by the difference in the workfunctions of the contacts [221]. Figure 7.13 demonstrates the two scenarios that can give rise to the determination of the open-circuit voltage.

The correct description of the temperature dependence of the open-circuit voltage is still under discussion. A common approach is based on the Shockley equation (7.13) with $j = 0$, see e.g. [222], yielding

$$V_{OC} = \frac{nkT}{q} \ln\left(\frac{j_{SC}}{j_0} + 1\right), \tag{7.15}$$

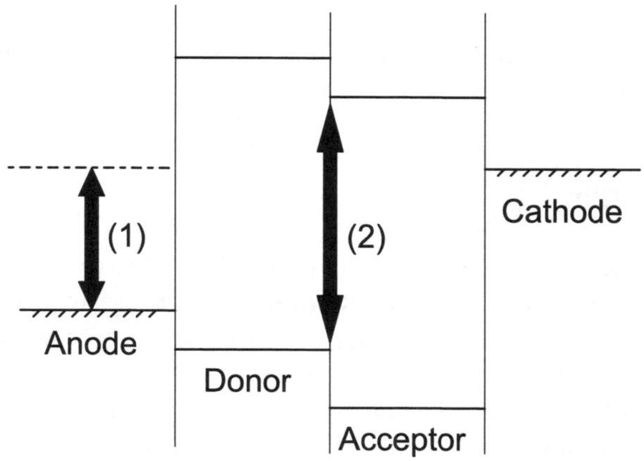

Figure 7.13 For ohmic contacts, the open-circuit voltage is given by the difference in energy between the HOMO of the donor and the LUMO of the acceptor. For nonohmic contacts, the V_{oc} is given by the difference in energy between the workfunctions of the device contacts

Koster *et al.* [223], however, have recently demanded a V_{oc} versus T dependence which does not rely on the ideality factor n, in order to explain their experimental data for PPV:PCBM solar cells. They propose the equation

$$V_{oc} = \frac{E_g}{q} - \frac{kT}{q}\ln\left(\frac{(1-P)\gamma N^2}{PG}\right), \tag{7.16}$$

where E_g denotes the energy difference between the HOMO of the donor and the LUMO of the acceptor, G is the generation rate of electron–hole pairs, P the dissociation probability of bound electron–hole pairs into free charges, γ is a Langevin (bimolecular) recombination constant, and N the effective density of states. Assuming a generation rate corresponding to white light of 80 mW/cm^2 and other reasonable parameters, they are able to achieve a good agreement between theory and experiment.

Studies on MDMO-PPV:PCBM [192] have shown that the solvent used can impact device efficiencies. The most efficient devices reported were made with chlorobenzene as the solvent and 80% PCBM content by weight. Measurements on PPV:PCBM blends via atomic force microscopy (AFM) and transmission electron microscopy (TEM) showed that phase separation in these blends (chlorobenzene) already begins to occur between 50 and 67% PCBM content by weight [224]. Measurements of the field-effect mobilities of electrons and holes in the blends revealed that the electron and hole mobilities are balanced at the 80% PCBM weight content [225].

The higher hole mobilities in P3HT makes it an attractive choice for polymer photovoltaics. A heat treatment is known to be essential for P3HT-based devices, such as OLEDs [226], OFETs [53], and solar cells [194, 227, 228]. Annealing some spin-coated organic semiconductor films can lead to a rearrangement or crystallization of the molecules and, in the case of spin-coated blends, can be used to improve the morphology [229–231]. Heat treatments have also been found to lead to oxygen dedoping, improved interchain interactions [232] and a reduction in the free volume and in the defects at the interface between contact and semiconductor due to evaporation of the solvent [226]. Still, understanding and optimizing the morphology of organic bulk heterojunction solar cells remains one of the major challenges for higher device efficiencies.

7.4.6 Organic lasers

The first laser was made using a ruby crystal in 1960 [233]. One year later it was recognized that conjugated organic molecules which were diluted and oriented within host crystals could be attractive materials for maser applications due to the energy systems in these materials; the triplet state was considered as the excited state [234]. Absorption and stimulated emission depend theoretically only on the relative populations of the two states between which the transition occurs [235, 236]. Reports of stimulated emission (singlet and triplet) were reported during the 1960s in organic molecules and dyes in host crystals or solution form (for some examples see [236–238] and for a review see [239]). The benefits of organic materials as the active material for lasers were recognized as: (i) the numerous possible device structures; and (ii) the wide spectral range that can be achieved simply by tuning the active laser chemically or mechanically.

Conjugated polymers in particular are good candidates for laser applications for several reasons. The first is that the inherent electronic structure of most conjugated polymers is a four-level system. Materials with four or more level systems are attractive for laser applications. The simplest structure for a laser is the three-level system, which means that excited electrons drop from a metastable excited state directly to the ground state. In a material with a four-level system, the electrons drop to a lower laser level above the ground state, leading to a more efficient system. Another advantage of conjugated polymers for laser applications is that the disorder of the system due to differences in conjugation length leads to regions with different bandgap widths which promotes the energy transfer from one region to another in the active layer. Conjugated polymers also typically have high luminescence efficiencies, and large cross-sections for stimulated emission, which leads to higher gain coefficients. Some conjugated polymers have the additional advantage that the emission and absorption spectra do not overlap, although this is a problem that delayed the observation of lasing from conjugated polymers until the 1990s. The absorption from the radiative state into higher energy levels and triplet absorption are issues for many laser materials, e.g., dyes. For a good review, see [240].

Until the 1990s, lasing had been observed in many organic materials, but not in conjugated polymers. Improvements in the chemistry, handling and processing of these materials, however, led to advances in the 1990s in this area [239]. In 1992, the first report of a conjugated polymer laser appeared [241]. The polymer MEH-PPV was used to replace the organic dye in the dye laser configuration. The work on polymer lasing that followed investigated stimulated emission and the excited state absorption on PPV via pump–probe experiments [242], and the results revealed interactions between neighboring molecules could prevent lasing in undiluted (or neat) films. In 1996, the first solid-state polymer lasing was reported [243]. The active layer consisted of MEH-PPV in a host matrix mixed with titanium dioxide nanoparticles. The nanoparticles serve to multiply the scattered photons within the active layer, allowing for stimulated emission if the media is optically pumped above a critical threshold, a technique which was found to be successful for dye lasers [244]. Late in 1996, four groups independently reported lasing from neat films of conjugated polymers [245–248], in particular, lasing was observed by optically pumping a neat PPV layer in an optical cavity [246]. The work on lasing in neat films of conjugated polymers opened the door for work toward polymer diode lasers.

Until now reports have focused on lasing observed from optically pumped organic materials. The transport and injection properties in organic materials are the limiting factor for electrically pumped organic lasers as of yet. Increasing knowledge about transport in disordered organic systems, new materials, and improved processing conditions, are again of the utmost importance for further advances in this area [239, 249].

7.5 CONCLUSIONS

In this chapter recent advances in the area of organic electronics are reviewed. Charge transport in amorphous organic semiconductors occurs via thermally activated and electric field dependent mechanisms. These charge transport properties are briefly described in the first section. A good understanding of them is necessary for the development of efficient devices. Experimental techniques that are commonly used by researchers in the field to determine the mobilities of charge carriers in organic semiconductors are presented, and the utility of the different experimental methods are discussed. The last section of the chapter deals with advances in the area of organic electronics. The nature of organic semiconductors, particularly those that can be processed in solution form, enable the economical production of electronic devices and allow for novel device applications. There have recently been many advances within the field of organic electronics which can be attributed to an increase in the knowledge of charge transport within organic semiconductors together with new materials and device structures.

REFERENCES

[1] H. Akamatsu, H. Inokuchi, and Y. Matsunaga, *Nature*, **173**, 168 (1954).

[2] M. Pope, and C.E. Swenberg, *Electronic Processes in Organic Crystals and Polymers*, 2nd edn, University Press, Oxford, 1999.

[3] H. Shirakawa, E.J. Louis, A.G. MacDiarmid, C.K. Chiang, and A.J. Heeger, *J. Chem. Soc. Chem. Comm.*, 579 (1977).

[4] R.A. Janssen, L. Smilowitz, N.S. Sariciftci, and D. Moses, *J. Chem. Phys.*, **101**, 1787 (1994).

[5] G. Hadziioannou and P. van Hutten, (eds) *Semiconducting polymers: chemistry, physics, engineering*, Wiley-VCH, Weinheim, 1999.

[6] H. Bässler, *Phys. Stat. Sol. B*, **175**, 15 (1993).

[7] A. Miller and E. Abrahams, *Phys. Rev.*, **120**, 745 (1960).

[8] O. Rubel, S.D. Baranovski, P. Thomas, and S. Yamasaki, Concentration dependence of the hopping mobility in disordered organic solids. *Phys. Rev. B*, **9**, 014206 (2004).

[9] M. Grunewald and P. Thomas, *Phys. Stat. Sol. B*, **94**, 125 (1979).

[10] F.R. Shapiro and D. Adler, *J. Non-Cryst. Solids*, **74**, 189 (1985).

[11] D. Monroe, *Phys. Rev. Lett.*, **54**, 146 (1985).

[12] S.D. Baranovski, T. Faber, F. Hensel, and P. Thomas, *J. Phys.: Condens. Matter*, **9**, 2699 (1997).

[13] S.D. Baranovski, H. Cordes, F. Hensel, and G. Leising, Charge-carrier transport in disordered organic solids. *Phys. Rev. B*, **62**, 7934 (2000).

[14] R. Schmechel, Gaussian disorder model for high carrier densities: Theoretical aspects and application to experiments. *Phys. Rev. B*, **66**, 235206 (2002).
[15] R. Schmechel, Hopping transport in doped organic semiconductors: A theoretical approach and its application to p-doped zinc-phthalocyanine. *J. Appl. Phys.*, **93**, 453 (2003).
[16] S. Narioka, H. Ishii, D. Yoshimura, M. Sei, Y. Ouchi, K. Seki, S. Hasegawa, T. Miyazaki, Y. Harima, and K. Yamashita, *Appl. Phys. Lett.*, **67**, 1899 (1995).
[17] H. Ishii, K. Sugiyama, E. Ito, and K. Seki, *Adv. Mater.*, **11**, 605 (1999).
[18] T. Shimada, K. Hamaguchi, A. Koma, and F.S. Ohuchi, *Appl. Phys. Lett.*, **72**, 1869 (1998).
[19] R. Schlaf, P.G. Schroeder, M.W. Nelson, B.A. Parkinson, P.A. Lee, K.W. Nebesny, and N.R. Armstrong, *J. Appl. Phys.*, **86**, 1499 (1999).
[20] J.M. Barathan and Y. Yang, *J. Appl. Phys.*, **84**, 3207 (1998).
[21] O.W. Richardson, *The Emission of Electricity from Hot Bodies*. Longmans Green and Company, London (1921).
[22] W. Schottky, *Z. Physik.*, **15**, 872 (1914).
[23] R.H. Fowler and L. Nordheim, *Proc. Roy. Soc.*, **121**, 626 (1928).
[24] I.D. Parker, *J. Appl. Phys.*, **75**, 1658 (1995).
[25] Z. Chiguvare, J. Parisi, and V. Dyakonov, *J. Appl. Phys.*, **94**, 2440 (2003).
[26] P.S. Davids, S.M. Kogan, I.D. Parker, and D.L. Smith, *Appl. Phys. Lett.*, **69**, 2270 (1996).
[27] J.G. Simmons, *Phys. Rev. Lett.*, **15**, 967 (1965).
[28] P.R. Emtage and J.J. O.'Dwyer, *Phys. Rev. Lett.*, **16**, 365 (1965).
[29] M.A. Baldo and S.R. Forrest, *Phys. Rev. B*, **64**, 085201 (2000).
[30] Y. Shen, W. Klein, D.B. Jacobs, J.C. Scott, and G.G. Malliaras, *Phys. Rev. Lett.*, **86**, 3867 (2001).
[31] M.A. Abkowitz, H.A. Mizes, and J.S. Facci, *Appl. Phys. Lett.*, **66**, 1288 (1995).
[32] E.M. Conwell and M.W. Wu, *Appl. Phys. Lett.*, **70**, 1867 (1997).
[33] V.I. Arkhipov, E.V. Emelianova, Y.H. Tak, and H. Bässler, *J. Appl. Phys.*, **84**, 848 (1998).
[34] U. Wolf, V.I. Arkhipov, and H. Bässler, *Phys. Rev. B*, **59**, 7507 (1999).
[35] V.I. Arkhipov, U. Wolf, and H. Bässler, *Phys. Rev. B*, **59**, 7514 (1999).
[36] S. Barth, U. Wolf, H. Bässler, P. Müller, H. Riel, H. Westweber, P.F. Seidler, and W. Rieß, *Phys. Rev. B*, **60**, 8791 (1999).
[37] J.C. Scott and G.G. Malliaras, *Chem. Phys. Lett.*, **299**, 115 (1999).
[38] W.R. Silveira and J.A. Marohn, *Phys. Rev. Lett.*, **93**, 116104 (2004).
[39] N.F. Mott and R.W. Gurney, *Electronic Processes in Ionic Crystals*. Second edn. Oxford University Press, Oxford, 1948.
[40] M.A. Lampert, Simplified theory of space-charge-limited currents in an insulator with traps. *Phys. Rev.*, **103**, 1648 (1956).
[41] M.A. Lampert and P. Mark, *Current injection in solids*. Academic Press, New York, 1970.
[42] P.N. Murgatroyd, *J. Phys. D: Appl. Phys.*, **3**, 151 (1969).
[43] H. Antoniadis, M.A. Abkowitz, and B.R. Hsieh, *Appl. Phys. Lett.*, **65**, 2030 (1994).
[44] P.W.M. Blom and M.C.J.M. Vissenberg, Electron and hole transport in poly(p-phenylene vinylene) devices. *Appl. Phys. Lett.*, **68**, 3308 (1996).
[45] L. Bozano, A. Cartera, J.C. Scott, G.G. Malliaras, and P.J. Brock, Temperature- and field-dependent electron and hole mobilities in polymer light-emitting diodes. *Appl. Phys. Lett.*, **74**, 1132 (1999).
[46] P.W.M. Blom, C. Tanase, D.M. de Leeuw, and R. Coehoorn, Thickness scaling of the space-charge-limited current in poly(p-phenylene vinylene). *Appl. Phys. Lett.*, **86**, 092105 (2005).
[47] V.I. Arkhipov, P. Heremans, E.V. Emelianova, and G.J. Adriaenssens, Space charge limited currents in materials with gaussian energy distributions of localized states. *Appl. Phys. Lett.*, **79**, 4145 (2001).
[48] V.I. Arkhipov, H.V. Seggern, and E.V. Emilianova, Charge injection versus space-charge-limited current in organic light-emitting diodes. *Appl. Phys. Lett.*, **83**, 5074 (2003).

[49] G. Horowitz, *Adv. Mater.*, **10**, 365 (1998).
[50] R.J.O.M. Hoofman, *Nature*, **392**, 54 (1998).
[51] P.W.M. Blom, M.J.M. de Jong, and M.G. van Munster, *Phys. Rev. B*, **55**, R656 (1997).
[52] H. Sirringhaus, P.J. Brown, R.H. Friend, M.M. Nielsen, K. Bechgaard, B.M.W. Lanveld-Voss, A.J.H. Spiering, R.A.J. Jannsen, E.W. Meijer, P. Herwig, and D.M. de Leeuw, *Nature*, **401**, 685 (1999).
[53] G. Wang, J. Swensen, D. Moses, and A.J. Heeger, *J. Appl. Phys.*, **93**, 6137 (2003).
[54] A. Salleo, T.W. Chen, A.R. Völkel, Y. Wu, P. Liu, B.S. Ong, and R.A. Street, *Phys. Rev. B*, **70**, 115311 (2004).
[55] D.M. Pai, *J. Chem. Phys.*, **52**, 2285 (1970).
[56] W.D. Gill, *J. Appl. Phys.*, **43**, 5033 (1972).
[57] P.W.M. Blom, M.J.M. de Jong, and J.J.M. Vleggaar, *Appl. Phys. Lett.*, **68**, 3308 (1996).
[58] E. Muller-Horsche, D. Haarer, and H. Scher, Transition from dispersive to nondispersive transport: photoconduction of polyvinylcarbazole. *Phys. Rev. B*, **35**, 1273 (1987).
[59] H. Meyer, D. Haarer, H. Naarmann, and H.H. Höhold, *Phys. Rev. B*, **52**, 2587 (1995).
[60] H.C.F. Martens, H.B. Brom, and P.W.M. Blom, *Phys. Rev. B*, **60**, R8489 (1999).
[61] J. Frenkel, *Phys. Rev.*, **54**, 647 (1937).
[62] S.V. Novikov, D.H. Dunlap, V.M. Kenkre, P.E. Parris, and A.V. Vannikov, Essential role of correlations in governing charge transport in disordered organic materials. *Phys. Rev. Lett.*, **81**, 4472 (1998).
[63] I.I. Fishchuk, A. Kadachuk, H. Bässler, and M. Abkowitz, Low-field chargecarrier hopping transport in energetically and positionally disordered organic materials. *Phys. Rev. B*, **70**, 245212 (2004).
[64] A.J. Mozer and N.S. Sariciftci, Negative electric field dependence of charge carrier drift mobility in conjugated, semiconducting polymers. *Chem. Phys. Lett.*, **389**, 438 (2004).
[65] G. Horowitz, P. Lang, M. Mottaghi, and H. Aubin, Extracting parameters from the current-voltage characteristics of organic field-effect transistors. *Adv. Funct. Mater.*, **14**, 1069 (2004).
[66] C. Tanase, E.J. Meijer, P.W.M. Blom, and D.M. de Leeuw, Unification of the hole transport in polymeric field-effect transistors and light-emitting diodes. *Phys. Rev. Lett.*, **91**, 216601 (2003).
[67] M.C.J.M. Vissenberg and M. Matters, *Phys. Rev. B*, **57**, 12964 (1998).
[68] B. Maennig, M. Pfeiffer, A.N. an X. Zhou, K. Leo, and P. Simon, Controlled p-type doping of polycrystalline and amorphous organic layers: Selfconsistent description of conductivity and field-effect mobility by a microscopic percolation model. *Phys. Rev. B*, **64**, 195208 (2001).
[69] W.F. Pasveer, J. Cottaar, C. Tanase, R. Coehoorn, P.A. Bobbert, P.W.M. Blom, D.M. de Leeuw, and M.A.J. Michels, Unified description of chargecarrier mobilities in disordered semiconducting polymers. *Phys. Rev. Lett.*, **94**, 206601 (2005).
[70] P.W.M. Blom and M.C.J.M. Vissenberg, Dispersive hole transport in poly(pphenylene vinylene). *Phys. Rev. Lett.*, **80**, 3819 (1998).
[71] V. Savvateev, A. Yakimov, and D. Davidov, Transient electroluminescence from poly(phenylenevinylene)-based devices. *Adv. Mater.*, **11**, 519 (1999).
[72] H.S. Majumdar, A. Bolognesi, and A.J. Pal, Relaxation dynamics in lightemitting devices based on a poly(3-alkylthiophene) derivative: transient capacitance and transient electroluminescence studies. *Synth. Met.*, **129**, 275 (2002).
[73] H.C.F. Martens, H.B. Brom, and P.W.M. Blom, Frequency-dependent electrical response of holes in poly(p-phenylene vinylene). *Phys. Rev. B*, **60**, R8489 (1999).
[74] A.J. Mozer, N.S. Sariciftci, A. Pivrikas, R. Österbacka, G.G. Jušska, L. Brassat, and H. Bässler, Charge carrier mobility in regioregular poly(3-hexylthiophene) probed by transient conductivity techniques: A comparative study. *Phys. Rev. B*, **71**, 035214 (2005).

[75] H. Scher and E.W. Montroll, *Phys. Rev. B*, **12**, 2455 (1975).

[76] G. Juška, K. Arlauskas, M. Viliūnas, K. Genevičius, R. Österbacka, and H. Stubb, Charge transport in π-conjugated polymers from extraction current transients. *Phys. Rev. B*, **62**, R16235 (2000).

[77] A.J. Mozer, N.S. Sariciftci, L. Lutsen, D. Vanderzande, R. Österbacka, M. Westerling, and G. Juška, Charge transport and recombination in bulk heterojunction solar cells studied by the photoinduced charge extraction in linearly increasing voltage technique. *Appl. Phys. Lett.*, **86**, 112104 (2005).

[78] R.W.I. de Boer, M. Jochemsen, T.M. Klapwijk, A.F. Morpurgo, J. Niemax, A.K. Tripathi, and J. Pflaum, Space charge limited transport and time of flight measurements in tetracene single crystals: A comparative study. *J. Appl. Phys.*, **95**, 1196 (2004).

[79] D. Poplavskyy and J. Nelson, Nondispersive hole transport in amorphous films of methoxy-spirofluorene-arylamine organic compound. *J. Appl. Phys.*, **93**, 341 (2003).

[80] T. van Woudenbergh, P.W.M. Blom, M.C.J.M. Vissenberg, and J.N. Huiberts, *Appl. Phys. Lett.*, **79**, 1697 (2001).

[81] S.M. Sze, *Physics of Semiconductor Devices.*, John Wiley & Sons, New York, 1981.

[82] G.G. Malliaras, J.R. Salem, P.J. Brock, and C. Scott, *Phys. Rev. B*, **58**, R13411 (1998).

[83] T.M. Brown, J.S. Kim, R.H. Friend, F. Cacialli, R. Daik, and W.J. Feast, *Appl. Phys. Lett.*, **75**, 1679 (1999).

[84] H.C.F. Martens, W.F. Pasveer, H.B. Brom, J.N. Huiberts, and P.W.M. Blom, Crossover from space-charge-limited to recombination-limited transport in polymer light-emitting diodes. *Phys. Rev. B*, **63**, 125328 (2001).

[85] B. Bohnenbuck, E. von Hauff, J. Parisi, C. Deibel, and V. Dyakonov, *J. Appl. Phys.*, **99**, 024506 (2006).

[86] G. Horowitz, *Adv. Mater.*, **13**, 53 (2003).

[87] S.F. Nelson, Y.Y. Lin, D.J. Gundlach, and T.N. Jackson, *Appl. Phys. Lett.*, **72**, 1854 (1998).

[88] R.W.I. de Boer, M.E. Gershenson, A.F. Morpurgo, and V. Podzorov, *Phys. Stat. Sol. A*, **6**, 1302 (2004).

[89] C. Waldauf, P. Schilinsky, M. Perisutti, J. Hauch, and C.J. Brabec, *Adv. Mater.*, **15**, 2084 (2003).

[90] E.J. Meijer, C. Tanase, P.W.M. Blom, E. van Veenendaal, B.H. Huisman, D.M. de Leeuw, and T.M. Klapwijk, *Appl. Phys. Lett.*, **80**, 3838 (2002).

[91] E.J. Meijer, C. Detcheverry, P.J. Baesjou, E. van Veenendaal, D.M. de Leeuw, and T.M. Klapwijk, Dopant density determination in disordered organic field-effect transistors. *J. Appl. Phys.*, **93**, 4831 (2003).

[92] P.V. Necliudov, M.S. Shur, D.J. Grundlach, and T.N. Jackson, *J. Appl. Phys.*, **88**, 6594 (2000).

[93] R.A. Street and A. Salleo, *Appl. Phys. Lett.*, **81**, 2887 (2002).

[94] G. Horowitz, R. Hajlaoui, D. Fichou, and A.E. Kassmi, *J. Appl. Phys.*, **85**, 3202 (1999).

[95] C. Tanase, P.W.M. Blom, D.M. de Leeuw, and E.J. Meijer, Charge carrier density dependence of the hole mobility in poly(p-phenylene vinylene). *Phys. Stat. Sol. A*, **201**, 1236 (2004).

[96] S. Luan and G.W. Neudeck, *J. Appl. Phys.*, **72**, 766 (1992).

[97] L. Bürgi, T.J. Richards, R.H. Friend, and H. Sirringhaus, *J. Appl. Phys.*, **94**, 6129 (2003).

[98] B.H. Hamadani and D. Natelson, *Appl. Phys. Lett.*, **84**, 443 (2004).

[99] R.J. Chesterfield, J.C. McKeen, C.R. Newman, C.D. Frisbie, P.C. Ewbank, K.R. Mann, and L.L. Miller, *J. Appl. Phys.*, **95**, 6396 (2004).

[100] E. von Hauff, J. Parisi, and V. Dyakonov, *Thin Solid Films*, **511–512**, 506–511 (2006).

[101] K. Seshadri and C.D. Frisbie, *Appl. Phys. Lett.*, **78**, 993 (2001).

[102] T. Hassenkam, D.R. Greve, and T. Bjornholm, *Adv. Mater.*, **13**, 631 (2001).

[103] L. Bürgi, H. Sirringhaus, and R.H. Friend, *Appl. Phys. Lett.*, **80**, 2913 (2002).

[104] V. Podzorov, V.M. Pudalov, and M.E. Gershenson, *Appl. Phys. Lett.*, **82**, 1739 (2003).

[105] S. Forrest, *Nature*, **428**, 911 (2004).
[106] C.J. Drury, C.M.J. Mutsaers, C.M. Hart, M. Matters, and D.M. de Leeuw, *Appl. Phys. Lett.*, **73**, 108 (1998).
[107] T. Shimoda, K. Morii, S. Seki, and H. Kiguchi, *MRS Bulletin*, **28**, 821 (2003).
[108] H. Sirringhaus, T. Kawase, R.H. Friend, T. Shimoda, M. Inbasekaran, W. Wu, and E.P. Woo, *Science*, **290**, 2123 (2000).
[109] S.E. Burns, P. Cain, J. Mills, J. Wang, and H. Sirringhaus, *MRS Bulletin*, **28**, 829 (2003).
[110] F. Pschenitzka and J.C. Sturm, *Appl. Phys. Lett.*, **74**, 1913 (1999).
[111] C.C. Wu, C. Yang, H.H. Chang, C.W. Chen, and C.C. Lee, *Appl. Phys. Lett.*, **77**, 794 (2000).
[112] G.B. Blanchet, Y.L. Loo, J.A. Rogers, F. Gao, and C.R. Fincher, *Appl. Phys. Lett.*, **82**, 463 (2003).
[113] M. Schtein, P. Peumans, J.B. Benziger, and S.R. Forrest, *J. Appl. Phys.*, **93**, 4005 (2003).
[114] A. Bernanose, M. Comte, and P. Vouaux, *J. Chim. Phys.*, **50**, 64 (1953).
[115] M. Pope, H. Kallman, and P. Magnante, *J. Chem. Phys.*, **28**, 2042 (1963).
[116] W. Helfrich and W.G. Schneider, *Phys. Rev. Lett.*, **14**, 229 (1965).
[117] P.S. Vincett, W.A. Barlow, R.A. Hann, and G.G. Roberts, *Thin Solid Films*, **94**, 171 (1982).
[118] C.W. Tang and S.A. VanSlyke, *Appl. Phys. Lett.*, **51**, 913 (1987).
[119] T. Tsutsui and K. Fujita, *Adv. Maters.*, **14**, 949 (2002).
[120] J.H. Burroughes, D.D.C. Bradley, A.R. Brown, R.N. Marks, K. Makay, R. Friend, P.L. Burns, and A.B. Holmes, *Nature*, **347**, 539 (1989).
[121] J.C. Scott, S.A. Carter, S. Karg, and M. Angelopoulos, *Synth. Met.*, **85**, 1197 (1997).
[122] S.A. Carter, M. Angelopoulos, S. Karg, P.J. Brock, and J.C. Scott, *Appl. Phys. Lett.*, **70**, 2067 (1997).
[123] F. Cacialli, R.H. Friend, N. Haylett, R. Daik, W.J. Feast, D.A. dos Santos, and J.L. Brédas, *Appl. Phys. Lett.*, **69**, 3794 (1996).
[124] J. Kido and Y. Iizumi, Fabrication of highlty efficient organic electroluminescent devices. *Appl. Phys. Lett.*, **73**, 2721 (1998).
[125] C.R. Newman, C.D. Frisbie, D.A. da Silva Filho, J.L. Brédas, P.C. Ewbank, and K.R. Mann, *Chem. Mater.*, **16**, 4436 (2004).
[126] L.S. Hung, C.W. Tang, and M.G. Madson, *Appl. Phys. Lett.*, **70**, 152 (1996).
[127] R. Schlaf, B.A. Parkinson, P.A. Lee, K.W. Nebesny, G. Jabbour, B. Kippelen, N. Peyghambarian, and N.R. Armstrong, Photoemission spectroscopy of LiF coated Al and Pt electrodes. *J. Appl. Phys.*, **84**, 6729 (1998).
[128] Y. Jin, X. Ding, J. Reynaert, V.I. Arkhipov, G. Borghs, P.L. Heremans, and M.V. der Auweraer, Role of LiF in polymer light-emitting diodes with LiF-modified cathodes. *Org. Electron.*, **5**, 271 (2004).
[129] Y. Yuan, D. Grozea, S. Han, and Z.H. Lu, Interaction between organic semiconductors and LiF dopant. *Appl. Phys. Lett.*, **85**, 4959 (2004).
[130] Y. Gao, K.T. Park, and B.R. Hsieh, *J. Appl. Phys.*, **73**, 7894 (1993).
[131] R.H. Friend, R.W. Gymer, A.B. Holmes, J.H. Burroughs, R.N. Marks, C. Taliani, D.D.C. Bradley, D.A.D. Santos, J.L. Brédas, W.R. Lögdlund, and M. Salaneck, *Nature*, **397**, 121 (1999).
[132] A.V. Slyke, C.H. Chen, and C.W. Tang, *Appl. Phys. Lett.*, **69**, 2160 (1996).
[133] R.H. Friend, *Pure Appl. Chem.*, **73**, 425 (2001).
[134] X. Gong, P. Iyer, D. Moses, G.C. Bazan, and A.J. Heeger, *Adv. Funct. Mater.*, **13**, 325 (2003).
[135] M. Mazzeo, V. Vitale, F.D. Sala, M. Anni, G. Barbarella, L. Favaretto, G. Sotgiu, R. Cingolani, and G. Gigli, *Adv. Funct. Mater.*, **17**, 34 (2005).
[136] X. Gong, S. Wang, D. Moses, G.C. Bazan, and A.J. Heeger, *Adv. Funct. Mater.*, **17**, 2053 (2005).

[137] M.A. Baldo, D.F.O. an dY. You, A. Shoustikov, S. Sibley, M.E. Thompson, and S.R. Forrest, Highly efficient phosphorescent emission from organic electroluminescent devices. *Nature*, **395**, 151 (1998).

[138] H. Yersin, Triplet emitters for OLED applications. mechanisms of exciton trapping and control of emission properties. *Top. Curr. Chem.*, **241**, 1 (2004).

[139] D. Kahng and M.M. Atalla, *IRE Solid-State Devices Research Conference, Carnegie Institute of Technology, Pittsburg, PA*, 1960.

[140] D.F. Barbe and C.R. Westgate, *J. Phys. Chem. Solids*, **31**, 2679 (1970).

[141] M.L. Petrova and L.D. Rozenshtein, *Fiz. Tverd. Tela. (Sov. Phys.—Solid State)*, **12**, 961 (1970).

[142] H. Koezuka, A. Tsumura, and T. Ando, *Synth. Met.*, **18**, 699 (1987).

[143] E.J. Meijer, D.M. de Leeuw, S. Setayesh, E. van Veenendaal, B.-H. Huisman, P.W.M. Blom, J.C. Hummelen, U. Scherf, and T.M. Klapwijk, Solution-processed ambipolar organic field-effect transistors and inverters. *Nat. Mater.*, **2**, 678 (2003).

[144] A. Dodabalapur, J. Laquindanum, H.E. Katz, and Z. Bao, *Appl. Phys. Lett.*, **69**, 4227 (1996).

[145] S. Steudel, K. Myny, V. Arkhipov, C. Deibel, S. de Vusser, J. Genoe, and P. Heremans, 50 MHz rectifier based on an organic diode. *Nat. Mater.*, **4**, 597 (2005).

[146] Y.Y. Lin, A. Dodabalapur, R. Sarpeshkar, Z. Bao, W. Li, K. Baldwin, V.R. Raju, and H.E. Katz, *Appl. Phys. Lett.*, **74**, 2714 (1999).

[147] P. Mach, S.J. Rodriguez, R. Nortrup, P. Wiltzius, and J.A. Rogers, *Appl. Phys. Lett.*, **78**, 3592 (2001).

[148] H.E.A. Huitema, G.H. Gelinck, J.B.P.H. van der Putten, K.E. Kuijk, C. Hart, E. Cantatore, P.T. Herwigm, A.J.J.M. van Breemen, and D.M. de Leeuw, *Nature*, **414**, 599 (2001).

[149] B.K. Crone, A. Dodabalapur, A. Gelperin, L. Torsi, H.E. Katz, A.J. Ovinger, and Z. Bao, *Appl. Phys. Lett.*, **78**, 2229 (2001).

[150] G.H. Gelinck, T.C.T. Geuns, and D.M. de Leeuw, *Appl. Phys. Lett.*, **77**, 1487 (2000).

[151] A. Hepp, H. Heil, W. Weise, M. Ahles, R. Schmechel, and H. von Seggern, *Phys. Rev. Lett.*, **91**, 157406 (2003).

[152] C.D. Dimitrakopoulos and P.R.L. Malenfant, *Adv. Mater.*, **14**, 99 (2002).

[153] Y. Sun, Y. Liu, and D. Zhu, *J. Mater. Chem.*, **15**, 53 (2005).

[154] D.M. de Leeuw, M.M.J. Simenon, A.R. Brown, and R.E.F. Einerhand, *Synth. Met.*, **87**, 53 (1997).

[155] S.A. Choulis, Y. Kim, J. Nelson, D.D.C. Bradley, M. Giles, M. Shkunov, and I. McCulloch, *Appl. Phys. Lett.*, **85**, 3890 (2004).

[156] T.D. Anthopoulos, C. Tanase, S. Setayesh, E.J. Meijer, J.C. Hummelen, P.W.M. Blom, and D.M. de Leeuw, *Adv. Mater.*, **16**, 2174 (2004).

[157] L.L. Chua, J. Zaumseil, J.F. Chang, E.C.W. Ou, P.K.H. Ho, H. Sirringhaus, and R.H. Friend, General observation of n-type field-effect behaviour in organic semiconductors. *Nature*, **434**, 194 (2005).

[158] V.C. Sundar, J. Zaumseil, V. Podzorov, E. Menard, R.L. Willett, T. Someya, M.E. Gershenson, and J.A. Rogers, *Science*, **303**, 1644 (2004).

[159] H. Klauk, M. Halik, U. Zschieschang, G. Schmid, W. Radlik, and W. Weber, *J. Appl. Phys.*, **92**, 5259 (2002).

[160] A.R. Brown, A. Pomp, D.M. de Leeuw, D.B.M. Klaassen, E.E. Havinga, P. Herwig, and K. Müllen, *J. Appl. Phys.*, **79**, 2136 (1996).

[161] P.T. Herwig and K. Müllen, *Adv. Mater.*, **11**, 480 (1999).

[162] A. Afzali, C.D. Dimitrakopoulos, and T.L. Breen, *J. Am. Chem. Soc.*, **124**, 8812 (2002).

[163] A. Afzali, C.D. Dimitrakopoulos, and T.O. Graham, *Adv. Mater.*, **15**, 2066 (2003).

[164] M. Halik, K. Hagen, U. Zschieschang, G. Schmid, C. Dehm, M. Schutz, S. Maisch, F. Effenberger, M. Brunnbauer, and F. Stellacci, *Nature*, **431**, 963 (2004).

[165] M. Panzer, C.R. Newman, and C.D. Frisbie, *Appl. Phys. Lett.*, **86**, 103603 (2005).
[166] H. Carchano, R. Lacoste, and Y. Segui, *Appl. Phys. Lett.*, **19**, 414 (1971).
[167] R.S. Potember, T.O. Poehler, and D.O. Cowan, *Appl. Phys. Lett.*, **34**, 405 (1979).
[168] T. Oyamada, H. Tanaka, K. Matsushige, H. Sasabe, and C. Adachi, *Appl. Phys. Lett.*, **83**, 1252 (2003).
[169] L.P. Ma, J. Liu, and Y. Yang, *Appl. Phys. Lett.*, **16**, 2997 (2002).
[170] L.P. Ma, Q.F. Xu, and Y. Yang, *Appl. Phys. Lett.*, **84**, 4908 (2004).
[171] J. He, L. Ma, J. Wu, and Y. Yang, *J. Appl. Phys.*, **97**, 064507 (2005).
[172] J.G. Simmons and R.R. Verderber, *Proc. R. Soc. London, Ser. A*, **301**, 77 (1967).
[173] L.D. Bozano, B.W. Kean, V.R. Deline, J.R. Salem, and J.C. Scott, *Appl. Phys. Lett.*, **84**, 607 (2004).
[174] S.H. Kang, T. Crisp, I. Kymissis, and V. Bulović, *Appl. Phys. Lett.*, **85**, 4666 (2004).
[175] C.P. Collier, E.W. Wong, M. Belohradsky, F.M. Raymo, J.F. Stoddart, P.J. Kuekes, R.S. Williams, and J.R. Heath, *Science*, **285**, 391 (1999).
[176] J. Chen, M.A. Reed, A.M. Rawlett, and J.M. Tour, *Science*, **286**, 1550 (1999).
[177] A. Bandyopadhyay and A.J. Pal, *Appl. Phys. Lett.*, **82**, 1215 (2003).
[178] F.L.E. Jakobsson, X. Crispin, and M. Berggren, *Appl. Phys. Lett.*, **87**, 063503 (2005).
[179] Y. Murata, K. Tsunashima, N. Koizumi, K. Ogami, F. Hosokawa, and K. Yokoyama, *Jpn. J. Appl. Phys.*, **32**, L849 (1993).
[180] R. Schroeder, L.A. Majewski, and M. Grell, *Adv. Mater.*, **16**, 633 (2004).
[181] G.H. Gelinck, A.W. Marsman, F.J. Touwslager, S. Setayesh, D.M. de Leeuw, R.C.G. Naber, and P.W.M. Blom, *Appl. Phys. Lett.*, **87**, 092903 (2005).
[182] C.J. Brabec, V. Dyakonov, J. Parisi, and N.S. Sariciftci, (eds) *Organic Photovoltaics*. Springer, Berlin, 2003.
[183] S.E. Shaheen, D.S. Ginley, and G.E. Jabbour, *MRS Bulletin*, **30**, 10 (2005).
[184] K. Zwiebel, B. von Roedern, and H. Ullal, *Photon Int.*, 48 (2004).
[185] S.E. Shaheen, R. Radspinner, N. Peyghambarian, and G.E. Jabbour, *Appl. Phys. Lett.*, **79**, 2996 (2001).
[186] J.M. Shaw and P.F. Seidler, *IBM J. Res. & Dev.*, **45**, 3 (2001).
[187] S. Wagner, E. Bonderover, W.B. Jordan, and J.C. Sturm, *IJHSES*, **12**, 391 (2002).
[188] H. Hoppe and N.S. Sariciftci, *J. Mater. Res.*, **19**, 1924 (2004).
[189] C.J. Brabec, J.A. Hauch, P. Schilinsky, and C. Waldauf, *MRS Bulletin*, **30**, 50 (2005).
[190] J.M. Kroon, M.M. Wienk, W.J.H. Verhees, and J.C. Hummelen, Accurate efficiency determination and stability studies of conjugated polymer/fullerene solar cells. *Thin Solid Films*, **403**, 223 (2002).
[191] F.C. Krebs, J.E. Carle, N. Cruys-Bagger, M. Andersen, M.R. Lilliedal, M.A. Hammond, and S. Hvidt, Lifetimes of organic photovoltaics: photochemistry, athmosphere effects and barrier layers in ITO-MEHPPV:PCBMaluminium devices. *Sol. Ener. Mat. Sol. Cells*, **86**, 499 (2005).
[192] S.E. Shaheen, C.J. Brabec, N.S. Sariciftci, F. Padinger, T. Fromherz, and J.C. Hummelen, *Appl. Phys. Lett.*, **78**, 15 (2001).
[193] C.J. Brabec, S.E. Shaheen, C. Winder, N.S. Sariciftci, and P. Denk, *Appl. Phys. Lett.*, **80**, 1288 (2002).
[194] F. Padinger, R. Rittberger, and N.S. Sariciftci, *Adv. Funct. Mater.*, **13**, 1 (2003).
[195] D.L. Morel, A.K. Gosh, T. Feng, E.L. Stogryn, P.E. Purwin, R.F. Shaw, and C. Fishman, *Appl. Phys. Lett.*, **32**, 495 (1978).
[196] A.K. Gosh and T. Feng, *J. Appl. Phys.*, **49**, 5982 (1978).
[197] C.W. Tang, *Appl. Phys. Lett.*, **48**, 183 (1986).
[198] M. Hiramoto, M. Suezaki, and M. Yokoyama, *Chem. Lett.*, **19**, 327 (1990).
[199] N.S. Sariciftci, L. Smilowitz, A.J. Heeger, and F. Wudl, *Science*, **258**, 1474 (1992).

[200] L. Smilowitz, N.S. Sariciftci, R. Wu, C. Gettinger, A.J. Heeger, and F. Wudl, *Phys. Rev. B*, **48**, 15425 (1993).

[201] N.S. Sariciftci, D. Braun, C. Zhang, V.I. Srdanov, A.J. Heeger, G. Stucky, and F. Wudl, Semiconducting polymer-buckmisterfullerene heterojunctions: diodes, photodiodes, and photovoltaic cells. *Appl. Phys. Lett.*, **62**, 585 (1993).

[202] J.J. Halls, K. Pichler, and R.H. Friend, Exciton diffusion and dissociation in a poly(p-phenylenevinylene)/C60 heterojunction photovoltaic cell. *Appl. Phys. Lett.*, **68**, 3120 (1996).

[203] G. Yu, J. Gao, J.C. Hummelen, F. Wudl, and A.J. Heeger, *Science*, **270**, 1789 (1995).

[204] M. Reyes-Reyes, K. Kim, and D.L. Carroll, High-efficiency photovoltaic devices based on annealed poly(3-hexylthiophene) and 1-(3-methoxycarbonyl)-propyl-1-phenyl-(6,6)C61 blends. *Appl. Phys. Lett.*, **87**, 083506 (2005).

[205] J.F. Ekert, J.F. Nicoud, J.F. Nierengarten, S.G. Liu, L. Echegoyen, F. Barigelletti, N. Armaroli, L. Ouali, V. Krasnikov, and G. Hadziioannou, *J. Am. Chem. Soc.*, **122**, 7467 (2000).

[206] F. Zhang, M. Svensson, M.R. Andersson, M. Maggini, S. Bucella, E. Menna, and O. Inganäs, *Adv. Mater.*, **13**, 1871 (2001).

[207] T. Aernouts, P. Vanlaeke, W. Geens, J. Poortmans, P. Heremans, S. Borghs, R. Mertens, R. Andriessen, and L. Leenders, Printable anodes for flexible organic solar cell modules. *Thin Solid Films*, **451**, 22 (2004).

[208] B.A. Gregg and M.C. Hanna, *J. Appl. Phys.*, **93**, 3605 (2002).

[209] D. Gebeyehu, C.J. Brabec, F. Padinger, T. Fromherz, J.C. Hummelen, D. Badt, H. Schindler, and N.S. Sariciftci, *Synth. Met.*, **118**, 1 (2000).

[210] A.C. Arango, L.R. Johnson, V.N. Bliznyuk, Z. Schlesinger, S.A. Carter, and H.H. Hörhold, *Adv. Mater.*, **12**, 1689 (2000).

[211] P. Schilinsky, C. Waldauf, J. Hauch, and C.J. Brabec, Simulation of light intensity dependent current characteristics of polymer solar cells. *J. Appl. Phys.*, **95**, 2816 (2004).

[212] D. Chirvase, Z. Chiguvare, M. Knipper, J. Parisi, V. Dyakonov, and J.C. Hummelen, Temperature dependent characteristics of poly(3 hexylthiophene)-fullerene based heterojunction organic solar cells. *J. Appl. Phys.*, **93**, 3376 (2003).

[213] I. Riedel, J. Parisi, V. Dyakonov, L. Lutsen, D. Vanderzande, and J.C. Hummelen, *Adv. Funct. Mater.*, **14**, 38 (2004).

[214] J. Gao, H. Wang, and F. Hide, *Synth. Met.*, **84**, 979 (1997).

[215] W.U. Huynh, J.J. Dittmer, N. Teclemariam, D.J. Milliron, and A.P. Alivisatos, *Phys. Rev. B*, **67**, 115326 (2003).

[216] G.G. Malliaras, J.R. Salem, P.J. Brock, and J.C. Scott, *J. Appl. Phys.*, **84**, 1583 (1998).

[217] C.J. Brabec, A. Cravion, D. Meissner, N.S. Sariciftci, T. Fromherz, M.T. Rispens, L. Sanchez, and J.C. Hummelen, *Adv. Funct. Mater.*, **11**, 374 (2001).

[218] J. Liu, Y. Shi, and Y. Yang, *Adv. Funct. Mater.*, **11**, 420 (2001).

[219] H. Frohne, D.C. Müller, and K. Meerholz, *Chem. Phys. Chem.*, **8**, 707 (2002).

[220] F.L. Zhang, M. Johansson, M.R. Anderson, J.C. Hummelen, and O. Inganäs, *Synth. Met.*, **137**, 1401 (2003).

[221] V.D. Mihailetchi, P.W.M. Blom, J.C. Hummelen, and M.T. Rispens, *J. Appl. Phys.*, **94**, 6849 (2003).

[222] P. Schilinsky, C. Waldauf, and C.J. Brabec, Recombination and loss analysis in polythiophene based bulk heterojunction photodetectors. *Appl. Phys. Lett.*, **81**, 3885 (2002).

[223] L.J.A. Koster, V.D. Mihailetchi, R. Ramaker, and P.W.M. Blom, Light intensity dependence of open-circuit voltage of polymer:fullerene solar cells. *Appl. Phys. Lett.*, **86**, 123509 (2005).

[224] J.K.J. van Duren, X. Yang, J. Loos, C.W.T. Bulle-Lieuwma, A.B. Sieval, J.C. Hummelen, and R.A.J. Janssen, *Adv. Func. Mater.*, **14**, 425 (2004).

[225] W. Geens, T. Martens, J. Poortmans, T. Aernouts, J. Manca, L. Lutsen, P. Heremans, S. Borghs, R. Mertens, and D. Vanderzande, *Thin Solid Films*, **451**, 498 (2004).

[226] T. Ahn, H. Lee, and S. Han, *Appl. Phys. Lett.*, **80**, 392 (2002).
[227] D. Chirvase, J. Parisi, J.C. Hummelen, and V. Dyakonov, *Nanotechnology*, **15**, 1317 (2004).
[228] I. Riedel and V. Dyakonov, *Phys. Stat. Sol. A*, **201**, 1332 (2004).
[229] J.J. Dittmer, E.A. Marseglia, and R.H. Friend, *Adv. Mater.*, **12**, 1270 (2000).
[230] W.U. Huynh, J.J. Dittmer, W.C. Libby, G.L. Whitting, and A.P. Alivisatos, *Adv. Funct. Mater.*, **13**, 73 (2003).
[231] T.J. Savenije, J.E. Kroeze, X. Yang, and J. Loos, The effect of thermal treatment on the morphology and charge carrier dynamics in a polythiophenefullerene bulk heterojunction. *Adv. Funct. Mater.*, **15**, 1260 (2005).
[232] P.J. Brown, D.S. Thomas, A. Köhler, J. Wilson, J.S. Kim, C.M. Ramsdale, H. Sirringhaus, and R.H. Friend, *Phys. Rev. B*, **67**, 064203 (2003).
[233] T.H. Maiman, *Nature*, **187**, 493 (1960).
[234] E.G. Brock, P. Csavinsky, E. Hormats, H.C. Nedderman, D. Tirpe, and F. Unterleitner, *J. Chem. Phys.*, **35**, 759 (1961).
[235] E.U. Condon and G.H. Shortley, *The Theory of Atomic Spectra*. Cambridge University Press, Cambridge, England, 1957.
[236] D.J. Morantz, B.G. White, and A.J.C. Wright, *J. Chem. Phys.*, **37**, 2041 (1962).
[237] A. Lempecki and H. Samelson, *Appl. Phys. Lett.*, **2**, 159 (1963).
[238] B.H. Soffer and B.B. McFarland, *Appl. Phys. Lett.*, **10**, 266 (1967).
[239] N. Tessler, *Adv. Mater.*, **11**, 363 (1999).
[240] M.D. McGehee and A.J. Heeger, *Adv. Mater.*, **12**, 1657 (2000).
[241] D. Moses, *Appl. Phys. Lett.*, **60**, 3215 (1992).
[242] L.J. Rothberg, M. Yan, F. Papadimitrakopoulos, M.E. Gavin, E.W. Kwock, and T.M. Miller, *Synth. Met.*, **80**, 41 (1996).
[243] F. Hide, B.J. Schwartz, M.A. Díaz-García, and A.J. Heeger, *Chem. Phys. Lett.*, **256**, 424 (1996).
[244] N.M. Lawandy, R.M. Balachandran, A.S.L. Gomes, and E. Sauvain, *Nature*, **368**, 436 (1994).
[245] W. Graupner, G. Leising, G. Lanzani, M. Nisoli, S.D. Silvestri, and U. Scherf, *Phys. Rev. Lett.*, **76**, 847 (1996).
[246] N. Tessler, G.J. Denton, and R.H. Friend, *Nature*, **382**, 695 (1996).
[247] F. Hide, M.A. Díaz-García, B.J. Schwartz, M. Andersson, Q. Pei, and A.J. Heeger, *Science*, **273**, 1833 (1996).
[248] S. Frolov, M. Ozaki, W. Gellerman, and V.Z.K. Yoshino, *Jpn. J. Appl. Phys.*, **35**, L1371 (1996).
[249] M.A. Baldo, R.J. Holmes, and S.R. Forrest, *Phys. Rev. B*, **66**, 035321 (2002).

8 Generation, Recombination and Transport of Nonequilibrium Carriers in Polymer–Semiconductor Nanocomposites

H. E. Ruda and Alexander Shik

Centre for Advanced Nanotechnology, University of Toronto, Toronto M5S 3E4, Canada

8.1 INTRODUCTION

Conducting polymers, as materials for light-emitting and photovoltaic devices, have been the subject of considerable interest over the last few decades (see, e.g., [1, 2]). Because of their light weight and excellent mechanical properties, they are ideal candidates for applica-

Charge Transport in Disordered Solids with Applications in Electronics Edited by S. Baranovski.

tions which require portability. The ease of processing and flexibility in thin-film polymers (e.g., fabricating using spin-on or inkjet-printed deposition [3]) contrasts with the stringent requirements of vacuum epitaxial growth of monocrystalline semiconductor structures. Flexible substrates lend themselves to innovative designs for end products and processing based on continuous coating of large-area surfaces. This physical flexibility is matched by functional versatility: a wide variety of compounds can be fashioned using synthetic methods, with energy gap and ionization potential being tuned by chemical modification of polymer chains. The physics of charge transport in conducting polymers and their device applications are described in this book in Chapter 6 by S.D. Baranovskii and O. Rubel and in Chapter 7 by E. von Hauff, C. Deibel, and V. Dyakonov, respectively.

Unfortunately, due to the low carrier mobility and low probability of radiative decay of excitons, the external quantum yield of optoelectronic devices in polymer structures is relatively small. This limits immediate and wide-ranging development of the devices. To overcome this drawback, a new degree of freedom has been introduced to the realm of conducting polymers: the incorporation of inorganic (i.e., mostly semiconductor) nanocrystals (NCs) in order to form composite structures [4–6]. The influence of these inclusions is at least two-fold:

- NCs create regions having high intensity of generation and recombination of nonequilibrium carriers. In photoelectric devices this may result in a considerable increase in the quantum yield and the additional possibility of tuning the spectral characteristics.
- NCs modify the characteristics of the surrounding polymer, which also may have a positive influence on the optical, photoelectrical and transport properties of the composite structures.

In this chapter, we present a theoretical description of nonequilibrium electronic phenomena in such nanocomposites, review a large amount of experimental data already existing in this dynamically developing area, and discuss possibilities of further optimization of light-emitting and photovoltaic devices based on composites comprised of conducting polymer and inorganic NCs.

This choice of materials excludes a great number of interesting studies on optical properties of composites using ordinary, nonconducting polymers (such as PMMA and many others), where electronic processes are confined to the embedded NCs [7, 8], while the polymer plays the role of an optically transparent passive matrix. We also will not consider polymers with ionic conductivity (see, e.g., [9]) where the physics of charge transfer differs essentially from the objects of our primary interest. Ionic conduction in disordered solids and device applications of such materials are described in this book in Chapter 10 by B. Roling and in Chapter 11 by P. Vinatier and Y. Hamon, respectively.

8.2 BASIC FEATURES OF POLYMER-SEMICONDUCTOR NANOCOMPOSITES

The present work deals with the behaviour of nonequilibrium carriers in polymer–semiconductor nanocomposites determined by the following processes:

- optical generation of nonequilibirum electrons and holes (both in polymer and in NCs) or their injection from contacts to the nanocomposite;
- diffusion-drift transport of carriers through the polymer matrix;
- carrier capture by and their release from NCs;
- both radiative and nonradiative recombination of carriers in polymer and in NCs.

At first glance, these phenomena are similar to those in lasers and photosensitive structures containing quantum dots, where a number of studies providing an adequate theoretical description, have already been published (see, e.g., [10–12]). However, comparison of polymer-based nanocomposites with usual quantum dot systems based on semiconductor heterostructures reveals a number of serious distinctions:

1. The electron affinity χ^{p} in most conducting polymers is essentially less than the affinity for most semiconductors χ^{s}. For instance, in the most popular conducting polymers—including polyphenylene vinylene (PPV), triphenyldiamine (TPD), polyvinyl carbazole (PVK), oxadiazole derivative (PBD)—χ^{p} ranges from 2.3 to 2.6 eV [1]. At the same time, χ^{s} is of the order of 4 eV for most semiconductors and even dielectrics, including such popular materials for NCs as CdS (3.8–4.8 eV [13]), CdSe (4.0 eV [13]), and TiO_2 (4.2 eV [14]). Thus the difference $\chi^{s} - \chi^{p}$ always exceeds 1 eV, which modifies noticeably the energy band diagram of nanocomposite, as compared with semiconductor quantum dot structures (see Section 8.3).
2. Nanocomposite components have a dramatic difference in dielectric constant, which in semiconductors usually exceeds that in polymers by a factor three or more. This results in a strong influence of image forces on some phenomena in nanocomposites.
3. NCs prepared for embedding into polymer are usually covered by a layer of organic molecules (ligands) preventing their aggregation in solution and playing the role of additional potential barriers for carriers at the interface.
4. Conducting polymer matrices are characterized by carrier mobilities many orders of magnitude lower than in semiconductors and noticeably different for electrons μ_e and holes μ_h. For most conducting polymers μ_h varies from ~10^{-1} cm^2 V^{-1} s^{-1} to the values several orders of magnitude less [15], and, as a rule, $\mu_e \ll \mu_h$ [16]. As we will show in Section 8.9, such values of mobility modify the picture of carrier transport not only quantitatively, but also qualitatively.

8.3 ENERGY BAND DIAGRAM AND OPTICAL ABSORPTION

In our further analysis, except for Section 8.12, we will assume that the energy gap of conducting polymers E_g^p (separation between the HOMO and LUMO molecular orbitals) exceeds the bandgap of semiconductor NCs E_g^s since in the opposite case, generation, recombination and transport of nonequilibrium carriers occurs mostly in the polymer matrix, and the role of NCs is minor. Thus, from the position of the energy band diagram, NCs represent an array of narrow-gap quantum dots in a wide-gap matrix. It is well known that in ideal heterojunctions the conduction band offset is equal to the difference in electron affinity for materials forming the junction: $\Delta E_c = \chi^s - \chi^p$. Based on the information contained in the previous section, we conclude that each NC represents a deep potential well for electrons, which in further calculations will be considered infinitely deep with the energy of ground state $E_{e0} = \pi^2\hbar^2/(2m_{es}a^2)$ (a is the NC radius and m_{es} is the electron effective mass in semiconductor) and the associated wavefunction, $\psi_{e0}(r) = (2\pi a)^{-1/2}\sin(\pi r/a)/r$. In most polymer–semiconductor systems, the difference in electron affinities is so much that $\chi^s - \chi^p > E_g^p - E_g^s$, which corresponds to so-called type-II heterojunctions where NCs are potential bumps, rather than wells, for holes (Figure 8.1). Moreover, even if the valence band offset $\Delta E_v = E_g^p - E_g^s - \chi^s + \chi^p > 0$ (dashed line in Figure 8.1) and NCs represent potential wells, for not very large a, these wells contain no bound states (at $\Delta E_v = 0.2\,\text{eV}$ and the hole effective mass in semiconductor $m_{hs} = 0.1m_0$ the corresponding condition $\Delta E_v < \pi^2\hbar^2/(8m_{hs}a^2)$ is fulfilled at $a < 3\,\text{nm}$) and all holes are delocalized.

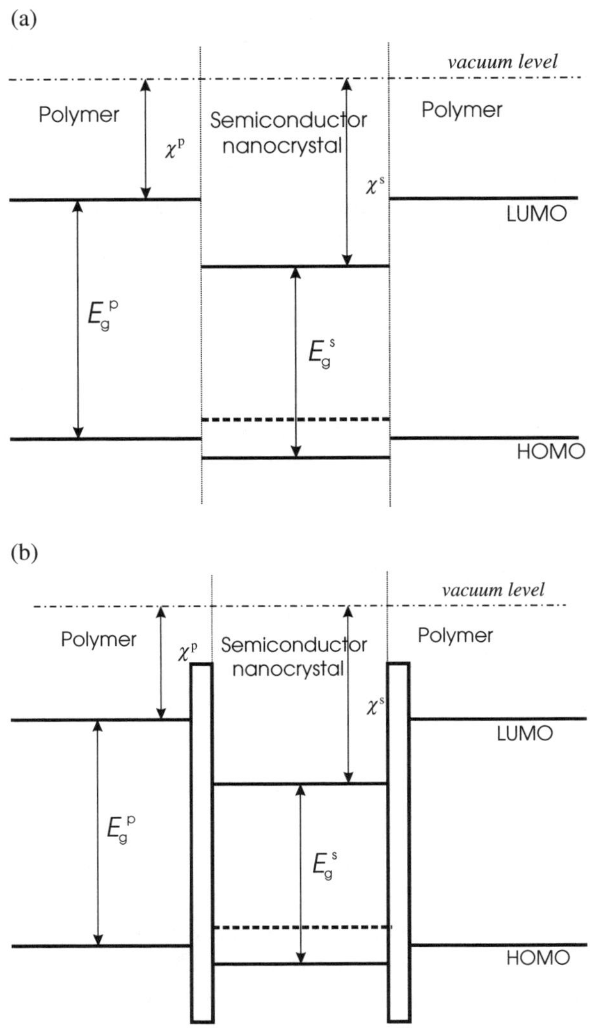

Figure 8.1 Schematic energy band diagram for a typical polymer-NC composite (solid line) and for a composite with positive ΔE_v (dashed line) without (a) and with (b) ligand shell

Thus we may claim that for most polymer–semiconductor nanocomposites holes, contrary to electrons, are not localized inside NCs, and this asymmetry strongly influences properties of a composite considered in the subsequent sections.

The presence of a potential barrier at the interface caused by the ligand layer modifies the picture so that holes inside NCs do not have a purely continuous, but a quasi-discrete energy spectrum, with the ground state energy E_{h0} determined by the NC radius a and the height of potential barrier for holes Δ_h created by the ligand shell. At high Δ_h, $E_{h0} \simeq \pi^2\hbar^2/(2m_{hs}a^2)$. This state has a final width δE (and hence the decay time $\tau_d \sim \hbar/\delta E$) determined by the barrier thickness d governing its tunnel transparency. The corresponding density of hole states $D(E)$ has a peak with the position and width determined, respectively, by E_{h0} and δE. Quantum mechanical description of such decaying states [17] results in the following expression for the wave function of holes with the energy E_h at $r < a$:

$$\begin{aligned}
\psi_h(r) &= \frac{1}{\sqrt{2\pi}}\left[(a-b)^*\sqrt{\frac{a+b}{a^*+b^*}}+\text{c.c.}\right]\frac{\sin(kr)}{r},\\
a &= \frac{\kappa}{4ik}e^{ikd+\kappa d}\left[(1+ik/\kappa)^2e^{-2\kappa d}-(1-ik/\kappa)^2\right],\\
b &= \frac{\kappa}{4ik}e^{ik(2a+d)+\kappa d}\left[(1-e^{-2\kappa d})(1+k^2/\kappa^2)\right],\\
k &= \frac{\sqrt{2m_{hs}E_h}}{\hbar}, \qquad \kappa = \frac{\sqrt{2m_{hs}(\Delta_h-E_h)}}{\hbar},
\end{aligned} \tag{8.1}$$

where ‘c.c.’ is the complex conjugate.

The described physical picture is presumably correct even in the absence of a ligand layer. The essentially different character of wavefunctions in polymers and semiconductors may result in partial hole reflection at the interface, even in the absence of a band offset. This can be interpreted as the presence of some effective interface barrier resulting in quasi-discrete states, similar to those mentioned above.

Now we can apply these results for calculating the specific character of the interband optical absorption of the system in the spectral region below the polymer bandgap [18]. While for nanocrystals in vacuum or in a crystalline dielectric where both electrons and holes are localized in NCs, absorption spectra consist of a number of discrete lines, in our case in the absence of exciton effects they must have continuous character with the absorption edge at $E_g^s + E_{e0}$. As an illustration, let us calculate the absorption spectrum for the simplest case $\Delta E_v = 0$. If we neglect the effective mass and wavefunction mismatch at the NC interface, the wavefunction of holes with the energy E_h can be written in the form of free particle spherical wave: $\psi_h(r) = (2/\pi)^{1/2}\sin\left(\sqrt{2m_{hs}E_h}\, r/\hbar\right)/r$ which is also the limit of Equation (8.1) at $d \to 0$. We have restricted ourselves to hole states with the orbital quantum number $l = 0$ because optical transitions from other states to the s-type ground electron state are forbidden. Calculation of the optical matrix element between ψ_0 and ψ_h followed by integration over E_h, results in the expression for the absorption spectrum:

$$\alpha(w) \sim \int_0^\infty \left|\int_0^a \psi_{e0}(r)\psi_h(r)r^2\mathrm{d}r\right|^2 \delta(\hbar w - E_g^s - E_{e0} - E_h)\frac{\mathrm{d}E_h}{\sqrt{E_h}} \sim \frac{\sin^2\sqrt{w}}{\sqrt{w}\,(w-\pi^2)^2}, \tag{8.2}$$

where $w = 2m_{hs}(\hbar\omega - E_g^s - E_{e0})a^2/\hbar^2$. Since Equation (8.2) contains summation only over hole states with $l = 0$, the one-dimensional density of states is used.

The absorption spectrum (Equation 8.2) is shown by curve 1 in Figure 8.2. Even for a purely continuous dispersion law of holes, the spectrum has a finite width with a maximum at $\hbar\omega - E_g^s - E_{e0} \sim E_{h0}$ though in this case, there are no quasi-discrete states and peaks in the hole density of states, and E_{h0} are absolutely fictitious. The origin of a maximum is the same as, say, for impurity absorption in a semiconductor: the Fourier transform of the electron wavefunction $\psi_{e0}(r)$ contains only wavevectors k not exceeding several a^{-1} so that vertical optical transitions from the hole states with larger k and, hence, larger E_h, are forbidden.

Similar calculations of $a(\omega)$ for NCs surrounded by a ligand shell, with $\psi_h(r)$ given by Equation (8.1), are also shown by solid lines in Figure 8.2.

Thus, in spite of the delocalized character of holes, the absorption spectrum of holes may contain a maximum with width determined by the character of polymer–semiconductor interface and the presence of a ligand layer at this interface. This agrees well with numerous experimental data for different polymer–semiconductor nanocomposites where the absorption spectra in the area of polymer transparency vary from noticeable narrow peaks to smooth monotonic curves with only a weak shoulder at $\hbar\omega$ close to $E_g^s + E_{e0} + E_{h0}$. In this connection, we draw our readers' attention to [19, 20] where the shape of absorption spectra at the light frequencies corresponding to interband transitions in NCs depended on the characteristics of ligand layers.

8.4 EXCITONS

So far, our considerations have ignored excitonic effects. Now we discuss how the electron–hole interaction will modify energy spectrum of the system discussed. Properties of excitons in quantum dots where both electron and hole are localized in the region $r < a$ have already been studied in detail (see, e.g., [21]). There are also theoretical works [22,

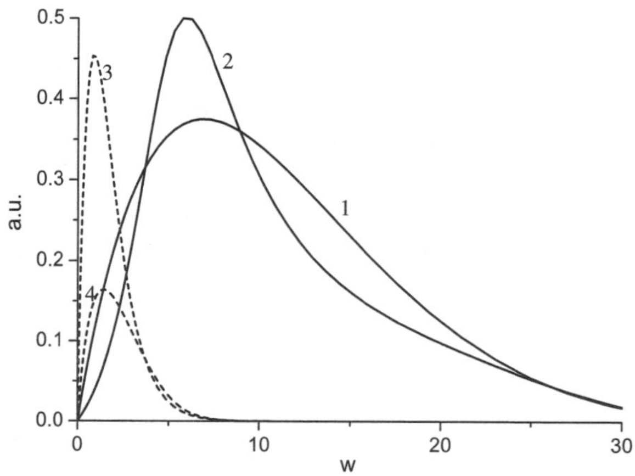

Figure 8.2 Spectral dependences of optical absorption (solid lines) and luminescence (dashed lines) for NCs without (1,3) and with (2,4) ligand shell. $w = [2m_{hs}(\hbar\omega - E_g - E_{e0})a^2]/\hbar^2$

23] devoted to excitons in type-II systems similar to ours, where electrons are localized within and holes—outside a NC. All these papers, however, deal with pure semiconductor heterojunctions where the difference in dielectric constant between NC and ambient is negligible. On the contrary, in the systems under consideration, the dielectric constant inside NC ε_s and outside it ε_p may differ by several times (the second factor mentioned in Section 8.1), resulting in a strong distortion of the electron–hole Coulomb interaction by image forces. Dramatic changes of exciton binding energies by these forces have been already analyzed for quantum wells [24, 25] and quantum wires [26]. One may expect that in our case of three-dimensional confinement, the effects can be even larger.

To find the binding energy of an exciton, we must solve the Schrödinger equation for a hole moving in the Coulomb field (plus corresponding image forces) of an electron localized in the region $r < a$. The corresponding electrostatic problem is straightforward in the case $\varepsilon_p << \varepsilon_s$ typical for polymer–semiconductor nanocomposites. In this case potential variations inside NC are irrelevant and the hole potential energy can be approximately written as

$$V(r) = \begin{cases} -\dfrac{e^2}{\varepsilon_p a} - \Delta E_v, & r < a \\ -\dfrac{e^2}{\varepsilon_p r}, & r > a \end{cases} . \tag{8.3}$$

Figure 8.3 shows the position of a ground (1*s*) state in this potential for different parameters of the problem, namely, the NC radius a, the valence band offset ΔE_v and the ratio of hole effective masses inside and outside NC m_{hs}/m_{hp}. For comparison, the dashed line in the same figure shows the depth of potential well in a polymer $-e^2/(\varepsilon_p a)$. One can see that for zero or negative ΔE_v (type-II band diagram) where the hole wave function is localized mostly in the polymer, the exciton binding energy decreases monotonically with the NC

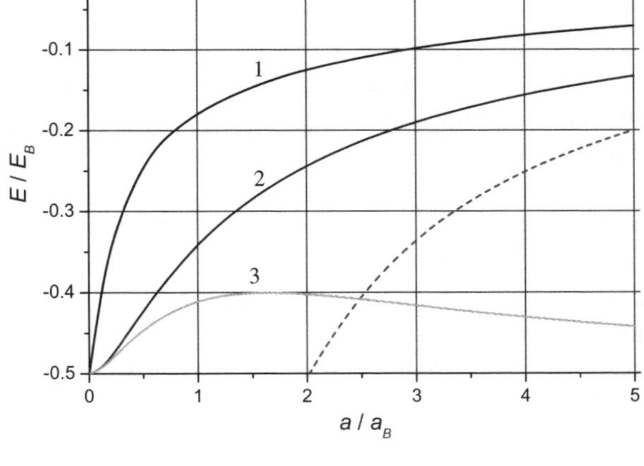

Figure 8.3 The energy of exciton for the case $m_{hs}/m_{hp} = 0.1$ and $\Delta E_v = -E_B$ (curve 1), $\Delta E_v = 0$ (curve 2), and $\Delta E_v = 0.5E_B$ (curve 3). $a_B = \varepsilon_p\hbar^2/(m_{hp}e^2)$ and $E_B = m_{hp}e^4/(\varepsilon_p^2\hbar^2)$ are the effective Bohr radius and energy at $r > a$. The dashed curve corresponds to $-e^2/(\varepsilon_p a)$

radius a. On the contrary, at positive ΔE_v, holes are mostly confined to NCs and the energy level is to a great extent determined by size quantization in the NC. In this case, the exciton properties are close to those of isolated nanocrystals already studied in detail [21]. At $a \to 0$ all curves naturally tend to -0.5, that is to the effective Bohr energy. Besides, since for large negative ΔE_v and not very small a, the wave function does not penetrate the NC and is very small at $r = a$, the corresponding energy (curve 1) for $a/a_B = 2$ is close to $-1/8$ (the second Bohr level) since the corresponding Bohr wave function has a node just at $r = 2a_B$. We also want to emphasize that for the typical situation $\varepsilon_p < \varepsilon_s$ and $m_{hp} > m_{hs}$ the exciton binding energy may be larger than in the case where both carriers are localized inside a NC.

The calculated energy spectrum determines the position of optical absorption lines. To find the transition intensity, we must know the amplitude of hole wavefunction in the region $r < a$ where electrons are localized. The answer depends dramatically on the band offset ΔE_v. For all positive (type-I band diagram) and not very large negative ΔE_v, a considerable part of hole wave function is also localized at $r < a$, resulting in a large interband matrix element. The criterion for this case can be easily formulated. It requires that the valence band in NC be classically accessible, i.e., to lie above the energy level of hole bound state. It means that $|\Delta E_v|$ must exceed the difference between the solid and dashed curves in Figure 8.2, which, as can be seen, in the given interval of parameters lies (in dimensional units) between $0.65e^2/(\varepsilon_p a)$ and $e^2/(\varepsilon_p a)$. Thus, the criterion can be formulated as

$$\Delta E_v > -e^2/(\varepsilon_p a) \tag{8.4}$$

(with an accuracy of about 1).

In the opposite case of large negative ΔE_v, the hole wavefunction exponentially decreases inside NC so that the matrix element drops dramatically and the influence of excitons on absorption and luminescence spectra is negligible.

8.5 POTENTIAL RELIEF AT HIGH EXCITATION LEVEL

We can describe the properties of nonequilibrium carriers in our system in terms of isolated, noninteracting excitons and use the results of Section 8.4 only for sufficiently weak optical excitation when the average number of such carriers per nanocrystal, Q/e, does not exceed one. At higher excitation, each negatively charged NC surrounded by equal positive charge of holes Q represents a many-body system, which at $Q/e >> 1$ can be analyzed using the methods of statistical physics.

We consider our nanocomposite under intensive interband optical excitation creating equal numbers of nonequilibrium electrons and holes and begin with the case of negative ΔE_v corresponding to the condition opposite to Equation (8.4). This means that each NC contains some number of localized electrons Q/e while equal number of holes are distributed outside NC in the self-consistent electrostatic potential $\phi(r)$ created by this charge separation (Figure 8.4). To find this potential, we should solve the Poisson equation at $r > a$ with the boundary conditions

$$\frac{d\phi}{dr}(a) = \frac{Q}{\varepsilon_p a^2}, \tag{8.5}$$

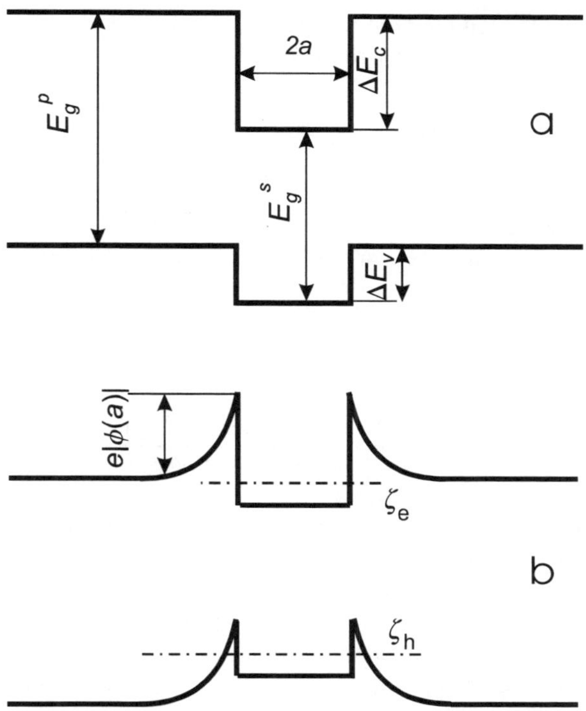

Figure 8.4 Band diagram of a nanocomposite in equilibrium (a) and under interband optical excitation (b)

and $\phi(\infty) = 0$. The form of the equation itself depends on the character of the hole distribution. It was shown [27] that even for very low carrier mobilities, typical for conducting polymers, the carrier distribution at sufficiently low excitation level has a quasi-equilibrium character and can be described by the electron ζ_e and hole ζ_h chemical potential quasi-levels, which will be measured from the HOMO level in polymer far from NC. For nondegenerate holes

$$\frac{1}{r^2}\frac{\mathrm{d}}{\mathrm{d}r}\left(r^2\frac{\mathrm{d}\phi}{\mathrm{d}r}\right) = -\frac{4\pi e N_\mathrm{v}^\mathrm{p}}{\varepsilon_\mathrm{p}}\exp\left(-\frac{\zeta_\mathrm{h}}{kT}\right)\left[\exp\left(-\frac{e\phi}{kT}\right)-1\right], \tag{8.6}$$

where N_v^p is the effective density of states in the HOMO band of the polymer. For $e|\phi(a)| >> kT$, Equation (8.6) can be integrated once. It gives the relation between $\phi(a)$ and $\mathrm{d}\phi/\mathrm{d}r(a)$, which, with the help of Equation (8.5), acquires the form

$$Q = a^2\sqrt{8\pi\varepsilon_\mathrm{p}N_\mathrm{v}^\mathrm{p}kT}\,\exp\left(\frac{-\zeta_\mathrm{h} - e\phi(a)}{2kT}\right), \tag{8.7}$$

As has already been mentioned, due to the condition $\varepsilon_\mathrm{p} << \varepsilon_\mathrm{s}$ we can neglect potential profile inside NCs and consider them as rectangular potential wells for electrons. In the quasi-classical approximation the requirement for the total number of nondegenerate electrons in a NC to be also Q/e can be written as

$$Q = \frac{4\pi e a^3 N_c^s}{3} \exp\left(\frac{\zeta_e - E_g^p + e\phi(a) + \Delta E_c}{kT}\right), \tag{8.8}$$

(N_c^s is the effective density of states in the conduction band of NC). Equations (8.7), (8.8) allow us to find separately the NC charge Q and the amplitude of potential relief $e|\phi(a)|$:

$$Q = 2a^2 \left(\frac{4\pi^2 e\varepsilon_p a N_c^s N_v^p kT}{3}\right)^{1/3} \exp\left(-\frac{E_g^p - \Delta E_c}{3kT}\right) \exp\left(\frac{\zeta_e - \zeta_h}{3kT}\right), \tag{8.9}$$

$$e|\phi(a)| = \frac{2\zeta_e + \zeta_h - 2E_g^p + 2\Delta E_c}{3} + \frac{kT}{3} \ln \frac{2\pi e^2 a^2 (N_c^s)^2}{9\varepsilon_p N_v^p kT}. \tag{8.10}$$

The formulae obtained depend on the positions of electron and hole quasi-levels ζ_e, ζ_h which, in turn, are determined by the intensity of optical carrier generation $G \equiv \alpha(\omega)I$ (I is the light intensity). The explicit dependences will be obtained later, but some qualitative features can already be seen from Equations (8.9), (8.10). The quasi-level difference $\zeta_e - \zeta_h$ and, hence, NC charge Q increases with I monotonically while the behavior of the potential relief is not so evident. Since an increase in I shifts ζ_e up and ζ_h down, $e|\phi(a)|$ may vary in a different way, depending on the ratio of densities of states N_c^s/N_v^p. It becomes physically clear if we consider two limiting cases. For $N_c^s/N_v^p \to \infty$ even a small deviation of ζ_e from the condiction band edge of a NC creates enough electrons to compensate the hole charge outside of it. For this reason, this conduction band edge is essentially pinned to $\zeta_e : \zeta_e \simeq E_g^p - \Delta E_c + e|\phi(a)|$ and as the intensity increases, $e|\phi(a)|$ increases as well. For $N_c^s/N_v^p \to 0$ the valence band (HOMO) edge at the interface is pinned to $\zeta_h : \zeta_h \simeq e|\phi(a)|$ and decreases with growing intensity, decreasing $e|\phi(a)|$.

At sufficiently high excitation intensity, quasi-levels ζ_e and ζ_h cross the band edges and at $\zeta_e - \zeta_h \simeq E_g^p - \Delta E_c$ both electrons and holes become degenerate. In this case the parameters Q and $e|\phi(a)|$ should be found from corresponding Thomas–Fermi equations. For electrons

$$Q = \frac{8\sqrt{2} e a^3}{9\pi\hbar^2} [m_{es}(\zeta_e - E_g^p + \Delta E_c)]^{3/2}. \tag{8.11}$$

For holes we have the Thomas–Fermi equation

$$\frac{1}{r^2}\frac{d}{dr}\left(r^2 \frac{d\phi}{dr}\right) = -\frac{8\sqrt{2}e}{3\pi\varepsilon_p}\left[\frac{m_{hp} e}{\hbar^2}|\phi|\right]^{3/2} \tag{8.12}$$

to be solved with the boundary condition (Equation 8.5) and a second condition, which can be obtained by excluding the unknown radius of space-charge region r_0 from the requirements $\phi(r_0) = 0$ and $d\phi/dr(r_0) = 0$. This procedure was performed numerically in [28] and some results are shown in Figure 8.5. As in the nondegenerate case, Q grows with $\zeta_e - \zeta_h$ monotonically while the behavior of $e|\phi(a)|$ is determined by the ratio of electron and hole

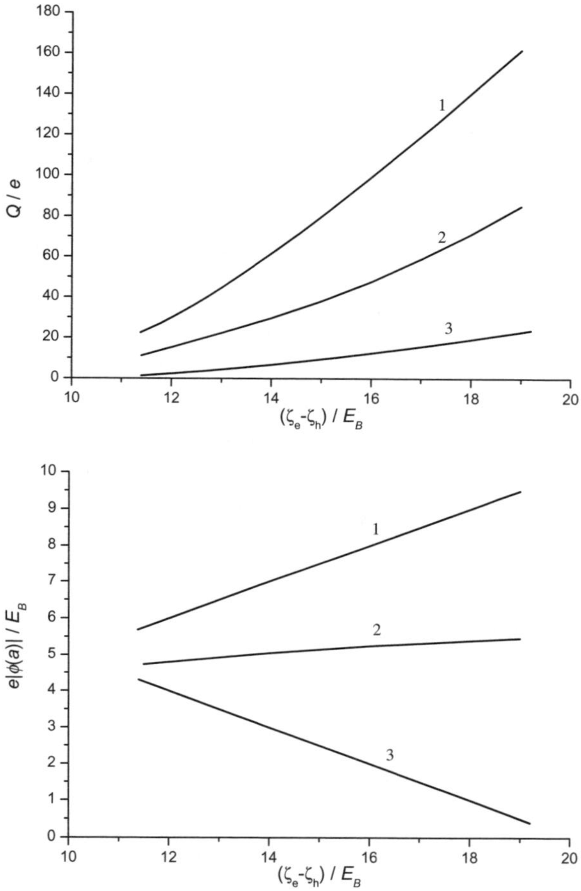

Figure 8.5 Dependences of the potential relief amplitude $e|\phi(a)|$ and of the NC charge Q on the difference of chemical potentials of degenerate electrons and holes in a system with $E_g^p = 20E_B$, $\Delta E_c = 10E_B$, and $a = 3a_B$ for the effective mass ratio $m_{es}/m_{hp} = 10$ (curve 1); 1 (curve 2); 0.1 (curve 3) (after [28] with permission from the Institute of Physics)

masses (that is densities of states). The same calculations show that the radius r_0 has a rather weak Q dependence, varying for all investigated parameters by not more than some tens of percent in the interval (6–10) a_B. All calculations of the present section remain adequate until this radius falls below the inter-NC distance.

To obtain an explicit dependence of Q and the potential relief amplitude on experimental parameters, such as the light intensity I, we must know the recombination rate of nonequilibrium carriers. Since for $\Delta E_v < 0$, electrons and holes are spatially separated, recombination requires activation at a potential barrier. For NCs with a lower bandgap than the polymer, $|\Delta E_v| < \Delta E_c$, we deal mostly with the activation of holes into a NC and recombination inside it. If it has a bimolecular character and can be described by the quadratic expression $\gamma_s np$, the total recombination rate per one NC is [6]

$$\Re = \frac{\gamma_s N_v^s Q}{e} \exp\left(\frac{e|\phi(a)| - \Delta E_v - \zeta_h}{kT}\right). \tag{8.13}$$

If $\hbar\omega < E_g^p$, optical generation, as well as recombination, takes place only in the NCs, so that in the stationary situation $G = \Re$. (The case $\hbar\omega > E_g^p$, when carriers are generated inside the whole polymer matrix, contains some additional problems related to carrier delivery to NCs and capture by them; they will be discussed in Section 8.7.2.) From Equations (8.9), (8.10), and (8.13) we obtain the dependence of Q on the light intensity

$$Q = \left(\frac{8\pi e \varepsilon_p a^4 kT N_v^p}{\gamma_s N_v^s} G\right)^{1/3} \exp\left(\frac{\Delta E_v}{3kT}\right). \tag{8.14}$$

The activation factors in Equations (8.13) and (8.14) are specific features of type-II structures in which nonequilibrium electrons and holes are spatially separated and in order to recombine they must overcome potential barriers. At low temperatures, this may occur by tunneling rather than thermal activation. The corresponding process was calculated in [28] and resulted in the formula for the recombination rate characterized by a very strong dependence on Q and on the NC size: $\Re \sim Q^{11/3} a^{-8}$. In this case Q is proportinal to $G^{3/11}$. We see that, independent of particular recombination mechanism, the Q versus light intensity dependence is essentially sublinear.

8.6 PHOTOCONDUCTIVITY

Since optical excitation causing interband transitions in NCs creates electrons in deep potential wells ΔE_c, photoconductivity in this spectral region is created exclusively by holes and thus has a monopolar character. It is very important to note that in the band diagram of nanocomposites corresponding to Figure 8.1 the spectrum of photoconductivity must coincide with that of optical absorption. This statement is far from being trivial and would be definitely false if NCs represented potential wells for both electrons and holes (as in type-I heterostructures) where light in the vicinity of absorption edge creates carriers localized in NCs and not contributing to the current (at least at low enough temperatures). Such coincidence of the absorption and photoconductivity spectra was observed experimentally in a number of polymer-NC composites [29–31]. Besides, the authors of [30] especially emphasize the observed independence of photoconductivity spectra on the specific polymer.

The dependence of photocurrent on the light intensity is linear at low intensities, and then acquires a sublinear character [32, 33]. This agrees with the results of Section 8.5 where we have shown that the number of nonequilibrium carriers per one NC, Q, is a sublinear function of the light intensity.

The amplitude of the photoconductivity demonstrates a strong dependence on the presence and type of ligand molecules. Removal of the capping layer [32], replacing it by a layer with shorter ligand molecules [33] or simply decreasing the relative amount of capping reagent [19] facilitates hole transport from the NC interior to the conducting polymer matrix, essentially increasing photoconductivity. As was discovered in [32], increase of photoconductivity in the structures with no capping layer is accompanied by a drop in photoluminescence. We will return to this question at the end of Section 8.11.

8.7 PHOTOLUMINESCENCE

8.7.1 Luminescence spectrum and Stokes shift

The main characteristic feature of luminescence observed in all polymer–NC nanocomposites is a strong Stokes shift between the absorption and luminescence spectra (see, e.g., [34–37]). This can be easily explained in terms of the band diagram shown in Figure 8.1 [18]. As was shown in Section 8.3, the maximum of $\alpha(\omega)$ corresponds to the generation of nonequilibrium holes with energy ~ E_{h0} above the valence band edge in NC. Due to continuity of the hole energy spectrum, these holes suffer very fast energy relaxation and for times much shorter than the recombination time, acquire a quasi-equilibrium distribution in the valence band.

For $\Delta E_v < 0$, quasi-equilibrium holes are concentrated mostly at the valence band edge in NC and at the edge of HOMO states (valence band) of a polymer near the interface with NC. The first region has a hole concentration a factor of ~$\exp(|\Delta E_v|/kT)$ lower than that of the second region, but at the same time has a considerably larger probability for optical transitions, which in this case represents direct transitions in the NC requiring no additional tunneling. At small $|\Delta E_v|$ (or at $\Delta E_v > 0$), these direct transitions dominate, and we have luminescence with a Stokes shift of ~E_{h0}. This situation is shown in Figure 8.2 where, besides the already mentioned absorption spectra in the absence and presence of a ligand shell, the corresponding photoluminescence spectra with a distinct Stokes shift, are shown as well. These are obtained as products of the corresponding transition probability calculated in Section 8.3 and the Boltzmann factor $\exp(-\hbar\omega/kT)$. Note that in the dimensionless frequency units w of Figure 8.2, the shift at E_{h0} in photon energies corresponds to the shift in w equal to π^2.

If the band offset is noticeable and the temperature is low enough, NCs contain a vanishingly small number of holes, and luminescence is due to spatially indirect transitions into HOMO or exciton states in a polymer. This is a tunneling process, resulting in a larger Stokes shift approximately equal to $E_{h0} + \Delta E_v$ (or, in other words, energy of emitted quanta $\hbar\omega = E_g^s + E_{e0} - |\Delta E_v| = E_g^p - \Delta E_c + E_{e0}$). To be more exact, the given energy represents maximal possible shift corresponding to electrons and holes exactly at the edges of corresponding bands. Since the density of states at the edge in the absence of quantization tends to zero, the whole luminescent line corresponds to larger $\hbar\omega$ (which are, nevertheless, less than $E_g^s + E_{e0}$ corresponding to the Stokes shift E_{h0}). Its shape can be easily calculated:

$$L(\omega) \sim \int d\varepsilon \rho_e(\varepsilon) f_e(\varepsilon) \int d\varepsilon' I^2(\varepsilon, \varepsilon') \rho_h(\varepsilon') f_h(\varepsilon') \delta(\hbar\omega - E_g^p + \Delta E_c - \varepsilon - \varepsilon'), \quad (8.15)$$

where $\rho_{e,h}$ and $f_{e,h}$ are densities of states and distribution functions for electrons and holes, $I(\varepsilon, \varepsilon')$ is the overlap integral of electron and hole wavefunctions.

The exact expression for $I(\varepsilon, \varepsilon')$ derived in [28] contains only a weak dependence on ε and ε', and the spectral shape $L(\varepsilon)$ is determined primarily by $\rho_{e,h}$ and $f_{e,h}$. In the quasi-classical approximation $\rho_e(\varepsilon) \sim \varepsilon^{1/2}$ while for holes confined in a triangular potential formed by the electric field of a NC equal to $Q/\varepsilon_p a^2$, $\rho_h(\varepsilon') \sim (\varepsilon')^{3/2}$. With these densities of states, Equation (8.15) for nondegenerate carriers gives $L(\omega) \sim (\Delta\omega)^4 \exp(-\Delta\omega/kT)$, where $\Delta\omega = \hbar\omega - E_g^p + \Delta E_c$. For a high excitation level I the shape of the emission spectrum depends on the quasi-levels ζ_e, ζ_h and, hence, becomes dependent on I. For completely degenerate carriers Equation (8.15) results in the following expression for the spectrum:

$$L(\omega) \sim (\Delta\omega)^4 \int_{\max\{0;1-\Delta\zeta_h/\Delta\omega\}}^{\min\{1;\Delta\zeta_e/\Delta\omega\}} x^{3/2}(1-x)^{3/2}\,dx, \tag{8.16}$$

where $\Delta\zeta_e = \zeta_e - E_g^p + \Delta E_c - e|\phi(a)|$, $\Delta\zeta_h = e|\phi(a)| - \zeta_h$ are the quasi-Fermi levels measured from the corresponding band edges. Figure 8.6 demonstrates this spectrum for several different ratios $\Delta\zeta_e/\Delta\zeta_h$ (determined by the effective mass ratio m_{es}/m_{hp}). It is seen to have a strongly asymmetric shape with the maximum slightly less than $\hbar\omega = \Delta\zeta_e + \Delta\zeta_h$ and, hence, suffering a blue shift proportional to the light intensity. Such a shift was experimentally observed in GaAs/GaSb quantum dot structures [11, 38, 39] characterized by a type-II band diagram.

An increase in the excitation intensity causes not only the described variations of the luminescence spectrum; at sufficiently high intensity, when the hole quasi-Fermi level approaches the valence band edge in NCs, a considerable increase of luminescence efficiency is expected as well.

It should be mentioned that the Stokes shift in luminescence was observed not only in nanocomposites, but in separate NCs in solution where it has typical values of 10–20 meV and is explained in terms of the so-called dark excitons in quantum dots [21]. However, this model presumably does not work in nanocomposites where the effect is an order of magnitude larger (e.g., 106 meV in PbS/PPV composites [37]), while our model seems quite natural.

8.7.2 Exciton capture by NCs

Describing photoluminescence phenomena in Section 8.7.1 we assumed that nonequilibrium electrons and holes were created by photoexcitation inside NCs. If the energy of exciting photons $\hbar\omega_e$ is less than the polymer bandgap E_g^p, it is really the main mechanism of photoexcitation, and the photogeneration rate of nonequilibrium carriers $G = \alpha(\omega_e)I$ where $\alpha(\omega_e)$ is the absorption coefficient of NCs with spectral dependence discussed in Section

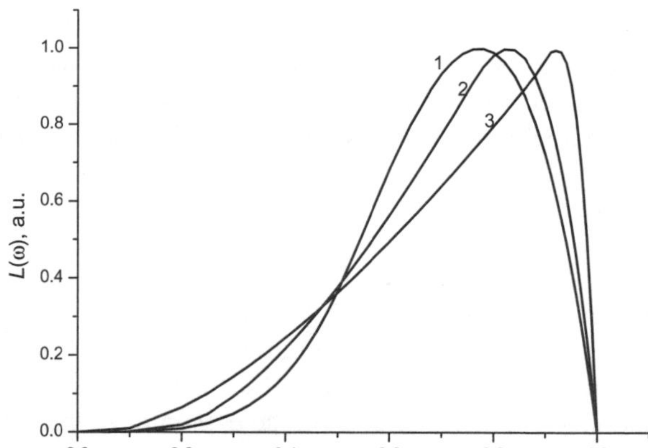

Figure 8.6 Spectra of tunnel luminescence for different rations of electron and hole densities of states. $\Delta\zeta_e/\Delta\zeta_h = 1$ (curve 1); $\Delta\zeta_e/\Delta\zeta_h = 3$ or 1/3 (curve 2); 20 or 1/20 (curve 3) (after [28] with permission from the Institute of Physics)

8.3. However, at $\hbar\omega_e > E_g^p$ most of the exciting light is absorbed in the polymer matrix. In this case, the effective G for carriers in NCs will be determined by their delivery to the NC interface and capture into its interior. Following [40], we consider these processes in more detail.

Excitations created by optical or electrical pumping in the polymer matrix exist, as a rule, in the form of Frenkel excitons, rather than free electron–hole pairs [41] and one of the most important mechanisms of carrier capture by a NC is the Förster one, which is the energy transfer from one system (in our case the Frenkel exciton in a polymer) to another (electron–hole pair in a NC) via dipole–dipole excitation. An alternative mechanism—separate capture of electron and holes—may be much weaker due to rather large binding energy of Frenkel excitons, typically 0.2–0.4 eV [42]. The Förster capture may be of special importance in the the presence of ligand molecules passivating NCs, which inhibit direct capture of carriers. Below we consider a theoretical description of the capture of nonequilibrium Frenkel excitons created in a polymer matrix by semiconductor NCs.

Consider a NC with center at the origin with radius a and a Frenkel exciton at location $\mathbf{r}_e(|\mathbf{r}_e| > a)$ in a polymer matrix. We wish to calculate the probability of exciton recombination with simultaneous generation of an electron–hole pair (or a Wannier–Mott exciton) in a NC via electrostatic interaction. Our theoretical approach is similar to that developed in [43, 44] for the reciprocal problem of energy transfer from a semiconductor quantum well or dot to a polymer matrix. The corresponding probability can be expressed in terms of classical Joule losses:

$$W = \frac{1}{2\pi\hbar}\int \mathrm{Im}\,\varepsilon(\mathbf{r},\omega)|\mathbf{E}(\mathbf{r})|^2\,\mathrm{d}^3\mathbf{r}. \tag{8.17}$$

Here $\mathbf{E}(\mathbf{r})$ is the electric field of an external dipole with the frequency ω in the medium to which energy is to be transferred, $\varepsilon(\mathbf{r}, \omega)$ is the dielectric function of this medium and integral is taken over its volume. In spite of the seemingly classical character of Equation (8.17), it is an exact expression if $\varepsilon(\mathbf{r}, \omega)$ is taken from strict quantum mechanical calculations considering all possible electron transitions at the frequency ω and is equivalent to calculations of the corresponding transition probabilities [43, 44]. In cited papers devoted to energy losses by quantum wells or dots, the role of the energy recipient medium was played by the matrix so that $\varepsilon(\mathbf{r}, \omega)$ refers to the polymer and the integration covered the whole matrix. In our case, energy is transferred to the NC and the integration region is the sphere $r < a$. The situation is quite opposite to that of [43, 44] and we are no longer interested in dielectric properties of a polymer. The energy-recipient medium is now the NC so that the function $\varepsilon(\mathbf{r}, \omega)$ is determined by its energy spectrum and the electron and hole wavefunctions.

If the energy spectrum of a NC were purely discrete with the frequency dispersion of Im ε consisting of a series of δ-peaks, this model would raise serious doubt. Since the Frenkel exciton is also characterized by a fixed discrete energy $E_{ex}(\omega = E_{ex}/\hbar)$, we could expect that the electron transition we are interested in can occur only at resonance when E_{ex} coincides with a gap between some electron and hole states or corresponding Wannier–Mott exciton energies in NC. Such coincidence (assumed in [45]) occurs only occasionally, so that for most ideal NCs the Förster-type transitions would be impossible. In other words, dependence of the transition probability on system parameters (e.g., on the NC radius a) should be represented by a series of δ-peaks. However, in the conditions of the type-II band diagram of Figure 8.1 with a continuous hole spectrum in NCs this problem does not exist and corresponding transitions are always possible. For further analytical calculations we will assume NCs to be not very small, ignore size quantization of electrons and use $\varepsilon(\mathbf{r}, \omega)$ for bulk semiconductors containing no coordinate dependence. For exact quantitative results one can use the dielectric functions obtained by numerical simulation (see, e.g., [46]) for particular quantum dots.

In the above-mentioned continuous approximation, Equation (8.17) is simplified. Using the connection between Im $\varepsilon(\omega)$ and the optical absorption coefficient $a(\omega)$, we have

$$W = \frac{n_r c\alpha(\omega)}{2\pi\hbar\omega} |\mathbf{E}(\mathbf{r})|^2 d^3\mathbf{r}, \tag{8.18}$$

where n_r is the NC refractive index. Thus for calculations of the Förster transition rate we need the spatial distribution of electric field created in the NC by a single Frenkel exciton in its immediate vicinity. To obtain a simple analytical formula, we ignore for the time being the difference in polarizability between the NC and the matrix, describing them by the same effective real dielectric constant ε_0 (this approximation also implies Im $\varepsilon << \varepsilon_0$). This allows us to ignore the image forces and to write for a dipole $\mathbf{d}$ at the origin, the squared electric field in the point $\mathbf{R}$ as $E^2 = (1 + 3\cos^2\Theta)d^2/\varepsilon_0^2 R^6$ where Θ is the angle between $\mathbf{d}$ and $\mathbf{R}$. If the dipole is located at the point with polar coordinates (r_e, 0, 0) and oriented along the polar direction (θ_e, 0), then at an arbitrary point (r, θ, ϕ)

$$E^2(r, \theta, \phi; r_e, \theta_e) = \frac{d^2\{r^2 + r_e^2 - 2rr_e\cos\theta + 3[\cos\theta_e(r\cos\theta - r_e) + r\sin\theta_e\sin\theta\cos\phi]^2\}}{\varepsilon_0^2(r^2 + r_e^2 - 2rr_e\cos\theta)^4}. \tag{8.19}$$

Substituting Equation (8.19) into (8.18), integrating over the whole NC volume and averaging over the dipole orientation θ_e, we obtain the resulting formula for the transition probability

$$W = \frac{4\eta_r c\alpha(w)d^2a^3}{3\hbar w\varepsilon_0^2(r_e - a)^3(r_e + a)^3}. \tag{8.20}$$

To estimate W given by Equation (8.20), we assume $d = 1$ Debye unit (10^{-18} CGSE), $n_r = 3$, $\varepsilon_0 = 5$, $\alpha = 10^5\,\text{cm}^{-1}$, $\hbar\omega = 2\,\text{eV}$. The value of W increases dramatically when the position of the Frenkel exciton r_e approaches the NC surface. For estimates, we assume a minimal possible value of $r_e - a$ to be of the order of interatomic distance in the polymer a_p. Taking $a_p = 0.5\,\text{nm}$, we obtain for the maximal transition probability $W \sim 1.5 \times 10^{11}\,\text{s}^{-1}$ (transition time ~ 7 ps). Note that in the limit $r_e - a << a$, when W is maximal, the answer does not depend on the NC size a and is inversely proportional to the cube of the distance between the exciton and NC surface: $W \sim (r_e - a)^{-3}$. This asymptotic answer could be also obtained by replacing NC with an infinite semi-space, which is possible due to the very fast decrease of E^2 with distance.

The formula of Equation (8.19) ignores modification of the dipole potential by the difference of polymer and semiconductor dielectric constants, ε_p and ε_s. In our system with a spherical dielectric nonuniformity this modification consists in the appearance of image charges so that the potential of a charge or a dipole is given by an infinite series [47]. The situation is, however, simplified for the case of NCs noticeably larger than a_p. In this case, the NC interface can be approximated by a plane and the potential of a dipole beyond the interface coincides with that in a uniform medium with dielectric constant $(\varepsilon_s + \varepsilon_p)/2$. Thus the capture probability is described, as before, by Equation (8.20) with the replacement $\varepsilon_0 \rightarrow (\varepsilon_s + \varepsilon_p)/2$.

The knowledge of the transition probability Equation (8.20) and its dependence on exciton distance from the NC center $W(r_e)$ allows us to find the net rate of carrier capture by NC. A complete description of the process must include generation of excitons in the polymer, their diffusion towards NCs, partial recombination in the matrix and capture by NCs. The specific feature of the Förster capture (Equation 8.20) is its very strong distance dependence so that when $r_e - a$ exceeds several a_p, W becomes negligibly small. For this reason, we can analyze the processes in the polymer matrix, considering NC capture as a phenomenon occuring almost directly at the NC interface and described by some effective surface recombination rate s. To obtain the expression for s, we note that if the exciton density in the vicinity of a NC is n, then the total rate of exciton capture is

$$\frac{\mathrm{d}n}{\mathrm{d}t} = 4\pi n \int_{a+a_p}^{\infty} W(r_e) r_e^2 \mathrm{d}r_e \simeq \frac{\pi n n_r c \alpha(\omega) d^2 a^2}{3\hbar\omega\varepsilon_0^2 a_p^2}. \tag{8.21}$$

The recombination flux $j_r = (4\pi a^2)^{-1}(\mathrm{d}n/\mathrm{d}t)$ and hence s, representing the proportionality factor between j_r and n, is equal to

$$s = \frac{n_r c \alpha(\omega) d^2}{12\hbar\omega\varepsilon_0^2 a_p^2}. \tag{8.22}$$

To describe the behaviour of Frenkel excitons in the matrix, we find the radial distribution of their density $n(r)$ ($r > a$ is measured from the NC center) from the continuity equation

$$D_{ex}\frac{\mathrm{d}}{r^2\mathrm{d}r}\left(r^2\frac{\mathrm{d}n}{\mathrm{d}r}\right) = \frac{n}{\tau_{ex}} - G, \tag{8.23}$$

where G, D_{ex} and τ_{ex} are the generation rate, the diffusion coefficient and the lifetime of Frenkel excitons in a bulk polymer. One of the boundary conditions to Equation (8.23) describes the capture:

$$D_{ex}\frac{\mathrm{d}n}{\mathrm{d}r}(a) = sn(a). \tag{8.24}$$

The other condition will be obtained by the approach similar to that used for description of nonequilibrium carriers in the low-temperature grown GaAs with arsenic clusters [48]. We assume that NCs form a periodic lattice with unit cells centered by a NC. By symmetry, the diffusion flux must vanish at the cell boundaries. By analogy with the Wigner–Seitz method of band structure calculations, we replace the unit cell by a sphere with radius $R_0 = [3/(4\pi N)]^{1/3}$ (N is the NC density), which results in the boundary condition

$$\frac{\mathrm{d}n}{\mathrm{d}r}(R_0) = 0. \tag{8.25}$$

With Equations (8.24), (8.25) we find the distribution $n(r)$ and, particularly, the value of $n(a)$ determining the capture rate by a NC. The final expression for the quantum yield η representing the fraction of all nonequilibrium excitons captured by NCs is:

$$\eta = \frac{4\pi a^2 sn(a)}{\frac{4\pi}{3}(R_0^3 - a^3)G} = \frac{3z}{y(x^3-1)}$$
$$\times \left[\frac{(xy+1)(y-1)\exp[-y(x-1)] - (xy-1)(y+1)\exp[y(x-1)]}{(xy+1)(y-1-yz)\exp[-y(x-1)] - (xy-1)(y+1+yz)\exp[y(x-1)]} \right],$$
$$L_{\mathrm{ex}} = \sqrt{D_{\mathrm{ex}}\tau_{\mathrm{ex}}}, \qquad x = \frac{R_0}{a}, \qquad y = \frac{a}{L_{\mathrm{ex}}}, \qquad z = \frac{sL_{\mathrm{ex}}}{D} = s\sqrt{\frac{\tau_{\mathrm{ex}}}{D_{\mathrm{ex}}}}, \tag{8.26}$$

containing three dimensionless parameters: x, y, z. The first of these characterizes the NC density (the relative volume occupied by NCs is $4\pi Na^3/3 \equiv x^{-3}$), y describes the properties of matrix while z is proportional to s and is the measure of the capture efficiency.

Let us discuss the requirements for these parameters to secure high values of η. The question has a high applied importance since, due to a large quantum yield of radiative recombination in semiconductor NCs (considerably higher than in a polymer matrix), large η is a condition of a noticeable increase of the light emitter effciency in organic–inorganic nanocomposites, as compared with that in a pure polymer. We note first of all that η increasing with z, saturates at the value

$$\eta_\infty = \frac{3}{y^2(x^3-1)} \left[\frac{(xy-1)(y+1)\exp[y(x-1)] - (xy+1)(y-1)\exp[-y(x-1)]}{(xy-1)\exp[y(x-1)] + (xy+1)\exp[-y(x-1)]} \right], \tag{8.27}$$

which at large $xy = R_0/L_{\mathrm{ex}}$ is considerably less than 1. This reflects the fact that at large s, the factor restricting NC capture is not the capture itself, but the exciton diffusion to the NC surface and η is determined by the interplay between diffusion and recombination in the matrix. To acquire high values of η, the sample must have not only effective Förster capture (large s), but also high enough diffusion length in the matrix, exceeding the NC separation. Quantitative estimates should answer two questions: (i) what values of s are high enough to use the asymptotic expression (Equation 8.27); and (ii) what are the requirements for the polymer matrix guaranteeing the corresponding values of η_∞ to be close to unity?

We begin with the second question. Figure 8.7 demonstrates the values of x and y corresponding to $\eta_\infty = 0.8$. To obtain higher η_∞, the x and y values should lie below the curve. Since for a fixed x (given NC density), η_∞ is determined by one single parameter $y = a/L$, small NCs are preferable for reaching high η. We take for estimates the values $L_{\mathrm{ex}} \simeq 5\,\mathrm{nm}$, $D_{\mathrm{ex}} \simeq 2 \times 10^{-4}\,\mathrm{cm^2/s}$ experimentally determined for the PPV polymer [41]. We see that for $x = 2$ (12% volume of NCs), one can reach $\eta_\infty = 0.8$ only for $a < 3\,\mathrm{nm}$ while for $x = 5$ (1% volume of NCs) the estimate gives $a < 0.5\,\mathrm{nm}$, which corresponds to a single interatomic distance and hence cannot be realized in a polymer with the given exciton diffusion length L_{ex}.

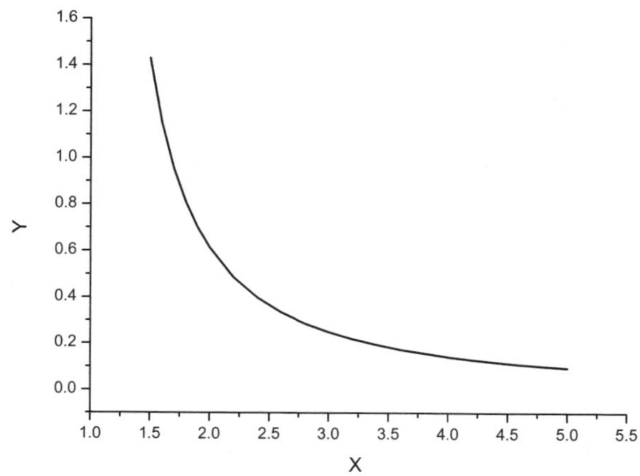

Figure 8.7 Relationship between dimensionless nanocomposite parameters x and y corresponding to a capture effciency $\eta_\infty = 0.8$. The region below the curve corresponds to structures with higher effciencies (after [18] with permission from the American Institute of Physics)

It can be easily shown from the general formula Equation (8.26) that for the x and y corresponding to Figure 8.7, η_∞ approximates the real value η with the 10% accuracy (in other words, $\eta \gtrsim 0.7$) if $z \gtrsim 10$ (at $x = 2$) or $z \gtrsim 30$ (at $x = 5$). For the L_{ex} and D_{ex} given above, it gives $s \gtrsim 10^3$ cm/s. Using the same values of parameters as in our previous estimates, we obtain from Equation (8.22) values of s several times larger. Thus, in most cases the main hurdle to high-effciency NC capture might be caused not by weakness of dipole–dipole interaction, but by low values of the exciton diffusion length in a polymer.

In this connection we point to a number of studies where the values of L_{ex} in polymers noticeably exceeding 5 nm (used in our estimates) were observed [49–52]. It allows us to look optimistically at the prospects of obtaining high quantum yield of photoluminescence in organic–inorganic nanocomposites.

We emphasize once more that the capture probability (Equation 8.20) depends drastically on the minimal exciton–NC distance a_p. Our estimate $a_p \sim 0.5$ nm assumes direct polymer–nanocrystal contact while in many real composites NCs are protected by a layer of ligands suppressing surface recombination, which may increase effective a_p up to 2–10 nm (depending on the length of ligand molecules) and suppress the capture effciency. This agrees with the experimentally observed dependence of the electroluminescence intensity in PbS/polymer nanocomposites on the length of ligand chain [53] and clearly displays the necessity of optimization the ligand material in light-emitting nanocomposites.

8.8 DIODE NANOCOMPOSITE STRUCTURES

So far we have discussed the main optical properties of polymer–semiconductor nanocomposites and their exploitation in photoluminescent and photoconducting devices. However, the most promising application prospects of these composites are connected with diode structures, such as light-emitting and photovoltaic devices. The principal distinction of diode structures consists in their bipolarity, that is in equal importance of the electron and hole contributions to the total current. In light-emitting diodes and lasers, nonequilibrium electrons and holes are injected from two opposite contacts into the polymer matrix, captured by nanocrystals and recombine, emitting radiation with the spectrum determined by the energy band diagram of NC and discussed above. In photodiodes and photovoltaics, nonequilibrium electrons and holes are created by light inside NCs, separated by the external or built-in electric field, and drift through the matrix to the contacts, giving rise to a current in the external circuit.

To give a complete picture of physical phenomena in these devices, we need an adequate description of all nonequilibrium processes listed in the beginning of Section 8.2. Characteristic features of optical transitions in NC–polymer system were discussed in Section 8.3. The next sections are devoted to other remaining problems: capture of carriers by NCs and drift carrier transport in the polymer matrix.

Capture of injected carriers by NC in a strong electric field is essentially different from that of light-induced carriers considered in Section 8.7.2. In the latter case nonequilibrium carriers in the matrix exist in the form of Frenkel excitons, while in the current-carrying situation we deal with electrons and holes moving and, hence, captured by NC separately. As in Section 8.7.2, NC capture is a two-stage process consisting of transport of carriers to a NC followed by capture itself, accompanied by phonon emission. But contrary to the exciton diffusion, transport of individual electrons and holes in a strong electric field has a drift, rather than diffusion, character, and the relative rate of the two above-mentioned stages is given by the parameter $A = \mu E/va^2N_s$ where a and N_s are the radius and areal density of NCs, μ and v are the mobility and velocity of corresponding carriers, E is the electric field in the structure. In crystalline semiconductor matrices usually $A >> 1$ so that the second stage represents the bottleneck for the combined capture process. For this reason, it has received detailed consideration in a number of theoretical works (see, e.g., [55–57]). In polymer matrices, where carrier mobilities are orders of magnitude lower, than in high-quality inorganic crystalline semiconductors, the opposite condition $A << 1$ is ofter realized for both types of carriers. In this case the particular details of carrier capture become irrelevant and all processes will be governed by carrier transport in the matrix. This allows us to formulate a general theory which describes nonequilibrium phenomena in polymer–NC composites while employing a minimum of unknown parameters.

8.9 CARRIER CAPTURE BY NANOCRYSTALS IN AN EXTERNAL ELECTRIC FIELD

Carrier capture in the physical system considered is dominated by drift of carriers to the NC. For this reason, the electric field distribution in the vicinity of NCs plays a crucial role. We treat this problem for a single NC, considering it as a sphere of radius a. It will be shown that in order to maintain stationary recombination process with equal fluxes of electrons and holes to NCs, the latter must acquire some charge q given by Equation (8.31) below. As a result, the potential distribution outside a NC will consist of three components: the potential of a uniform external field $\mathbf{E}$, the potential of image forces caused by the difference of dielectric constants inside and outside NC, and the potential of the charge q. Using the well-known expressions for the first two components (see, e.g., [58]) and the Coulomb formula for the last one, we have

$$\varphi(r,\theta) = -E\left[r - \frac{(\varepsilon_s - \varepsilon_p)a^3}{(2\varepsilon_p + \varepsilon_s)r^2}\right]\cos\theta + \frac{q}{\varepsilon_p r}. \tag{8.28}$$

Here θ is the angle between $\mathbf{r}$ and $\mathbf{E}$. The normal component of electric field at the NC boundary $E_n = -(\partial\varphi/\partial r)_{r=a}$ is positive for some range of θ and negative elsewhere.

Electrons are transported accross the polymer–NC interface and almost instantly captured by NCs in the regions of positive E_n, while holes are captured in the complementary regions of θ. The corresponding carrier fluxes will be proportional to the electric field fluxes calculated separately for the regions $E_n > 0$ and $E_n < 0$:

$$F_+ = 2\pi a^2 \int E_n \Theta(E_n)\sin\theta d\theta = \frac{2\pi q}{\varepsilon_p} + \frac{3\pi\varepsilon_s E a^2}{2\varepsilon_p + \varepsilon_s} + \frac{\pi q^2(2\varepsilon_p + \varepsilon_s)}{3\varepsilon_s\varepsilon_p^2 E a^2}, \tag{8.29}$$

$$F_- = -2\pi a^2 \int E_n \Theta(-E_n)\sin\theta d\theta = -\frac{2\pi q}{\varepsilon_p} + \frac{3\pi\varepsilon_s E a^2}{2\varepsilon_p + \varepsilon_s} + \frac{\pi q^2(2\varepsilon_p + \varepsilon_s)}{3\varepsilon_s\varepsilon_p^2 E a^2}, \tag{8.30}$$

where $\Theta(x)$ is the unit step function. As required, the total flux $F_+ - F_-$ is equal to $4\pi q/\varepsilon_p$.

If electron and hole concentrations in the vicinity of a NC are n and p, and their mobilities respectively, μ_e and μ_h, then the number of electrons and holes captured by a NC per unit time will be equal if $n\mu_e F_+ = p\mu_h F_-$. This condition determines the stationary charge q of a NC for given n, p, and E:

$$\frac{q(2\varepsilon_p + \varepsilon_s)}{3\varepsilon_s\varepsilon_p E a^2} = \frac{\sqrt{\dfrac{p\mu_h}{n\mu_e}} - 1}{\sqrt{\dfrac{p\mu_h}{n\mu_e}} + 1}. \tag{8.31}$$

The charge q vanishes in the symmetric situation $n\mu_e = p\mu_h$ and tends to $\pm 3\varepsilon_s\varepsilon_p E a^2/(2\varepsilon_p + \varepsilon_s)$ for extremely strong asymmetry. The recombination rate is given by $R = n\mu_e F_+ = p\mu_h F_-$ [6]:

$$\Re(n,p,E) = \frac{12\pi\varepsilon_s E a^2 np\mu_n\mu_p}{(2\varepsilon_p + \varepsilon_s)\left(\sqrt{n\mu_e} + \sqrt{p\mu_h}\right)^2}. \tag{8.32}$$

The expressions obtained remain valid at suffciently low concentration of nonequilibrium carriers. At high deviation from equilibrium (intensive illumination of photodiodes or high currents in light-emitting structures) high concentration of nonequlibrium carriers makes the physical picture much more complicated. In this case we have the NC itself with the negative charge of localized electrons $-Q$ surrounded by the hole cloud with the charge $Q + q$. Here Q is determined by the intensity of excitation similarly to Section 8.5 while the net charge q, as above, is to be found from the condition of stationary recombination. In this case, the conditions of capture are different for electrons and holes. For holes to be captured, they should reach the external boundary of the hole-containing region having the radius $r_0 > a$ discussed in Section 8.5. Thus, in the expression for F_- (Equation 8.30) we should replace a by r_0. On the contrary, electrons must reach the interface of NC itself, which has the charge $-Q$. Thus in Equation (8.29) we should retain a as the radius of the capturing sphere, but replace q by $-Q$. This will result in more cumbersome expressions for q and R which are not shown here. Their qualitative meaning is, however, clear: with the growth of Q capture of electrons becomes relatively more diffcult, as compared with holes. In other words, with the increase of light (or injection) intensity, in the formulae to be derived below we must assume that the effective ratio $n\mu_e/p\mu_h$ decreases smoothly.

8.10 THEORY OF NANOCOMPOSITE LIGHT EMITTERS

8.10.1 Basic equations

To obtain a complete theoretical description of polymer–NC light emitters, we must complement the formulae of the previous sections by equations describing electron and hole transport in the polymer matrix [6]. We consider a composite layer of thickness L provided with an electron-injecting contact at $x=0$ and a hole-injecting contact at $x=L$. We represent the influence of these contacts by fixing the concentrations of corresponding carriers:

$$n(x=0)=n_0, \qquad p(x=L)=p_0. \tag{8.33}$$

The density of NCs inside the layer is, in general, nonuniform and described by some function $N(x)$. To find the electronic and luminescence characteristics of the system, we must know the distributions of carrier concentrations n, p and the electric field E in the layer. Given a random spatial distribution of NCs, all of these characteristics will be coordinate dependent. We will average them in the yz-plane so that in our model n, p, and E are functions of only the x-coordinate. In agreement with the previous section, the charge of NC is a given function of n, p, and E and, hence, is also x dependent.

In addition to considering carrier capture by and recombination in the NCs, we consider carrier recombination in the matrix. We assume that its rate has a quadratic character associated with a bimolecular process, hence γnp. In this case the continuity equations for electrons and holes are given by

$$\mu_n \frac{d(nE)}{dx} = -\Re(n, p, E) - \gamma np, \tag{8.34}$$

$$\mu_p \frac{d(pE)}{dx} = \Re(n, p, E) + \gamma np. \tag{8.35}$$

Adding Equations (8.34) and (8.35), we obtain

$$(\mu_n n + \mu_p p)E = j/e = \mathrm{const}(x), \tag{8.36}$$

where j is the total current through the structure.

If the carrier concentrations and NC density are large enough, the electric field E in Equations (8.34), (8.35) is not a constant, but varies across the structure due to space-charge effects. Since a thin layer between x and $x + dx$ has an areal charge density $[q(x)N(x) + ep(x) - en(x)]$, $E(x)$ is found from the Poisson equation

$$\frac{\mathrm{d}E}{\mathrm{d}x} = -\frac{4\pi}{\varepsilon_\mathrm{p}}(qN + ep - en), \tag{8.37}$$

with the boundary condition $\int_0^L E(x)\mathrm{d}x = V$, where V is the voltage applied to the electrodes.

To obtain particular quantitative results, we must specify the distribution of NCs in the structure. We assume a nanocrystal density profile with a Gaussian shape,

$$N(x) = N_0 \exp\left[-\frac{(x - x_0)^2}{l^2}\right], \tag{8.38}$$

where N_0 represents the volume density, l the thickness, and x_0 the position of the NC layer.

8.10.2 Current–voltage characteristic

To obtain an analytical description of the current–voltage characteristic (CVC) with the account of space charge effects, and to discuss its main properties, we begin with the limiting case of large density of NCs localized in a very thin layer $x = x_0$ [6, 59]. In this case, due to strong recombination in NCs, concentrations of carriers injected from two opposite electrodes decrease sharply after crossing this layer. As a result, the current given by Equation (8.36), and the field given by Equation (8.37) will be dominated by n for $x < x_0$ and by p for $x > x_0$. Since the areas with high electron and high hole density are spatially separated, the term γnp in Equations (8.34), (8.35) becomes negligibly small. This allows us to integrate Equation (8.37) separately in the two regions mentioned: $E^2 = 8\pi jx/(\varepsilon_\mathrm{p}\mu_\mathrm{n}) + j^2(e\mu_\mathrm{n}n_0)^2$ for $x < x_0$ and similarly for $x > x_0$. From the requirement $\int_0^L E(x)\mathrm{d}x = V$ we obtain thereby the CVC:

$$V = \frac{2}{3}\sqrt{\frac{8\pi j}{\varepsilon_\mathrm{p}}}\left\{\frac{1}{\sqrt{\mu_\mathrm{n}}}\left(x_0 + \frac{j\varepsilon_\mathrm{p}}{8\pi e^2 n_0^2 \mu_\mathrm{n}}\right)^{3/2} - \left(\frac{j\varepsilon_\mathrm{p}}{8\pi e^2 n_0^2 \mu_\mathrm{n}}\right)^{3/2}\right.$$
$$\left. + \frac{1}{\sqrt{\mu_\mathrm{p}}}\left(1 - x_0 + \frac{j\varepsilon_\mathrm{p}}{8\pi e^2 n_0^2 \mu_\mathrm{n}}\right)^{3/2} - \left(\frac{j\varepsilon_\mathrm{p}}{8\pi e^2 n_0^2 \mu_\mathrm{n}}\right)^{3/2}\right\}. \tag{8.39}$$

For small applied voltages, the CVC has a quadratic character which is characteristic of space-charge-limited currents [60]. As V grows, the role of space-charge effects diminishes and the CVC becomes almost linear. The important role played by space-charge effects in polymer–NC composites has been suggested in [35] on the basis of experimental data.

To present the complete analysis of the problem, we introduce dimensonless variables by measuring length in units of L, concentrations in $(4\pi La^2)^{-1}$, current density in $e^2\mu_\mathrm{n}/(4\pi\varepsilon_\mathrm{p}La^4)$, voltage in $eL/(\varepsilon_\mathrm{p}a^2)$ and by characterising matrix recombination using the parameter $g = \varepsilon_\mathrm{p}\gamma/(4\pi e\mu_\mathrm{n})$. We present in Figure 8.8 the results for structures with symmetric ($n_0 = \mu_p p_0/\mu_n = 1$) and asymmetric ($n_0 = 1$, $\mu_p p_0/\mu_n = 3$) injection and for two $N(x)$ profiles: thin layer ($l = 0.1$) and uniform density ($l >> 1$).

By comparing curve 1 with 3 and curve 2 with 5, we conclude that for uniform density of NCs the conduction mechanism is less effective than for a thin layer with the same total number $N_i = \int_0^L N(x)\mathrm{d}x$ of NCs, and the current is lower. The position of a thin NC layer x_0 is unimportant in symmetric structures, but plays an important role in the case of asymmetric junction. In the symmetric case, results obtained for different x_0 differ by less than 10% and for this reason only one curve, 3, is presented in the figure. For asymmetric structures, the current increases when x_0 approaches the electrode with weaker injection which in our example ($n_0 = 1$, $\mu_p p_0/\mu_n = 3$) lies at $x = 0$. This effect is demonstrated by a series of curves 4–6 in Figure 8.8. The intensity of matrix recombination in a wide interval of g has only a minor influence on the CVC.

It is interesting to compare the results of numerical calculations (dotted lines) with the approximate formula of Equation (8.39) (solid lines). For a thin layer of NCs (which was assumed while deriving Equation (8.39) the methods give almost the same results, as demonstrated by the pair of curves 3 and 5. For uniform NC density Equation (8.39) is inadequate and curves 1,2 are the result of numerical calculations.

8.10.3 Quantum yield of NC electroluminescence

For light-emitting structures, the most important characteristic is the quantum yield of electroluminescence from NCs η. This characteristic is determined by the ratio of total carrier flux inside NCs to the electric current in the external circuit:

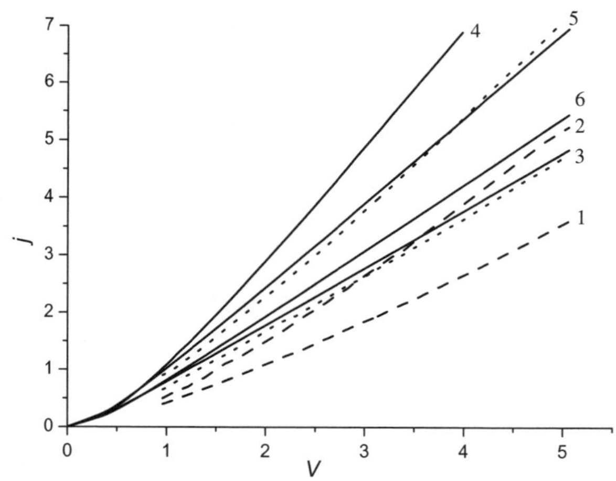

Figure 8.8 Current–voltage characteristic of a polymer–NC composite with $g = 25$ and coverage factor $\pi a^2 N_i = 1$. Curves 3–6 correspond to a thin layer, curves 1,2 to a uniform distribution of NCs. Curves 2,3 correspond to a symmetric structure with $n_0 = \mu_h p_0/\mu_e = 1$ and $x_0 = 0.5$; curves 1,4,5,6, to an asymmetric structure with $n_0 = 1$, $\mu_h p_0/\mu_e = 3$ and $x_0 = 0.2$ (curve 2), 0.5 (curve 5), 0.8 (curve 6). Dashed and dotted curves are obtained by numerical calculations; solid curves are given by the approximate analytical expression (Equation 8.39). j and V are measured in dimensionless units defined in the text (after [6] with permission from American Scientific Publishers)

$$\eta = \beta \frac{e\int_0^L \Re[n(x), p(x), E(x)]\mathrm{d}x}{j}, \tag{8.40}$$

where $\beta \leq 1$ is the internal quantum yield, the ratio of the radiative to the total recombination rate in the NCs. There are two factors which prevent η from acquiring its maximal possible value β:

- incomplete recombination when some electrons reach the hole electrode $x = L$ and/or some holes reach $x = 0$;
- recombination in the polymer matrix described by the last term in Equations (8.34), (8.35).

The necessary (but not sufficient, see below) condition for eliminating incomplete recombination is that of a high enough NC concentration:

$$a^2 \int_0^L N(x)\mathrm{d}x \gg 1, \tag{8.41}$$

since the cross-section of carrier capture by NCs is $\sim a^2$.

Another important issue relates to the spatial distribution of NCs described by the $N(x)$ function. If the total number of NCs, $N_i = \int_0^L N(x)\mathrm{d}x$, is fixed, we wish to know whether it is better to have NCs distributed over the whole volume or concentrated in a thin layer. The results of numerical solution of the system (Equations 8.33–38) shown in Figure 8.9 demonstrate that, in the presence of a strong matrix recombination, the thin-layer geometry has an advantage (solid vs dashed curves). If NCs whose total number satisfies Equation

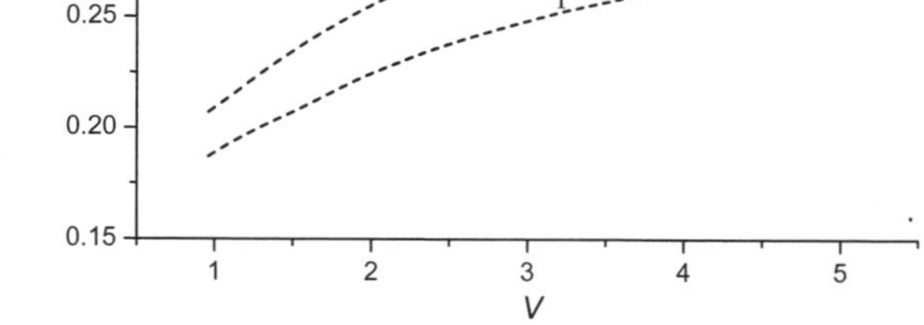

Figure 8.9 Quantum yield of NC electroluminescence in a polymer–NC composite with $g = 25$ and coverage factor $\pi a^2 N_i = 1$. Curves 3–6 correspond to a thin layer, curves 1,2 to a uniform distribution of NCs. Curves 2,3 correspond to a symmetric structure with $n_0 = \mu_h p_0/\mu_e = 1$ and $x_0 = 0.5$; curves 1,4,5,6, to an asymmetric structure with $n_0 = 1$, $\mu_h p_0/\mu_e = 3$ and $x_0 = 0.2$ (curve 2), 0.5 (curve 5), 0.8 (curve 6). V is measured in dimensionless units defined in the text (after [6] with permission from American Scientific Publishers)

(8.41) lie in a single plane, this plane provides complete recombination so that $n = 0$ to the right of this plane and $p = 0$ to the left. Everywhere outside of this single plane, the product $np = 0$ and matrix recombination is zero. If NCs are distributed over a layer of finite thickness, in this layer $np \neq 0$ and matrix recombination competes with NC recombination, decreasing η.

If NCs are concentrated in a single layer, we wish to know the position of this layer x_0 which will maximize η. The answer depends on the injection asymmetry characterized by the ratio $\mu_n n_0/\mu_p p_0$. In symmetric structures in which this ratio is close to 1, η is almost independent of x_0 and, to several percent accuracy, the quantum yield for any x_0 is described by curve 3. In asymmetric structures the yield increases when the layer is moved toward the electrode with weaker injection (compare curves 4,5,6). This is accompanied by the increase in current described in the previous subsection since a considerable part of the current has a recombination character.

In the absence of space charge effects, there would be a dramatic difference in η between symmetric and asymmetric structures. This is connected with the constancy of the electric field E throughout the system. For $E = \text{const}(x)$, a difference between the contact concentrations n_0 and p_0 results in a difference between injected electron and hole currents $j_n = e\mu_n n_0 E$ and $j_p = e\mu_p p_0 E$. Under these conditions the recombination current connected with luminescence cannot exceed $\min\{j_n, j_p\}$ whereas the current in the external circuit is not less than $\max\{j_n, j_p\}$. As a result, $\eta \leq \beta \min\{j_n, j_p\}/\max\{j_n, j_p\}$ and for highly asymmetric structures is very small, contrary to what is required for light-emitting devices. Fortunately, space charge effects redistribute electric field in the system so that the electron and hole currents become equal (the calculations in Section 8.10.2 show that such redistribution is really possible). Eventually, the decrease in η caused by asymmetry and shown in Figure 8.9 (compare curve 3 with 4,5, and 6) does exist, but is not dramatic.

The dependence of η on the applied voltage V is also of interest. At first glance, Equations (8.34), (8.35) suggest that η should grow with V since the capture rate R increases with E and, hence, with V (see Equation 8.32), whereas γ is assumed to be field independent. In symmetric structures this prediction is uniformly true, but in the presence of asymmetry, account must be taken of an additional factor. From the properties of the CVC discussed in the previous subsection, we know that the influence of space-charge effects decreases with applied voltage. These effects cause an increase of η in asymmetric structures (see the previous paragraph). As a result, the η vs V characteristic of asymmetric structures is influenced by at least two different factors and one cannot predict its behavior from simple qualitative arguments. The results of numerical calculations presented in Figure 8.9 show that in asymmetric structures the η vs V dependence is also an increasing one, but weaker than in symmetric structures.

From these theoretical calculations, we may elaborate the main qualitative conditions to be fulfilled for optimization of the quantum yield [59, 61]:

- the 'coverage factor' of NCs $N_i a^2$ must considerably exceed unity;
- NCs should be concentrated in a thin layer shifted toward the electrode with worse injection characteristics;
- the electron and hole injection effciencies must be approximately equal;

- the applied voltage must be large enough, preferably beyond the initial quadratic region of the CVC.

In real polymers electron and hole mobilities often differ dramatically and in most cases $\mu_n << \mu_p$. In this situation, to satisfy the third condition, one should choose the contact material so that the condition $n_0 >> p_0$ is fulfilled.

These recommendations are in agreement with numerous experimental observations. The increase of η with the 'coverage factor' was revealed [36] by varying the thickness of the layer with a constant volume density of NCs. The dependence of η on nanocrystal migration toward a cathode was observed in [5] and the fast increase of η with the applied voltage followed by saturation was reported in [35].

8.11 ELECTROLUMINESCENCE vs PHOTOLUMINESCENCE

The results of the previous section allow us to determine the intensity of electroluminescence and its dependence on the parameters of nanocomposite structure and applied voltage. Now we discuss briefly the spectrum of emitted light. A number of authors observing electroluminescence in polymer–semiconductor nanocomposites drew attention to the difference between electro- and photoluminescence spectra. This difference, however, did not have a regular character. In [35, 53] the line of electroluminescence was red-shifted, related to the photoluminescence one, in [4, 62, 63] it was blue-shifted, while in [34, 36] the positions of these lines almost coincided. It was also reported [64] that electroluminescent spectra had a considerably larger width that those of photoluminescence; however, the authors actually compared electroluminescence of polymer-based nanocomposites with photoluminescence of NCs in solution, that is systems with principally different energy diagrams. To explain these facts, we must realize the difference in excited states created by illumination and injection. In the first case, optical pumping creates electron–hole pairs directly in NCs (at least until the exciting photon energy is less than the energy gap in polymer matrix). As shown in Section 8.5, electrons remain localized inside NCs while most holes are concentrated outside them, though their total numbers are equal so that the whole region of radius r_0 remains neutral. In the case of electroluminescence due to difference in concentrations and mobilities of injected electrons and holes in the matrix NCs acquire some net charge q, calculated in Section 8.9. This deviation from neutrality modifies the spectrum of holes and hence the energy of emitted light. This effect can also be responsible for a voltage-induced shift of the electroluminescence spectrum observed by some authors [4, 62].

Another important distinction between electro- and photoluminescence is the much stronger dependence of the former on the presence and the type of a ligand capping layer. Such a strong dependence for another effect—photoconductivity—has been already discussed in Section 8.6. For electroluminescence the situation is similar. For instance, in [53] the replacement of oleate ligands covering PbS NCs by the octylamine, with essentially less carbon atoms in the molecular chain, caused a noticeable increase in electroluminescence without serious changes in photoluminescence.

Moreover, the authors of [32] point to a noticeable quenching of photoluminescence after removal of a capping level.

8.12 POLYMER–DIELECTRIC NANOCOMPOSITES

In conclusion, we will discuss briefly properties of nanocomposites formed by embedding dielectric, rather than semiconductor, NCs into a conducting polymer. Contrary to the band diagram of Figure 8.1, such NCs represent potential maxima for both electrons and holes and hence do not capture nonequilibrium carriers. Nevertheless they may modify electronic properties of composites, presumably via two main mechanisms.

First, they create essential inhomogeneities in current and hence electric field distribution in light-emitting and photovoltaic devices. Such a redistribution creates 'hot spots' with a high local current density which, due to a superlinear luminance–current dependence in conducting polymers, not only modifies current–voltage characteristics, but also increases the effciency of polymer electroluminescence [65–67]. The effect is practically independent of the nature of NCs, as long as they are electronically inactive.We should add to this list studies [68, 69] in which the increase of effciency in polymer light-emitting diodes caused by embedding SiO_2 nanoparticles in the vicinity of the anode was attributed to a redistribution of internal electric field by the charge of carriers trapped at the nanoparticle surface.

Second, dielectric NCs may influence the properties of a polymer and, particularly, its luminescence spectrum in the immediate vicinity of their interface. These nanoparticles change the exciton energy in the polymer by influencing the polymer conjugation length [70–72] and/or modifying the polarization component of energy [73, 74], which eventually changes the intensity and spectrum of luminescence.

More detail and references regarding polymer–dielectric nanocomposites can be found in the review paper [6].

8.13 CONCLUDING COMMENTS

It is important that a number of results obtained are applicable not only to polymer–semiconductor nanocomposites but also to pure semiconductor quantum dot structures forming type-II heterostructures, e.g., InGaAs/GaSb, Si/Ge, ZnTe/ZnSe, CdS/PbS and some others. While a great number of publications are devoted to the type-I quantum dots with simultaneously confined electrons and holes, the properties of type-II dot remain almost uninvestigated.

Furthermore, we want to point at several problems which have been left beyond the scope of this chapter.

1. We have always assumed that the polymer matrix is not intentionally doped (or has a p-doping) so that the equilibrium Fermi level lies below LUMO at the energy exceeding $x^s - x^p - E_{e0}$. If this requirement is not fulfilled, which can be the case in n-doped polymers, then even in equilibrium NCs acquire some charge and the band diagram appears like Figure 8.4b and is discussed in more detail in [28]. The only difference consists in the fact that the compensating positive charge outside NC is formed not by holes, but by ionized donor centers.

2. We have not considered properties of composites at very high (some tens of percent) volume density of NCs when direct percolation of electrons through NCs becomes possible. Experimental data available show a number of interesting phenomena in

these systems: essentially nonlinear current–voltage characteristics [75], variable-range hopping conduction [76], dramatic increase in conductivity due to doping [77], strong increase of photovoltaic effciency [32], formation of collective electronic states [78, 79], stimulated optical emission [80]. The role of polymer in such composites is essentially suppressed and in many cases is reduced to simple mechanical binding. Electronic phenomena in them should be analyzed in a model different from ours and paying special attention to the properties of direct contacts between NCs.

ACKNOWLEDGEMENTS

The authors are deeply grateful to L. Bakueva and S. Musikhin who actively discussed the main ideas presented above and kindly offered their experimental results for comparison with theoretical assumptions. The paper would not have been written without S. Baranovski who suggested the idea and stimulated authors in the process of work.

REFERENCES

[1] *Organic Electroluminescent Materials and Devices*, S. Miyata and H.S. Nalwa (Eds), Gordon & Breach, New York, 1997.
[2] R.N. Friend, R.W. Gymer, A.B. Holmes, J.H. Burroughes, R.N. Marks, C. Taliani, D.D.C. Bradley, D.A. Dos Santos, J.L. Bredas, M. Logdlund, and W.R. Salaneck, *Nature*, **397**, 121 (1999).
[3] T.R. Hebner, C.C. Wu, D. Marcy, M.H. Lu, and J.C. Sturm, *Appl. Phys. Lett.*, **72**, 519 (1999).
[4] V.L. Colvin, M.C. Schlamp, and A.P. Alivisatos, *Nature*, **370**, 354 (1994).
[5] H. Mattoussi, L.H. Radzilowski, B.O. Dabbousi, D.E. Fogg, R.R. Schrock, E.L. Thomas, M.F. Rubner, and M.G. Bawendi, *J. Appl. Phys.*, **86**, 4390 (1999).
[6] L. Bakueva, S. Musikhin, E.H. Sargent, H.E. Ruda, and A. Shik, in: *Handbook of Organic-Inorganic Hybrid Materials and Nanocomposites*, H.S. Nalwa (ed.), American Scientific Publishers, 2003, p.181.
[7] S.V. Gaponenko, *Optical Properties of Semiconductor Nanocrystals*, Cambridge University Press, 1998.
[8] A.L. Efros, D.J. Lockwood, and L. Tsybeskov (eds), *Semiconductor Nanocrystals: from Basic Principles to Applications*, Kluwer Academic/Plenum Publishers, New York, 2003.
[9] Y.-S. Shon, W.P. Wuelfing, and R.W. Murray, *Langmuir*, **17**, 1255 (2001).
[10] L.V. Asryan and R.A. Suris, *IEEE J. Quant. Electr.*, **34**, 841 (1998).
[11] D. Bimberg, M. Grundman, and N.N. Ledentsov, *Quantum Dot Heterostructures*, John Wiley & Sons, Ltd, Chichester, 1999.
[12] V.V. Mitin, V.I. Pipa, A.V. Sergeev, M. Dutta, and M. Stroscio, *Infrared Phys. Technol.*, **42**, 467 (2001).
[13] I.S. Grigoriev and E.Z. Meilikhov (eds), *Handbook of Physical Quantities*, CRC, Boca Raton, 1997.
[14] A.C. Arango, S.A. Carter, and P.J. Brock, *Appl. Phys. Lett.*, **74**, 1698 (1999).
[15] Z. Bao, A. Dodabalapur, and A.J. Lovinger, *Appl. Phys. Lett.*, **69**, 4108 (1996).
[16] P.W.M. Blom, M.J.M. de Jong, and J.J.M. Vleggaar, *Appl. Phys. Lett.*, **68**, 3308 (1996).
[17] A.M. Perelomov and Y.B. Zel'dovich, *Quantum Mechanics—Selected Topics*, World Scientific, 1998.

[18] A. Shik, L. Bakueva, and H.E. Ruda, *Phys. Stat. Sol., (b)* **242**, 1183 (2005).

[19] J.G. Winiarz, L. Zhang, M. Lal, C.S. Friend, and P.N. Prasad, *Chem. Phys.*, **245**, 417 (1999).

[20] L. Sun, C. Liu, C. Liao, and C. Yan, *Solid State Comm.*, **111**, 483 (1999).

[21] Al.L. Efros and M. Rosen, *Annu. Rev. Mater. Sci.*, **30**, 475 (2000).

[22] J.M. Rorison, *Phys. Rev. B*, **48**, 4643 (1993).

[23] U.E.H. Laheld, F.B. Pedersen, and P.C. Hemmer, *Phys. Rev. B*, **52**, 2697 (1995).

[24] N.S. Rytova, *Vestnik Mosk. Univ. Fiz.*, **3**, 13 (1967) (in Russian).

[25] L.V. Keldysh, *JETP Lett.*, **29**, 658 (1979).

[26] A. Shik, *J. Appl. Phys.*, **74**, 2951 (1993).

[27] A. Shik, H. Ruda, and E.H. Sargent, *J. Appl. Phys.*, **88**, 3448 (2000).

[28] A. Shik, H. Ruda, and E.H. Sargent, *Nanotechnology*, **12**, 523 (2001).

[29] Y. Wang and N. Herron, *J. Lumin.*, **70**, 48 (1996).

[30] K.S. Narayan, A.G. Manoj, J. Nanda, and D.D. Sarma, *Appl. Phys. Lett.*, **74**, 871 (1999).

[31] D.V. Talapin, S.K. Poznyak, N.P. Gaponik, A.L. Rogach, and A. Eychmüller, *Physica. E*, **14**, 237 (2002).

[32] N.C. Greenham, X. Peng, and A.P. Alivisatos, *Phys. Rev. B.*, **54**, 17628 (1996).

[33] S.A. McDonald, P.W. Cyr, L. Levina, and E.H. Sargent, *Appl. Phys. Lett.*, **85**, 2089 (2004).

[34] B.O. Dabboussi, M.G. Bawendi, O. Onitsuka, and M.F. Rubner, *Appl. Phys. Lett.*, **66**, 1316 (1995).

[35] M.C. Schlamp, X. Peng, and A.P. Alivisatos, *J. Appl. Phys.*, **82**, 5837 (1997).

[36] H. Matoussi, L.H. Radzilowski, B.O. Dabbousi, E.L. Thomas, M.G. Bawendi, and M.F. Rubner, *J. Appl. Phys.*, **83**, 7965 (1998).

[37] L. Bakueva, G. Konstantatos, L. Levina, S. Musikhin, and E.H. Sargent, *Appl. Phys. Lett.*, **84**, 3459 (2004).

[38] E.R. Glaser, B.R. Bennett, B.V. Shanabrook, and R. Magno, *Appl. Phys. Lett.*, **68**, 3614 (1996).

[39] F. Hatami *et al.*, *Appl. Phys. Lett.*, **67**, 656 (1995).

[40] A. Shik, G. Konstantatos, E.H. Sargent, and H.E. Ruda, *J. Appl. Phys.*, **94**, 4066 (2003).

[41] M. Pope and C.E. Swenberg, *Electronic Processes in Organic Crystals and Polymers*, Oxford University Press, 1999, p. 852.

[42] E.M. Conwell, *Synth. Metals*, **83**, 101 (1996).

[43] D. Basko, G.C. La Rocca, F. Bassani, and V.M. Agranovich, *Eur. Phys. J. B.*, **8**, 353 (1999).

[44] D.M. Basko, V.M. Agranovich, F. Bassani, and G.C. La Rocca, *Eur. Phys. J. B*, **13**, 653 (2000).

[45] A. Engelmann and V.I. Yudson, P. Reineker, *Phys. Rev. B.*, **57**, 1784 (1998).

[46] L.-W. Wang and A. Zunger, *Phys. Rev. B*, **53**, 9579 (1996).

[47] V.V. Batygin and I.N. Toptygin, *Problems in Electrodynamics*, Academic Press, 1978.

[48] H. Ruda and A. Shik, *Phys. Rev. B*, **63**, 085203 (2001).

[49] T. Stübinger and W. Brütting, *J. Appl. Phys.*, **90**, 3632 (2001).

[50] A. Haugeneder, M. Neges, C. Kallinger, W. Spirkl, U. Lemmer, J. Feldman, U. Scherf, E. Harth, A. Gügel, and K. Müllen, *Phys. Rev. B*, **59**, 15346 (1999).

[51] V. Choong, Y. Park, Y. Gao, T. Wehrmeister, K. Müllen, B.R. Hsieh, and C.W. Tang, *Appl. Phys. Lett.*, **69**, 1492 (1996).

[52] H.R. Kerp, H. Donker, R.B.M. Koehorst, T.J. Schaafsma, and E.E. van Faassen, *Chem. Phys. Lett.*, **298**, 302 (1998).

[53] L. Bakueva, S. Musikhin, M.A. Hines, T.-W.F. Chang, M. Tzolov, G.D. Scholes, and E.H. Sargent, *Appl. Phys. Lett.*, **82**, 2895 (2003).

[54] H.E. Ruda and A. Shik, *Phys. Rev. B*, **67**, 235309 (2003).

[55] R. Ferreira and G. Bastard, *Appl. Phys. Lett.*, **74**, 2818 (1999).

[56] Y. Toda, O. Moriwaki, M. Nishioka, and Y. Arakawa, *Phys. Rev. Lett.*, **82**, 4114 (1999).

[57] S. Marcinkevicius and R. Lean, *Physica Scripta T*, **79**, 79 (1999).

[58] L.D. Landau and E.M. Lifshits, *Electrodynamics of Continuous Media*, Pergamon, New York, 1984.
[59] A. Shik, S. Yu, E. Johnson, H.E. Ruda, and E.H. Sargent, *Solid-State Electr.*, **46**, 61 (2002).
[60] M.A. Lampert and P. Mark, *Current Injection in Solids*, Academic Press, San Diego, 1970.
[61] L. Bakueva, A. Shik, S. Yu, and E.H. Sargent, in: *197th Meeting of the Electrochemical Society*, Toronto, 2000, Book of Abstracts, # 1027.
[62] N.P. Gaponik, D.V. Talapin, A.L. Rogach, and A. Eychmüller, *J. Mater. Chem.*, **10**, 2163 (2000).
[63] L. Bakueva, I. Gorelikov, S. Musikhin, X.S. Zhao, E.H. Sargent, and E. Kumacheva, *Adv. Mater.*, **16**, 926 (2004).
[64] N.P. Gaponik, D.V. Talapin, and A.L. Rogach, *Phys. Chem. Chem. Phys.*, **1**, 1787 (1999).
[65] S.A. Carter, J.C. Scott, and P.J. Brock, *Appl. Phys. Lett.*, **71**, 1145 (1997).
[66] L. Bozano, S.E. Tuttle, S.A. Carter, and P.J. Brock, *Appl. Phys. Lett.*, **73**, 3911 (1998).
[67] P.W.M. Blom, H.F.M. Schoo, and M. Matters, *Appl. Phys. Lett.*, **73**, 3924 (1998).
[68] V. Bliznyuk, B. Ruhstaller, P.J. Brock, U. Scherf, and S.A. Carter, *Adv. Mater.*, **11**, 1257 (1999).
[69] X.M. Ding, L.M. Hung, L.F. Cheng, Z.B. Deng, X.Y. Hou, C.S. Lee, and S.T. Lee, *Appl. Phys. Lett.*, **76**, 2704 (2000).
[70] W.-P. Chang and W.-T. Whang, *Polymer*, **37**, 4229 (1996).
[71] P.K.H. Ho, J.-S. Kim, N. Tessler, and R.H. Friend, *J. Chem. Phys.*, **115**, 2709 (2001).
[72] L. Bakueva, S. Musikhin, and E.H. Sargent, *J. Nanosci. Nanotech.*, **1**, 457 (2001).
[73] S. Musikhin, L. Bakueva, E.H. Sargent, and A. Shik, *J. Appl. Phys.*, **91**, 6679 (2002).
[74] L. Bakueva, S. Musikhin, E.H. Sargent, and A. Shik, *Surf. Sci.*, **532–535**, 1051 (2003).
[75] M. Drndic, M.V. Jarocz, N.Y. Morgan, M.A. Kastner, and M.G. Bawendi, *J. Appl. Phys.*, **92**, 7498 (2002).
[76] D. Yu, C. Wang, B.L. Wehrenberg, and P. Guyot-Sionnest, *Phys. Rev. Lett.*, **92**, 216802 (2004).
[77] D. Yu, C. Wang, and P. Guyot-Sionnest, *Science*, **300**, 1277 (2003).
[78] C.R. Kagan, C.B. Murray, and M.G. Bawendi, *Phys. Rev. B.*, **54**, 8633 (1996).
[79] M.V. Artemyev, A.I. Bibik, L.I. Gurinovich, S.V. Gaponenko, and U. Woggon, *Phys. Rev. B*, **60**, 1504 (1999).
[80] V.I. Klimov, A.A. Mikhailovsky, Su Xu, A. Malko, J.A. Hollingsworth, C.A. Leatherdale, H.-J. Eisler, and M.G. Bawendi, *Science*, **290**, 314 (2000).

9 AC Hopping Transport in Disordered Materials

Igor Zvyagin

Faculty of Physics, Moscow State University, 119899 Moscow, Russia

9.1 INTRODUCTION

It is well known that a characteristic feature of the kinetics of relaxation and response of disordered materials to external perturbations is the nonexponential time variation of some material properties (polarization, conductivity, etc.). Nonexponential dispersive time depen-

Charge Transport in Disordered Solids with Applications in Electronics Edited by S. Baranovski.

dences are generally characteristic of transients in disordered solids [1]. Such dependences are observed not only when switching an electric field on or off, but also at optical excitation, in particular, nonuniform. Thus, in time-of-flight experiments in a number of disordered materials, the transients have distinctive features of the dispersive transport [2].

A general expression for the current density is

$$j(t) = \sigma_D E(t) + dP(t)/dt, \tag{9.1.1}$$

where $E(t)$ is the electric field, σ_D is the electrical conductivity describing the fast (instantaneous) response to the electric field, and the second term on the right-hand side is the displacement current related to the delayed polarization $P(t)$ of the medium. Equation (9.1.1) can describe the situation where there are different contributions to the conductivity (parallel conduction channels). The first term on the right-hand side may correspond to free-carrier band conduction (the Drude conductivity), whereas the second term may describe the delayed polarization produced by localized charge carrier hopping. As usually, for the electric displacement $D(t)$, we may write

$$D(t) = \kappa_0 \kappa_\infty E(t) + P_h(t), \tag{9.1.2}$$

where κ_0 is the permittivity of free space, $\kappa_\infty = 1 + \chi_\infty$ is the high-frequency relative permittivity characterizing the instantaneous polarization $P_\infty(t) = \kappa_0 \chi_\infty E(t)$ related to the fast polarization processes, χ_∞ is the instantaneous susceptibility, $P(t) = P_\infty(t) + P_h(t)$, and $P_h(t)$ is the contribution of the delayed (e.g., hopping) polarization. Generally, a linear relation between $P_h(t)$ and $E(t)$ has a non-Markovian form [3]

$$P_h(t) = \kappa_0 \int_0^\infty \chi_h(t') E(t - t') dt', \tag{9.1.3}$$

where $\chi_h(t')$ is the delayed susceptibility.

Expressions (9.1.1) and (9.1.2) may be written in the frequency domain. Assuming that the applied field is proportional to $\exp(i\omega t)$, $E(t) = E(\omega)\exp(i\omega t)$, one obtains

$$D(\omega) = \kappa_0 \kappa(\omega) E(\omega), \tag{9.1.4}$$

where

$$\kappa(\omega) = \kappa_\infty + \chi_h(\omega) \tag{9.1.5}$$

is the complex permittivity,

$$P_h(\omega) = \kappa_0 \chi_h(\omega) E(\omega), \tag{9.1.6}$$

and

$$\chi_h(\omega) = \int \Phi(t') \exp(-i\omega t') dt' \tag{9.1.7}$$

is the complex susceptibility that describes the delayed polarization. The real and imaginary parts of the susceptibility $\chi_h(\omega) = \kappa(\omega) - \kappa_\infty = \chi_h'(\omega) - i\chi_h''(\omega)$ are related by the Kramers–

Kronig relations [3]. The quantity $\Delta\kappa = \kappa(0) - \kappa_\infty = \chi_h(0)$ is called the dielectric strength of the material. From Equation (9.1.1) we have

$$j(\omega) = \sigma(\omega) E(\omega), \tag{9.1.8}$$

where the electrical conductivity $\sigma(\omega)$ contains real and imaginary parts: $\sigma(\omega) = \sigma'(\omega) + i\sigma''(\omega)$.

Generally, the conductivity is a sum of the so-called Drude contribution σ_D and the contribution of the delayed polarization σ_h:

$$\sigma(\omega) = \sigma_D + \sigma_h(\omega), \tag{9.1.9}$$

where the polarization part is determined by the corresponding susceptibility

$$\sigma_h(\omega) = i\omega\kappa_0\chi_h(\omega). \tag{9.1.10}$$

The real part of the conductivity $\sigma_h'(\omega)$ related to the imaginary part of the susceptibility by Equation (9.1.10) determines the absorption of the wave power (the dielectric loss),

$$P = \sigma_h'(\omega) E^2(\omega)/2. \tag{9.1.11}$$

The standard exponential Debye-type relaxation with a single relaxation time τ is

$$\chi_\tau(t) = \chi_\tau(0)\exp(-t/\tau). \tag{9.1.12}$$

In this case, the resulting frequency dependence of the complex susceptibility is

$$\chi_\tau(\omega) = \frac{\Delta\kappa}{1 + i\omega\tau}. \tag{9.1.13}$$

Note that the Drude conductivity determined from the conventional Boltzmann transport theory is dependent on frequency at frequencies comparable to the inverse momentum relaxation time τ_p

$$\sigma_D(\omega) = \frac{\sigma_D}{1 + i\omega\tau_p}, \tag{9.1.14}$$

where σ_D is the DC Drude conductivity. We are mainly interested in the low-frequency region, where the frequency is small compared with the attempt-to-jump frequency which is of the order of the characteristic phonon frequency $\nu_{ph} \simeq 10^{12} - 10^{13}\,\mathrm{s}^{-1}$ [4]. In this region, we have $\omega\tau_p \ll 1$ and the free-carrier conductivity determined by Equation (9.1.14) is virtually independent of frequency.

In many disordered dielectric materials, the relaxation of material characteristics is well described by the fractional-power time dependence

$$\chi_h(t) = At^{-s}, \tag{9.1.15}$$

where s is a parameter that is usually smaller than unity and A is a coefficient. The kinetics described by Equation (9.1.15) is substantially different from the standard exponential Debye-type relaxation (Equation 9.1.12). For the case of the power-law delayed response (9.1.15), we have

$$\chi_h(\omega) = A\Gamma(1-s)(i\omega)^{-(1-s)} = A\Gamma(1-s)\left(\sin\frac{\pi s}{2} - i\cos\frac{\pi s}{2}\right)\omega^{-(1-s)}, \tag{9.1.16}$$

where Γ is the gamma function. A power-law relaxation described by Equation (9.1.15) corresponds to the power-law frequency dependence of the conductivity. One obtains from Equation (9.1.16) the tangent of the loss angle δ in the form

$$\tan\delta = \frac{\sigma'_h(\omega)}{\sigma''_h(\omega)} = \frac{\chi''_h(\omega)}{\chi'_h(\omega)} = \cot\frac{\pi s}{2}. \tag{9.1.17}$$

For the real and imaginary parts of the conductivity $\sigma_h(\omega)$, we obtain the frequency dependences

$$\sigma'_h(\omega) = \tilde{A}\omega^s, \tag{9.1.18}$$

$$\sigma''_h(\omega) = \tilde{A}\tan\frac{\pi s}{2}\omega^s \tag{9.1.19}$$

with some coefficient $\tilde{A}$.

The polarization contribution to the conductivity $\sigma_h(\omega)$ does not necessarily vanish as $\omega \to 0$. Indeed, the polarization due to *hopping* between localized states can produce a finite DC current corresponding to the susceptibility $\chi_h(\omega)$ (and permittivity) divergent as $\omega \to 0$. According to Equation (9.1.9), the conductivity in this case is written as a sum of the Drude component (the contribution from the electrons in the extended states of the conduction band) and the contribution (9.1.10) from the electrons in localized states determined by the hopping polarization,

$$\sigma(0) = \sigma_D + \sigma_h(0), \tag{9.1.20}$$

where $\sigma_h(0) = i\kappa_0 \lim_{\omega\to 0} \omega\chi_h(\omega)$ is the DC hopping contribution to the conductivity. If $\sigma_D \gg \sigma_h(0)$ and at not too low frequencies the conductivity has a power-law dependence described by Equations (9.1.18) and (9.1.19). Then the crossover from the DC to a hopping power-law conductivity is accompanied by the change in the conduction mechanism. In this case, using Equations (9.1.9) and (9.1.18), we can present the real part of the conductivity as a superposition of the Drude and hopping contributions,

$$\sigma'(\omega) = \sigma_D\left[1 + (\omega/\omega_{cr})^s\right]. \tag{9.1.21}$$

Here ω_{cr} is the crossover frequency from the power-law region to a virtually constant conductivity σ_D. Using Equation (9.1.18), we obtain

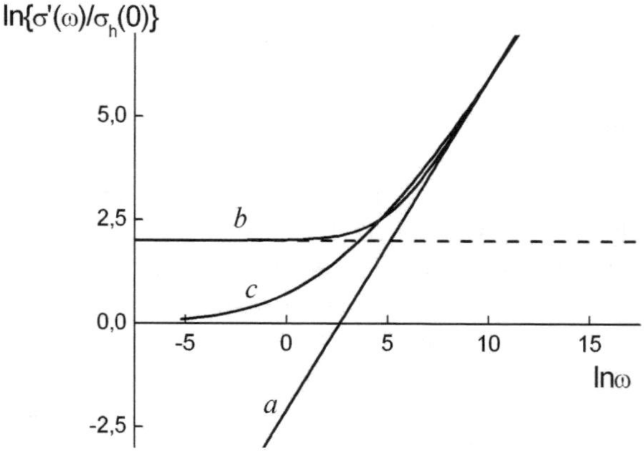

Figure 9.1 Real part of the conductivity $\sigma_h'(\omega)$ as a function of frequency (schematically): curve *a* fractional-power law dependence; (curve *b*) real part of the total conductivity $\sigma'(\omega)$ (Equation 9.1.21) for the case where $\sigma_D > \sigma'(0)$ (crossover from the power-law dependence to constant DC conductivity); curve *c* real part of the total conductivity $\sigma'(\omega)$ defined by Equation (9.5.16) describing the multiple hopping regime

$$\omega_{cr} = \left(\sigma_D / \tilde{A}\right)^{1/s}. \tag{9.1.22}$$

In Figure 9.1, curve *a* shows the fractional-power frequency dependence of the real part of the hopping conductivity $\sigma_h'(\omega)$, provided it can be described by Equation (9.1.18) in the entire frequency range. Curve *b* illustrates the variation $\sigma'(\omega)$ described by Equation (9.1.21) for the case $\sigma_D >> \sigma_h(0)$. If, however, $\sigma_D << \sigma_h(0)$, the Drude contribution may be disregarded and the total conductivity is $\sigma(\omega) \simeq \sigma_h(\omega)$ (curve *c* in Figure 9.1). In this case, the conductivity in the crossover region and the crossover frequency are determined by the features of the hopping mechanism in the multiple hopping regime (see Section 9.5).

9.2 UNIVERSALITY AND SCALING

Dispersive non-Debye behavior of the permittivity is characteristic of many disordered dielectrics (see, e.g., [1]). A frequency dependence of the AC conductivity in the form of Equation (9.1.18), where the exponent s is usually in the range $0.7 < s < 1$, was observed in numerous conducting disordered materials in wide frequency ranges covering many orders of magnitude where $\sigma'(\omega) >> \sigma(0)$. Usually, the parameters $\tilde{A}$ and s depend only weakly on frequency and temperature. The parameter s slightly increases with frequency and with decreasing temperature, approaching unity. Such a behavior was reported for compensated crystalline semiconductors [5, 6], elementary amorphous semiconductors and semiconductor glasses, including chalcogenides (see, e.g., [7–13]), a-Si:H [14–16], oxides [17], $SrTiO_3$-based perovskite oxides [18], organic disordered semiconductors [19], polymer blends and composites [20, 21], and many other electronic materials. Such a frequency dependence of the AC conductivity was also observed for ionic conductors [22, 23]. Note that the fractional-power frequency dependence of the conductivity is characteristic not only of electronic conductors; it is a very general effect observed in a wide range of disordered materials of different structures with different conduction mechanisms under different conditions (*universality*).

In the region of universal fractional-power law, the temperature dependence of the conductivity at high frequencies, $\omega > \omega_{cr}$, is determined mainly by that of the parameter $\tilde{A}$ and is usually substantially weaker than at low frequencies (for $\omega < \omega_{cr}$), where it is closely related to the variation of the DC conductivity. For an activated Drude conduction, the temperature dependence of the crossover frequency ω_{cr} determined by Equation (9.1.21) is also activated; however, the activation energy ε_a/s may be somewhat different from that for the Drude conductivity ε_a. The correlation between the crossover frequency and the DC conductivity was currently observed.

The experimental data on the AC conductivity can usually be merged onto a single master curve by plotting the reduced conductivity $\sigma(\omega)/\sigma_D$ as a function of the reduced frequency ω/ω_s, where ω_s is a properly chosen scaling frequency. The master curve corresponds to the equation [24–27]

$$\sigma(\omega)/\sigma_D = F(\omega/\omega_s), \tag{9.2.1}$$

where $F(x)$ is a master function of the type shown in Figure 9.1. If the AC conductivity is described by Equation (9.1.22), we have

$$F(u) = 1 + u^s \tag{9.2.2}$$

and

$$\omega_s = \omega_{cr}. \tag{9.2.3}$$

An example of scaling is shown in Figure 9.2, where the AC conductivity master curve for the scaled conductivity of amorphous silicon a-Si measured at three different temperatures is plotted as a function of a properly scaled frequency [16]. Apart from the above correlation between the scaling frequency ω_s and the DC conductivity, the choice of ω_s remains to some degree arbitrary. In some systems (in particular, in alkali oxide ionic glasses), the scaling frequency appears to depend on glass composition. Thus for sodium borate glasses with varying sodium oxide content x, it has been shown that for any given value of x, the curves of the conductivity frequency variation corresponding to different temperatures, when both the conductivity and frequency are scaled by $T\sigma(0)$, collapse onto a common master curve; however, the master curves for different compositions are shifted to higher values of ω/ω_s as x decreases. This is approximately described by assuming that $\omega_s \simeq \omega_{s0}/x$ so that the master curves at different x merge into a single master curve if x is introduced into the scaling factor [22],

$$\sigma(\omega)/\sigma_D = F(x\omega/\omega_{s0}), \tag{9.2.4}$$

where $\omega_{s0} = kT\sigma(0)$. The form of the scaling relation provides some information about the charge transfer mechanism. Thus, the deviations from the scaling relation (9.2.4) observed in alkali germanate glasses with alkali content varying by more than an order of magnitude

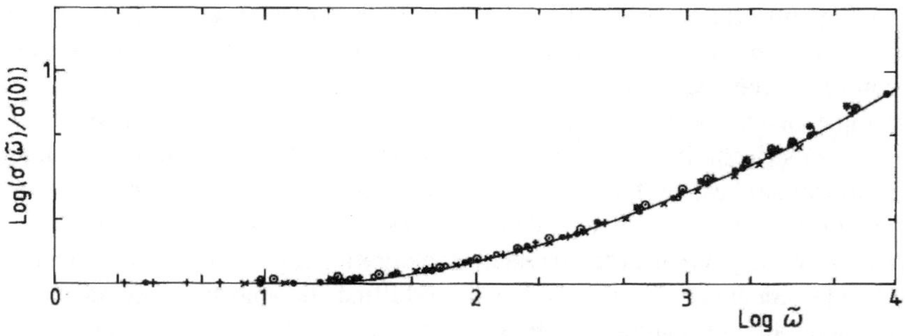

Figure 9.2 Universality of the frequency dependence of the conductivity of an a-Si film measured at different temperatures

were attributed to the composition dependence of the AC hopping distance and were accounted for by choosing the scaling frequency in the form $\omega_s = \sigma(0)/\kappa_0\Delta\kappa$ [27], since the change $\Delta\kappa$ in the permittivity (the dielectric strength) is a direct consequence of the relaxation of hopping cations and incorporates the composition dependence of the hopping distance.

In some cases, however, when the frequency ω exceeds ω_{cr} by several orders of magnitude, the fractional power law (Equation 9.1.18) is no longer sufficient to describe the frequency-dependent conductivity. At high frequencies and/or low temperatures (where ω_{cr} is low), the slope s of log–log conductivity frequency plots becomes close to the value of 1.0 (rather than gradually approaches unity) is and virtually independent of frequency in a large frequency range, implying a negligible frequency dependence of the imaginary part of the permittivity $\kappa_h'' = \sigma_h'(\omega)/\omega \simeq$ constant. This type of frequency dependence of the imaginary part of the permittivity is called *nearly constant* (*or flat*) *loss*; it was observed in some amorphous semiconductors and glasses [13, 14, 28], evaporated Al_2O_3, Al-doped $CaTiO_3$ single crystals, low-loss polymers [1], and ionic conductors [29]. A crossover from the fractional-power to the nearly constant loss behavior was observed with increasing frequency and decreasing temperature. It was suggested to describe the contribution from the nearly constant loss by introducing the nearly constant loss contribution into Equation (9.1.21),

$$\sigma(\omega) = \sigma_D\left[1 + (\omega/\omega_{cr})^s\right] + B\omega, \tag{9.2.5}$$

where B is constant. Such a superposition of the fractional-power and nearly constant loss contributions provided a good description of the AC conductivity of ion-containing glasses in a wide frequency range from MHz to GHz at different temperatures and compositions [30]. The origin of the nearly constant loss is discussed in Section 9.6.3.

The ubiquitous character of the highly dispersive behavior of the conductivity and transient characteristics appearing for different conduction mechanisms makes it difficult to obtain detailed information on specific features of transport mechanisms from the measurements of the frequency dependence of the conductivity $\sigma(\omega)$. Therefore, the analysis of the *deviations from universality* is important for extracting information about the transport mechanism and the structural characteristics of the system. Such nonuniverasal behavior can be related to the temperature and composition dependence of the slope s in the high-frequency region, of the behavior of the conductivity near the crossover frequency, or to the analysis of the nearly constant loss contribution.

In some materials, information on the transport mechanism is available from other measurements (including DC conductivity, photoconductivity, thermopower, magnetoresistance, optical studies, etc.), thus making it possible to advance reasonable assumptions on the type of charge carriers and basic charge transfer mechanisms. One class of such relatively well-investigated disordered materials exhibiting highly dispersive properties includes disordered semiconductors such as doped crystalline and amorphous semiconductors (most often, silicon and germanium). The charge carriers in these materials are often known to be of electronic origin and the low-temperature DC transport mechanism is hopping (inelastic tunneling between localized states). The same transport mechanism is also expected to control the AC conductivity in the low-frequency range. Another wide class of disordered solids that are extensively studied experimentally is formed by ion-conducting glasses. Although the details of the ionic transport mechanism are not yet established in final form, one can state some reasonable models providing a good description of both DC and AC properties. In this chapter, we concentrate on the frequency dependence of the AC hopping conduction in disordered materials and on various concepts and models used to describe the AC transport.

9.3 PHONONLESS AC CONDUCTIVITY

In order to describe electronic transport in the region of localized states, one must know the form of the electron wavefunctions of the states giving the main contribution to transport. Generally, the wavefunctions of electrons in disordered semiconductors depend on the atomic structure of a disordered material and the properties of the random field, which are usually not known sufficiently well. Therefore, an approach is often used assuming the electron states to be known and expressing the transport coefficients in terms of such characteristics of the system as the density of states, asymptotic behavior of the wavefunctions at large distances from the localization center (site), etc.. The parameters through which kinetic coefficients are expressed can often be calculated approximately or determined from independent experiments. It should be noted that the shape of localized wavefunctions is generally more important for AC than for DC transport. Indeed, the average distance between sites contributing to the conductivity decreases with increasing frequency and the knowledge of long-range asymptotic behavior of the wavefunctions may appear to be insufficient. In particular, for close sites, hybridization of electron states can modify the wavefunctions and substantially affect the conductivity.

AC hopping transport, likewise DC transport may be due to inelastic tunneling (hopping) of charge carriers between localized states with different energies and different positions in space. To ensure energy conservation, energy must be released or absorbed at hopping transitions usually involving phonons. At $\omega \neq 0$, however, transitions are possible even in the absence of the interaction with phonons, due to the absorption of the AC field quanta $\hbar\omega$. Indeed, in a large system of centers with random energies of localized states, one can always find a pair of centers for which the energy difference is exactly equal to the field quantum $\hbar\omega$ (*a resonance pair*). In spite of small number of such pairs, their contribution (*resonance, or phononless*) to the AC conductivity at $\omega \neq 0$ is finite.

The contribution of phononless transitions to the low-field AC conductivity can be calculated using the Kubo formula or the density matrix approach (see [4, 31]). If the current density is expressed in terms of the off-diagonal elements of the one-particle density matrix, one arrives at the expression

$$j(\omega)=\frac{ie\omega}{\Omega}\sum_{\substack{\lambda\lambda'\\(\lambda\neq\lambda')}}\langle\lambda|x|\lambda'\rangle\langle\lambda|V|\lambda'\rangle\frac{n_F(\varepsilon_\lambda)-n_F(\varepsilon_{\lambda'})}{\varepsilon_\lambda-\varepsilon_{\lambda'}+\hbar\omega+i\eta}, \tag{9.3.1}$$

where Ω is the volume of the system, λ and λ' denote the exact single-particle localized eigenstates, ε_λ and $\varepsilon_{\lambda'}$ are the exact energy eigenvalues for these states, $\langle\lambda|x|\lambda'\rangle$ and $\langle\lambda|V|\lambda'\rangle$ are the off-diagonal matrix elements of the coordinate x (the ox-axis is directed along the external electric field $\mathbf{E}$) and of the potential energy V of the external field, $n_F(\varepsilon_\lambda)$ is the equilibrium occupancy of a state with energy ε_λ, and η is an infinitesimal positive number, $\eta\to+0$.

For simplicity, a standard simplifying approximation assuming that the applied field is uniform in space is used

$$V=-e\mathbf{E}\mathbf{r}. \tag{9.3.2}$$

After a straightforward transformation of Equation (9.3.1), we obtain for the real part of the conductivity

$$\begin{aligned}\mathrm{Re}\,\sigma(\omega)=&\frac{\pi e^2\omega}{3\Omega}\{1-\exp(-\hbar\omega/kT)\}\sum_{\lambda\lambda'}|\langle\lambda|\mathbf{r}|\lambda'\rangle|^2\\&\times n_F(\varepsilon_\lambda)\{1-n_F(\varepsilon_{\lambda'})\}\delta(\varepsilon_\lambda-\varepsilon_{\lambda'}+\hbar\omega).\end{aligned} \tag{9.3.3}$$

For sufficiently low concentrations of the centers, the conductivity can be calculated assuming that the main contribution to the conductivity comes from isolated pairs of centers (*the pair, or two-site approximation*). For hydrogenic states, the matrix element $\langle\lambda|\mathbf{r}|\lambda'\rangle$ in Equation (9.3.3) can be calculated explicitly [32],

$$|\langle\lambda|\mathbf{r}|\lambda'\rangle|^2=r^2I^2(r)/(\varepsilon_\lambda-\varepsilon_{\lambda'})^2, \tag{9.3.4}$$

here $I(r)$ is the overlap integral exponentially decreasing with increasing distance r between the centers,

$$I(r)=I_0\exp(-r/\alpha), \tag{9.3.5}$$

and α is the localization radius.

We may pass from summation over states λ and λ' in Equation (9.3.3) to integration with respect to site positions and energies ε, ε'

$$\begin{aligned}\sigma'(\omega)=&\frac{4\pi e^2g_F^2\omega}{3\Omega}\{1-\exp(-\hbar\omega/kT)\}\int d\varepsilon\int d\varepsilon'\int_0^\infty dr r^4\Psi(\varepsilon,\varepsilon',r)\frac{I^2(r)}{(\varepsilon-\varepsilon')^2}\\&\times n_F(\varepsilon_\lambda)\{1-n_F(\varepsilon_{\lambda'})\}\delta(\varepsilon-\varepsilon'+\hbar\omega).\end{aligned} \tag{9.3.6}$$

Here, due to the presence of the factor $n_F(\varepsilon_\lambda)\{1 - n_F(\varepsilon_{\lambda'})\}$ that decreases exponentially with increasing separation of the energy levels ε and ε' from the Fermi level ε_F, only centers with energies lying within the layer of width of about kT contribute to Equation (9.3.6); we have assumed the density of states to be a slowly varying function and set it equal to its value g_F at the Fermi level. The factor $\Psi(\varepsilon, \varepsilon', r)$ in Equation (9.3.6) is the correlation function of energy levels related to the conditional probability that a center located at a distance r from a given center with energy ε has an energy ε' (see [31, 33, 34]). Correlation between the levels at short distances is due mainly to quantum hybridization of electronic states of the centers and corresponds to the level repulsion. The correlation function $\Psi(\varepsilon, \varepsilon', r)$ is readily calculated for an isolated pair of sites (we may consider a close pair of centers as isolated if we may disregard the overlap of the wavefunctions of the centers of the pair with the wavefunctions of the other centers of the system). For a pair of centers, we can find electron energies with allowance for hybridization. The variational calculation gives [32]

$$\varepsilon^{\pm} = \frac{\varepsilon_0 + \varepsilon_0'}{2} \pm \frac{1}{2}\sqrt{(\varepsilon_0 - \varepsilon_0')^2 + 4I^2(r)}, \tag{9.3.7}$$

where $\varepsilon^+ = \varepsilon'$, $\varepsilon^- = \varepsilon$, and ε_0 and ε_0' are the energies of an electron on the isolated centers of the pair (without taking into account the overlap of the wave functions). The level correlation function $\Psi(\varepsilon, \varepsilon', r)$ is just the Jacobian of the transformation from the variables ε_0 and ε_0' to the variables ε and ε', i.e.,

$$\Psi(\varepsilon, \varepsilon', r) = \frac{\varepsilon' - \varepsilon}{\sqrt{(\varepsilon' - \varepsilon)^2 - 4I^2(r)}}. \tag{9.3.8}$$

Then Equation (9.3.6) has the form

$$\sigma'(\omega) = \frac{4\pi e^2 g_F^2}{3\hbar}\left\{1 - \exp\left(-\frac{\hbar\omega}{kT}\right)\right\}$$
$$\int d\varepsilon n_F(\varepsilon)\{1 - n_F(\varepsilon + \hbar\omega)\}\int_{r_\omega}^{\infty} dr r^4 \frac{I^2(r)}{\sqrt{(\hbar\omega)^2 - 4I^2(r)}}, \tag{9.3.9}$$

where

$$r_\omega = \alpha \ln(2I_0/\hbar\omega) \tag{9.3.10}$$

is the distance at which $\hbar\omega = 2I(r)$.

The integrals in Equation (9.3.9) are readily calculated at $T \to 0\,\mathrm{K}$. We have

$$\int d\varepsilon n_F(\varepsilon)\{1 - n_F(\varepsilon + \hbar\omega) =$$
$$\int \frac{d\varepsilon \exp\left(\dfrac{\varepsilon + \hbar\omega - \varepsilon_F}{kT}\right)}{\left\{1 + \exp\left(\dfrac{\varepsilon - \varepsilon_F}{kT}\right)\right\}\left\{1 + \exp\left(\dfrac{\varepsilon + \hbar\omega - \varepsilon_F}{kT}\right)\right\}} \xrightarrow[T\to 0\mathrm{K}]{} \hbar\omega, \tag{9.3.11}$$

and

$$\int_{r_\omega}^{\infty} \mathrm{d}r r^4 \frac{I^2(r)}{\sqrt{(\hbar\omega)^2 - 4I^2(r)}} \simeq \frac{1}{4}\alpha r_\omega^4 \hbar\omega. \tag{9.3.12}$$

Here, we have used the fact that the function r^4 varies slowly compared with the last factor in the integrand which decays exponentially at $r > r_\omega$ at a distance of about α. Therefore, we may take the factor r^4 at the point $r = r_\omega$ out of the integral. Substituting Equations (9.3.11) and (9.3.12) into (9.3.9), we arrive at the following expression for the real part of the phononless contribution to conductivity at low temperatures:

$$\sigma'(\omega) = \frac{\pi^2 e^2 g_F^2 \alpha\hbar}{3} r_\omega^4 \omega^2. \tag{9.3.13}$$

This expression has been obtained in [33]. (In [33], the numerical factor differs, however, from that in Equation (9.3.13); the correct value of this factor has been obtained in [34].) Expression (9.3.13) for the phononless contribution varies almost quadratically with frequency, is proportional to the square of the density of states at the Fermi level, and depends weakly on temperature at low temperatures.

Note that the functions of the form $\omega^m \ln^n(\omega_0/\omega)$ appearing in (9.3.13) are well approximated (at $\omega_0 >> \omega$) by a power law $A\omega^s$ with the exponent

$$s = \frac{\mathrm{d}\ln\sigma(\omega)}{\mathrm{d}\ln\omega} = m - \frac{n}{\ln(\omega_0/\omega)}. \tag{9.3.14}$$

This exponent depends weakly on temperature and frequency, decreasing with ω.

Since the main contribution to the conductivity comes from close pairs of centers, on calculating a conductivity, it can be important to take account of the Coulomb interaction between the electrons simultaneously occupying both centers of the pair [34]. The Coulomb repulsion of electrons results in a decrease in the equilibrium probability of double occupation of the pair; in other words, the conditional occupation probability of the second center of the pair already containing one electron decreases. Due to the increase in the number of singly occupied pairs, the intra-pair carrier interaction increases the AC conductivity.

Coulomb interaction between the electrons of a pair is readily incorporated into the above calculation. It amounts to replacing the factor $n_F(\varepsilon)\{1 - n_F(\varepsilon')\}$ by the equilibrium conditional probability $\tilde{f}_\lambda^{(\lambda')}$ of the occupation of center λ provided that the other center of the pair λ' is empty. This conditional probability can be written down at once, using the Gibbs principle

$$\tilde{f}_\lambda^{(\lambda')} = \frac{\exp\left(-\frac{\varepsilon_\lambda - \varepsilon_F}{kT}\right)}{1 + \exp\left(-\frac{\varepsilon_\lambda - \varepsilon_F}{kT}\right) + \exp\left(-\frac{\varepsilon_{\lambda'} - \varepsilon_F}{kT}\right) + \exp\left(-\frac{\varepsilon_\lambda + \varepsilon_{\lambda'} - 2\varepsilon_F + v_{\lambda\lambda'}}{kT}\right)}. \tag{9.3.15}$$

Here

$$v_{\lambda\lambda'} = e^2/(4\pi\kappa_0\kappa r_{\lambda\lambda'}) \tag{9.3.16}$$

is the energy of Coulomb interaction between the electrons which occupy centers λ, λ' of the pair, and ε_λ is the energy of an electron on the center λ with allowance for its interaction with electrons on the other centers of the system which do not belong to the pair considered. At low temperatures, the quantity described by Equation (9.3.15) (with $\varepsilon_{\lambda'} = \varepsilon_\lambda + \hbar\omega$) is almost unity within the interval $\varepsilon_F - (v_{\lambda\lambda'} + \hbar\omega) < \varepsilon_\lambda < \varepsilon_F$, and integral (9.3.11) can be replaced by

$$\int d\varepsilon_{\lambda'} \tilde{f}_\lambda^{(\lambda')} \big|_{\varepsilon\lambda'=\varepsilon\lambda+\hbar\omega} \xrightarrow[T\to 0]{} e^2/(4\pi\kappa_0\kappa r_{\lambda\lambda'}) + \hbar\omega. \tag{9.3.17}$$

At high temperatures, where $kT >> e^2/(4\pi\kappa_0\kappa r_{\lambda\lambda'})$, one obtains

$$\int d\varepsilon_{\lambda'} \tilde{f}_\lambda^{(\lambda')} \big|_{\varepsilon\lambda'=\varepsilon\lambda+\hbar\omega} \to kT. \tag{9.3.18}$$

Accordingly, in the presence of Coulomb interaction between the electrons of the pair, instead of Equation (9.3.13), we find at low temperatures

$$\sigma'(\omega) = \frac{\pi^2 e^2 g_F^2 \alpha}{3} r_\omega^4 \omega\left(e^2/(4\pi\kappa_0\kappa r_\omega) + \hbar\omega\right). \tag{9.3.19}$$

At high temperatures, we have

$$\left\{1 - \exp\left(-\frac{\hbar\omega}{kT}\right)\right\} \int d\varepsilon_{\lambda'} \tilde{f}_\lambda^{(\lambda)} \big|_{\varepsilon_\lambda=\varepsilon_\lambda+\hbar\omega} \approx \hbar\omega$$

and formula (9.3.13) for the conductivity is retrieved.

Thus, at low temperatures, the frequency dependence of the conductivity is determined by formula (9.3.19), rather than by (9.3.13), i.e., at low frequencies [$\hbar\omega << e^2/(4\pi\kappa_0\kappa r_{\lambda\lambda'})$], the frequency dependence is almost linear. With increasing frequency [at $\hbar\omega \simeq e^2/(4\pi\kappa_0\kappa r_\omega)$], a crossover from an almost linear to an almost quadratic frequency dependence is predicted by Equation (9.3.19). This crossover is discussed in Section 9.4.4.

9.4 PHONON-ASSISTED AC CONDUCTIVITY IN THE PAIR APPROXIMATION

9.4.1 Model

The mechanism of the DC transport in the region of localized states is known to be phonon-assisted hopping; accordingly, the contribution of phonon-assisted hopping to the AC conductivity must be important, at least, at low frequencies. In this case, a resonance condition of the type $\varepsilon_\lambda - \varepsilon_{\lambda'} = \hbar\omega$ is no longer required, since the energy variation at transitions is compensated due to emission or absorption of phonons. Phonon-assisted hopping can be described using the rate equation; a standard linearized form of this equation in the presence of weak external electric field is [31, 35, 37]

$$\frac{d\delta f_\lambda}{dt} = -\sum_{\lambda'} \frac{\Gamma_{\lambda\lambda'}}{kT} \left\{ V_\lambda - V_{\lambda'} + \frac{kT\delta f_\lambda}{n_F(\varepsilon_\lambda)(1-n_F(\varepsilon_\lambda))} - \frac{kT\delta f_{\lambda'}}{n_F(\varepsilon_{\lambda'})(1-n_F(\varepsilon_{\lambda'}))} \right\}, \tag{9.4.1}$$

where V_λ is the potential energy averaged over the localized state (site) λ, $\delta f_\lambda = f_\lambda - n_F(\varepsilon_\lambda)$ is the field-induced variation of the occupancy f_λ of site λ, $n_F(\varepsilon_\lambda)$ is the equilibrium occupancy of site λ,

$$\Gamma_{\lambda\lambda'} = W_{\lambda\lambda'} n_F(\varepsilon_\lambda)(1 - n_F(\varepsilon_{\lambda'})) \tag{9.4.2}$$

is the transition rate,

$$W_{\lambda\lambda'} = w_{\lambda\lambda'} \exp(-2r_{\lambda\lambda'}/\alpha)\exp\left(\frac{\varepsilon_\lambda - \varepsilon_{\lambda'}}{2kT}\right)\sinh^{-1}\left(\frac{|\varepsilon_\lambda - \varepsilon_{\lambda'}|}{2kT}\right), \tag{9.4.3}$$

is the probability of transition from λ to λ', ε_λ and $\varepsilon_{\lambda'}$ are the energies of the initial and final states, and $w_{\lambda\lambda'}$ is a factor expressed in terms of the matrix element of electron–phonon interaction. For transitions between the hydrogenic impurity states in doped semiconductors, the factor $w_{\lambda\lambda'}$ is explicitly calculated [32, 36],

$$W_{\lambda\lambda'} = \frac{\Xi_1^2 |\varepsilon_\lambda - \varepsilon_{\lambda'}|}{\pi^3 s^5 \hbar^4 \mathrm{d}}\left(\frac{e^2}{6\kappa_0 \kappa \alpha}\right)^2 \frac{(r_{\lambda\lambda'}/\alpha)^2}{\left\{1 + \left(\alpha \dfrac{\varepsilon_\lambda - \varepsilon_{\lambda'}}{2\hbar s}\right)^2\right\}^4}, \tag{9.4.4}$$

where Ξ_1 is the deformation potential constant, s is the velocity of sound, d is the density of the material, $r_{\lambda\lambda'}$ is the distance between sites λ and λ', and α is the localization radius. At large distances between the impurity centers λ and λ' ($r_{\lambda\lambda'} >> \alpha$), the dependence of $W_{\lambda\lambda'}$ on the intersite distance is mainly determined by the exponentially decreasing function corresponding to an exponentially small overlap of the wave functions of localized states of different centers. Expression (9.4.3) was derived for single-phonon transitions between the hydrogenic states; the main exponential factors that determine the dependence of the transition probability on energies and intersite separations remain, however, the same for multiphonon intersite transitions as well. Usually, the dependence of the prefactor $w_{\lambda\lambda'}$ on $r_{\lambda\lambda'}$ and site energies is weak compared with that of the exponential factor in Equation (9.4.3) and may be disregarded; thus we may set $w_{\lambda\lambda'} \simeq w_0$. The transport coefficients are expressed in terms of the quantities δf_λ obtained by solving Equation (9.4.1) [31].

In the DC case, the solution of Equation (9.4.1) is known to be equivalent to finding the overall resistance of the Miller–Abrahams random resistance network, where the resistances

$$R_{\lambda\lambda'} = \{(e^2/kT)\Gamma_{\lambda\lambda'}\}^{-1} \tag{9.4.5}$$

connect every site λ with every other site λ' [32]. In the presence of an AC field, the rate equation (9.4.1) corresponds to a generalized equivalent network [37] that contains both random resistors (Equation 9.4.5) and the capacitors

$$C_{\lambda\lambda'} = (e^2/kT)\Gamma_{\lambda\lambda'} n_F(\varepsilon_\lambda)(1 - n_F(\varepsilon_{\lambda'})). \tag{9.4.6}$$

Using Equations (9.4.5) and (9.4.6), we can present Equation (9.4.1) in the form describing the distribution of capacitor charges $q_\lambda = e\delta f_\lambda$

$$\frac{dq_\lambda}{dt} = -\sum_{\lambda'} R^{-1}_{\lambda\lambda'} \left\{ \frac{V_\lambda}{e} - \frac{V_{\lambda'}}{e} + \frac{q_\lambda}{C_\lambda} - \frac{q_{\lambda'}}{C_{\lambda'}} \right\}. \tag{9.4.7}$$

The solution of the AC conductivity problem (Equations 9.4.1 or 9.4.7) can be substantially simplified at not too low frequencies using the pair (two-site) approximation which assumes that the current is determined by the sum of the contributions from isolated pairs of sites [5]. The pair approximation holds at sufficiently high frequencies of the applied field such that, for any of these pairs, the transitions to other sites outside the pair may be disregarded. Then the rate equation is reduced to independent equations describing the transitions inside the pairs. We generalize this equation, taking account of the correlation of the site occupancies inside the pairs by introducing the quantities $f_\lambda^{(\lambda')}$ such that $f_\lambda^{(\lambda')}$ is the nonequilibrium probability of finding state λ occupied provided that state λ' is empty, i.e., the probability of a single occupancy of the pair with electron located on site λ. In the presence of a weak external field (slowly varying over the localization radius), the linearized equation describing the field-induced transitions inside a pair of sites λ, λ' is [38]

$$\frac{d\delta f_\lambda^{(\lambda')}}{dt} = -\frac{\Gamma_{\lambda\lambda'}}{kT} \left\{ V_\lambda - V_{\lambda'} + \frac{kT\delta f_\lambda^{(\lambda')}}{\tilde{f}_\lambda^{(\lambda')}} - \frac{kT\delta f_{\lambda'}^{(\lambda)}}{\tilde{f}_{\lambda'}^{(\lambda)}} \right\}, \tag{9.4.8}$$

where $V_{\lambda'}$ is the potential energy averaged over state λ, $\delta f_\lambda^{(\lambda')} = f_\lambda^{(\lambda')} - \tilde{f}_\lambda^{(\lambda')}$, $\tilde{f}_\lambda^{(\lambda')}$ is the equilibrium occupancy, and $\Gamma_{\lambda\lambda'} = W_{\lambda\lambda'}\tilde{f}_\lambda^{(\lambda')} = W_{\lambda'\lambda}\tilde{f}_{\lambda'}^{(\lambda)}$. If the occupancies of sites λ and λ' are uncorrelated, we may set

$$f_\lambda^{(\lambda')} \simeq f_\lambda (1 - f_{\lambda'}). \tag{9.4.9}$$

Note that using the functions $f_\lambda^{(\lambda')}$ instead of the occupancies f_λ allows us to take account both of the Coulomb correlations and of hybridization of electronic states inside a pair.

Equation (9.4.8) at a fixed total occupancy of the pair ($f_\lambda^{(\lambda')} + f_{\lambda'}^{(\lambda)} \simeq \tilde{f}_\lambda^{(\lambda')} + \tilde{f}_{\lambda'}^{(\lambda)}$, where $\tilde{f}_\lambda^{(\lambda')} + \tilde{f}_{\lambda'}^{(\lambda)}$ is the equilibrium single occupancy of the pair) describes the relaxation to the state in which the ratio of the occupancies of the sites is equilibrium. Accordingly, equations of the type (9.4.8) are valid only at not too long times (not too low frequencies), as long as it is possible to disregard the transitions outside the pair.

According to Equation (9.1.1), the contribution of an isolated pair to the current density is expressed in terms of the time derivative of the polarization of the pair. For the current density, we have

$$j_h = \frac{dP(t)}{dt} = \frac{e}{\Omega} \sum_{(\lambda\lambda')} \frac{d}{dt} \left(x_\lambda f_\lambda^{(\lambda')} + x_{\lambda'} f_{\lambda'}^{(\lambda)} \right) = \frac{e}{\Omega} \sum_{(\lambda\lambda')} (x_\lambda - x_{\lambda'}) \frac{df_{\lambda'}^{(\lambda)}}{dt}, \tag{9.4.10}$$

where, as above, Ω is the volume of the system, x_λ is the coordinate of site λ in the direction of the field, and the sum is taken over all isolated pairs.

After performing the Fourier transformation in Equation (9.4.8), we obtain

$$i\omega\delta f_\lambda^{(\lambda')}(\omega) = -\frac{\Gamma_{\lambda\lambda'}}{kT} \left\{ V_\lambda(\omega) - V_{\lambda'}(\omega) + \frac{kT\delta f_\lambda^{(\lambda')}(\omega)}{\tilde{f}_\lambda^{(\lambda')}} - \frac{kT\delta f_{\lambda'}^{(\lambda)}(\omega)}{\tilde{f}_{\lambda'}^{(\lambda)}} \right\}. \tag{9.4.11}$$

In the high-frequency limit, the last two terms in the braces on the right-hand side of Equation (9.4.11) are small and we obtain

$$\delta f_{\lambda}^{(\lambda')}(\omega)=\frac{i}{\omega kT}\Gamma_{\lambda\lambda'}\left(V_{\lambda}(\omega)-V_{\lambda'}(\omega)\right). \tag{9.4.12}$$

For a uniform field E, (Equation 9.3.2), we find the asymptotic high-frequency value of the conductivity

$$\sigma_{\infty}=\frac{e}{\Omega kT}\sum_{(\lambda\lambda')}\left(x_{\lambda}-x_{\lambda'}\right)^{2}\Gamma_{\lambda\lambda'}. \tag{9.4.13}$$

Thus, in the high-frequency limit, the conductivity becomes independent of frequency. At lower frequencies, the conductivity can be found, using the pair approximation (Equation 9.4.10) where the main contribution comes from the transitions between close sites located at distances much shorter than the average site-to-site separation. Using the Fourier representation and the condition $\delta f_{\lambda}^{(\lambda')}(\omega)=-\delta f_{\lambda'}^{(\lambda)}(\omega)$, which is valid for an isolated pair, we can write the solution of Equation (9.4.11) as

$$\delta f_{\lambda}^{(\lambda')}(\omega)=\frac{eE}{4kT}\left(x_{\lambda}-x_{\lambda'}\right)\frac{F\left(\varepsilon_{\lambda},\varepsilon_{\lambda'}\right)}{1+i\omega\tau_{\lambda\lambda'}}, \tag{9.4.14}$$

where

$$\tau_{\lambda\lambda'}^{-1}=W_{\lambda\lambda'}+W_{\lambda'\lambda}, \tag{9.4.15}$$

and

$$F\left(\frac{\varepsilon_{\lambda}}{kT},\frac{\varepsilon_{\lambda'}}{kT}\right)=4\left\{\left(\tilde{f}_{\lambda}^{(\lambda')}\right)^{-1}+\left(\tilde{f}_{\lambda'}^{(\lambda)}\right)^{-1}\right\}^{-1}. \tag{9.4.16}$$

Note that, for noninteracting electrons, the function $F(u, u')$, defined by Equation (9.4.16), does not depend on $r_{\lambda\lambda'}$; however, generally, $r_{\lambda\lambda'}$ must be included in the arguments of $F(u, u')$, since, if the electron–electron interaction is taken into account, the equilibrium single-electron occupancy of pairs actually depends on $r_{\lambda\lambda'}$.

9.4.2 AC conductivity for noninteracting electrons in the pair approximation

Fourier transforming Equation (9.4.10), substituting Equation (9.4.14) into (9.4.10), and passing from summation to integration, we obtain the following expression for the real part of the conductivity:

$$\sigma'(\omega)=\frac{\pi e^{2}g_{\mathrm{F}}^{2}\omega}{3kT}\int d\varepsilon\int d\varepsilon'\int_{r_{\omega}}^{\infty}dr\,r^{4}\Psi(\varepsilon,\varepsilon',r)F(\varepsilon/kT,\varepsilon'/kT)\frac{\omega\tau}{1+\omega^{2}\tau^{2}}. \tag{9.4.17}$$

Here the density of states is assumed to be constant and equal to its value g_F at the Fermi level, $\Psi(\varepsilon, \varepsilon', r)$ is the quantum level correlation function (9.3.8), and we have integrated over the solid angle with account of the orientations of the dipole moments of the pair. With regard to Equation (9.4.3), neglecting the nonexponential dependence of τ, determined by Equation (9.4.15), we may write

$$\tau^{-1} = \tau_0^{-1} \exp(2r/\alpha), \tag{9.4.18}$$

where $\tau_0 \simeq \nu_{\text{ph}}$ and ν_{ph} is the phonon attempt-to-jump frequency, which is typically $\nu_{\text{ph}} \simeq 10^{12}\,\text{s}^{-1}$ [4].

In Equation (9.4.17), we pass from integration with respect to the variables ε, ε' to integration with respect to $u = (\varepsilon + \varepsilon' - 2\varepsilon_F)/2kT$ and $\delta = (\varepsilon - \varepsilon')/kT$

$$\sigma'(\omega) = \frac{\pi e^2 g_F^2 \omega T k}{3kT} \int_{r_\omega}^{\infty} \mathrm{d}r\, r^4 \int_{-\infty}^{\infty} d\delta \int_{-\infty}^{\infty} \mathrm{d}u \Psi(u+\delta/2, u-\delta/2, r), \tag{9.4.19}$$
$$F(u+\delta/2, u-\delta/2)\frac{\omega\tau}{1+\omega^2\tau^2}.$$

If the phonon-assisted hopping distance $\tilde{r}_\omega$ is large compared with the characteristic distance in Equation (9.3.10) that determines the quantum correlation range, we may set $\Psi(\varepsilon, \varepsilon', r) \simeq 1$ (the correlation is taken into account in [34, 41]). In the absence of the interaction between electrons, function (9.4.16) is

$$F\left(u - \frac{\delta}{2}, u + \frac{\delta}{2}\right) = \frac{1}{\cosh(\delta/2)(\cosh(\delta/2) + \cosh(u))} \tag{9.4.20}$$

and we may directly perform the energy integrations in Equation (9.4.19),

$$\psi_0 = \int_{-\infty}^{\infty} \mathrm{d}u \int_{-\infty}^{\infty} \mathrm{d}\delta F\left(u - \frac{\delta}{2}, u + \frac{\delta}{2}\right) \simeq 9.84. \tag{9.4.21}$$

Since in the frequency range under study we have $\omega\tau_0 << 1$, we can integrate with respect to r. Indeed, for $\omega\tau_0 << 1$, the function $\dfrac{\omega\tau_0 \exp(2r/\alpha)}{1+\omega^2\tau_0^2 \exp(4r/\alpha)}$ is sharply peaked at $r = \tilde{r}_\omega$, where $\tilde{r}_\omega$ is the value of r at which $\omega\tau = 1$

$$\tilde{r}_\omega = (\alpha/2)\ln\left(\frac{1}{\omega\tau_0}\right). \tag{9.4.22}$$

We have $\tilde{r}_\omega >> \alpha$ and the width of the peak is of the order of α [5]. This fact makes it possible to replace r by $\tilde{r}_\omega$ in the slowly varying functions of r and perform the r-integration in Equation (9.4.19). Integrating, we arrive at the Austin–Mott formula [4, 40]

$$\sigma'(\omega) = Be^2\alpha^5 g_F^2 kT\omega \ln^4(\nu_{\text{ph}}/\omega), \tag{9.4.23}$$

where $B = \pi^2\psi_0/96 \simeq 0.25$. The condition $\omega\tau_0 << 1$, which is the condition for the validity of expression (9.4.23), is satisfied in a wide frequency range, and the high-frequency limit (Equation 9.4.13) is virtually not reached in the frequency range of interest.

9.4.3 Pair approximation for interacting electrons

Coulomb interaction between the electrons can be taken into account by using expression (9.3.15) for the equilibrium conditional probability $\tilde{f}_{\lambda}^{(\lambda')}$ of filling site λ provided that site λ' is empty. Electron–electron interaction can modify the result obtained in the pair approximation, due to reduction of the number of doubly occupied pairs of close sites [41]. With allowance for electron–electron correlation, we can find the probability of single occupancy of a pair

$$F\left(u-\frac{\delta}{2},u+\frac{\delta}{2},z\right)=\frac{1}{\cosh(\delta/2)(\cosh(\delta/2)+(1/2)\{\exp u+\exp(-(u+z))\})}, \tag{9.4.24}$$

where $z = v_{\lambda\lambda'}/kT$ and $v_{\lambda\lambda'} = e^2/(4\pi\kappa_0\kappa r_{\lambda\lambda'})$ is the intra-pair Coulomb interaction energy (Equation 9.3.16). When calculating the inner integral in Equation (9.4.19), we obtain

$$\int_{-\infty}^{\infty} \mathrm{d}u F\left(u-\frac{\delta}{2},u+\frac{\delta}{2},z\right)=$$
$$\frac{1}{\cosh(\delta/2)\sqrt{\cosh^2(\delta/2)-\exp(-z)}}\ln\frac{1+\sqrt{1-\cosh^{-2}(\delta/2)\exp(-z)}}{1-\sqrt{1-\cosh^{-2}(\delta/2)\exp(-z)}}. \tag{9.4.25}$$

Introducing the function

$$\psi(z)=\int_{-\infty}^{\infty}\mathrm{d}u\int_{-\infty}^{\infty}\mathrm{d}\delta F\left(u-\frac{\delta}{2},u+\frac{\delta}{2},z\right)$$
$$=4\int_{1}^{\infty}\frac{\mathrm{d}y}{y\sqrt{(y^2-1)(y^2-\exp(-z))}}\ln\frac{y+\sqrt{y^2-\exp(-z)}}{y-\sqrt{y^2-\exp(-z)}} \tag{9.4.26}$$

we obtain, instead of Equation (9.4.23),

$$\mathrm{Re}\,\sigma(\omega)=\frac{\pi^2e^2\alpha^5\rho_F^2}{96}kT\psi(v_\omega/kT)\omega\ln^4(v_{\mathrm{ph}}/\omega), \tag{9.4.27}$$

where $v_\omega = e^2/(4\pi\kappa_0\kappa\tilde{r}_\omega)$.

We have $\psi(z) \approx \psi_0$ for $z << 1$ and $\psi(z) \simeq 4z$ for $z >> 1$. For intermediate values of z, the interpolation expression $\psi(z) \approx \psi_0 + 4z$ may be used, giving an error not exceeding 10% in the entire interval of z. Thus, for the real part of the conductivity, we find [41]

$$\sigma'(\omega)=Be^2\alpha^5g_F^2\omega\ln^4(v_{\mathrm{ph}}/\omega)\left(kT+\frac{2e^2}{\pi\psi_0\kappa_0\kappa\alpha}\ln^{-1}(v_{\mathrm{ph}}/\omega)\right). \tag{9.4.28}$$

We see that the frequency dependence of the real part of the conductivity is only slightly changed by the interactions, while the temperature dependence of $\sigma(\omega)$ at low temperatures

$$\left(kT << \frac{2e^2}{\pi\psi_0\kappa_0\kappa\alpha} \ln^{-1}\left(\nu_{\text{ph}}/\omega\right) \right)$$

becomes appreciably weaker due to interactions.

9.4.4 Crossover from phonon-assisted to phononless regime

The ratio of phononless contribution (Equation 9.3.19) to that of phonon-assisted processes (9.4.28) is

$$\frac{\sigma'_{\text{phononless}}(\omega)}{\sigma'_{\text{phonon-assisted}}(\omega)} = \frac{\pi^2}{48B}\left(\frac{r_\omega}{\tilde{r}_\omega}\right)^4 \frac{\hbar\omega + e^2/(4\pi\kappa_0\kappa r_\omega)}{kT + e^2/(\pi\psi_0\kappa_0\kappa\tilde{r}_\omega)}. \tag{9.4.29}$$

At not too high frequencies, the contribution of phonon-assisted hopping to the AC conductivity usually substantially exceeds the contribution of phononless processes. Indeed, the frequency ν_{T} defined by $h\nu_{\text{T}} = kT$ is about 20 GHz at $T = 1$ K. Therefore, for most of the measurements of the AC conductivity, which are made at frequencies below the GHz range, the ratio (9.4.29) is small and phonon-assisted hopping dominates.

On the other hand, at high frequencies and low temperatures, the phononless contribution may become greater than the contribution of the phonon-assisted hopping. Moreover, as discussed in Section 9.3, the phononless contribution is expected to exhibit a crossover from linear to quadratic dependence with increasing frequency.

Recent experiments in the quantum limit ($\hbar\omega > kT$) were performed on Si:B [42] and Si:P [43] samples at high frequencies and low temperatures (from 0.1 to 20 GHz in the range of tens of K for Si:B and from tens of GHz to 1 THz at temperatures of several K for Si:P). In agreement with the approach described in Section 9.3, a crossover from an almost linear to almost quadratic dependence of the AC conductivity was observed. However, it appeared that the crossover from one power to the other was much sharper than that predicted by the theory. A possible reason is the effect of the long-range Coulomb interactions (in particular, responsible for the appearance of the Coulomb gap) [42]. The incorporation of these effects is complicated by the fact that the role of the long-range interactions depends on the hopping distance (this is clear, e.g., from the calculations of the low-temperature heat capacity due to electronic excitations of the localized carriers [45]). A crossover from linear to quadratic dependence was also reported for metallic nanocomposites, where the AC conductivity was measured in a large frequency range (5 Hz–3 THz) at room temperature [44]; this crossover was attributed to the transition from phonon-assisted to phononless regime.

9.4.5 Different tunneling mechanisms

Until now, when considering tunneling hopping, we specifically assumed the transport to be due to inelastic electron tunneling between localized states. The details of the mechanism of the phonon-assisted electron tunneling have been discussed in a number of studies [4, 32]. One of the possible mechanisms suggested by Holstein is polaron hopping [4, 47–49];

he assumed that in the presence of strong electron–phonon interactions, electron transitions between sites occur when the energies of initial and final states (including the lattice polarization energy) coincide due to thermal fluctuations of the atomic configuration. A modification of Holstein's model to describe hopping in the impurity band of doped compensated semiconductors was considered in [50, 51]. These models assume that the energy levels of localized states fluctuate due to lattice vibrations or to a fluctuating Coulomb potential created by hopping charge carriers, and tunneling transitions between localized states occur when the localized levels come in resonance.

Electron tunneling is not the only mechanism capable of explaining a nearly linear frequency dependence of the conductivity, which is quite generally observed in numerous disordered solids. In many systems, the nature of charge carriers may be different (in this context, localized electron pairs [28], bipolarons [52], etc., were discussed). In principle, AC conductivity of glasses can also be due to tunneling transitions of ions [53]; indeed, tunneling of atoms or groups of atoms between close equilibrium states, almost degenerate in energy, are usually involved in the interpretation of low-temperature anomalies of thermal capacity and heat conductivity of glasses [54, 55]. In [56], it has been shown that the structure of chemical bonds in chalcogenide glasses implies a possibility of the existence of such 'tunneling states', related not only to structural defects, but also to a significant fraction of the host atoms (chalcogenes) containing lone pairs. Of course, atoms are assumed to be at least partially ionized so that their motion produces a varying dipole moment and contributes to the current. A power-law frequency dependence of the conductivity is quite generally obtained for the systems in which the relaxation times of the polarizable centers are exponential functions of some random variables (such as hopping distance and energy) with a reasonably smooth and wide distribution.

9.5 MULTIPLE HOPPING REGIME

9.5.1 Frequency-dependent cluster construction

The pair approximation applies only at sufficiently high frequencies where the hopping distance r_ω at a frequency ω is small compared with the characteristic hopping distance r_{h} that determines the DC conductivity. Clearly, at low frequencies we must take into account the possibility that during a half-period of the AC field, the electron makes several hops [39, 57–60]. A regime at which the electrons giving the main contribution to the current make more than one hop during a half-period π/ω is called the *multiple hopping regime.*

In contrast to DC conduction, for $\omega \neq 0$, the electrons have time to be displaced only over finite distances. Accordingly, in this case, the problem is reduced to finding the optimum paths that correspond to a finite hopping displacement of electrons during a finite time interval. Therefore, the shape of the optimum paths depends on the field frequency, i.e., the AC current arising in the system after the application of an AC electric field of frequency ω at different frequencies is determined by the displacements of charge carriers along different paths.

It is convenient to find the displacements of charge carriers under the action of an electric field in a nonstationary hopping problem by reducing the problem to the corresponding bond percolation problem (see [57–61]). In this approach, the hopping current at different frequencies is determined by different characteristic clusters of bonds; hence both the cluster construction and the characteristic cluster size appear to be dependent on frequency. In [57, 58] the ideas of the finite-size percolation theory were used to evaluate the frequency dependence of the hopping current for a system with a large scatter in the transition rates. A somewhat different approach was used in [59, 60], where the network impedance was calculated using a variational procedure for the disconnected network regime, which divides all resistances into two groups, those of the first one being finite and equal to each other and those of the second one being infinite.

To see how the standard arguments of the percolation theory should be modified in the nonstationary case, we consider the linear response of the system to an applied AC electric field described by the rate equation (9.4.1). The rate (9.4.2) of transitions between localized states can be written in the form

$$\Gamma_{\lambda\lambda'} = \Gamma_0 \exp(-\eta_{\lambda\lambda'}), \tag{9.5.1}$$

where

$$\eta_{\lambda\lambda'} = \frac{2r_{\lambda\lambda'}}{\alpha} + \frac{|\varepsilon_\lambda - \varepsilon_F| + |\varepsilon_{\lambda'} - \varepsilon_F| + |\varepsilon_\lambda - \varepsilon_{\lambda'}|}{2kT} \tag{9.5.2}$$

and $\Gamma_0 \simeq w_0$ is the preexponential factor whose dependence on ε_λ, $\varepsilon_{\lambda'}$, and $r_{\lambda\lambda'}$ is weak compared with that of the exponential factor and is usually disregarded. As usual, choosing some value of Γ, we assume that sites λ and λ' are bonded if

$$\Gamma_{\lambda\lambda'} > \Gamma. \tag{9.5.3}$$

For an exponentially large distribution of $\Gamma_{\lambda\lambda'}$, the DC conductivity of the system is determined by the percolation threshold Γ_c corresponding to the appearance of the infinite percolation cluster of conjugated bonds [36, 62].

Consider a finite cluster of bonds and let the minimum transition rate within kth cluster be $\Gamma_k^* > \Gamma > \Gamma_c$. Due to large disorder, the equilibration time for this cluster τ is determined by Γ_k^* (see Appendix 9.1), i.e., $\tau = \tau(\Gamma_k^*)$. Therefore, for clusters with

$$\tau(\Gamma_k^*) < \omega^{-1} \tag{9.5.4}$$

the site occupancies inside the cluster are equilibrated (the ratios of the occupancies of different sites are the same as in equilibrium, although the occupancies themselves are nonequilibrium). We call the clusters for which condition (9.5.4) is satisfied *quasi-equilibrium* clusters. If we define the bonds by condition (9.5.3), all clusters in the system are finite if $\Gamma > \Gamma_c$. Let $\Gamma = \Gamma(\omega)$ be defined by

$$\tau(\Gamma(\omega)) = \omega^{-1}; \tag{9.5.5}$$

then the clusters of bonds are quasi-equilibrium (a frequency-dependent cluster construction). The cluster size increases with decreasing frequency ω. However, at low frequencies (at $\omega < \omega_c$, where ω_c is the crossover frequency from the multiple hopping regime to the regime of conduction over large clusters), the concept of isolated quasi-equilibrium clusters is no longer justified. Indeed, this concept is valid only if the cluster size is not too large so that the minimum rate of transitions inside the cluster substantially exceeds Γ, and the maximum rate of transitions from any of the centers of the cluster to any other site of the system is much lower than Γ. On the other hand, at frequencies higher than some frequency ω_0, the displacement of a charge carrier during a half-period of the field becomes small, and the crossover from the multiple hopping regime to the regime of pair hopping occurs. Thus, in the frequency range [38, 61]

$$\omega_c < \omega < \omega_0 \tag{9.5.6}$$

the dipole polarization of the system, and hence the AC current, is determined by the geometry of quasi-equilibrium clusters; the crossover frequencies ω_c and ω_0 are discussed in Section 9.5.3.

9.5.2 AC current and conductivity

The current at a frequency ω can be expressed in terms of the frequency-dependent size and the number of quasi-equilibrium clusters defined with respect to $\Gamma = \Gamma(\omega)$ in accordance with Equation (9.5.4). Indeed, we may write for the hopping polarization current density

$$j_h(\omega) = i\omega P(\omega) = \frac{i\omega}{\Omega}\sum_k \delta D_k(\omega), \tag{9.5.7}$$

where the sum is taken over the quasi-equilibrium clusters and δD_k is the field-induced variation of the dipole moment of the kth cluster estimated in Appendix 9.1. Using Equation (9.A1.12) and writing the current density as a sum over the cluster sizes, we obtain for the real part of the conductivity

$$\sigma_h'(\omega) = \omega \sum_s n(s) s C_0 l^2(s), \tag{9.5.8}$$

where C_0 is the cluster capacitance per site, $l(s)$ is the linear cluster size, and $n(s)$ is the cluster size distribution function, i.e., the number of clusters of s sites per unit volume (for a given Γ) (the properties of the distribution function $n(s)$ are discussed in Appendix 9.2).

The sum on the right-hand side of Equation (9.5.8) is a sum of the type $\sum_s s^p n(s)$. According to (9.A2.1), the number of large clusters in the system is always small compared with the number of the clusters consisting of a small number of sites; it might seem that, in the absence of the infinite cluster, pairs always give the main contribution to the conductivity. Actually this is not the case, since for $s < s_c$, where s_c is the critical number of sites (Equation 9.A2.2) that diverges at the percolation threshold as $\eta \to \eta_c$, $n(s)$ decays with increasing s as a power-law function, $n(s) \sim s^{-\tau}$, and for $s >> s_c$, it decays exponentially (see Appendix 9.2). Thus, the main contribution to the sum $\sum_s n(s) s^p$ comes from the clusters with small numbers of sites for $p < \tau - 1$; on the other hand, for $p > \tau - 1$, the main contribution comes from the critical sites with $s \simeq s_c$ corresponding to the cutoff, where the power-law decrease of the function $n(s)$ is succeeded by its exponential decay; the sum considered diverges at the threshold as $s_c \to \infty$. Therefore, the number of sites that belong to the finite clusters $\phi = (\eta - \eta_c/\eta_c \to 0$ remains finite at the critical point where $\phi = (\eta - \eta_c)/\eta_c \to 0$ and the function $S(\phi)$ defined by (9.A2.3) behaves as $S(\phi) \simeq \phi^{-\gamma}$. For $p > \tau - 1$, we estimate

$$\frac{\sum_s s^p n(s)}{\sum_s sn(s)} \simeq S(\phi) s_c^{p-2} \approx \phi^{-\gamma-(p-2)\Delta}. \tag{9.5.9}$$

The sum in Equation (9.5.8) has a form (9.5.9) with $p = 1 + 2\nu/\Delta > \tau - 1$ and, using (9.5.9), we find (up to a constant factor of the order of unity)

$$\sigma_h'(\omega) = \omega C_0 \alpha^{-1} \eta_c^{\xi-1} (\eta_c - \eta)^{-\xi}, \tag{9.5.10}$$

where

$$\xi = \gamma + 2\nu - \Delta = \Delta - \nu \tag{9.5.11}$$

(with regard to Equation 9.A2.8). Substituting here the indices Δ and ν ($\Delta = 2.2$ and $\nu = 0.9$ [36]), we estimate $\xi = 1.3$.

According to Equation (9.5.1), we have $\eta_c - \eta = \ln(\Gamma/\Gamma_c)$, where $\Gamma_c = \Gamma_0 \exp(-\eta_c)$ is the transition rate corresponding to the percolation threshold that determines the DC conductivity. The quantity Γ can be expressed in terms of the conductivity,

$$\Gamma(\omega) = \Gamma_c = \sigma_h'(\omega)/\sigma_h(0), \tag{9.5.12}$$

where $\sigma_h(0) = (e^2\Gamma_c/(kTl_{c0})) \exp(-\eta_c)$ [36, 38] and Equation (9.5.10) assumes the form

$$\sigma_h'(\omega) \ln^{\xi}(\sigma_h'(\omega)/\sigma_h(0)) = C_0 \alpha^{-1} \eta_c^{\xi-1} \omega. \tag{9.5.13}$$

Note that C_0 may depend on temperature; thus for Mott's variable range hopping regime, the sites that belong to a cluster have energies in the range of the width of the order of $kT\eta_c = kT(T_0/T)^{1/4}$ and the sites that contribute to the capacitance Equation (9.A1.7) lie in the layer kT near the Fermi level so that $C_0 \sim \eta_c^{-1} = (T/T_0)^{1/4}$.

We see that in the multiple hopping regime, the frequency dependence of $\sigma_h'(\omega)$ follows a sublinear fractional power dependence, just as in the regime of pair hopping. Thus, although the power exponents in the two cases may be slightly different, it is quite difficult to experimentally establish the crossover between these regimes.

9.5.3 Frequency range for the multiple hopping regime

Expression (9.5.13) for the conductivity becomes inapplicable at low and high frequencies outside the range $\omega_c < \omega < \omega_0$, (9.5.6). At low frequencies, it fails and does not give a correct value of the conductivity at $\omega \to 0$, since the assumption of the key role of one bond in a cluster is no longer valid for large clusters of size exceeding $l \sim l_{c0} \simeq l_0 \eta_c^{\nu}$, i.e., for $\eta_c - \eta < 1$ [58]. On the other hand, the scaling description near the percolation threshold can be used only at not too high frequencies where $\eta_c - \eta < \zeta\eta_c$ (here ζ is a number of order unity). Thus the frequency interval where expression (9.5.12) is valid is determined by the conditions

$$1 << \eta_c - \eta < \zeta\eta_c. \tag{9.5.14}$$

It is asymptotically satisfied only for $\eta_c >> 1$ (*the extreme disorder limit*, see [26]).

From (9.5.14), we can estimate the frequencies ω and ω_c which determine the frequency range where the multiple hopping regime can be effective. For the frequency ω_0 of crossover from the multiple hopping regime to pair hopping, we find (in agreement with [61])

$$\omega_0 \simeq \Gamma_0 \exp\{-(1-\zeta)\eta_c\}, \tag{9.5.15}$$

where, for Mott hopping, $\eta_c = (T_0/T)^{1/4}$ and T_0 is the Mott parameter. The frequency ω_c of the crossover to the low-frequency region is estimated as $\omega_c \simeq a\Gamma_c$, where a is a number of the order of unity.

We can write Equation (9.5.13) in the form of the scaling relation (9.2.1) [58],

$$\left(\frac{\sigma'(\omega)}{\sigma(0)}\right)\ln^{\xi}\left(\frac{\sigma'(\omega)}{\sigma(0)}\right) = \frac{\omega}{\omega_s}, \tag{9.5.16}$$

where

$$\omega_s = \frac{\alpha}{C_0\eta_c^{\xi-1}}\sigma(0) \tag{9.5.17}$$

is the scaling frequency.

Equation (9.5.16) predicts that if we approximate the conductivity by a power law (9.1.18), we find that the exponent s appears to be frequency and temperature dependent, although for $\sigma(\omega) >> \sigma(0)$ this dependence is weak. Indeed, we find

$$s = \frac{d\ln\sigma}{d\ln\omega} = \left(1 + \frac{\xi}{\ln[\sigma/\sigma(0)]}\right)^{-1}. \tag{9.5.18}$$

It follows that in the range where $\sigma(\omega) >> \sigma(0)$, s is slightly less than unity, slowly (logarithmically) increasing with frequency (due to the increase in σ) and with decreasing temperature approaching unity. This behavior is in qualitative agreement with the experimental data for virtually all disordered materials studied.

Equations of the type (9.5.16) (with different values of ξ) appear in many descriptions of the hopping conductivity; in particular, they were obtained using different versions of cluster theories [58, 59], effective-medium models [63] and the related self-consistent approach [64], percolation path approximation [65], etc. One might expect that cluster theories based on percolation theory are better suited for the description of the frequency dependence of the conductivity at intermediate frequencies for systems with extremely large disorder, whereas effective-medium theories provide a better description if the scatter of the transition rates in the system is not very large. However, the results for the frequency dependence of the conductivity for different models are quite similar and are weakly sensitive to specific features of the systems considered. Figure 9.3 shows the scaled real part of the conductivity $\sigma_h'(\tilde{\omega})/\sigma_h(0)$ as a function of the scaled frequency $\tilde{\omega} = \omega/\omega_s$ in the intermediate frequency region for various theoretical models (reproduced from [15]). The scaling frequencies are: $\omega_s = \tau_b^{-1}$, where τ_b is the bulk conductivity relaxation time (the bulk model with high-conductivity inclusions [66]); $\omega_s = W_c N^{-1/3}\alpha$, where W_c is the critical hopping probability at the percolation threshold and N is the site density (effective medium approximation for the r-percolation problem [63]); ω_s is the minimum hopping rate (continuous-time random walk model [67]); $\omega_s = 100\sigma'(0)kT/e^2\alpha$ (extended pair approximation [68]); and ω_s is given by Equation (9.5.17) (cluster model [58]). Thus, different models can provide a reasonably good description of the experimental data in the intermediate frequency domain. An example of the comparison of the experimental data for amorphous hydrogenated silicon in the low-frequency region, where the multiple hopping regime is expected to dominate, with Equation (9.5.16) is shown in Figure 9.2 [16]. After the subtraction of the contribution of delocalized carriers, the frequency dependence of the conductivity in the frequency range of 10^{-2}–10^4 Hz is well described by the universal equation (9.5.16).

Expression (9.5.17) for the scaling frequency ω_s is obtained for the case where the transition from the power-law to DC conductivity is not accompanied by the change in the conduction mechanism (e.g., from hopping to band conduction). We see that ω_s is proportional to $\sigma(0)$, the other factors in (9.5.17) being weakly (nonexponentially) temperature dependent, and its temperature dependence is expected to be almost the same as that of the DC conductivity. In Section 9.2 we argued that if the conduction due to two parallel channels is described by (9.1.21), the main temperature dependence of the crossover frequency (9.1.22) is somewhat different from that of the DC conductivity (the exponent s is usually not very different from unity). Note that the transition from the power-law to DC conduction is somewhat smoother in the case where there is no change in the conduction mechanism. This is schematically illustrated by curve c in Figure 9.1 plotted using the scaling equation (9.5.16).

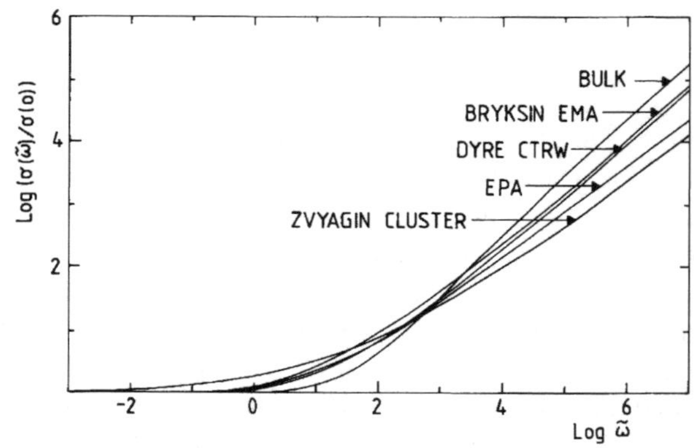

Figure 9.3 Frequency dependence of the conductivity for different theoretical models (redrawn from [15] with permission from Elsevier)

9.6 CLASSICAL HOPPING

9.6.1 Pike's model

The mechanism of charge transfer in the case of heavy charge carriers (ions, paired electrons at the valence-alternation pairs in chalcogenides, polarons and bipolarons, etc.) is expected to be classical hopping over the potential barriers, rather than tunneling. A model for classical hopping over the barriers created by overlapping Coulomb-like potentials centered at a distance R from one another was suggested by Pike [69] (this situation is schematically shown in Figure 9.4). The lowering of the potential barrier due to the overlap of the potential energy curves is

$$\varepsilon_0 - \varepsilon_{\rm m} = \frac{e^2}{\pi\kappa_0\kappa R}, \tag{9.6.1}$$

where ε_0 is the energy level corresponding to free carrier motion and $\varepsilon_{\rm m}$ is the maximum of the potential energy in the region between the Coulombic potential wells (sites). We see that the barrier height in the double-well system depends on the intersite distance R and it randomly fluctuates for random positions of the Coulomb centers (sites). Let ε_λ and $\varepsilon_{\lambda'}$ be the ground state energies in the two wells (sites) and let $\varepsilon = (\varepsilon_\lambda + \varepsilon_{\lambda'})/2$. In this model, the contribution of an isolated double well to the conductivity (after averaging over the angles) can be written as ([5], compare with Section 9.4)

$$\sigma_{\rm i}'(\omega) = \frac{e^2R^2\omega}{12kT\cosh^2(\xi_{\lambda\lambda'}/2kT)}\frac{\omega\tau_{\lambda\lambda'}}{1+\omega^2\tau_{\lambda\lambda'}^2} \tag{9.6.2}$$

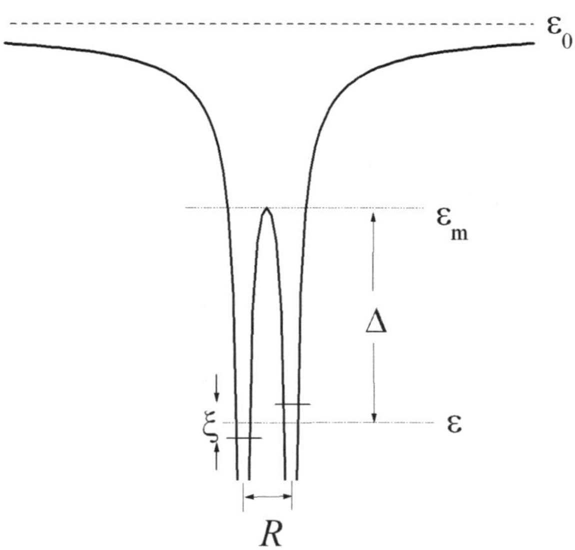

Figure 9.4 Double potential well corresponding to overlapping Coulomb-type potentials

where $\xi_{\lambda\lambda'} = \varepsilon_\lambda - \varepsilon_{\lambda'}$ is the asymmetry of the double well (the difference between the ground-state energies of the wells),

$$\tau_{\lambda\lambda'} = (W_{\lambda\lambda'} + W_{\lambda'\lambda})^{-1} = (\tau_0/2)\exp(u)\cosh^{-1}(\xi_{\lambda\lambda'}/2kT) \tag{9.6.3}$$

and $u = (\varepsilon_m - \varepsilon)/kT$ (this expression holds for $u \pm \xi/2kT >> 1$). Generally, the factor τ_0 may depend weakly on the intersite distance R, but this dependence is usually disregarded.

The conductivity is given by the sum of the contributions Equation (9.6.2) of separate double-wells,

$$\sigma_h'(\omega) = \frac{e^2 N_p \omega}{12kT} \int_{-\infty}^{\infty} \frac{d\xi}{\cosh^2(\xi/2kT)} \int_0^{\infty} dR R^2 p(R,\xi) \frac{\omega\tau}{1+\omega^2\tau^2}, \tag{9.6.4}$$

where N_p is the concentration of singly occupied double-wells and $p(R,\xi)$ is the probability density of having an interwell distance R and the energy separation between the levels ξ (see Figure 9.4). In Equation (9.6.4), τ depends on u and, according to (9.6.1), u is uniquely related to the intersite distance R. Following Pike, we assume that $p(R,\xi) = p(R)p(\xi)$ and, since the distribution of Δ is not known, we take a uniform distribution

$$p(\xi) = (2\xi_0)^{-1}\theta(\xi_0 - |\xi|), \tag{9.6.5}$$

where $\theta(\xi)$ is a step function. Neglecting correlation in the spatial distribution of the sites (wells), we have that the probability of having the nearest neighbor at a distance in the interval R, $R + dR$ is $p(R)dR = 4\pi NR^2\exp(-4\pi NR^3/3)$, where N is the density of wells. We assume that the distance between the sites of the pairs that give the main contribution to the conductivity is smaller than the average intersite separation $N^{-1/3}$ so that $\exp(-4\pi NR^3/3) \simeq 1$. According to Equation (9.6.1), we may write

$$R^4 dR = \left(\frac{e^2}{\pi\kappa_0\kappa kT}\right)^5 (u_0 - u)^{-6}\, du, \tag{9.6.6}$$

where $u_0 = \Delta/kT$ and $\Delta = \varepsilon_0 - \varepsilon$. In order to estimate the integral in (9.6.4), it is convenient to use the relation $(1 - x)^{-6} \simeq \exp(6x)$, which is satisfied for $x << 1$. Then Equation (9.6.4) yields

$$\sigma_h'(\omega) = \left(\frac{e^2}{\pi\kappa_0\kappa\Delta}\right)^5 \frac{\pi^2 N N_p b(\xi_0/2kT)}{6(2\tau_0)^{\xi}\Delta\sin(\pi s/2)}\omega^s, \tag{9.6.7}$$

where,

$$s = 1 - \zeta, \tag{9.6.8}$$

$$\zeta = 6kT/\Delta \tag{9.6.9}$$

and

$$b(x)=\frac{1}{x}\int_0^x \frac{dt}{\cosh^{2-\zeta} t}. \tag{9.6.10}$$

Thus, the exponential distribution in transition probabilities due to the variation of barrier heights yields a fractional-power-law frequency dependence of the conductivity with an exponent somewhat smaller than unity. However, it follows from Equation (9.6.7) that a frequently observed weak temperature dependence of the AC conductivity can be reconciled with Pike's theory only if $\xi_0/2 << 1$, i.e., if the asymmetry of the double-wells giving the main contribution to the conductivity is small. Pike's theory also predicts that, with decreasing temperature, the exponent s, determined by Equations (9.6.8) and (9.6.9), linearly approaches unity.

One of the characteristic features of Pike's model is a linear temperature dependence of the exponent s. Such a dependence was indeed observed for some disordered materials (e.g., for evaporated silicon monoxide films [17]. On the other hand, for other materials (e.g., for silver vanadate glasses [70]), the results disagree with Pike's model.

Pike's model is attractive, since it explicitly specifies the origin of disorder in barrier heights and allows one to calculate the barrier height distributions. However, it is expected to hold for systems where the interwell distances are sufficiently large so that the interaction of an electron with the Coulomb potential can be described macroscopically. Thus it can hardly be directly applied to ionic conduction, where the distance between the wells is of the order of the interatomic spacing. In this case, the appearance of random potential barriers is intimately related to the local structure (bond configuration) in the material. A number of methods can provide information about the local structure in the neighborhood of the hopping ions (EXAFS, XPS, Raman spectroscopy, NMR, etc.). Thus, in [71], using the information on the inter-ion distances, it was possible to estimate Coulomb barrier heights; for alkali silicate and alkali allumosilicate glasses, they appeared to be about 0.1–0.2 eV. These values are manifestly below the measured activation energies for alkali diffusion (about 0.6 eV), indicating that for these structures variations in the Coulomb energies are not the main factor responsible for creation of random barriers.

9.6.2 Random barrier models for ionic conduction

Classical activated ion hoping over random potential barriers is often assumed to describe the properties of ionically conducting solids (or of solids with other heavy charge carriers of nonelectronic type), since ion tunneling probabilities, at least at not too low temperatures, are extremely small. A standard model used to describe transport in disordered solids with ionic-type conduction assumes the existence of a random potential landscape with ions randomly distributed over the local minima ('sites') of the potential energy, and the conduction is due to activated transitions of ions over the barriers separating these sites (Figure 9.5). A random barrier model was earlier applied to polycrystalline semiconductors and ferroelectric ceramics (see, e.g., [72, 73], where a percolation approach to DC conduction was used). The model in which the site energies are assumed to be equal whereas the barrier heights are random is often called the *random barrier model*; this model is also often referred to as a symmetric hopping model, since the barrier heights for forward and backward hopping between two neighboring sites are the same [26]). A more general model that allows for the fluctuations of both the site energies and the barrier heights is called the *random energy model* [74, 75]; in the following, we admit this terminology. The random barrier model [76] was successfully used to describe both the DC effects (mixed alkali effect, i.e., nonlinear dependence of the conductivity when replacing one alkali oxide by another, and highly nonlinear dependence of the DC conductivity on the fraction x of modifying ions in single-modifier glasses [74, 75, 77]) and the AC conductivity in ionic conductors [74, 75]. Different models for the conductivity of ionic solids were suggested assuming substantial structural relaxation that accompanies ion hopping transitions (the dynamic structure model [78] and related jump relaxation model [23]). However, there are indications (in particular, the absence of the mixed-alkali effect in ion-exchanged glasses) that structural relaxations usually play a minor role [74] and the rigid barrier model provides an adequate basis for understanding the above effects.

Accordingly, we discuss the model where the interaction of mobile ions with the glass matrix can be characterized by a rigid random potential landscape schematically shown in Figure 9.5 [74, 75]. Although, as noted above, some information on the properties of the potential landscape in ionic glasses can be obtained, actually little is known about its exact form, its statistical characteristics and its relation to the composition and structure of the material. In particular, the landscape can be strongly affected by the modifying ions introduced into the glass matrix during preparation. Therefore, it is reasonable to introduce some *ad hoc* assumptions about the properties of the potential (to be checked by comparing the predictions of the theory with experimental observations). A simplest assumption corresponds to the random barrier model, where all minima of the potential energy landscape (sites) are the same, whereas the heights of the barriers have a broad distribution. For the random energy model, one should take into account the statistics of site population, whereas for the simple random barrier model in thermal equilibrium, the site occupancy is simply equal to the fraction x of the sites filled by mobile ions. For the random energy model, the equilibrium site occupancies are determined by a Fermi-type distribution with the Fermi level ε_F. A modification of expression (9.5.2) for the exponent of the transition rate (9.5.1) is

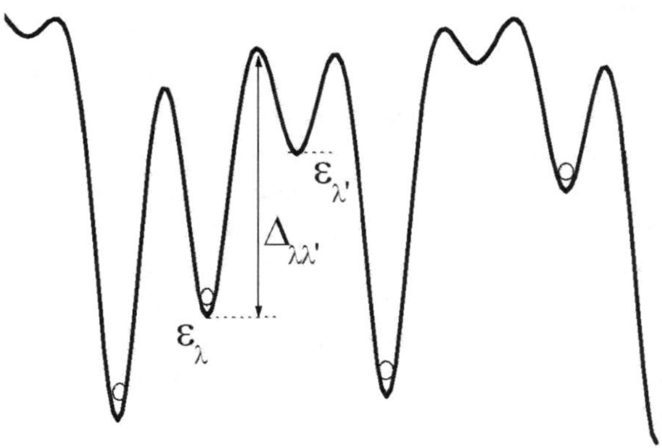

Figure 9.5 Potential landscape for mobile ions along an arbitrary direction in the random energy model (schematic)

$$\eta_{\lambda\lambda'} = \frac{|\varepsilon_\lambda - \varepsilon_F| + |\varepsilon_{\lambda'} - \varepsilon_F| + |\varepsilon_\lambda - \varepsilon_{\lambda'}|}{2kT} + \frac{\Delta_{\lambda\lambda'}}{kT}, \tag{9.6.11}$$

where ε_λ and ε_λ are the energies of the minima and $\Delta_{\lambda\lambda'}$ are the potential barrier heights (Figure 9.5). This has a form similar to that for tunneling problems; accordingly, for the DC case and in the low-frequency region, a standard percolation approach [36, 38] may be used.

Assuming that the conduction mechanism is the same in the entire frequency domain and applying the random energy model, one can consistently explain the main features of both DC and AC conduction of ionic solids [74, 75]. In this model, a highly nonlinear dependence of the DC conductivity on the fraction x of modifying ions in single-modifier glasses may be due the variation of the activation energy related to the shift of the Fermi energy ε_F. Thus, for an exponential density of localized states

$$g(\varepsilon) = (1/\varepsilon_t)\exp(\varepsilon/\varepsilon_t) \tag{9.6.12}$$

with energy scale ε_t we have

$$x \cong \int_{-\infty}^{\varepsilon_F} g(\varepsilon)\mathrm{d}\varepsilon, \tag{9.6.13}$$

and the DC conductivity is

$$\sigma(0) \sim x^{\varepsilon_t/kT}. \tag{9.6.14}$$

A specific form of this dependence may be different for a different density of states, but it is clear that this dependence may be highly nonlinear [74].

Furthermore, the mixed-alkali effect in DC conduction can be also explained by the random energy model [77]. It follows from the fact that ions of different sizes can be incorporated into a random glass structure more easily, geometrically, than ions of the same size. Hence, the binding energy is greater, and the Fermi level ε_F is lower in the mixed-alkali system than in the single-alkali one. Thus the activation energy appears to be greater for a mixed system, provided that the structure of barriers and the value of ε_c are the same as in the single-alkali glass.

It should be noted that the mixed-alkali effect is also present in the AC conduction [79, 80], although its magnitude decreases with increasing frequency. The presence of this effect in the AC conductivity at low frequencies is probably due to the same mechanisms as for the DC case which indicates that the conductivity is controlled by hopping in large clusters, i.e., that the multiple hopping regime postulated in [74] for ionic conductors is actually realized.

The origin of the composition-dependent scaling (9.2.4) also follows from the theory [74] using the energy barrier model in the region of multiple hopping. If the DOS function $g(\varepsilon)$ is steep enough (e.g., like Equation 9.6.12), then most of the carriers occupy localized states with energies close to the Fermi energy ε_F. Then one obtains that in (9.5.13), (9.5.17), the capacitance per site C_0, defined by (9.A1.7), (9.A1.14), is $C_0 \sim x$. This gives rise to the scaling relation of the form (9.2.4).

9.6.3 Nearly constant loss

The random barrier model makes it also possible to understand the origin of the nearly constant loss discussed in Section 9.2. Let us consider a system, where the disorder is due to random variations of barrier heights, and assume that the displacement in each hop does not vary much (this is the energy disorder in the absence of appreciable variations of site positions). In Section 9.4, we have seen that the logarithmic frequency factors responsible for sublinear frequency dependence of the conductivity appear due to the fact that the magnitude of the hopping distance (factor r^4 in the integrand in Equation 9.4.17) is determined by the peak of the last factor in the integrand and is dependent on frequency. The situation is different for the case of energy disorder in the absence of appreciable variations of the hopping distance (*fixed-range hopping*). In this case, $r \simeq R_0$ and expression (9.6.4) is replaced by (cf. [53])

$$\sigma_{\rm h}'(\omega) = \frac{e^2 N_{\rm p} R_0^2 \omega}{12kT} \int_{-\infty}^{\infty} \frac{d\xi}{\cosh^2(\xi/2kT)} \int_0^{\infty} d\Delta p(\Delta,\xi) \frac{\omega\tau}{1+\omega^2\tau^2}, \tag{9.6.15}$$

where τ is given by Equation (9.6.3). The exponential dependence of τ on the random variable Δ makes the integrand in Equation (9.6.15) sharply peaked. As in Section 9.4, we see that the main contribution comes from the double-wells with asymmetries $\xi\delta 2kT$ and with barrier heights determined by the condition $\omega\tau \simeq 1$, i.e., with

$$\Delta_\omega = kT \ln(1/\omega\tau_0). \tag{9.6.16}$$

The result of integration in (9.6.15) is virtually independent of frequency and we obtain a conductivity proportional to the first power of frequency, i.e., just the dependence characteristic of the nearly constant loss behavior. The nearly constant loss contribution described by this approach is expected to become important at high frequencies and low temperatures (see Section 9.6.2), where the regime of near-neighbor fixed-range hopping is established.

The model discussed is essentially very similar to the model of asymmetric double-well potentials with distributed parameters [30], often used for the interpretation of the nearly constant loss not only in glassy ionic conductors, but in glasses without mobile ions [14] as well. It should be noted, however, that in our model, the asymmetry of double-wells is not essential, the contribution (Equation 9.6.15) corresponding to the nearly constant loss remains finite at low δ. Other models were also used to describe the nearly constant loss in disordered solids involving the processes of fast structural relaxation with relaxation times shorter than the ionic hopping time scale; one of these models is the cage model, assuming polaron-like trapping of the carrier after a hop [81].

Expression (9.2.5) is based on the assumption that the dispersive fractional-power behavior of the conductivity and nearly constant loss are due to different dynamic processes. An alternative standpoint is that both are due to hopping motion of the mobile ions, and one expects a continuous increase in the apparent in exponent to 1.0, in accordance with the above discussion of the random energy model that would predict a transition from distant-neighbor to fixed-range ion hopping with increasing frequency and decreasing temperature. This standpoint is confirmed by the observations on numerous ion-conducting glasses in a wide range of compositions [82]; in particular, a direct correlation between the characteristic frequencies ω_{cr} and $\omega_{NCL} = B/\sigma_D$ in Equation (9.2.5) was established in high-temperature measurements. At the same time, measurements at temperatures below 100 K, where ion-conducting glasses also exhibit a nearly constant loss behavior of the AC conductivity, but the slope of the conductivity curves and the temperature dependence of the conductivity vary with glass composition, indicate the presence of a different mechanism of nearly constant loss [83] (at temperatures below approximately 100 K, the frequency and temperature dependence of the conductivity cannot be described by Equation (9.2.5) [84]).

9.7 CONCLUSIONS

Fractional-power dispersive behavior of the AC conductivity at relatively low frequencies (usually below the GHz range) is a universal feature of disordered solids, which has been extensively studied in various systems for many years. A striking similarity in the AC behavior is observed for many different classes of disordered solids having different structures, different DC transport properties (both semiconducting and insulating), different charge carriers and charge transport mechanisms, etc. The universality makes it possible to describe the observed effects in a unified manner using scaling relations and collapsing numerous data onto a single master curve. Due to this universality, the understanding of the related processes in solids is of fundamental interest.

On the other hand, the universality implies that the observed effects are weakly sensitive to specific microscopic features of the systems under study, as well as to the assumptions of the models used for the interpretation of the experimental data. Therefore, the elucidation of the underlying mechanisms and the corresponding processes from the studies of the frequency response appears to be quite difficult. Quite often, different models based on fundamentally different assumptions about the nature of the system provide an equally satisfactory description of the observed effects. The result is that there may exist various models which, although apparently different, predict a similar behavior of the AC response, relaxations, etc., indicating a fundamental similarity of time-dependent processes in disordered systems based on extremely large scatter in the characteristic times of the elementary processes. To obtain specific information about a system, one must study finer features of the behavior of the AC response, in particular, the nonunversality (or deviations from universality).

We have described some of the existing models of the AC conduction in disordered solids disregarding some of the aspects that might appear to be important for real materials. Thus we took account of the electron–electron interaction only by introducing a direct interaction of the charge carriers in isolated pairs; it was virtually disregarded when considering the multiple hopping regime. However, the role of long-range Coulomb interactions that give rise to the Coulomb gap effects [36] in the low-temperature hopping conductivity is not yet quite clear, although one may assume that these effects are of importance in the low-frequency range at low temperatures. In addition, charge transfer is also possible by correlated transitions of several charge carriers (e.g., sequential or simultaneous transitions of several electrons or tunneling transitions of atomic configurations). These multi-electron effects deserve further study. Nevertheless, even the models based on a very simplified description of interaction effects appear to be quite effective. Let us summarize some of the conclusions of these models.

The main features of the frequency dependence exhibited by the quasi-universal shape of the scaled master curves are described by most of the models. In most cases, an increase in the AC conductivity with frequency indicates that conduction is by charge carrier hopping between localized states. The above discussion has shown, however, that the AC conductivity of disordered systems (impurity semiconductors; amorphous semiconductors; semiconducting, ion-conducting, and insulating glasses; polymers; organic solids; granular and composite materials) is not always described by a simple superposition of the DC contribution and a power-law contribution of the type described by Equation (9.1.18). In Figure 9.6 we schematically show various conductivity regimes corresponding to different frequency regions (not all of them are necessarily observed for the same material). In the region of very low frequencies the conductivity differs only slightly from the DC conductivity (depending on the system, it may be extended-state or hopping conduction). In this frequency range, extended-state (Drude) conductivity is virtually frequency independent, and the conductivity increase with frequency is related to the hopping contribution. If the DC conductivity is due to variable-range hopping, the flat region corresponds to the conditions where the carrier displacement during a half-period exceeds the characteristic scale l_{c0} (the characteristic value of the correlation length) of the percolation cluster of bonds for the DC problem) [36]. Actually, finer features of frequency dependence of the real part of the conductivity $\sigma'(\omega)$ in the low-frequency domain are important, since they determine the behavior of the dielectric loss $\kappa_0\kappa''(\omega) = \sigma'(\omega)/\omega$. If the increase in $\sigma'_h(\omega)$ in this range is superlinear, $\kappa''(\omega)$ increases with frequency giving rise to a loss peak at a frequency of about ω_c, where the crossover to a highly dispersive behavior occurs. It should be noted that the function $\sigma'_h(\omega)$ is nonanalytic at $\omega = 0$ and cannot be expanded in powers of ω; this situation has been analyzed in [85], where the relation $\sigma'_h(\omega) - \sigma'_h(0) \sim \omega^{3/2}$ has been obtained, predicting the existence of the loss peak. Further analysis is, however, needed to establish the role of the fractal structure of the percolation cluster.

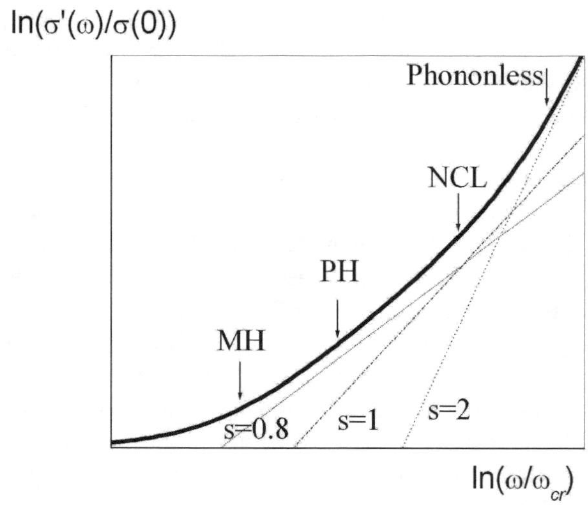

Figure 9.6 Various regimes for the AC conductivity, schematically (MH: multiple hopping, PH: pair hopping, NCL: nearly constant loss)

At frequencies where the carrier displacement becomes smaller than the value of the correlation length l_{c0} that establishes the scale of the backbone cluster [36], a crossover to the multiple hopping regime (indicated by the MH arrow) occurs. We may expect that this crossover is smoother if there is no change in the conduction mechanism in the crossover region. With further increase in frequency, crossover from the multiple hopping to pair hopping regime (PH in the figure) occurs. In both regions corresponding to multiple and pair hopping, the frequency dependence of the conductivity obeys a sublinear fractional-power law with slightly different exponents, so that the experimental proof of this crossover is quite difficult (the multiple hopping regime is characterized by a somewhat different, (although nonexponential, see Equation 9.5.13), temperature dependence of the conductivity). At higher frequencies corresponding to shorter hopping distances (smaller than the localization radii of the states), the conductivity controlled by tunneling between localized states may saturate Equation (9.4.8); this is shown by the dotted curve in Figure 9.6. Furthermore, a region of the linear frequency dependence of the AC conductivity (the nearly constant loss, NCL in the figure), which can extend over many decades, is observed in many materials. This region is usually associated with fixed-range hopping corresponding to nearest-neighbor transitions between localized states with transition rates randomly distributed over a very broad range. It is expected that this situation may be realized for classical hopping, where the hopping distance is determined by the structure of the material and a broad distribution of the transition rates is due to the fluctuations of the classical barrier heights. The nearly constant loss is indeed usually observed in ion-conducting solids (and in some nonionic glasses), where the conduction is expected to be due to classical hopping; on the other hand, this mechanism is not expected for tunneling impurity conduction in doped semiconductors. At still higher frequencies (in the MHz and THz range, and at low temperatures, see Section 9.3) a crossover from phonon-assisted to phononless hopping accompanied by resonance absorption of the quanta of the AC field can be observed. Note that the frequency dependence of the phononless contribution to the conductivity can change from approximately linear to approximately quadratic with increasing frequency.

Thus we see that many different regimes give rise to a power-law frequency dependence of the conductivity with power exponents that typically lie in the range $0.7 < s \leq 1$. It follows that in order to obtain information about the nature of charge carriers and the dominant processes, to the studies of the frequency dependence of the conductivity should not be restricted to establishing the power-law dependence and simple universality. The deviations from the universality such as the variation of the exponent s with temperature and composition, the shape of the curves of the scaled conductivity near the crossover to the low frequency region, and the variation of the crossover frequency ω_{cr} should be analyzed in detail together with other effects that can shed light on the nature of charge carriers and the mechanisms of charge relaxation (e.g., the NMR data for ionically conducting solids).

APPENDIX 9.1 FREQUENCY RESPONSE OF A FINITE ISOLATED CLUSTER

Let us consider a finite cluster of bonds with a large scatter of transition rates corresponding to different bonds. As a first approximation, the relaxation of site occupancies can be considered, having set all transition rates in the cluster, except for the minimum one Γ^*, equal to infinity [38, 58]. In such a way we arrive at the equivalent circuit of a finite cluster consisting of the two parts Λ_1 and Λ_2 connected by the key bond Γ^*. Let a weak electric field E of frequency ω be applied to the system. Since inside each of the subsystems Λ_1 and Λ_2 the equilibration times are short compared with the equilibration time for the whole cluster, the subsystems can be characterized by quasi-Fermi levels ε_{F1} and ε_{F2}. Accordingly, the variation of average occupancy of site λ belonging to subsystem Λ_i (i = 1,2) is

$$\delta f_\lambda^{(i)} = n_F(\varepsilon_\lambda + V_\lambda - \delta\varepsilon_{F,i}) - n_F(\varepsilon_\lambda) \approx \frac{dn_F(\varepsilon_\lambda)}{d\varepsilon_\lambda}(V_\lambda - \delta\varepsilon_{F,i}). \qquad (9.A1.1)$$

The frequency dependence of the quantities $\delta f_\lambda^{(i)}$ is determined by the shift of the corresponding quasi-Fermi level, and change in the number of particles δn_i in subsystem Λ_i is

$$\delta n_i = \sum_{\lambda\in\Lambda_i}(V_\lambda - \delta\varepsilon_{F,i})\frac{dn_F(\varepsilon_\lambda)}{d\varepsilon_\lambda}. \qquad (9.A1.2)$$

Since the total number of particles in an isolated cluster is conserved, we have $\delta n_1 = -\delta n_2$, and the quantity δn_1 is determined by the rate equation for transitions through the key bond,

$$\frac{d\delta n_1}{dt} = -\frac{\Gamma^*}{kT}\left\{V_a - V_b + \frac{kT\delta f_a}{n_F(\varepsilon_a)(1-n_F(\varepsilon_a))} - \frac{kT\delta f_b}{n_F(\varepsilon_b)(1-n_F(\varepsilon_b))}\right\}. \qquad (9.A1.3)$$

Here the indices a and b denote the centers lying at the ends of the key bond, a $\in \Lambda_1$ and b $\in \Lambda_2$. Using the relation

$$\delta f_a = \frac{dn_F(\varepsilon_a)}{d\varepsilon_a}(V_a - \delta\varepsilon_{F,1}), \qquad (9.A1.4)$$

and a similar relation for δf_b, we can express δf_a, δf_b in terms of $\delta\varepsilon_{F,1}$, $\delta\varepsilon_{F,1}$ and then, using (9.A1.2), in terms of δn_1. The equation obtained for δn_1 has the form

$$\frac{d\delta n_1}{dt} = \frac{eE\bar{x}}{kT}\Gamma^* - \frac{\delta n_1}{\tau}, \qquad (9.A1.5)$$

where $\bar{x} = \bar{x}_1 - \bar{x}_2$,

$$\bar{x}_i = \frac{e^2}{kTC_i}\sum_{\lambda\in\Lambda_i} x_\lambda n_F(\varepsilon_\lambda)(1-n_F(\varepsilon_\lambda)) \qquad (9.A1.6)$$

is the position of subcluster i,

$$C_i = \frac{e^2}{kT}\sum_{\lambda\in\Lambda_i} n_F(\varepsilon_\lambda)(1-n_F(\varepsilon_\lambda)) \qquad (9.A1.7)$$

is the subcluster capacitance

$$\tau = \frac{kT}{e^2}\frac{C}{\Gamma^*}, \qquad (9.A1.8)$$

is the relaxation time for the cluster considered and C is the total cluster capacitance

$$C^{-1} = C_1^{-1} + C_2^{-1}. \qquad (9.A1.9)$$

In the frequency domain, the solution of Equation (9.A1.6) assumes the form

$$\delta n_1(\omega) = \frac{eE\bar{x}}{kT}\Gamma^* \frac{\tau}{1+i\omega\tau}. \qquad (9.A1.10)$$

Using the representation of the random resistor–capacitor network, we can say that the charge carrier relaxation time in a finite cluster is equal to the time constant of the equivalent circuit containing the series connection of the key resistance and the capacitances of the parts of the cluster located on different sides of the key bond.

The variation of the dipole moment of the cluster at frequency ω is

$$\delta D(\omega) = e\sum_\lambda x_\lambda \delta f_\lambda(\omega) = -\frac{e}{kT}\sum_i \sum_{i\in\Lambda_i} x_\lambda (V_\lambda - \delta\varepsilon_{F,i}) n_F(\varepsilon_\lambda)(1-n_F(\varepsilon_\lambda)) \qquad (9.A1.10)$$
$$= \frac{1}{e}\sum_i C_i \left(eE\Delta_i^2 + \delta\varepsilon_{F,i}\bar{x}_i\right)$$

where Δ_i is the average linear cluster size,

$$\Delta_i^2 = \sum_{i\in\Lambda_i} x_\lambda^2 n_F(\varepsilon_\lambda)(1-n_F(\varepsilon_\lambda)) \Big/ \sum_{i\in\Lambda_i} n_F(\varepsilon_\lambda)(1-n_F(\varepsilon_\lambda)). \qquad (9.A1.11)$$

Expressing $\delta\varepsilon_{F,i}$ in terms of E, we obtain (to a factor of order unity)

$$\delta D(\omega) = e\bar{x}\delta n_1(\omega) \qquad (9.A1.12)$$

(we have omitted here the frequency-independent contributions corresponding to instantaneous response of the subclusters). The response of a system of clusters of s sites is obtained by summing over all such clusters characterized by a random logarithmic distribution of key transition rates Γ^* (or by averaging over them). The result for the average value of the real part of frequency response of clusters of s sites is estimated to be

$$\mathrm{Re}(i\omega\delta D(\omega)) \simeq \omega l^2(\varepsilon)C(\varepsilon)E. \qquad (9.A1.13)$$

here the main contribution comes from clusters with $\omega\tau \simeq 1$ and $\bar{x} \simeq l(\varepsilon)$ is the linear cluster size. If $C_1 \sim C_2$, the capacitance of the equivalent circuit is of the order of the cluster capacitance and is proportional to the number of sites in the cluster,

$$C \sim C_0 s, \qquad (9.A1.14)$$

where C_0 is the capacitance per site.

APPENDIX 9.2 SIZE DISTRIBUTION OF FINITE CLUSTERS

Statistical properties of systems of finite-size clusters have been widely studied for different percolation problems on a lattice [36, 86, 87]. In connection with the problem of time-dependent hopping processes, we are interested in the cluster size distribution function $n(s)$ that gives the average number of clusters of s sites per unit volume. The function $n(s)$ was studied both analytically and by numerical methods. It follows from scaling invariance that in the critical region near the percolation threshold $\Gamma_c = \Gamma_0 \exp(-\eta_c)$, the function $n(s)$ at large s has a form

$$n(s) = s^{-\tau} F\left(\phi s^{1/\Delta}\right), \tag{9.A2.1}$$

where $\phi = (\eta - \eta_c/\eta_c)$ $F(x)$ is an analytical function, finite at $x = 0$ and exponentially decaying at large $|x|$, and τ and Δ are the critical indices. Estimations based on the results of simulation for three-dimensional lattices yield $\tau \simeq 2.2$ and $\Delta \simeq 2.4$ [36, 86].

We see that, for $\phi s^{1/\Delta} << 1$, i.e., for $s << s_c$, where

$$s_c = \phi^{-\Delta} \tag{9.A2.2}$$

$n(s)$ is a power-law decreasing function. For $s >> s_c$, the function $n(s)$ decreases exponentially. Clusters with $s \simeq s_c$ sites are called critical. It follows from definition (9.A2.2) that the size of critical clusters increases as we approach the percolation threshold.

The following characteristics of the system of finite clusters are expressed in terms of the function $n(s)$: the average number of sites in a cluster

$$S(\phi) = \frac{\sum_s s^2 n(s)}{\sum_s s n(s)}, \tag{9.A2.3}$$

the number of sites (per unit volume) that belong to the infinite cluster

$$P(\phi) = N - \sum_s s n(s), \tag{9.A2.4}$$

where N is the number of sites, $\sum_s s n(s)$ is the number of sites that belong to the all finite clusters, and $P(\phi)$ is the order parameter for the problem under study; and the number of finite clusters per unit volume,

$$N_{cl} = \sum_s n(s), \tag{9.A2.5}$$

in Equations (9.A2.3)–(9.A2.5), the summation is performed over the finite clusters.

It follows from the form of the function $n(s)$ that the quantity (9.A2.4) vanishes at the percolation threshold and the sum (9.A2.5) remains finite as $\phi \to 0$, i.e., the main contribution to the sum comes from small-size clusters with $s << s_c$. At the same time, the quantity (9.A2.3) diverges as $\phi \to 0$

$$S(\phi) \simeq \phi^{-\gamma} \tag{9.A2.6}$$

(the main contribution to it comes from critical clusters). In (9.A2.6), γ is the critical index, equal to 1.7 for three-dimensional systems. Assuming that the critical clusters closely fill the entire space (see Section V.19 in [37]), we obtain

$$S(\phi)\sum_s sn(s) \simeq (l_c/l_0)^{-3} s_c^2, \tag{9.A2.7}$$

where l_c is the linear cluster size (the correlation length) diverging at the threshold as $l_c \approx l_0\phi^{-\nu}$ ($l_{c0} \approx l_0\eta_c^\nu$ is the correlation length at $\eta - \eta_c \simeq 1$, i.e., the scale of the critical backbone cluster for the DC conduction). It follows that the critical indices are related by

$$\gamma = 2\Delta - 3\nu. \tag{9.A2.8}$$

Further, the main contribution to the sum (9.A2.3), diverging as $\phi \to 0$, gives clusters of about the critical size; therefore, in the critical region, we have $S \sim s_c^3 n(s_c) \sim s_c^{3-\tau}$. Comparing this expression with (9.A2.6) (with allowance for 9.A2.2), we find one more relation between the critical indices,

$$(3-\tau)\Delta = \gamma. \tag{9.A2.9}$$

Simulation for two-dimensional systems (see, e.g., [88]) have shown that, for the systems under study, the scaling hypothesis is valid, at least, for $\phi \le 0.1$.

ACKNOWLEDGEMENT

This study was supported by the Russian Foundation for Basic Research.

REFERENCES

[1] A.K. Jonscher, *Universal Relaxation Law*, Chelsea Dielectric Press, London, 1996.
[2] G. Pfister and H. Scher, *Adv. Phys.*, **27**, 747 (1978).
[3] L.D. Landau and E.M. Lifshitz, *Electrodynamics of Continuous Media*, Pergamon Press, Oxford, 1982.
[4] N.F. Mott and E.A. Davis, *Electron Processes in Non-Crystalline Materials*, 2nd edn, Clarendon Press, Oxford, 1979.
[5] M. Pollak and T.H. Geballe, *Phys. Rev.*, **122**, 1742 (1961).
[6] I.V. Klyatskina and I.S. Shlimak, *Fiz. Techn. Poluprov.*, **12**, 134 (1978).
[7] S.K. Bahl and A.L. Chopra, *J. Appl. Phys.*, **41**, 2196 (1970).
[8] M.H. Gilbert and C.J. Adkins, *Phil. Mag.*, **34**, 143 (1976).
[9] M. Abkowitz, P.G. Le Comber, and W.E. Spear, *Commun. Phys.*, **1**, 175 (1976).
[10] E.V. Ivkin and B.T. Kolomiets, *J. Non-Cryst. Solids*, **3**, 41 (1970).
[11] A.E. Owen and J.M. Robertson, *J. Non-Cryst. Solids*, **2**, 40 (1970).
[12] A. Lakatos and M. Abkowitz, *Phys. Rev. B*, **3**, 1791 (1971).
[13] X. Le Cleac'h, *J. de Physique*, **40**, 417 (1979).
[14] A. Long, *Adv. Phys.*, **31**, 553 (1982).
[15] A. Long, *in Hopping Conduction in Solids*, M. Pollak and B.I. Shklovskii (eds.), North Holland, Amsterdam, 1991, p. 207.

[16] V.O. Abramov, Yu.A. Zarifyants, and I.P. Zvyagin, *Fiz. Techn. Poluprov.*, 16, 717 (1982).
[17] M. Jourdain, A. de Polighac, and J. Despujols, *J. Phys. C*, **12**, 4999 (1979).
[18] Ch. Ang, Zh. Yu, and Zh. Jing, *Phys. Rev. B*, **61**, 3922–3926 (2000).
[19] M. Sugi, I. Nembach, D. Möbius, and H. Kuhn, *Solid State Commun.*, **13**, 603 (1973).
[20] P. Dutta, S. Biswas, M. Ghosh, S.K. De, and S. Chatterjee, *Synthetic Metals*, **122**, 455–461 (2001).
[21] G.C. Psarras, E. Manolakaki, and G.M. Tsangaris, *Composites A: Applied Science and Manufacturing*, **34**, 1187 (2003).
[22] B. Roling, A. Happe, K. Funke, and M.D. Ingram, *Phys. Rev. Lett.*, **78**, 2160 (1997).
[23] K. Funke, *Solid State Ionics*, **94**, 27 (1997).
[24] H.E. Taylor, *Trans. Faraday Soc.*, **52**, 873 (1956).
[25] J.O. Isard, *J. Non-Cryst. Sol.*, **4**, 357 (1970).
[26] J.C. Dyre and Th.B. Schrøder, *Rev. Mod. Phys.*, **72**, 873 (2000); *Phys. Stat. Sol. (b)*, **230**, 5 (2002).
[27] D.L. Sidebottom, *Phys. Rev. Lett.*, **82**, 3653 (1999).
[28] S.R. Elliott, *Adv. Phys.*, **36**, 135 (1987).
[29] A. Rivera, C. León, J. Sanz, J. Santamaria, C.T. Moynihan, and K.L. Ngai, *Phys. Rev. B*, **65**, 224302 (2002).
[30] A.S. Nowick, B.S. Lim, and A.V. Vaysleyb, *J. Non-Cryst. Solids*, **172–174**, 1243 (1994).
[31] V.L. Bonch-Bruevich, R. Enderlein, B. Esser, R. Keper, A.G. Mironov, and I.P. Zvyagin, *Elektronentheorie ungeordneter Halbleiter*, Deutscher Verlag der Wissenschaften, Berlin, 1984.
[32] A. Miller and E. Abrahams, *Phys. Rev.*, **120**, 745 (1960).
[33] N.F. Mott, *Phil. Mag.*, **22**, 7 (1970).
[34] B.I. Shklovskii and A.L. Efros, *JETP*, **81**, 406 (1981).
[35] I.P. Zvyagin, *Phys. Stat. Sol. (b)*, **101**, 9 (1980).
[36] B.I. Shklovskii and A.L. Efros, *Electronic Properties of Doped Semiconductors*, Springer, Berlin, 1984.
[37] M. Pollak, *in Proc. 5th Int. Conf. on Amorphous and Liquid Semiconductors,* Garmisch-Partenkirchen, 1973, London, 1974, p. 127.
[38] I.P. Zvyagin, *Transport Phenomena in Disordered Semiconductors*, Moscow State University Press., Moscow, 1984 (in Russian).
[39] H. Böttger and V.V. Bryksin, *Fizika Tverdogo Tela*, **18**, 88, 1888 (1976).
[40] I.G. Austin and N.F. Mott, *Adv. Phys.*, **18**, 41 (1969).
[41] A.L. Efros, *Phil. Mag. B*, **43**, 829 (1981).
[42] M. Lee and M.L. Stutzmann, *Phys. Rev. Lett.*, **87**, 056402 (2001).
[43] E. Helgren, N.P. Armitage, and G. Grüner, *Phys. Rev. Lett.*, **89**, 246601 (2002).
[44] S.D. Baranovskii, B.I. Shklovskii, and A.L. Efros, *JETP*, **51**, 199 (1980).
[45] J.A. Reedijk, L.J. Adriaanse, H.B. Brom, L.J. de Jongh, and G. Schmid, *Phys. Rev. B*, **57**, R15116 (1998).
[46] H. Böttger and V.V. Bryksin, *Hopping Conduction in Solids*, Akademie-Verlag, Berlin, 1985.
[47] T. Holstein, *Ann. Phys.*, **281**, 706 (2000).
[48] D. Emin, *Adv. Phys.*, **24**, 305 (1975).
[49] D. Emin, *Phys. Rev. B*, **46**, 9419 (1992).
[50] A.L. Burin and L.A. Maksimov, *JETP*, **95**, 1345 (1989).
[51] N. A. Poklonskii, S.A. Vyrko, and A.G. Zabrodskii, *Fizika Tverdogo Tela*, **47**, 1195 (2005).
[52] M. Foygel, A.G. Petukhov, and A.S. Andreev, *Phys. Rev. B*, **48**, 17018 (1993).
[53] M. Pollak and G.E. Pike, *Phys. Rev. Lett.*, **28**, 1449 (1972).
[54] P.W. Anderson, B.I. Halpenn, and C.M. Varma, *Phil. Mag.*, **25**, 1 (1972).
[55] W.A. Phillips, *J. Low Temp. Phys.*, **7**, 351 (1972).
[56] M.I. Klinger and Yu.N. Belianinov, *Solid State Commun.*, **37**, 505 (1978).
[57] I.P. Zvyagin, *Vestnik Moskovsk. Univ.. Fiz./Astr.*, **19**, 82 (1978).

[58] I.P. Zvyagin, *Phys. Stat. Sol. (b)*, **143**, 97 (1980).
[59] H. Böttger, V.V. Bryksin, and G.Yu. Yashin, *Fizika Tverdogo Tela*, **21**, 1256 (1979).
[60] H. Böttger, V.V. Bryksin, and G.Yu. Yashin, *J. Phys. C*, **12**, 3951, (1979).
[61] A. Hunt, *Phil. Mag. B*, **81**, 875 (2001).
[62] V. Ambegaokar, B.I. Halperin, and J.S. Langer, *Phys. Rev. B*, **4**, 2612 (1971).
[63] V.V. Bryksin, *Fizika Tverdogo Tela*, **22**, 1421 (1980); O. Bleibaum and H. Böttger, and V.V. Bryksin, *Phys. Rev. B*, **54**, 5444 (1996).
[64] B. Movaghar, *J. de Physique*, **42**, suppl. au No. 10, C4 (1981).
[65] J.C. Dyre, *Phys. Rev. B*, **48**, 12511 (1993).
[66] A. Long, *Phil. Mag. B*, **59**, 377 (1989).
[67] J.C. Dyre, *Phys. Lett. A*, **108**, 457 (1985).
[68] S. Summerfield and P.N. Butcher, *Phil. Mag. B*, **49**, L65 (1984).
[69] G.E. Pike, *Phys. Rev. B*, **6**, 1572 (1972).
[70] S. Bhattacharya and A. Ghosh, *Phys. Rev. B*, **68**, 224202 (2003).
[71] G.N. Greaves and K.L. Ngai, *Phys. Rev. B*, **52**, 6358 (1995).
[72] I.P. Zvyagin and W. Kinase, *J. Phys. Soc. Japan.*, **54**, 4776 (1985).
[73] W. Kinase and I.P. Zvyagin, *J. Phys. Soc. Japan.*, **64**, 2988 (1985).
[74] S.D. Baranovski and H. Cordes, *J. Chem. Phys.*, **111**, 7546 (1999).
[75] H. Cordes and S.D. Baranovskii, *Phys. stat. sol. (b)*, **218,** 133 (2000).
[76] I. Svar, F. Borsa, D.R. Torgeson, and S.W. Martin, *Phys. Rev. B*, **48**, 9336 (1993).
[77] S. Sen, A.M. George, and J.F. Stebbins, *J. Non-Cryst. Solids*, **197**, 53 (1996).
[78] A. Bunde, M.D. Ingram, and P. Maass, *J. Non-Cryst. Solids*, **172–174**, 1222 (1994).
[79] P. Maass, A. Bunde, and M.D. Ingram, *Phys. Rev. Lett.*, **68**, 3064 (1992).
[80] H. Jain and X. Lu, *J. Non-Cryst. Solids*, **196**, 285 (1996).
[81] A. Rivera, C. León, J. Sanz, J. Santamaria, C.T. Moynihan, and K.L. Ngai, *Phys. Rev. B*, **68**, 224302 (2002).
[82] B. Roling, C. Martiny, and S. Murugavel, *Phys. Rev. Lett.*, **87**, 085901 (2001).
[83] C.H. Hsieh and H. Jain, *J. Non-Cryst. Solids*, **203**, 293 (1996).
[84] C. Leon, A. Rivera, A. Varez, J. Sanz, J. Santamaria, and K.L. Ngai, *Phys. Rev. Lett.*, **86**, 1279 (2001).
[85] A. Hunt, *J. Phys. CM*, **4**, 6957 (1992).
[86] D. Stauffer & A. Aharony. *Introduction to Percolation Theory*, Taylor & Francis, London, 1992.
[87] D. Stauffer, *Phys. Rev. Lett.*, **35**, 394 (1975).
[88] H. Nakanishi and H.E. Stanley, *J. Phys. A*, **11**, L189 (1978).

10 Mechanisms of Ion Transport in Amorphous and Nanostructured Materials

Bernhard Roling

Fachbereich Chemie, Philipps-Universität Marburg, Hans-Meerwein-Str., Marburg, Germany

Charge Transport in Disordered Solids with Applications in Electronics Edited by S. Baranovski.

10.1 INTRODUCTION

Solid ionic conductors have found many applications in the fields of energy storage and conversion technology, chemical sensors and electrochromic devices. Examples are solid electrolytes in microbatteries, supercapacitors, fuel cells, electrochemical gas sensors and smart windows. For these applications, an ionic conductivity of the electrolyte exceeding 10^{-3} S/cm is desirable. Materials with ambient temperature conductivities above this limit are refered to as 'fast ion conductors' or 'superionic conductors'. Examples are the crystalline compounds α-$RbAg_4I_5$ and β-alumina [1, 2].

High ionic conductivity is, however, by no means the only prerequisite for technical applicability. General drawbacks of crystalline ionic conductors with regard to applications are often high grain boundary resistances, a strong influence of impurities and nonstoichiometries on the ionic conductivity, and a low electrochemical stability. The two former drawbacks can be avoided by preparing amorphous materials. Therefore, in particular in the 1980s and 1990s, many research groups have worked on the preparation of glasses with high ionic conductivities. Improvements in conductivity were achieved by replacing oxide glasses by chalcogenide glasses and by adding alkali halides to the glasses [3, 4]. In 1996, Takada and co-workers succeeded in integrating a highly conducting $LiPO_3{\cdot}Li_2S{\cdot}SiS_2$ glass as electrolyte into a lithium microbattery [5]. However, frequent drawbacks of such fast ion-conducting glasses are high sensitivity to moisture and low electrochemical stability. An alternative approach for reducing the electrical resistance is the preparation of thin films. For instance, Bates and co-workers prepared $Li_{2.9}PO_{3.3}N_{0.4}$ glass films by sputtering of Li_3PO_4 in the presence of nitrogen [6]. These films exhibit a high chemical and electrochemical stability and were successfully integrated into lithium batteries. The research interest in thin-film glass electrolytes has rapidly increased during the last decade [7–9].

Another important class of materials are polymer electrolytes. They have been studied since 1975 when Wright discovered that polyethylene oxide (PEO) is capable of dissolving large amounts of alkali salts, resulting in ionic conduction in the polymer matrix [10]. Due to their high mechanical flexibility, such salt-in-polymer electrolytes are attractive materials for applications. However, these electrolytes also have several drawbacks, in particular a high tendency to crystallize and a small transference number of the alkali ions [11]. An alternative are gel electrolytes. These are polymer networks with moderate crosslink density which can be swollen with plasticizers, such as propylene carbonate. These plasticizers are excellent solvents for alkali salts [12]. Another alternative are polymer electrolytes based on 'hairy rod molecules'. Hairy rod molecules are polymers with a stiff backbone and flexible side chains. The stiff backbone guarantees a high mechanical stability of the material, while the flexible side chains are important for ion transport. Oxygen atoms in the side chains are capable of binding alkali ions, and assisted by motions of the side chains, the alkali ions possess a long-range mobility [12].

During the last decade, the interest in ionic conductors with complex structures on nano- and mesoscales is rapidly increasing. Examples are nanocrystalline composites, nanoscale heterostructures consisting of different ion conducting layers, glass ceramics containing nanocrystallites, nanocomposite and nanogel polymer electrolytes, and inorganic–organic hybrid electrolytes. The mechanisms of ionic conduction in such materials will be discussed in Sections 10.5–10.9 of this chapter. After this introduction, the basic prerequisites for ionic conduction in solid materials are described in Section 10.2. Subsequently, in Sections 10.3 and 10.4, new experimental and theoretical results regarding conduction mechanisms in glasses and amorphous polymer electrolytes are discussed. Device applications of ion transport in disordered solids are described in Chapter 11 of this book.

10.2 PREREQUISITES FOR IONIC CONDUCTION IN SOLIDS

A basic prerequitsite for ionic conduction in solids is the existence of *empty sites* for the mobile ions. For instance, in classical defect crystals with Frenkel or Schottky disorder, the number of empty sites is thermodynamically controlled and increases with increasing temperature. In contrast, in materials with disordered structures, the number of empty sites is generally determined by the structure. In structurally disordered crystals, such as α-$RbAg_4I_5$ and β-alumina, the number of ionic sites is much larger than the number of ions. Therefore, all ions are mobile and contribute to the electrical conductivity. As an example, Figure 10.1 illustrates the structure of the fast ion conductor α-$RbAg_4I_5$. On the average, the occupation number of the silver sites forming percolative pathways is 1/3. Together with low energy barriers between the sites, this leads to a room temperature conductivity of 0.27 S/cm.

In amorphous materials, such as glasses and polymer electrolytes, the ions move in a highly disordered matrix, and it is not possible to distinguish between ions on regular sites and defects. Furthermore, due to the lack of information about the structure, it is not clear how many sites exist for the ions.

In all solid ionic conductors, the movement of an ion to a neighboring site involves the breaking of chemical bonds to the matrix and the formation of new chemical bonds at the

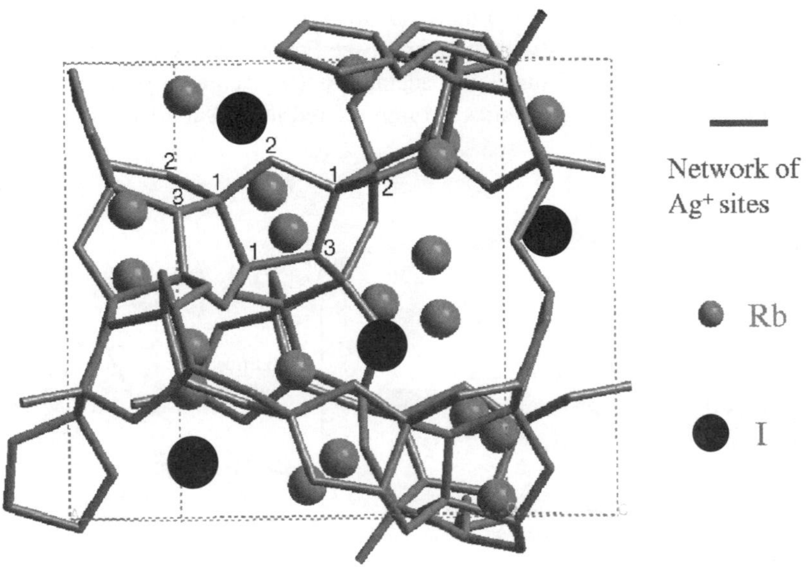

Figure 10.1 (See also colour plates.) Structure of the silver fast ion conductor α-$RbAg_4I_5$. The average occupation number of the silver sites forming a percolative network is 1/3. Together with low energy barriers between the sites this leads to fast ion conduction (reproduced with permission from Stefan Adams, Habilitation Thesis, Goettingen 2000)

new position. Therefore, ion transport in solid materials is a thermally activated hopping process, and a low activation energy for bond breaking is an important prerequisite for a high hopping rate of the ions. If the structure of the crystalline or amorphous matrix itself does not depend on temperature, then the temperature-dependent hopping rate can be described by an Arrhenius equation.

10.3 GLASSES

Due to the lack of knowledge about the structure of ion conducting glasses, the theoretical understanding of the ion dynamics and transport is a challenging task. Important issues concern the universality of the transport mechanisms, the spatial extent of the subdiffusive short-time ion dynamics, the heterogeneity of the ion dynamics, the number of empty sites for the mobile ions, and the structure of the transport pathways. Recent experimental and theoretical results have shed new light on these issues.

10.3.1 Spatial extent of subdiffusive ion dynamics

It is well known that the electrical conductivity spectra $\sigma(\nu)$ of ion-conducting materials, including glasses, exhibit a quasi-univeral shape [13]. Below a characteristic crossover frequency ν^*, the frequeny-dependent conductivity is characterised by a plateau, the plateau value being identical to the DC conductivity σ_{DC}, see Figure 10.2. The crossover frequency ν^* can be defined as: $\sigma(\nu^*) \equiv 2\sigma_{DC}$. According to linear response theory, this implies that on time scales $> 1/(2\pi\nu^*)$, the ionic movements in thermal equilibrium are diffusive, i.e., the mean square displacement $\langle r^2(t)\rangle$ increases linearly with time. At frequencies $\nu > \nu^*$, the conductivity $\sigma(\nu)$ increases with frequency, see Figure 10.2. This implies that on time scales $< 1/(2\pi\nu^*)$, the mean square displacement $\langle r^2(t)\rangle$ increases *sublinearly* with time. Therefore, the short-time ion dynamics is often refered to as 'subdiffusive'.

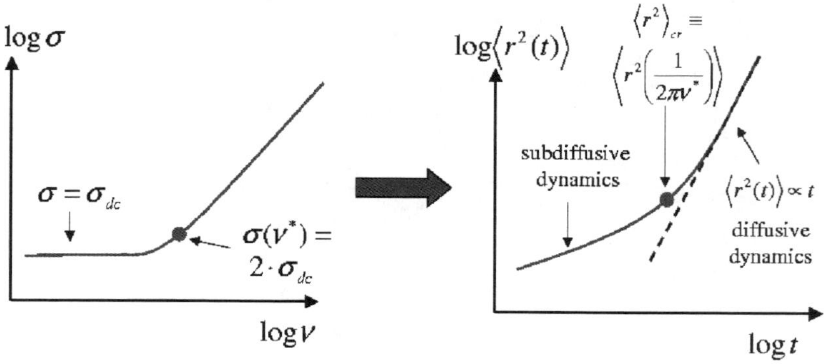

Figure 10.2 By means of linear reponse theory, the time-dependent mean square displacement of the mobile ions, $\langle r^2(t)\rangle$, can be calculated from the frequency-dependent ionic conductivity, $\sigma(\nu)$. At the time $1/(2\pi\nu^*)$, the short-time subdiffusive ion dynamics passes over into the long-time diffusive ion dynamics

In the case of ion conducting glasses, the network structure and the number density of mobile ions have virtually no effect on the spectral shape of the conductivity, especially in a frequency range below a few MHz [14–16]. Therefore, one common viewpoint is the existence of a universal microscopic ion transport mechanism. An alternative assumption is that the macroscopic conductivity displays a much more universal behavior than the microscopic ion dynamics, since a lot of information about differences in the microscopic dynamics is lost due to the macroscopic averaging over all ions in the sample. Recent experimental results provide strong indications in favor of the second viewpoint [17, 18]. This indication is based on an analysis of the mean square displacement where the subdiffusive ion dynamics passes over into diffusive dynamics. In the following, this 'crossover mean square displacement' will be denoted by $\langle r^2 \rangle_{cr} \equiv \left\langle r^2\left(\frac{1}{2\pi\nu^*}\right)\right\rangle$, see Figure 10.2, and its square root $\sqrt{\langle r^2 \rangle_{cr}}$ will be called the 'spatial extent of the subdiffusive ion dynamics'.

This spatial extent $\sqrt{\langle r^2 \rangle_{cr}}$ was determined for various network-forming oxide glasses, such as germanate and borate glasses, with variable amounts of alkali oxide. For these glasses, $\sqrt{\langle r^2 \rangle_{cr}}$ is independent of temperature, but depends strongly on glass composition [17, 18]. In Figure 10.3, $\sqrt{\langle r^2 \rangle_{cr}}$ is plotted versus the alkali oxide content of germanate and borate glasses, x. As seen from the figure, $\sqrt{\langle r^2 \rangle_{cr}}$ increases with decreasing alkali oxide content. In highly modified glasses, $\sqrt{\langle r^2 \rangle_{cr}}$ is smaller than 1 Å. In contrast, a value of approximately 9 Å is obtained for a sodium germanate glass with 0.5 mol% sodium oxide. Thus, a variation of the spatial extent of the subdiffusive ion dynamics by a factor of about 15 is observed.

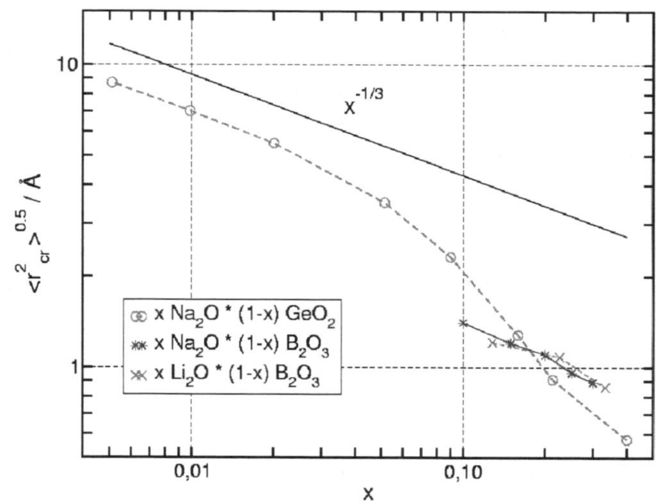

Figure 10.3 Spatial extent of subdiffusive ion dynamics $\sqrt{\langle r^2 \rangle_{cr}}$ versus alkali oxide content x of sodium germanate, sodium borate and lithium borate glasses (reproduced with permission from [16], Copyright 2001 by the American Physical Society)

These findings suggest that, despite the quasiuniveral spectral shape of the conductivity, the microscopic mechanisms of the ion transport depend strongly on glass composition. If one assumes that the distance for a jump of an ion to a neighboring site is in the range 2–3 Å, the finding that $\sqrt{\langle r^2 \rangle_{cr}} < 1\text{Å}$ for glasses with *high* alkali oxide contents implies that at the crossover time from subdiffusive to diffusive dynamics, $t^* = 1/(2\pi\nu^*)$, only a small fraction of ions have left their original sites. On the other hand, in glasses with low alkali oxide content, the length scale $\sqrt{\langle r^2 \rangle_{cr}}$ is considerably larger than a typical jump distance. This implies that at the crossover time t^*, most ions have moved over several jump distances.

These results demonstrate that the spatial extent of the subdiffusive ion dynamics contains valuable information about the microscopic ion dynamics and is, therefore, an important quantity to consider when microscopic mechanisms of the ion transport in glasses are discussed.

10.3.2 Dynamic heterogeneities probed by multidimensional NMR techniques

Many experimental quantities that characterize ion dynamics and ion transport in solid ionic conductors are based on two-time correlation functions. Examples are electrical conductivity, diffusion coeffcients, and neutron scattering functions. For these quantities, the positions of the ions at two different points in time are relevant, the time difference between these points being determined by the time scale or the inverse frequency of the experiment. By means of modern multidimensional NMR techniques, additional information about three-time and four-time correlation functions have become available. Three-time correlation functions yield direct information about heterogeneties in the ion dynamics. During the NMR experiment, subensembles of ions moving on a predetermined time scale are selected, and the movements of these ions on a second predetermined time scale are tracked. By means of four-time correlation functions, additional information about the lifetimes of dynamic heterogeneities can be obtained. The lifetimes specify how long it takes until a fast ion becomes slow and a slow ion becomes fast, respectively.

Vogel *et al.* studied the silver ion dynamics in AgI-doped silver phosphate glasses by means of multi-time correlation functions [19]. Since the ^{109}Ag nucleus has a spin $I = 1/2$, chemical shift interactions are strong as compared with dipole–dipole interactions, and thus single-particle correlation functions are accessible, providing information about the depopulation of silver sites occupied at the start of the experiment. Vogel *et al.* proved, for the first time, the existence of dynamic heterogenities in ion-conducting glasses [19]. Their results allow the authors to put severe constraints on successful models for ion conduction in glasses.

10.3.3 New information about ion transport pathways from reverse Monte Carlo modeling and bond valence calculations

A new approach for elucidating structure–dynamics relations in ion conducting glasses has been developed by Adams and Swenson [20, 21]. In a first step, they generate structural models of glasses by applying the reverse Monte Carlo technique to neutron and X-ray diffraction data. In a second step, the bond valence method is applied to the structural models in order to obtain information about the position of the mobile cations. The bond valence concept is widely used in crystal chemical considerations, e.g., to assess equilibrium position of atoms in crystal structures. The modified bond valence approach by Adams and Swenson takes into account the softness of the bonds between the cations and the glass network. For the determination of the transport pathways, the structural model is divided into several million subelements. Such subelement is considered as accessible for a cation, if the bond valence of the cation in the center of the subelement differs by less than a treshold valence mismatch from the ideal equilibrium bond valence [20]. Accessible volume elements that share edges or faces are considered to belong to the same pathway cluster. Only percolating clusters are assumed to contribute to the ionic DC conductivity.

Adams and Swenson analyse the 'pathway volume', i.e., the volume fraction of the percolating clusters, for glasses with ionic conductivities differing by many orders of magnitude, and they find a strong positive correlation between pathway volume and ionic conductivity [20]. This correlation can be used to predict the ionic conductivity of glasses from structural data.

For fast ion-conducting oxide glasses containing AgI, the bond valence analysis does not provide any evidence for transport pathways in an entire iodine environment, but the Ag^+ ions are generally moving in a mixed oxgen–iodine coordination [21]. Furthermore, the bond valence approach provides an explanation for the mixed alkali effect, i.e., the strong drop of the ionic conductivity, when one type of alkali ions is partially replaced by a second type. Swenson and Adams find that the two types of alkali ions are randomly mixed and have distinctively different conduction pathways with low dimensionality. Thus, the main reason for the mixed alkali effect is that the ions of one type block the pathways of the other type [22, 23]. The structure of the conduction pathways in $Li_xRb_{1-x}PO_3$ glasses is illustrated in Figure 10.4. The random mixing of the alkali ions in mixed-alkali glasses has been confirmed by recent NMR studies of Eckert and co-workers [24].

10.3.4 New information about empty sites and transport mechanisms from molecular dynamics simulations

With the increase in computing power, detailed molecular dynamics (MD) simulations of ion-conducting glasses have become feasible. By means of these simulations considerable progress has been made in the understanding of the relation between structure and dynamics. Heuer *et al.*, Vogel, and Habasaki and Hiwatari carried out simulations on lithium silicate and lithium phosphate glasses and analyzed the number of available sites for the mobile Li^+ ions [25–27]. They found that the number of empty sites is only 5–10% larger than the number of ions. Thus, only few empty sites are available. This is reminiscent of the situation in crystals with low concentrations of defects. Consequently, Heuer, Vogel and Habasaki postulated a vacancy-type transport mechanism in the sense that the small number of empty sites has a strong influence on the hopping dynamics of the ions. Such a vacancy-type mechanism was also postulated by Cormack *et al.* based on an analysis of the Na^+ ion trajectories in a sodium silicate glass [28].

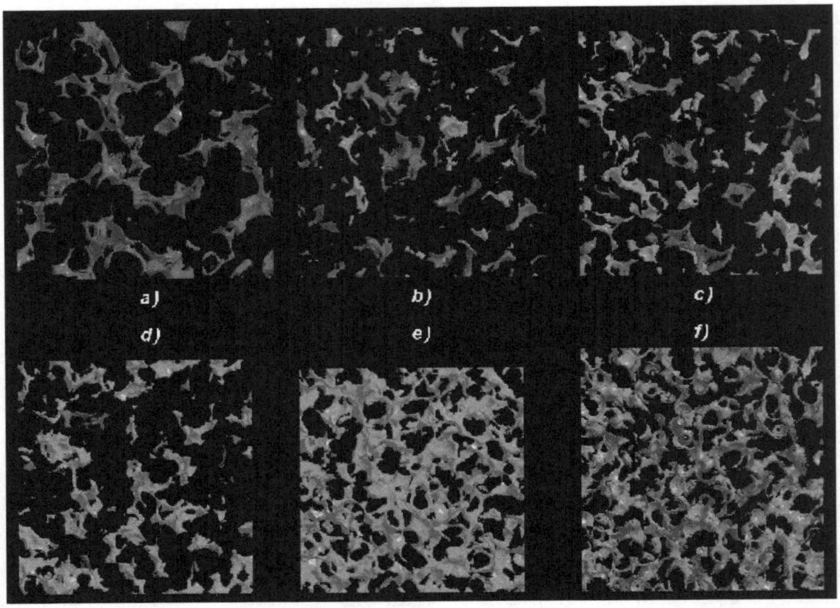

Figure 10.4 (See also colour plates.) Slices through conduction pathways for Li^+ and Rb^+ ions in $Li_xRb1_{1-x}PO_3$ glasses. The picture are from bond valence calculations applied to reverse Monte Carlo models of the glasses. (a) $x = 0$; (b) $x = 0.25$; (c) $x = 0.5$; (d) $x = 0.75$; (e) $x = 1$. The pathways for the Li^+ and the Rb^+ ions are shown as blue and red isosurfaces, respectively. (f) Li^+ pathways (blue) in a $Li_{0.5}Rb_{0.5}PO_3$ glass and those regions that are blocked by Rb^+ ions (pink), but otherwise would have a matching bond valence and would therefore be conduction pathways for the Li^+ ions (reproduced with permission from [22], Copyright 2003 by the American Physical Society)

Meyer *et al.* [29] and Sunyer *et al.* [30] carried out MD simuations on sodium silicate glasses and analyzed the spatial distribution of the mobile Na^+ ions. They found that sodium-rich channels exist in the silicate matrix, the channels acting as preferred transport pathways for the Na^+ ions. In these channels, the number of empty sites is small, leading to a vacancy-type transport mechanism [29]. This result is consistent with the findings by Heuer, Vogel, and Habasaki [25–27].

10.3.5 Field-dependent conductivity of thin glass samples

For many applications, thin-film electrolytes with thicknesses in the range of 100 nm and below are desirable. When a thin film is subjected to voltages of several volts, the resulting electrical field strengths in the film are very high. For instance, the application of 5 V to a sample with a thickness of 100 nm leads to a field strength of $E = 500\,\text{MV/cm}$. Many solid electrolytes show a significant increase of their conductivity at field strengths exceeding 50 kV/cm [31–35]. Thus, when the thickness of a sample is reduced, its electrical resistance is not simply proportional to thickness, but the field-induced conductivity enhancement leads to a further reduction of the resistance. Thus, for the application of thin solid electrolyte samples, it is important to characterize the field dependence of the conductivity. On the other hand, the field dependence is also of interest from a basic scientific point of view, since it contains important information about basic mechanisms of ion transport.

The simplest model for an ionic conductor is a random walk model with independent ions moving in a periodic potential landscape. In the framework of this model, the field dependence of the current density j is given by:

$$j \propto \sinh\left(\frac{qaE}{2k_BT}\right). \tag{10.1}$$

Here, q and a denote the charge and the jump distance of the ions, respectively. The jump distance a corresponds to the distance between neighboring potential minima. Thus, for a given temperature T, the field dependence contains direct information about the jump distance a.

Although this model is too simple for providing a realistic description of a disordered ion-conducting material, the $j(E)$ curves of many real materials can, in a good approximation, be fitted by means of Equation (10.1). However, the jump distances obtained thereby are much larger than what is expected for typical distances between neighboring sites. In the case of ion conducting glasses, values in the range 15–30 Å were derived [31, 32].

The underlying field-dependent measurements were carried out by applying DC electric fields. In this case, a major problem is the potential influence of Joule heating on the conductivity. Heating of the sample due to the dissipation of electrical energy may lead to an increase of the sample temperature, which in turn results in an increase of the conductivity. Thus, Joule heating may pretend a field-dependent conductivity.

In order to avoid this problem, Murugavel and Roling studied the field dependence of the ion transport by means of nonlinear AC conductivity spectroscopy [36]. The application of ac electric fields allows for an umambiguous differentiation between field-dependent ion transport and Joule heating effects. This is easily rationalized when the current density is expressed in terms of a power series of the electric field:

$$j = \sigma_1 E + \sigma_3 E^3 + \sigma_5 E^5 + \ldots. \tag{10.2}$$

Here, σ_1 denotes the low-field conductivity, while σ_3, σ_5 etc. are higher-order conductivity coeffcients. Inserting a sinusoidal electrical field with angular frequency ω into this equation results in higher harmonic contributions to the current density spectrum at 3ω, 5ω, etc. [36]. In contrast, Joule heating leads to an increase of σ_1, but does not generate higher harmonics. By means of nonlinear ac conductivity spectroscopy, independent information about the higher-order conductivity coeffcients, including their respective frequency dependence, is obtainable, while nonlinear DC conductivity measurements yield only one $j(E)$ curve.

The DC values (obtained at low frequencies) of the low-field conductivity and of the third-order conductivity coeffcient, σ_{1DC} and σ_{3DC}, can be used to define an apparent jump distance [36]:

$$a_{app} \equiv \sqrt{\frac{24\,\sigma_{3,DC}(k_BT)^2}{\sigma_{1,DC}q^2}}. \tag{10.3}$$

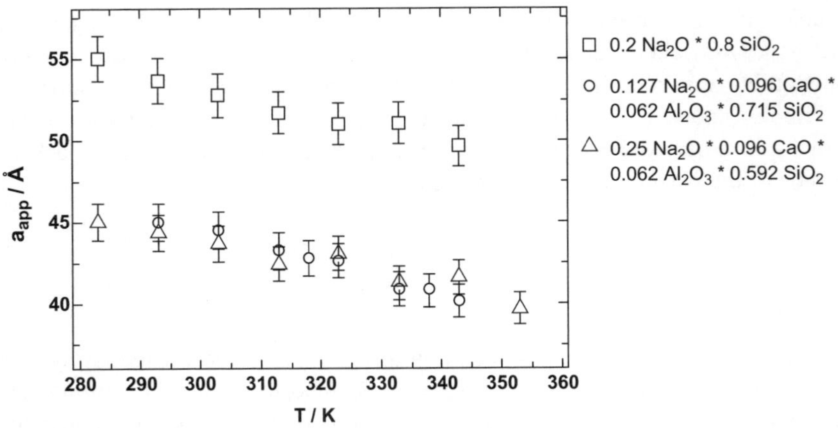

Figure 10.5 Apparent jump distances of mobile Na^+ ions, a_{app} in different silicate glasses plotted versus temperature *T*. The apparent jump distances were determined by means of nonlinear conductivity spectroscopy (reproduced from [36] with permission from Elsevier)

This definition of a_{app} implies that for the random walk model, the simple relation $a_{app} = a$ holds. In Figure 10.5, the apparent jump distance a_{app} obtained for three different sodium ion-conducting glasses is plotted versus temperature [36]. Remarkably, the values are in the range 40–55 Å and are thus considerably higher than typical values obtained from field-dependent DC conductivity measurements. From this it can be concluded that Equation (10.1) does not provide an exact description of the nonlinear conductivity of ionic conductors. This is confirmed by the observation of negative values for $\sigma_{5,DC}$ [36], whereas the validity of Equation (10.1) would imply positive values for $\sigma_{5,DC}$.

In order to learn more about the physical meaning of the apparent jump distances, extensive theoretical studies on nonlinear ion transport in disordered potential landscapes are currently underway [37].

10.4 AMORPHOUS POLYMER ELECTROLYTES

10.4.1 Salt-in-polymer electrolytes

In salt-in-polymer electrolytes, the alkali ions are bound to oxygen or nitrogen donor atoms in the polymer chains. As an example, Figure 10.6 (upper part) illustrates the structure of crystalline $(PEO)_6 \cdot LiAsF_6$. The lithium ions are bound to ether oxygens of the PEO chains, while the AsF_6^- ions interact electrostatically with the Li^+ ions, but do not form strong chemical bonds to the PEO chains [38]. In amorphous polymer electrolytes, which are generally more suitable for technical applications, the nature of the chemical bonds between cations, anions and polymer chains is similar to their crystalline counterparts. Due to the weak bonds of the anions to the polymer chain, the anions are generally more mobile than the cations. Furthermore, the diffusion of the ions is strongly coupled to motions of the polymer chains. This illustrated schematically in Figure 10.6 (lower part). Consequently,

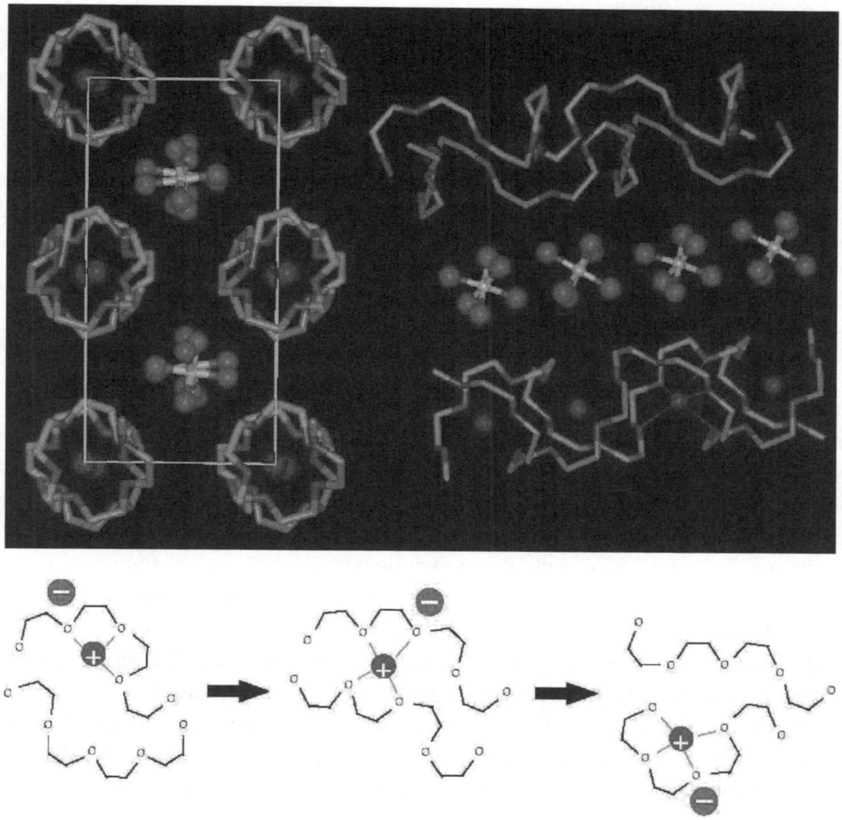

Figure 10.6 (See also colour plates.) Upper: structure of the crystalline polymer electrolyte $(PEO)_6 \cdot LiAsF_6$. Magenta spheres: Li; white spheres: As; light magenta spheres: F; light green: C in chain 1; dark green: O in chain 1; light red: C in chain 2; dark red: O in chain 2 (reproduced with permission from [38], Copyright 2001 by Nature Publishing Group)

Lower: schematic illustration of the coupling between ion dynamics and polymer chain dynamics in an amorphous polymer electrolyte based on PEO

below the glass transition temperature where the dynamics of the polymer chains are frozen in, the mobility of the ions is very low. Above the glass transition temperature, the ions move in a dynamic polymer matrix, in contrast to the situation in inorganic crystals and glasses with an essentially static matrix. Measurements of both diffusion coeffcients of the ions and ionic conductivity for salt-in-polymer electrolytes provide evidence that there are correlations between the motions of cations and anions in the sense that both types of ions move preferably in the same direction [39, 40]. This is sometimes called the 'ion pairing' effect. However, the exact nature of these ionic pairs is unknown at present.

The interplay between ion dynamics and polymer chain dynamics at different ion concentrations, temperatures and pressures can be studied in detail by carrying out Monte Carlo simulations on simple model systems. Snyder *et al.* employed a dynamic bond percolation model and found that a weak temperature dependence of the polymer chain dynamics leads to an Arrhenius-type temperature dependence of the ionic conductivity, while a strong temperature dependence of the polymer chain dynamics results in a Vogel–Fulcher–Tammann-type temperature dependence of the conductivity [41]. Duerr *et al.* used a lattice model with polymer chains occupying nearest-neighbor points of the lattice and with point-like charges representing the ions. They showed that at low ion concentrations, the ion transport is *completely* coupled to the polymer dynamics, while at high ion concentrations, a decoupling takes place, i.e., ions are able to move without assisting movements of the polymer chains [42].

10.4.2 Gel electrolytes

The crystallization of simple salt-in-polymer systems can be suppressed by means of crosslinking. In particular, chemical and photochemical techniques are suitable for this purpose [12]. The resulting polymer networks exhibit good mechanical properties and good processibility to form thin films. However, the crosslinking slows down the polymer chain dynamics, resulting in a decrease of the ionic conductivity with increasing crosslink density. In order to overcome this drawback, it is convenient to use polymer networks with a moderate crosslink density and to swell them by means of a plasticizer, i.e., a low-molecular-weight liquid, such as propylene carbonate. Such liquids are excellent solvents for alkali salts, and the alkali ions are highly mobile in the plasticizer phase. Therefore, the ambient-temperature conductivity of gel electrolytes may well exceed 10^{-3} S/cm [43]. Several groups found that a low molecular weight and a high polarity of the plasticizer are beneficial for a high ionic conductivity [44–46]. This can be explained by the high self-diffusivity of low-molecular-weight liquids in the polymer matrix [46] and the increased dissociation of cation–anion pairs in polar liquids [44, 45].

Due to their high ionic conductivities, gel electrolytes are attractive materials for battery applications. However, major drawbacks are the high chemical reactivity of many plasticizers towards the electrode materials in lithium batteries [12] and a possible leakage of plasticizers.

10.4.3 Polymer-in-salt electrolytes

In salt-in-polymer electrolytes, the conductivity becomes maximum when the ratio of alkali ions to ether oxygens is about 1:16. Larger salt concentrations cause ion pairing and a strong increase of the glass transition temperature, resulting in a decrease of the ionic mobility. However, Angell and co-workers discovered that the reverse approach of dissolving small amounts of polymers in low-melting-point mixtures of lithium salts leads to materials with high conductivities at ambient temperatures [47]. These materials are refered to as 'polymer-in-salt electrolytes'. The mixture of lithium salts can be supercooled below ambient temperatures while retaining its high ionic conductivity. The reason is that in the salt mixture, the transport of the lithium ions is decoupled from the structural relaxation determining the mechanical properties. The addition of a small amount of polymer leads to an increase of the glass transition temperature, but when the amount is small enough, the glass transition temperature remains below ambient temperatures. Thus, the resulting materials exhibit a rubbery behavior at room temperature. Angell and co-workers showed that, although the addition of the polymer leads to a small drop of the conductivity, the conductivities are higher than those of usual salt-in-polymer electrolytes. Furthermore, the transference number of the lithium ions is close to unity [47]. Thus, polymer-in-salt electrolytes combine the electrical characteristics of glassy electrolytes (decoupled ion transport, high cationic transference number) with the mechanical properties of rubbery polymers. A drawback of these materials with regard to an integration into lithium batteries is, however, a high sensitivity to moisture and a high reactivity of the salt anions, in particular of the perchlorate ions, towards metallic lithium.

A new polymer-in-salt system prepared by Angell and co-workers is based on the polyanionic electrolyte poly (lithium oligoetherato mono-oxalato orthoborate), called PolyMOB [48]. Pure PolyMOB electrolytes show single lithium ion conductivities up to 10^{-5} S/cm at room temperature and a good electrochemical stability up to 4.5 V versus Li^+/Li [49]. The addition of small amounts of PolyMOBs to lithium salts leads to materials with high ionic conductivities and excellent mechanical properties [48].

In order to elucidate the ion transport mechanisms in polymer-in-salt electrolytes, Forsyth *et al.* investigated the free volume behavior of a polyacrylonitrile/lithium triflate system by means of positron annihilation lifetime spectroscopy (PALS) [50]. They found that in the composition range between 45 and 75 wt% salt, the free volume is virtually independent of the salt content, while the ionic conductivity increases strongly with the salt content. This strong increase cannot be explained in terms of an increased number of salt ions, but must be mainly caused by an increase of the ionic mobilities. Since this increase in mobility is not caused by an increase in free volume, Forsyth *et al.* propose a percolation-type model. They suggest that the salt ions form clusters which are isolated at low salt content, but form continuous pathways for the mobile ions at high salt concentrations.

Zalewska *et al.* studied the structure of polyacrylamide-based polymer-in-salt electrolytes by means of infrared spectroscopy and Maldi-TOF mass spectroscopy [51]. Their results indicate that most of the ions are not complexed by the polymer, but are present in the form of clusters. This is in good agreement with the model proposed by Forsyth *et al.* [50]. However, Zalewska *et al.* suggest an additional possibility for ion transport, that is, the dissociation of ions from the clusters and the transport via complexation with the polyacryamide [51].

10.4.4 'Hairy-rod' polymer electrolytes

Another possibility to decouple ion transport from the motions of the polymer chains is the synthesis of 'hairy-rod molecules' with a stiff polymer backbone and flexible side chains. An example is poly (*p*-phenylene) (PPP) with oligo (ethylene oxide) (EO) side chains. This material has a strong tendency to self-organize in a liquid-crystalline superstructure with the PPP backbones arranged in parallel layers [52]. Due to the stiff backbones, the material is a hard rubber at room temperature. However, the glass transition temperature of the EO side chains is about –50°C, i.e., their behavior is liquid-like at room temperature. Lithium salts are dissolvable in this material, since the oxygen atoms in the EO chains form chemical bonds to the Li^+ ions. The Li^+ ions possess a long-range mobility assisted by motions of the side chains. The ionic conductivity depends strongly on the length of the EO side chains. With increasing length, the volume fraction of the EO matrix in which the ion transport takes place increases. However, at very long chain lengths, the side chains crystallize, and the superstructure of the material is destroyed. This leads to a strong drop of the conductivity.

From an application point of view, hairy-rod electrolytes exhibit excellent mechanical properties, however, their ionic conductivities are generally still too low [12, 53].

10.5 NANOCRYSTALLINE MATERIALS AND COMPOSITES

One method for improving the ionic conductivity of crystalline materials is the systematic creation of internal interfaces by means of nanostructuring. A simple approach is the preparation of well-known materials in a nanocrystalline form. Example is nanocrystalline oxide ion-conducting ceramics based on CeO_2 [54] or ZrO_2 [55] for the use as electrolytes in fuel cells. A possible explanation for the increased conductivity is an increased number of ionic defects close to the interfaces. Remarkably, the conductivity of nanocrystalline ionic conductors can often be improved by adding nanocrystalline insulators. Indris *et al.* studied composites consisting of lithium ion conducting Li_2O crystallites and insulating B_2O_3 crystallites with an average grain size of about 30 nm [56]. They found that the conductivity becomes maximum when the amount of both types of crystallites is similar. This apparently paradox behaviour can be explained on the assumption that the interfaces between Li_2O and B_2O_3 crystallites act as fast conduction pathways for the Li^+ ions. This is illustrated schematically in Figure 10.7 [57]. The left-hand illustration shows a material consisting of ion-conducting nanocrystallites. On the addition of insulating nanocrystallites (middle illustration), highly conducting pathways at the interfaces are formed. If these pathways percolate, the ionic conductivity of the material is high. When the insulator content is further increased (right-hand illustration), the percolation of the pathways disappears, and the ionic conductivity of the material becomes low. The same qualitative behavior has been found by Knauth and co-workers in Cu^+-conducting CuBr–TiO_2 and CuBr–Al_2O_3 composites [58, 59].

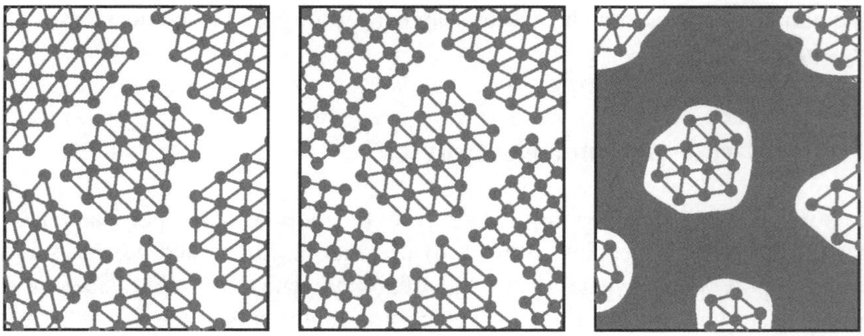

Figure 10.7 (See also colour plates.) Schematic illustration of the structure of a conductor–insulator nanocomposite.

Left-hand: Material composed of ion conducting nanocrystallites. Middle: Composite prepared by mixing conductor (blue) and insulator (red) crystallites. The interfaces between dissimilar crystallites provide fast conduction pathways for the ions. A percolation of these pathways leads to a high ionic conductivity. Right-hand: At high concentrations of insulator crystallites (red area), the pathways are interrupted, and the ionic conductivity is low (reproduced with permission from [57], Copyright 2003 by Institute of Physics Publishing Ltd)

A general description of ion transport in such conductor–insulator composites can be achieved in the framework of three-phase percolation models [56, 58, 60]. In these models, three phases are distinguished, namely the conductor phase, the insulator phase, and the regions close to the interfaces. It is assumed that the conductivity of the interfacial regions is much higher than that of the other two phases. In this case, the overall ionic conductivity of the sample is determined by the percolation behavior of the interfacial regions. In the framework of these models, it is not necessary to specify whether the high interfacial conductivity is caused by a high number density of mobile defects or by a high mobility of the defects.

If the number density of defects in the conductor phase is low, space-charges in the interfacial regions may lead to an enhanced number of defects. In the framework of the space-charge concept by Maier [61], a typical scenario is the seggegration of mobile ions to the interfaces, a process that is driven by gradients in the chemical potential between the bulk of the conductors phase and the interfaces. This seggregation leads to the formation of space-charge regions close to the interfaces where oppositely charged defects are accumulated. The width of these space-charge regions is of the order of the Debye length [61], and the increased number density of defects in the space charge regions is responsible for the conductivity enhancement.

10.6 HETEROSTRUCTURES

Another possibility to make use of interfacial effects is the preparation of heterostructures consisting of thin layers of different ionic conductors. Sata *et al.* prepared heterolayered films composed of BaF_2 and CaF_2 by means of molecular-beam epitaxy [62, 63]. By varying the thickness of the alternating BaF_2 and CaF_2 layers they were able to show that the electrical conductivity is determined by fluoride ion conduction at the interfaces between the layers. When the thickness of the individual layers exceeds 50 nm, the conductivity is proportional to the interface density in the heterostructures. This can be explained by the formation of space-charge regions close to the interfaces leading to enhanced defect concentrations, see Figure 10.8. However, at layer thicknesses below 50 nm, the conductivity shows an anomalous increase with decreasing layer thickness, which can be attributed to the overlap of space charge regions, see Figure 10.8. Due to this overlap, the individual layers lose their identity, and an artifical ionic conductor with anomalous transport properties is formed [62].

10.7 NANO- AND MESOSTRUCTURED GLASS CERAMICS

In a glass ceramic, crystallites are embedded in an amorphous matrix. When the chemical composition of a glass is identical to that of a crystalline compound, this glass can be crystallized partially or completely with the formation of only one crystalline phase having the same chemical composition as the amorphous phase. For an ion-conducting glass ceramic, ion transport in the crystallites, in the amorphous matrix and at the interfaces has to be considered. Usually, the ion transport in an amorphous phase is faster than in a crystalline phase of the same chemical composition [64–66]. This should lead to a decrease of the ionic conductivity with increasing degree of crystallinity. However, several systems are

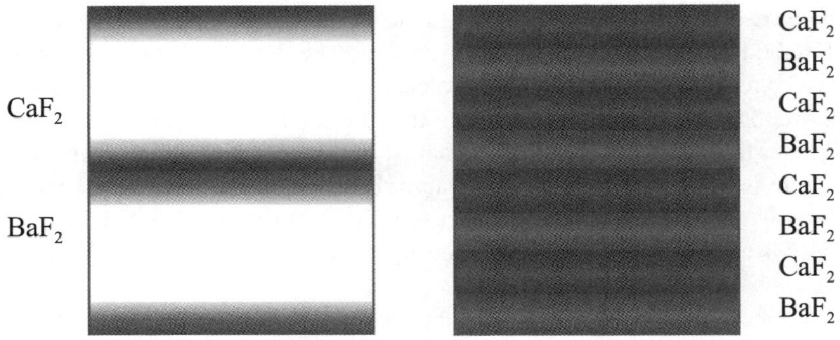

Figure 10.8 (See also colour plates.) Schematic illustration of space charge regions with enhanced ionic conductivity in BaF_2/CaF_2 heterostructures.

Left-hand: When the thickness of the layers is large, individual space charge regions close to the interfaces are formed, and the ionic conductivity is proportional to the interface density.

Right-hand: At small layer thicknesses, the space charge regions overlap, the individual layers loose their identity, and an artifical ionic conductors with anomalous transport properties is formed (reproduced with permission from [62], Copyright 2000 by the Nature Publishing Group)

known where a partial crystallization leads even to an increase of the ionic conductivity. An example is the enhancement of the silver ion conductivity in a 0.57 AgI · 0.29 Ag_2O · 0.14 V_2O_5 glass during partial crystallization [67]. By comparing conductivity and X-ray diffraction data, Adams and co-workers showed that the excess conductivity of the glass ceramics with respect to the conductivity of the glass is proportional to the interfacial area between the glassy phase and the nanocrystallites, which are formed in the early stages of the crystallization process. This indicates a high ionic conductivity in the interfacial regions. The overall conductivity becomes maximum, when the average diameter of the crystallites is about 80 nm.

Similar effects were observed by Roling and Murugavel for $LiAlSiO_4$ glass ceramics. Below 42% crystallinity, the lithium ion conductivity increases with increasing crystallinity [68]. This increase can also be ascribed to fast ion conduction at the interfaces between crystallites and amorphous phase. The dependence of the conductivity on the crystallinity could be reproduced in the framework of a three-phase (crystallites, amorphous phase, interfacial regions) continuum percolation model. However, a remarkable outcome of the model are anomalous interfacial widths of about one-third of the crystallite radii. For instance, in the case of the glass ceramic with 42% crystallinity, the average crystallite radius is about 300 nm, resulting in an interfacial width of about 100 nm. A speculative explanation for the large interfacial widths are mechanical stresses between crystallites and glassy phase, leading to the formation of interfacial layers with structures distinct from both the crystalline and the glassy phase [68].

A limiting factor hindering a better theoretical understanding of ion transport in nano- or mesostructured materials is the traditional characterization of the ion dynamics by means of macroscopic techniques, such as conductivity spectroscopy. These techniques average over the ion dynamics in different phases and at interfaces, leading to a loss of information about the microscopic and nanoscopic mechanisms of the ion transport. Therefore, an experimental method capable of probing ion transport on nanoscopic length scales would

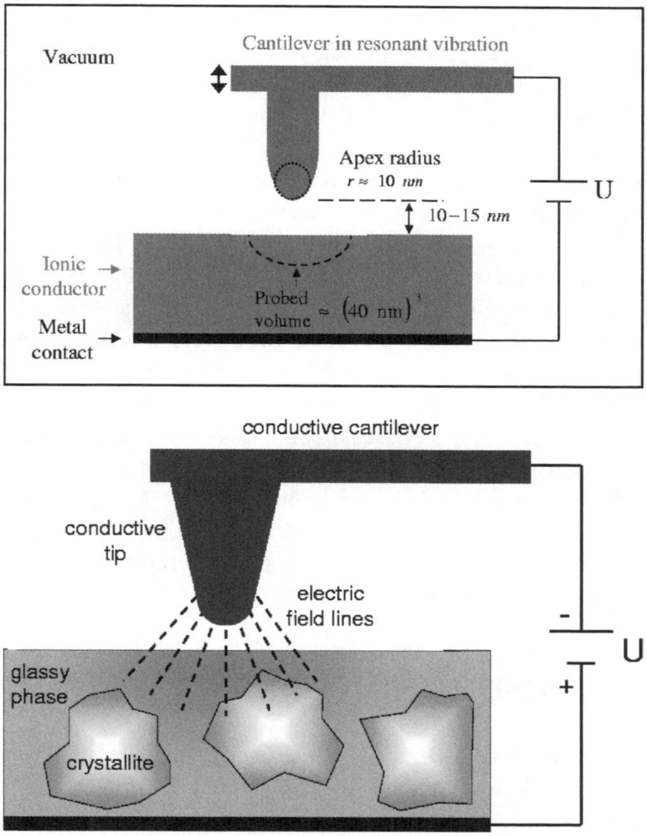

Figure 10.9 Upper: Schematic illustration of the experimental setup for electrostatic force spectroscopy (EFS) measurements on solid ionic conductors. Due to the radial decay of the electric field around the AFM tip, the voltage decay in the sample occurs mainly in a small subvolume below the tip. Thus, the electrical properties of the sample can be probed with a spatial resolution of the order of the tip diameter (reproduced with permission from [69], Copyright 2004 by the American Institute of Physics)

Lower: When EFS is applied to a partially crystallized glass ceramic, the amount of glassy phase and crystalline phase present in the probed volume depend on the position of the tip. Thus, the spatial variation of the electrical force signals provides information on the nanostructure of the glass ceramic and on the diffusion pathways for the mobile ions (reproduced with permission from [64], Copyright 2005 by the PCCP Owner Societies)

be highly desirable. Schirmeisen, Roling, Bracht and co-workers have shown that electrostatic force spectroscopy (EFS) by means of an atomic force microscope (AFM) is well suited for this purpose [64, 69]. The measuring principle is illustrated in Figure 10.9 (upper part). In the dynamic AFM mode, a voltage is applied between a conducting AFM tip and a sample attached to a metal contact, at typical tip–sample distances of 10–15 nm. In this case, the voltage drop in the sample occurs mainly in a nanoscopic subvolume below the surface. This subvolume is of the order of $(40\,\text{nm})^3$ [69]. Mobile ions in the subvolume diffuse into the direction of the electric field, leading to an increase of the electrostatic forces acting between tip and sample and thus to a decrease of the resonant frequency of the AFM cantilever. Since the ionic movements create space-charges, which inhibit a further diffusion, the resonant frequency of the cantilever becomes constant after some time. From the decay time of the resonant frequency, the ionic conductivity in the subvolume can be estimated [69].

When EFS is applied to nano- or mesostructured glass ceramics, the relative amounts of crystalline and amorphous phases being present in the probed subvolume depend on the position of the tip above the surface, see schematic illustration in Figure 10.9 (lower part) [64]. If the ionic conductivities of these phases are different, the relaxation curves of the cantilever resonant frequency as a function of the tip position should contain information about the nanostructure of the material and about the diffusion pathways of the ions. Roling, Schirmeisen, Bracht and co-workers applied EFS to a $LiAlSiO_4$ glass ceramic with 42% crystallinity and detected two different dynamic processes with different activation energies [64]. In contrast, the macroscopic electrical conductivity of the glass ceramic is exclusively determined by the faster process, which is caused by long-range ion transport through the percolating amorphous phase and along the interfaces. The slow process detected by EFS is most likely related to localized movements of Li^+ ions in isolated crystallites. The activation energy of this process is similar to the activation energy for the Li^+ ion conductivity in a completely crystallized $LiAlSiO_4$ ceramic [64, 68].

10.8 NANOCOMPOSITE AND NANOGEL ELECTROLYTES

The properties of polymer electrolytes can be improved by dispersing ceramic nanoparticles into the polymer matrix. Scrosati and co-workers studied the influence of Al_2O_3, TiO_2 and SiO_2 particles with diameters of the order of 10 nm on the properties of lithium salt-in-PEO electrolytes [70, 71]. They found that the dispersion of the nanoparticles improves the conductivity by several orders of magnitude. On the one hand, this conductivity enhancement is related to the suppression of crystallization by the presence of the nanoparticles. As already mentioned, amorphous polymer electrolytes are usually better conductors than their crystalline counterparts. On the other hand, it is likely that fast conduction pathways occur at the interfaces between nanoparticles and polymer matrix [71]. This is confirmed by recent quantum chemical calculations on the local structure around nanoparticles in model polymer electrolytes. Johansson and Jacobsson studied the structure close to a TiO_2 cluster in a dimethyether with dissolved $LiBF_4$ salt [72]. They found a preferred adsorption of BF_4^- ions on the surface of the cluster. The adsorption leads to a dissociation of cation–anion pairs and to the formation of diffusion pathways for the Li^+ ions at the interfaces. This should result in an increase of the transport number of the Li^+ ions and to a decoupling of the Li^+ ion transport from the motions of the polymer chains, in good agreement with experimental results [71]. In addition to the improved electrical properties, nanocomposite electrolytes exhibit enhanced mechanical stability and processibility as compared with standard polymer electrolytes. Therefore, they are promising materials for applications and are currently attracting a lot of research interest.

Another interesting approach is the preparation of nanogel electrolytes. These materials are synthesized by incorporating layered nanoparticles into a gel electrolyte. This is illustrated schematically in Figure 10.10 (upper part) [73]. Figure 10.10 (middle) is a snapshot from a MD simulation of PEO intercalated between the individual layers of a layered

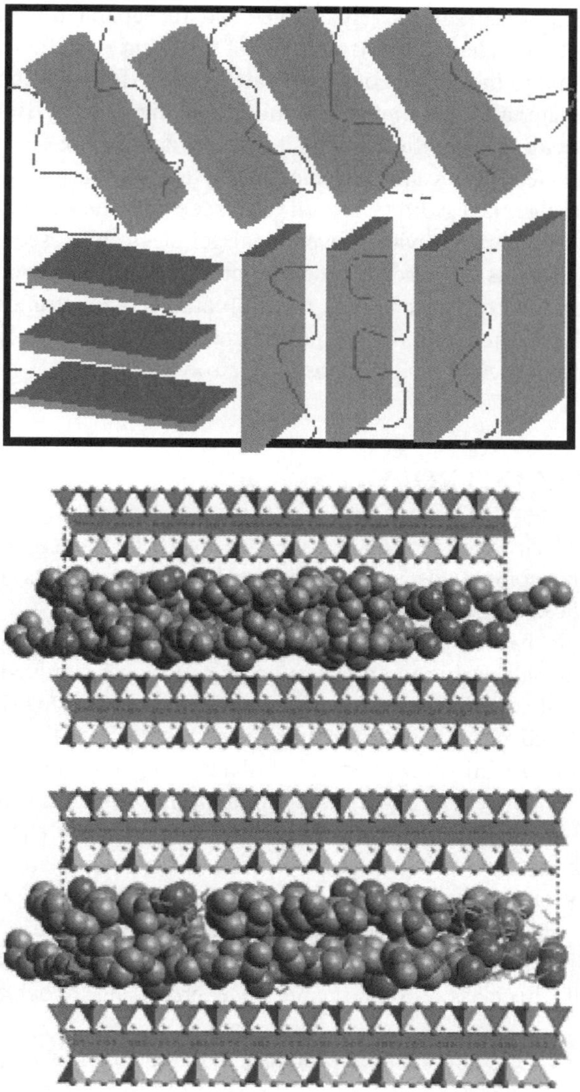

Figure 10.10 (See also colour plates.) Upper: Schematic illustration of a nanogel electrolyte with ordered layers of nanoparticles.

Middle: PEO chains (gray spheres: C; red spheres: O) between nanocrystals of layered silicates. The nanocrystals are negatively charged, and the Li^+ ions (blue spheres) are located close the surfaces of the crystals (results from Monte Carlo and molecular dynamics simulations).

Lower: After addition of a solvent (green molecules), such as propylene carbonate, most Li^+ ions are coordinated by solvent molecules and are highly mobile (reproduced with permission from [73], Copyright 2003 by the Royal Society of Chemistry) silicate. The silicate layers are negatively charged and are separated by interlayers containing Li^+ ions and polymer chains. In the absence of a solvent (middle illustration), the Li^+ ions are mainly bound to the surface of the silicate layers. After addition of a solvent, such as propylene carbonate, a large number of lithium ions move away from the surface and bind to solvent molecules (lower illustration). These solvated Li^+ ions exhibit relatively high mobilities. Nanogel electrolytes are characterised by a high thermal stability and a high chemical and electrochemical stability towards electrode materials in lithium batteries. This enhanced stability as compared with conventional gel electrolytes seems to be caused by the nanoparticles acting as a barrier for the evaporation or diffusion of the solvent molecules. Thus, nanogel electrolytes combine the high ambient-temperature conductivities of conventional gel electrolytes with the high chemical and electrochemical stability and the good mechanical properties of nanocomposite electrolytes [73].

10.9 HYBRID ELECTROLYTES

The term 'hybrid electrolyte' is used, on the one hand, for materials synthesized from inorganic and organic compounds, and, on the other hand, for materials conducting different types of ionic species.

Inorganic–organic hybrid electrolytes are often prepared via sol–gel methods. The addition of a lithium salt during the preparation leads to ionic conductivities up to 10^{-5} S/cm. Examples are silicate/polymer [74] and aluminate/polymer hybrids [75]. Bronstein and co-workers carried out detailed structural and dynamic studies on composites prepared from polyether-functionalized methoxysilanes and aluminum alkoxides blended with PEO [76–78]. Their results suggest that an amorphous aluminosilicate network with nanoscopic pores is formed [76]. This network stabilizes the polymer and suppresses its crystallization without suppressing local movements of polymer chain units [77]. Furthermore, the tendency of cations and anions to segregate seems to be much lower than in simple salt-in-PEO electrolytes, and the large interfacial area between inorganic network and polymer promotes ionic conduction. Lithium ion transference numbers up to 0.7 were obtained [78]. Thus, this class of hybrid materials is promising for achieving single-cation conduction.

A hybrid electrolyte conducting both protons and oxide ions was prepared by Zhu *et al.* [79]. The composite material $BaCe_{0.8}Y_{0.2}O_{3-\delta}\cdot n\ H_2O$ (proton conductor) + Y_2O_3 / Sm_2O_3 (oxide ion conductor) is of interest as a separator in hydrogen/oxygen fuel cells. Due to the mixed conduction of the composite, water can be formed at the anode as well as at the cathode of the fuel cell.

10.10 SUMMARY AND CONCLUSIONS

Among the amorphous materials covered in this chapter, the best understanding of the ion transport mechanisms has been achieved for glasses. Recently, new insights have been obtained into the spatial extent of the subdiffusive ion dynamics, into dynamic heterogeneities, into the number of empty sites and the structure of the diffusion pathways and into the electric field dependence of the ion transport. For instance, the strong composition dependence of the spatial extent of subdiffusive ion dynamics provides a strong indication that there is no universal mechanism of the ion transport in glasses. Nevertheless, a lot of work remains to be done. In particular, it would be desirable to obtain comprehensive information about the potential landscapes in which the ions are moving. This kind of information should be extractable from the results of molecular dynamics simulations. Linking the shape of the potential landscape to experimentally available information about the ion dynamics would then be the ultimate goal. Very little is known up to now about differences in the ion transport mechanisms in bulk and thin-film glasses. The latter materials are a subject of Chapter 11 of this book. Although a number of studies on different thin-film materials have been carried out [7–9], there is not much knowledge about the influence of thickness on the ion transport properties. Thickness-dependent properties may be expected at thicknesses below 100 nm.

In the case of amorphous polymer electrolytes, considerable information is available about the ion transport mechanisms in salt-in-polymer electrolytes, in gel electrolytes and in hairy-rod electrolytes. In contrast, little is known about the structure and the ion transport mechanisms in polymer-in-salt complexes. Here, more structural data would be desirable, for instance by combining small-angle diffraction, NMR, IR, Raman, PALS and mass spectroscopy.

The ion transport in nanocrystalline materials and composites can be well described by using multi-phase percolation models. However, there remain open questions concerning the influence of the interfacial regions on the number density of mobile ions or defects and on their mobilities. For many composites, the space-charge concept of Maier [61] accounts for a high conductivity in interfacial regions. However, in the case of nanocrystalline Li_2O–B_2O_3 composites, Indris *et al.* assume a width of the interfacial regions of only 1 nm [56]. This suggests that a high mobility of the ions in these regions is responsible for the conductivity enhancement. However, the origin of the high mobility is unclear at present. The ion transport in the BaF_2/CaF_2 heterostructures studied by Sata *et al.* [62] is well explained by Maier's space-charge concept. Here, it would be interesting to check the prediction of this concept by means of electrostatic force spectroscopy.

The conductivity enhancement due to the formation of nano- or mesoscale crystallites in glass ceramics can also be explained in the framework of multi-phase percolation models. However, since the number density of mobile ions is high in these materials, the space-charge concept does not seem appropriate for explaining a high conductivity in the interfacial regions. Here, an electrostatic force spectroscopy study should be helpful for elucidating the nature of the interfacial ion transport.

The exact role of the interfaces in nanocomposite and nanogel electrolytes for ion transport is also not clear at present. The quantum chemical calculations by Johansson and Jacobsson on a model nanocomposite system indicate that anions adsorb preferentially at the oxidic nanoparticles leading to a dissociation of cation–anion pairs and to a formation of diffusion pathways for the cations close to the interfaces [72]. These theoretical results could be checked experimentally by combining a broad range of methods, such as NMR, diffusion measurements and electrostatic force spectroscopy. The same applies to inorganic–organic hybrid electrolytes where the interfaces between inorganic and organic phases seem to play an important role for ion transport.

ACKNOWLEDGEMENTS

I thank Steven S. Pas for carefully reading the manuscript and for many helpful suggestions. Financial support by the German Science Foundation (DFG) is also gratefully acknowledged.

REFERENCES

[1] R. Vargas, M.B. Salamon, and C.P. Flynn, *Phys. Rev. Lett.*, **37**, 1550 (1976).
[2] D.P. Almond, A.R. West, and R.J. Grant, *Solid State Common*, **44**, 1277 (1982).
[3] E.F. Hairetdinov, N.F. Uvarov, H.K. Patel, and S.W. Martin, *Phys. Rev. B.*, **50**, 13259 (1994).
[4] J. Kincs and S.W. Martin, *Phys. Rev. Lett.*, **76**, 70 (1996).
[5] K. Takada, N. Aotani, K. Iwamoto, and S. Kondo, *Solid State Ionics*, **86–88**, 877 (1996).
[6] X.H. Yu, J.B. Bates, G.E. Jellison, and F.X. Hart, *J. Electrochem. Soc.*, **144**, 524 (1997).
[7] K.H. Joo, P. Vinatier, B. Pecquenard, A. Levasseur, and H.J. Sohn, *Solid State Ionics.*, **160**, 51 (2003).
[8] S.J. Lee, J.H. Bae, H.W. Lee, H.K. Baik, and S.M. Lee, *J. Power Sources*, **123**, 61 (2003).
[9] M. Yamashita and H. Yamanaka, *Solid State Ionics*, **158**, 151 (2003).
[10] P.V. Wright, *Brit. Polym. J.*, **7**, 319 (1975).
[11] P.G. Bruce, M.T. Hardgrave, and C.A. Vincent, *Solid State Ionics*, **53–56**, 1087 (1992).
[12] W.H. Meyer, *Adv. Mater.*, **10**, 439 (1998).
[13] A.K. Jonscher, *Nature*, **267**, 673 (1977).
[14] B. Roling, A. Happe, K. Funke, and M.D. Ingram, *Phys. Rev. Lett.*, **78**, 2160 (1997).
[15] D.L. Sidebottom, *Phys. Rev. Lett.*, **82**, 3653 (1999).
[16] D.L. Sidebottom, B. Roling, and K. Funke, *Phys. Rev. B*, **63**, 024301 (2001).
[17] B. Roling, C. Martiny, and K. Funke, *J. Non-Cryst. Solids*, **249**, 201 (1999).
[18] B. Roling, C. Martiny, and S. Brueckner, *Phys. Rev. B*, **63**, 214203 (2001).
[19] M. Vogel, C. Brinkmann, H. Eckert, and A. Heuer, *J. Non-Cryst. Solids*, **307**, 971 (2002).
[20] S.Adams and J. Swenson, *Phys. Rev. Lett.*, **84**, 4144 (2000).
[21] J. Swenson and S. Adams, *Phys. Rev. B*, **64**, 024204 (2001).
[22] J. Swenson and S. Adams, *Phys. Rev. Lett.*, **90**, 155507 (2003).
[23] A. Hall, S. Adams, and J. Swenson, *Ionics*, **10**, 396 (2004).
[24] E. Ratai, J.C.C. Chan, and H. Eckert, *Phys. Chem. Chem. Phys.*, **4**, 3198 (2002).
[25] H. Lammert, M. Kunow, and A. Heuer, *Phys. Rev. Lett.*, **90**, 215901 (2003).
[26] M. Vogel, *Phys. Rev. B*, **70**, 094302 (2004).
[27] J. Habasaki and Y. Hiwatari, *Phys. Rev. B*, **69**, 144207 (2004).
[28] A.N. Cormack, J. Du, and T.R. Zeitler, *Phys. Chem. Chem. Phys.*, **4**, 3193 (2002).
[29] A. Meyer, J. Horbach, W. Kob, F. Kargl, and H. Schober, *Phys. Rev. Lett.*, **93**, 027801 (2004).
[30] E. Sunyer, P. Jund, and R. Jullien, *J. Phys. Condens. Mat.*, **15**, S1659 (2003).
[31] J.M. Hyde and M. Tomozawa, *Phys. Chem. Glasses*, **27**, 147 (1986).
[32] J.L. Barton, *J. Non-Cryst. Solids*, **203**, 280 (1996).
[33] J.O. Isard, *J. Non-Cryst. Solids*, **202**, 137 (1996).
[34] Y. Tajitsu, *J. Mat. Sci.*, *31*, 2081 (1996).
[35] Y. Tajitsu, *J. Electrostat.*, **43**, 203 (1998).
[36] S. Murugavel and B. Roling, *J. Non-Cryst. Solids*, **351**, 2819 (2005).
[37] A. Heuer, S. Murugavel, and B. Roling, *Phys. Rev. B*, **72**, 174304 (2005).
[38] Z. Gadjourova, Y.G. Andreev, D.P. Tunstall, and P.G. Bruce, *Nature*, **412**, 520 (2001).
[39] N.A. Stolwijk and S. Obeidi, *Phys. Rev. Lett.*, **93**, 125901 (2004).
[40] C.A. Vincent, *Electrochim. Acta.*, **40**, 2035 (1995).

[41] J.F. Snyder, M.A. Ratner, and D.F. Shriver, *Solid State Ionics*, **147**, 249 (2002).
[42] O. Duerr, W. Dieterich, and A. Nitzan, *Solid State Ionics*, **149**, 125 (2002).
[43] Y. Tada, M. Sato, N. Takeno, T. Kameshima, Y. Nakacho, and K. Shigehara, *Macromol. Chem. Phys.*, **195**, 571 (1994).
[44] D.R. MacFarlane, J. Sun, P. Meakin, P. Fasoulopoulos, J. Hey, and M. Forsyth, *Electrochim. Acta.*, **40**, 2131 (1995).
[45] A. Reiche, T. Steurich, B. Sandner, P. Lobitz, and G. Fleischer, *Electrochim. Acta*, **40**, 2153 (1995).
[46] A. Reiche, J. Tubke, K. Siury, B. Sandner, G. Fleischer, S. Wartewig, and S. Shashkov, *Solid State Ionics*, **85**, 121 (1996).
[47] C.A. Angell, C. Liu, and E. Sanchez, *Nature*, **362**, 137 (1993).
[48] W. Xu, L.-M. Wang, and C.A. Angell, *Electrochim. Acta.*, **48**, 2037 (2003).
[49] W. Xu and C.A. Angell, *Solid State Ionics*, **147**, 295 (2002).
[50] J.S.M. Forsyth, D.R. MacFarlane, and A.J. Hill, *J. Polymer Sci. B*, **38**, 341 (2000).
[51] A. Zalewska, I. Pruszczyk, E. Sulek, and W. Wieczorek, *Solid State Ionics*, **157**, 233 (2003).
[52] U. Lauter, W.H. Meyer, and G. Wegner, *Macromolecules*, **30**, 2092 (1997).
[53] C.T. Imrie, M.D. Ingram, and G.S. McHattie, *Adv. Mater.*, **11**, 832 (1999).
[54] S.K. Tadokoro and E.N.S. Muccillo, *J. Alloys and Compounds*, **374**, 190 (2004).
[55] S.K. Tadokoro and E.N.S. Muccillo, *J. Alloys and Compounds*, **344**, 186 (2002).
[56] S. Indris, P. Heitjans, H.E. Roman, and A. Bunde, *Phys. Rev. Lett.*, **84**, 2889 (2000).
[57] P. Heitjans and S. Indris, *J. Phys. Cond. Mat.*, **15**, R1257 (2003).
[58] G. Albinet, J.M. Debierre, P. Knauth, C. Lambert, and L. Raymond, *Eur. Phys. J. B*, **22**, 421 (2001).
[59] P. Knauth, J.M. Debierre, and G. Albinet, *Solid State Ionics*, **121**, 101 (1999).
[60] A. Bunde and W. Dieterich, *J. Electroceram.*, **5**, 81 (2000).
[61] J. Maier, *Prog. Solid State Chem.*, **23**, 171 (1995).
[62] N. Sata, K. Eberman, K. Eberl, and J. Maier, *Nature*, **408**, 946 (2000).
[63] N. Sata, N.Y. Jin-Phillipp, K. Eberl, and J. Maier, *Solid State Ionics*, **154**, 497 (2002).
[64] B. Roling, A. Schirmeisen, H. Bracht, A. Taskiran, H. Fuchs, S. Murugavel, and F. Natrup, *Phys. Chem. Chem. Phys.*, **7**, 1472 (2005).
[65] R. Winter, K. Siegmund, and P. Heitjans, *J. Non-Cryst. Solids*, **212**, 215 (1997).
[66] O. Kanert, R. Kuechler, D. Suter, G.N. Shannon, and H. Jain, *J. Non-Cryst. Solids*, **274**, 202 (2000).
[67] S. Adams, K. Hariharan, and J. Maier, *Solid State Ionics*, **86–88**, 503 (1996).
[68] B. Roling and S. Murugavel, *Z. Phys. Chem.*, **219**, 23 (2005).
[69] A. Schirmeisen, A. Taskiran, H. Fuchs, B. Roling, S. Murugavel, H. Bracht, and F. Natrup, *Appl. Phys. Lett.*, **85**, 2053 (2004).
[70] F. Croce, G.B. Appetecci, L. Persi, and B. Scrosati, *Nature*, **394**, 456 (1998).
[71] F. Croce, R. Curini, A. Martinelli, L. Persi, F. Ronci, and B. Scrosati, *J. Phys. Chem. B*, **103**, 10632 (1999).
[72] P. Johansson and P. Jacobsson, *Solid State Ionics*, **170**, 73 (2004).
[73] M. M.E. Jacob, E. Hackett, and E.P. Giannelis, *J. Mat. Chem.*, **13**, 1 (2003).
[74] K. Nishio, K. Okubo, Y. Watanabe, and T. Tsuchiya, *J. Sol–Gel Sci. Technol.*, **19**, 187 (2000).
[75] V.D. Noto, V. Zago, G. Pace, and M. Fauri, *J. Electrochem. Soc.*, **151**, A224 (2004).
[76] L.M. Bronstein, C. Joo, R. Karlinsey, A. Ryder, and J.W. Zwanziger, *Chem. Mater.*, **13**, 3678 (2001).
[77] C.G. Joo, L.M. Bronstein, R.L. Karlinsey, and J.W. Zwanziger, *Solid State Nucl. Magn. Res.*, **22**, 235 (2002).
[78] L.M. Bronstein, E. Ashcraft, P. DeSanto, R.L. Karlinsey, and J.W. Zwanziger, *J. Phys. Chem. B*, **108**, 5851 (2004).
[79] B. Zhu, X.R. Liu, and T. Schober, *Electrochem. Commun.*, **6**, 378 (2004).

11 Applications of Ion Transport in Disordered Solids: Electrochemical Micro-ionics

Philippe Vinatier and Yohann Hamon

Institut de Chimie de la Matière Condensée de Bordeaux ICMCB-CNRS and ENSCPB, Université Bordeaux I, 87, Av. Dr A. Schweitzer, 33608 PESSAC Cedex (France)

11.1 INTRODUCTION

Ion transport occurs in a wide range of systems. It plays a vital role, notably in many biological processes. In a completely different field, alkali ions, present in almost any piece of glass, are mobile. Humans have been fabricating glasses for thousands of years without directly making use of this property, which was demonstrated at the end of the nineteenth

Charge Transport in Disordered Solids with Applications in Electronics Edited by S. Baranovski.

century [1]. The first direct application of ion transport in glass rapidly followed: the pH glass electrode was introduced one hundred years ago [2]. The development of solid-state ionic sensors has continued, and today, many ions or gases can be detected and quantified using amorphous electrolytes [3]. New applications of ion transport in disordered solids are also emerging. In this chapter we will discuss three applications related to electronics: micro-sources of power, nonvolatile memories and sensors.

Recent years have seen many advances in lithium batteries for consumer electronics and electrical vehicles [4]. Most of these batteries utilize crystallized electrode materials and organic liquids or polymers as electrolytes. Great effort is currently being devoted to research on fuel cells in which the electrolyte is either an oxide ion or a proton-conducting ceramic or polymer [5]. However, in this chapter, we intend to put the emphasis on a new class of rechargeable power sources, thin-film micro-sources of power. Because of their small size and of their fabrication technologies, these devices are likely to be more intimately related to electronics. In the microbattery, a solid-state thin-film battery about to be marketed, the electrolyte is an oxide thin-film lithium ion-conducting glass, and mixed conductivity (electronic and ionic) amorphous electrodes are sometimes utilized [6]. Two other kinds of electrochemical microsystems are very close to microbatteries. Microsupercapacitors are designed to release their electrical charge very rapidly [7]; electrochromic systems are microbatteries in which transparency is controlled by the state of charge [8].

It has been proposed that microbatteries be deposited directly onto silicon chips in order to create nonvolatile memory. However, within this field, a more specific device is being developed: the solid-state ionic memory, or programmable metallization cell memory (PMCm or PMC-RAM). Industrial competition for the next generation of memory technology is very severe, and various systems are possible candidates: magnetic memory (MRAM), ferroelectric memory (FeRAM), and two kinds of memories based on chalcogenide glasses: the phase-change memory (PC-RAM) and the PMCm [9]. The latter is based on thin-film electrolytes, which have a low intrinsic electronic conductivity, but can be reversibly switched to a metallic conductivity state upon application of a small bias, creating a second stable, easily readable, state. The silver chalcogenide amorphous thin films are, at present, the best candidates as active materials [10].

While microbatteries and solid-state ionic memories are still under development, sensors, on the contrary, have long been on the market. The discovery of pH glass electrodes has been followed by the introduction of ion-selective electrodes, gas gauges (oxygen, etc.), and, more recently, ion-selective field-effect transistors (ISFET) [11]. These analytical tools have applications in a wide range of industrial sectors. Many of them utilize a thin-film chalcogenide glass as an electrolyte and a reference electrode to form an electrochemical cell with the medium to be analyzed, which can be a solution or a mixture of gases. In the simplest systems, the mobile ion of the cell is the ion to be analyzed. The voltage across the cell depends on the chemical potentials of this ion in both the reference electrode (generally fixed and known) and the medium. The voltage is therefore directly linked to ion concentration in the medium.

Miniaturization is a central issue in electronics: it is not surprising that the materials in the previous applications are in the form of thin films. These applications of ion transport are clearly related to electrochemistry, so that they may be grouped under the name 'electrochemical micro-ionics'. This is simply because an ion flux is conveniently controlled or read by an easily measurable electron flux. In the case of microbatteries, the active materials, which transform chemical energy into electrical energy, are the electrodes. The electrolyte is simply used to transfer the ions from one electrode to another. On the contrary, in solid-state ionic memories or sensors, the electrolyte is at the heart of the device, either storing information according to its conductivity state, or creating a bridge between a reference electrode and the medium to be analyzed. In all three cases, the electrolyte is often a disordered solid, prepared as a thin film, and ion transport is always the central phenomenon.

11.2 MATERIALS AND IONIC CONDUCTIVITY

In all the above applications, the relevant materials for this chapter are mostly inorganic amorphous thin-film electrolytes. In this discussion, we restrict ourselves to this type of material, even if, for instance, ion-conducting polymers are used in lithium batteries [4]. However, until now, miniaturization has been easier using inorganic materials. Most of these are homogeneous glasses, but nanostructured disordered materials are emerging, and could become more important if synthesis methods are improved to control their organization. These are oxides, chalcogenides, nitrides or mixed-anion thin-film compounds. The best conducting ions are lithium, because of its small size, and silver, because of its high deformability and polarizability. However, other cations are also utilized, such as copper in solid-state ionic memories [12], or the proton in electrochromic systems [8]. In addition, anion-conducting materials are also being widely investigated—oxide ion-conducting materials for fuel cells, for instance,—though these are often ceramic or polymer electrolytes [5].

11.2.1 Glasses

Amorphous thin-film electrolytes are mostly glasses, and ionic conduction in glasses has been widely investigated as described in Chapter 10 of this book. The advantages of glasses over crystals are: isotropic conductivity; an absence of grain boundaries; and the fact that they are readily obtainable as thin films with no need for annealing. In addition, the existence of a continuous glassy domain enables a continuous variation of the properties. The chemical composition of a typical glass consists of a network former (B_2X_3, SiX_2, GeX_2, As_2X_3, P_2X_5 . . .), a network modifier (Li_2X, Ag_2X, Cu_2X . . .), and sometimes a doping salt (LiI, Li_2SO_4 . . . , AgI . . .); X stands for O, S or Se. The classical structure of a glass is shown in Figure 11.1 [13]. A glass is made of a rigid covalent skeleton without long-range order, constituted by the former cation and the anion. Modifier cations occupy sites between the ramified chains, along with the doping salt anions. The addition of the modifier results in significant structural modifications: the depolymerization of the glassy skeleton into smaller and smaller entities, and the formation of nonbridging anions. The doping salt is supposed to be dissolved in the glassy matrix without affecting its structural organization. However, the role of each constituent is not always so clear. This is particularly crucial in silver chalcogenides as we will see. In the B_2O_3–Li_2O–Li_2SO_4 system, the effect of the addition of doping salt on the covalent skeleton has been investigated using vibrational spectroscopies [14]. It has been found that, even if the doping salt does not participate in the formation of the covalent network, it may influence the equilibrium between three-coordinate and four-coordinate boron atoms.

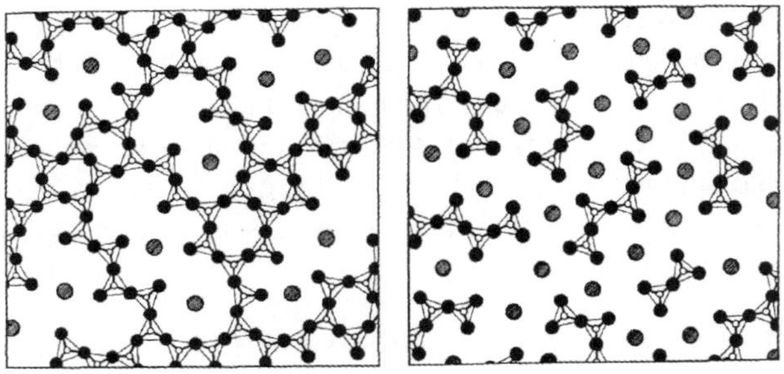

Figure 11.1 Classical structure of a glass, schematized in two dimensions. Small white disks: former cations, large black disks: anions, large grey disks: modifier cations or doping salt counter-anions. The modifier amount is higher for the right-hand picture (reproduced from [13] with permission from The American Ceramic Society)

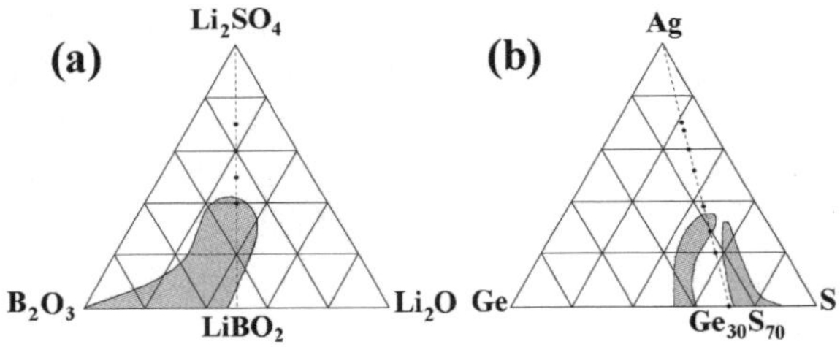

Figure 11.2 (a) Glasses in the B_2O_3–Li_2O–Li_2SO_4 system. Grey area: vitreous domain for glasses prepared by melt quenching [18]; black circles along the $LiBO_2$–Li_2SO_4 line: amorphous thin films prepared by sputtering (redrawn from [15] with permission from Elsevier); (b) Glasses in the Ag–Ge–S system. Grey area: vitreous domain for glasses prepared by melt quenching; black circles along the dashed line of composition $(Ge_{30}S_{70})_{1-y}Ag_y$: amorphous thin film prepared by evaporation (redrawn from [16] with permission from The American Institute of Physics)

A glass exists as a homogeneous material for a wide range of compositions, termed the vitreous domain or glass-forming region. For the B_2O_3–Li_2O–Li_2SO_4 system, the vitreous domain is shown in Figure 11.2a [15]. These glasses can therefore be represented by a formula such as $(1-y)[(1-x)B_2O_3\ xLi_2O]yLi_2SO_4$, where y can vary continuously from 0 to 0.4 for $x = 0.4$. Whereas oxide glasses are mostly made of mixtures of stoichiometric oxides and salts, chalcogenide glasses depart from this simplicity. This is why a typical formula is then written as $(Ge_xSe_{1-x})_{1-y}Ag_y$, and the vitreous domain (Figure 11.2b) is represented in a ternary diagram with atoms at the corners instead of chalcogenides [16]. In other words, there are homogeneous glasses apart from the $GeX_2 - Ag_2X$ line if X = S or Se (indeed, these form the majority), but not for X = O. Neutron diffraction has revealed that homopolar Ge–Ge bonds are formed in $0.5Ag_2S\ 0.5GeS_2$ glasses [17]; indeed their energy is not far from the energy of heteropolar Ge–S bonds. Also S and Se may easily form chains: a specific medium-range order then appears.

The continuous random network model of the glass structure depicted in Figure 11.1 is basically a homogeneous model. On the contrary, recent results for chalcogenide glasses suggest a heterogeneous structure [19, 20]. For $(Ge_{30}X_{70})_{1-x}Ag_x$($X$ = S, Se) thin films, for instance, a phase separation model has been proposed [16]. At low silver concentration, the glass is homogeneous; the small dots in Figures 11.3a and 11.3b represent the silver concentration, and do not signify phase separation. The amount of silver can be increased without any phase separation until it reaches the proportions shown in Figure 11.3c. Upon further silver addition, a silver-rich phase ensues (Figure 11.3d), and eventually the entire material is occupied (Figure 11.3e). The role of silver has been also discussed with respect to the Ge/Se ratio: it acts as a network modifier in Se-rich glasses, but as a network former in Ge-rich glasses [21]. In homogeneous Ge-rich glasses, Ag atoms replace available Ge sites of the network to be tricoordinated by Se. It is worth noting that some authors have considered heterogeneity as an essential characteristic of glass structure. For instance, the cluster-tissue model envisages glass as made of crystallized or amorphous clusters dispersed in a connecting phase [22, 23]. However, heterogeneity as described for silver chalcogenide glasses has not been observed for the corresponding oxide systems.

Any material can, in principle, be amorphized, but it may not be stable at room temperature. And the quenching speed required may exceed the fastest currently available, probably provided by the physical vapor deposition methods. Nevertheless, the vitreous domain is larger for higher quenching speeds. In Figures 11.2a and 11.2b, the glassy domains for melt quenching methods are the grey areas. They have been greatly extended in both cases using evaporation [16] or sputtering [15]. These methods are therefore likely to provide new disordered solids, as we will discuss.

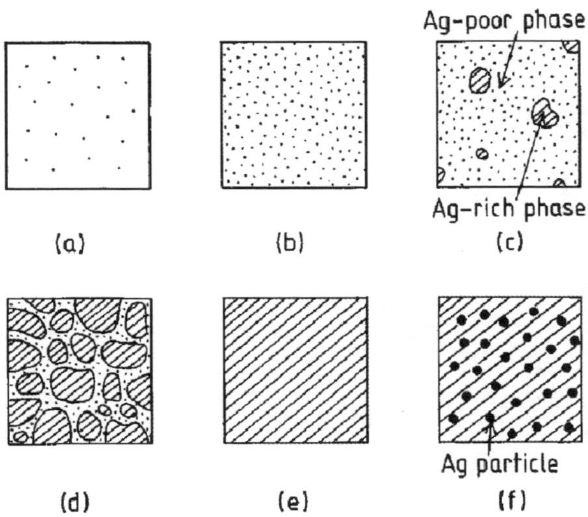

Figure 11.3 Schematic illustration of the change in structure of silver chalcogenide thin film with an increase in Ag content. The Ag content increases from (a) to (f). The small dots in (a)–(d) display schematically the Ag concentration, i.e., a high density means a high Ag concentration. The Ag concentration of the Ag-rich phase (hatched area) is close to the maximum Ag concentration. The large dots in (f) represent fine Ag particles formed on the film during preparation (reproduced from [16] with permission from The American Institute of Physics)

11.2.2 Ionic conductivity in glasses

The existence of a large number of equivalent sites for each modifier cation (clearly visible in Figure 11.1), and its nondirectional ionic bond with the anions, are responsible for its mobility. However, such features do not exist only in glass. For example, they also occur in Na^+-β-alumina, α-AgI and $RbAg_4I_5$, Nasicon and Lisicon [24]. The existence of a large number of sites per mobile ion has been called 'structural disorder'. Its signature is detected in frequency-dependent conductivity or NMR spin–lattice relaxation time. The conduction mechanism associated with structural disorder is still debated as described in Chapter 10 of this book.

For a given mobile cation, the ionic conductivity of a glass depends, foremost, on two factors: its concentration and the polarizability of the amorphous network, which is related to its mobility. Thus, lithium sulfide glasses have a conductivity three orders of magnitude higher than oxide glasses [25]. This is due to the higher polarizability of sulfur, which makes the hop of a mobile ion from one site to the other easier. Similarly, silver selenide glasses are better ionic conductors than their sulfide counterparts. This effect has not been fully explained or quantified. On the contrary, the variation of ionic conductivity as a function of the mobile ion concentration is well documented and demonstrated by a power law, which has been both empirically proven and reproduced by recent models: $\sigma = Ac^{\gamma}$, where σ is the ionic conductivity, c is the concentration, and A and γ depend on the vitreous system [26]. An example of this relationship is given in Figure 11.4a [26], and the variation of the silver diffusion coefficient versus a wide range of silver concentrations in chalcogenide

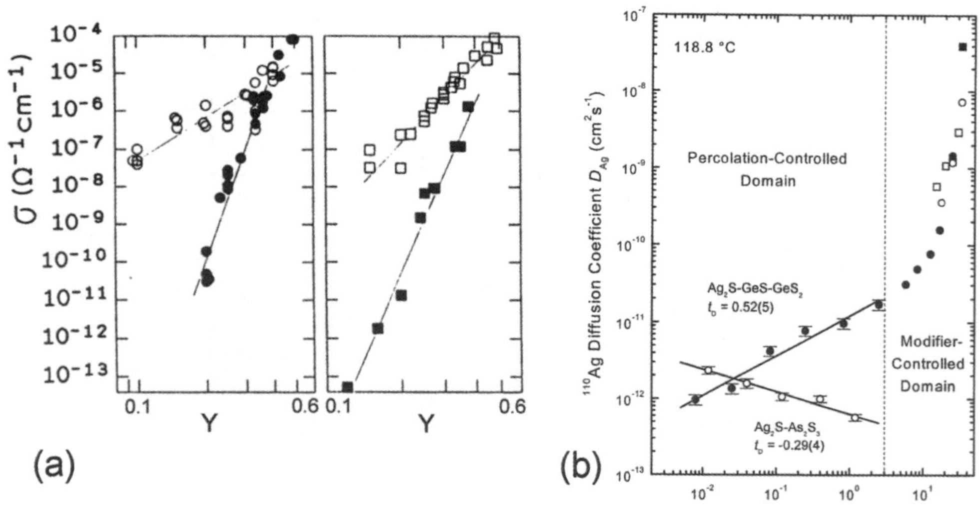

Figure 11.4 (a) DC conductivities in several alkali oxide glasses: $yLi_2O(1-y)SiO_2$ (open circles), $yLi_2O(1-y)B_2O_3$ (closed circles), $yNa_2O(1-y)SiO_2$ (open squares), $yNa_2O(1-y)B_2O_3$ (closed squares) (reproduced from [26] with permission from Elsevier). (b) Silver diffusion coefficient as a function of silver concentration in Ag_2S–As_2S_3 and Ag_2S–GeS–GeS_2 glasses (reproduced from [27] with permission from Elsevier)

glasses is reported in Figure 11.4b [27]. Other factors are also important, such as density and free volume, as well as the presence of more than one former cation (mixed former effect) or mobile ion (the still-debated mixed-alkali effect). Increasing the modifier amount depolymerizes the glassy network and creates more and more nonbridging anions, which contribute to the increase of a network polarizability. Similarly, increasing the doping salt amount introduces anions such as I^-, which are also likely to increase the network polarizablity.

Such modifications result in higher mobile ion concentration, and they have consequently been utilized to improve ionic conductivity. However, they should not pull the system out of the vitreous domain; the conductivity generally decreases when a crystalline phase is formed in a glassy matrix. For instance, in the B_2S_3–Li_2S–LiI system, conductivity increases with the doping salt amount until LiI aggregates are formed in the material [28]. For lithium sulfide glasses, the fully depolymerized compositions corresponding to the maximum modifier amount can be achieved using melt quenching. For instance, in the $(1 - x)B_2S_3\ xLi_2S$ system, the glass-forming region extends to $x = 0.75$, i.e., Li_3BS_3, which consists of Li^+ ions and BS_3^{3-} anions [29]. This is not the case for the oxide homolog $(1 - x)B_2O_3\ xLi_2O$ system, for which the vitreous domain obtained using melt quenching ends around $x = 0.5$. On the contrary, fully depolymerized silver oxide glasses are available, such as AgI–Ag_3PO_4 [30].

Ionic conduction has been investigated mainly in electrolytes (i.e., purely ionic conductors), even though it is also a crucial phenomenon in electrode materials (i.e., mixed ionic and electronic conductors). Moreover amorphous and crystalline intercalation materials display interesting features for the fundamental study of ion transport. In crystallized materials, the concentration of mobile ions can be adjusted without changing the structure of the material. The role of dimensionality in ion transport can be studied in layered and tunnel crystallized electrodes. Such studies are difficult, however, because they require separating the ionic and electronic components. In this context, quasi-elastic neutron scattering could be an efficient tool, as already demonstrated for the mixed conductor Ag_2S [31].

The levels of ionic conductivity at room temperature in different materials are spread within orders of magnitude. The ionic conductivity of a lithium-ion-conducting oxide glass is typically 10^{-7} S/cm, whereas that of a liquid electrolyte used in lithium batteries is about 10^{-2} S/cm. Surprisingly, some solid electrolytes almost reach this value, which is comparable to the electronic conductivity of semiconductors. Lithium sulfide glasses [32], as well as silver oxide glasses [30], have a conductivity close to 10^{-3} S/cm. An optimized amorphous ionic conductor has been designed, with silver as the conducting cation, sulfur as the anion, and boron and silicon are used to provide a mixed former effect [33]. However, an even better ionic conductor is a $70Li_2S$–$30P_2S_5$ glass–ceramic material prepared by mechanical milling followed by annealing. The room temperature ionic conductivity of this material is 3.2×10^{-3} S/cm, and the activation energy of the conductivity is 0.19 eV. These outstanding values are attributed to the crystallization of a highly conductive phase in the glassy matrix [34].

11.2.3 Thin-film preparation

Thin-film preparation methods are classified as either physical or chemical vapor deposition (PVD and CVD). Here, we will briefly discuss two PVD methods which are well established in microelectronics: sputtering and evaporation. Laser ablation is another interesting PVD method, though it has currently no industrial applications. CVD methods may play an important role in the future, for example in the fabrication of three-dimensional micro-batteries [35]. Such microbatteries could be built directly on a silicon wafer, which would be used as the negative electrode. The wafer would be etched with an array of ditches, then lined with the electrolyte and the positive electrode to form an interdigitated plate array of cathodes and anodes. The geometry of such ditches may prevent the utilization of PVD to deposit a thin film on them. The great advantage of this design is that it considerably increases the (geometrical) surface capacity and utilizes an approach which has proven efficient for three-dimensional capacitors.

In a sputtering chamber, a target—the source of material to be deposited—is placed in front of a substrate. The target is generally a single-phase pressed-powder pellet. Voltage is applied between the target and the substrate. In certain pressure conditions (typically 1 Pa), the gas in the chamber forms a plasma, and the target is bombarded by the ions. Atoms are ejected from the target and eventually deposited on the substrate. The thin film grows at a rate of a few tenths of a micrometer per hour. If one wishes to synthesize a new material, a second source of atoms has to be introduced. The target can be made of a mixture of two phases, or a small disk of a second phase can simply be placed on the main target. Better control of the composition can be obtained using co-sputtering from two (or more) single phase targets. The composition of the film then depends on the relative power applied to each target. Finally, a reactive gas can be introduced into the plasma. Oxides can be prepared from a metallic target sputtered in a plasma containing oxygen. More exotic materials can be obtained by using nitrogen to prepare oxynitrides from an oxide target [36], or oxygen to prepare oxysulfides from a sulfide target [37]. Using sputtering, the thin-film composition is, in many cases, very close to the sputtered target composition, which is more difficult to obtain using evaporation. This is why co-evaporation has been utilized to prepare $(Ge_{30}Se_{70})_{1-y}Ag_y$ thin films, using two different tungsten boats containing $Ge_{30}Se_{70}$ glass and metallic silver [16]. The deposition rates of each material were adjusted to obtain thin films with appropriate silver content.

The new compositions obtained using the methods described can lead to new single-phase thin-film amorphous or crystallized materials, as well as to phase-separated materials. Sputtering and evaporation can both be considered as ultra-high-quenching-rate synthesis methods. The temperature of the film during the deposition does not typically exceed 150°C. This does not leave enough energy to provoke the diffusion of the incoming atoms on the film, and thus the atoms do not reach a stable thermodynamic state of equilibrium. However, some materials require an even faster quenching speed to be homogeneously amorphized, or their glass transition temperature may be below room temperature. In this case, original microstructures are obtained, either amorphous or crystallized. In the latter case, the crystallite size can be as small as about 10 nm. Preferred orientations are obtained as well as typical morphologies. Controlling the nanostructure is quite difficult. It involves substrate heating or cooling, application of a small negative bias on the substrate to let it be bombarded by ions from the plasma, or bombardment by an ion gun. Not all the possibilities presented by these techniques have yet been explored, and new metastable materials are to be expected.

Characterizing amorphous nanostructured thin films is obviously a difficult task, requiring a wide combination of experimental techniques. The case of amorphous silicon monoxide thin films is an interesting example. Utilization of diffraction, electronic microscopy, spectroscopy and magnetometry has revealed a disproportionation: the material is made of amorphous SiO_2 clusters and of amorphous Si clusters dispersed in a suboxide matrix [38]. With investigation methods becoming more and more sensitive, it is likely to expect that new lower-scale inhomogeneities will be discovered more often in amorphous thin films.

11.3 LITHIUM-ION-CONDUCTING OXIDE GLASSES IN MICRO-SOURCES OF POWER

The miniaturization of electronic circuitries and devices has not kept pace with the evolution of their power sources. However, in the field of lithium batteries, research in miniaturization has led to the development of a new device, which is about to be launched on the market by several companies: the microbattery. It is an all-solid-state thin-film rechargeable battery, whose fabrication has required replacing the liquid electrolyte with a solid electrolyte—a lithium-ion-conducting glass. The thickness of a microbattery, represented in Figure 11.5 [39], is about $10\,\mu m$ for the active layers, and less than 0.1 mm with the flexible packaging included. The surface area can be adjusted on demand from a few mm^2 to more than $10\,cm^2$. The typical capacity is about $100\,\mu A\,h\,cm^{-2}$, and the microbattery can deliver a current up to a few mA/cm^2 at a voltage of 2–4 V. The most promising applications may be related to security issues and new functions in smart cards. Many other low-power devices could make use of microbatteries, such as sensors, radiofrequency identification tags, and micro-electromechanical systems (MEMs) [6].

11.3.1 Principle of lithium microbatteries and related systems

A lithium microbattery is made of three layers deposited successively (in addition to current collectors): the positive electrode; the electrolyte; and the negative electrode for which metallic lithium is used. During the discharge, lithium is oxidized in the negative electrode. The lithium ions move towards the positive electrode through the electrolyte, whereas the electrons are transferred via the external circuit. The lithium ions are intercalated into the positive electrode material, which is simultaneously reduced by the electrons. These spontaneous processes are schematized in Figure 11.6a [40] for a lithium-ion battery and a typical cycling behavior is represented in Figure 11.6b. The reverse reactions are forced

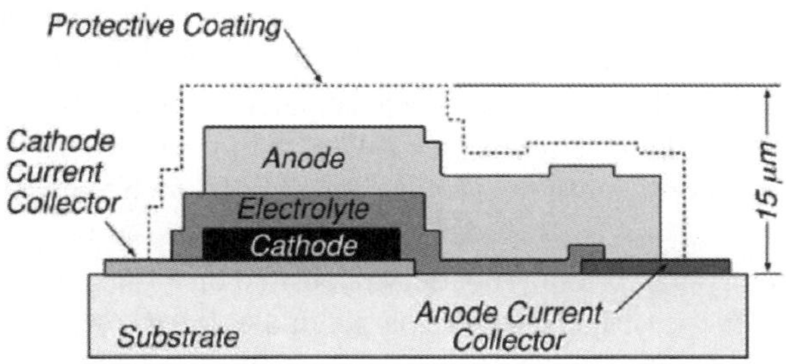

Figure 11.5 A lithium thin-film microbattery (reproduced from [39] with permission from Elsevier)

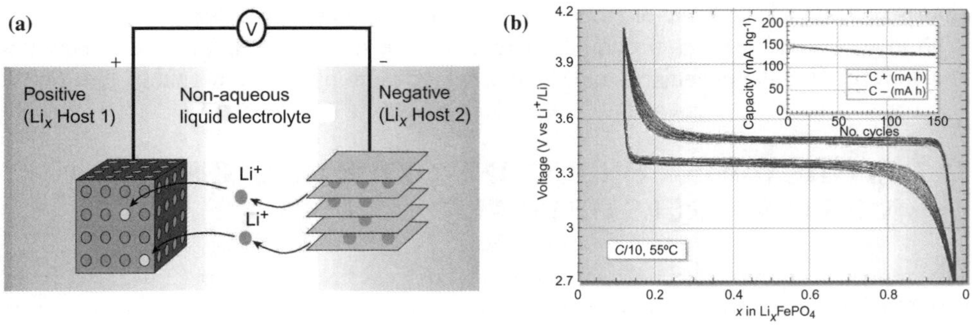

Figure 11.6 (a) Schematic representation and operating principles of rechargeable Li-ion batteries. (b) Cycling behaviour of an optimized $LiFePO_4$/C composite electrode. The evolution of voltage has been plotted as a function of x, the amount of lithium intercalated in the host material (reproduced from [40] with permission from Nature Publishing Group)

during charge. A lithium-ion battery is a battery whose negative electrode is also made of an intercalation material. This has the advantage of preventing the formation of lithium dendrites that cause short-circuits in bulk batteries, and to require a less severe encapsulation than the highly hygroscopic metallic lithium in microbatteries. Also, lithium melts at 181°C, which prevents the performance of soldering operations on lithium microbatteries.

The choice of the mobile ion is based on two factors: high conductivity in the solid electrolyte and in the intercalation materials is required to achieve a high current density; and the electrochemical couples involved should give a high voltage in order to obtain a high power. Lithium has the great advantage of providing such a high voltage (up to 4.5–5 V [41]), and of diffusing easily in certain electrolyte and electrode materials. In alkaline batteries, the electrolyte is aqueous, which limits the voltage to 1.2 V. Solid-state batteries have been made using the very fast silver ion as a mobile ion, but their voltage is low [42]. However, with electronics requiring lower and lower voltage (see the solid-state ionic memories disussed later in this chapter), this problem could disappear in the near future. And the intrinsic advantages of silver are interesting: a very high conductivity in solid electrolytes, and an excellent chemical stability.

Even though microbatteries have a high energy density, their power may be insufficient to start some electronic devices. Microsupercapacitors can satisfy this power specification. They consist of a metal oxide/electrolyte/metal oxide electrochemical cell. Lithium ions are used instead of protons utilised in bulk supercapacitors since proton-conducting oxide films are difficult to grow [43, 44]. The energy storage mechanisms in these devices do not involve only the storage of charges in the double-layer at or near the electrolyte/electrode interface, but also a so-called redox pseudocapacitance mechanism, in which redox reactions take place between the ions in the electrolyte and the electrodes [7, 45]. Their use in hybrid systems with microbatteries has been considered to be promising for 'on-chip' applications.

An electrochromic system is a microbattery deposited on a sheet of glass, which is colorless and transparent in one of its states of charge, and colored, opaque or reflecting in the other [8]. Electrochromic systems mostly utilize the proton or the hydroxide ion as a mobile ion, since voltage is not relevant. A very rapid diffusion of the mobile ion is required to effect a rapid color change. Most of the electrochromic systems developed up to now use a polymer electrolyte [46]. However, an inorganic thin-film electrolyte may be interesting in this context. It is worth noting that proton- and hydroxide-ion-conducting thin-films could also prove applicable to micro-fuel cells.

11.3.2 Requirements of thin-film electrolytes for electrochemical microsystems

A lithium-ion-conducting electrolyte for microbatteries should satisfy the following criteria:

- a very high ionic conductivity for a rapid transfer of the lithium ions from one electrode to the other, and therefore a high current. In the following the values of conductivity are given at room temperature;
- a very low electronic conductivity to avoid self-discharge, i.e., a transport number close to 1;
- a wide electrochemical stability window;
- complete compatibility with the electrode materials, both chemically and mechanically;
- the ability to be deposited efficiently using sputtering, which is the most commonly used deposition technique.

For the last 25 years, much effort has been devoted to enhancing solid electrolyte ionic conductivity. The conductivity of a liquid electrolyte is about 10^{-2} S/cm. The standard thin-film electrolyte material used in microbatteries, a lithium phosphorus oxynitride called Lipon (typically $Li_{2.9}PO_{3.3}N_{0.36}$), has a conductivity of 2×10^{-6} S/cm [47]. This difference is reduced by three orders of magnitude if the thickness is taken into account: 1 μm for a microbattery compared with 1 mm for a battery. However, the surface area is likely to be smaller; for example, a microbattery can be made as small as 15 mm^2. In this case, a 1-μm-thick electrolyte will induce an internal resistance of 200 Ω if its ionic conductivity is similar to Lipon. A 5 mA current leads to an unmanageable ohmic drop of 1 V. Instead, a low ohmic drop of 0.1 V is obtained if an electrolyte with a conductivity one order of magnitude higher is used. Any enhancement of the ionic conductivity is reflected in the maximum current available, and therefore in the discharge power and the charge duration, provided that the main cause of the internal resistance is electrolyte resistivity. This may not always be the case, since the electronic and ionic resistivities of the electrode materials and the interface resistances play also important role.

The electronic conductivity of the electrolyte should be low enough to avoid self-discharge. Unfortunately, this is not always the case. For instance, thin films prepared from a SiO_2–V_2O_5–Li_2O target are very good lithium ion conductors; their ionic conductivity is 10^{-6} S/cm [48]. But their electronic conductivity, 2×10^{-10} S/cm, leads to a significant self-discharge, corresponding to a 4 μA/cm^2 current [49]. The introduction of transition elements in the chemical composition of an electrolyte is likely detrimental to the microbattery performances.

The electrolyte has to be able to withstand the potential difference between the two electrodes. It is about 3 V in a TiS_2//Li battery, but it can reach up to 5 V with some recent cathode materials [41]. It is worth noting that liquid electrolytes can hardly withstand such a high voltage. On the contrary, the electrochemical stability window of the Lipon electrolyte is about 5.5 V, as shown in Figure 11.7 [47]. The chemical stability of an electrolyte in contact with lithium has also to be investigated, because lithium is a highly reductive agent. The stability of lithium borate glasses compared with lithium has been checked at 100°C for several days using impedance spectroscopy [50]. On the contrary, the stability of the P_2S_5–Li_3PO_4–LiI thin-film electrolyte is too fragile. A protective LiI layer must be deposited between the lithium electrode and the electrolyte, according to an approach developed for sulfide glasses [51]. The ionic conductivity of the oxysulfide glass, 2×10^{-5} S/cm, leads eventually to an overall value of 10^{-6} S/cm [52].

The mechanical properties of thin-film electrolytes have not been studied. Nevertheless, this parameter is crucial for their industrial application, and especially their compatibility with the other layers of the system. There are further requirements for the industrial application of thin-film electrolytes. The target from which they are fabricated using sputtering should be hard enough to prevent breakage under the heating caused by its bombardment. Moreover, the target should be able to withstand air conditions for its mounting in the sputtering machine: highly hygroscopic materials should be avoided. Finally, as always, cost and environmental issues must be considered, even if the quantity of materials is quite low.

11.3.3 Electrolyte materials used in electrochemical microsystems

We are now going to discuss thin-film electrolytes, focusing on the choice of composition in order to obtain a high ionic conductivity. The main components are, in order of importance: (i) a high lithium concentration, using either highly modified or highly doped glasses; (ii) a very polarizable anion; (iii) a mixed former composition.

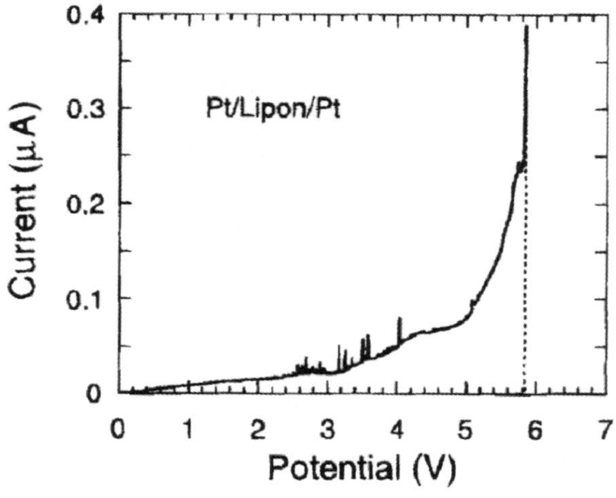

Figure 11.7 The electrochemical stability window of Lipon thin films (reproduced from [47] by permission of The Electrochemical Society, Inc.)

In the early 1980s, the first thin-film electrolytes for lithium microbatteries were designed. The easiest approach for the first step was to reproduce the existing bulk materials, in other words, oxide glasses. For example, a $0.69[0.55B_2O_3\ 0.45Li_2O]\ 0.31Li_2SO_4$ electrolyte has been deposited and used in a TiS_2//Li microbattery [53]. This glassy composition is characterised by a high lithium concentration. Its ionic conductivity is 2×10^{-7} S/cm, the electronic conductivity is about 10^{-13} S/cm. This material is stable in contact with lithium, and has an electrochemical stability window of 5 V. These data are very close to those of the corresponding bulk glass. Simultaneously, a radically new approach was initiated. Amorphous thin films were prepared from a $0.4Li_3PO_4\ 0.6Li_4SiO_4$ crystallized target [54]. This composition is characterized by two features: it is a mixed former material (it can be written $0.78Li_2O\ 0.22(0.5P_2O_5\ 0.5SiO_2)$), and both components, Li_3PO_4 and Li_4SiO_4 are fully modified. This composition stands outside the vitreous domain obtained using melt quenching, but is amorphized using sputtering. The presence of two former cations probably makes the occurrence of crystallization more difficult. The ionic conductivity reaches a very high value, 5×10^{-6} S/cm. Unfortunately, the long-term stability of this electrolyte in contact with lithium is not satisfactory [6].

At the same time, bulk sulfide glasses were discovered, which have an ionic conductivity as high as 10^{-3} S/cm [25, 55]. Because of their hygroscopicity, lithium sulfide thin films have received only little attention. It seems that there have been only two successful attempts, using the P_2S_5–SiS_2–Li_2S–LiI system [51] and the GeS_2–Ga_2S_3–Li_2S system [56]. Oxysulfide glasses with an oxide content of about 5%, for instance in the SiS_2–Li_2S–Li_4SiO_4 system, have been prepared by mechanochemical synthesis and used in bulk solid-state batteries [57]. Their conductivity is similar to sulfide glasses, 10^{-4} S/cm. The introduction of a small proportion of an oxide into a sulfide glass has been found to stabilize it without a detrimental effect on conductivity. Prior to these recent studies, oxysulfide glasses had been used in the form of thin films. A very efficient TiS_2//Li microbattery with a P_2S_5–Li_3PO_4–LiI glass has been fabricated [52]. This electrolyte, however, is not stable in contact with lithium; it must be protected by a LiI layer. This line of research seems to have been abandoned.

A decisive breakthrough was made in the 1990s. It had been shown that, in bulk oxide glasses, the incorporation of nitrogen improves the ionic conductivity [58]. This is partly due, once again, to the higher polarizability of nitrogen, and probably also to some structural modifications. Nitrogen incorporation is quite easy using sputtering: N_2 can simply be introduced into the discharge gas. In this way, Lipon glassy thin films have been prepared from a crystallized Li_3PO_4 target sputtered in an N_2 plasma [36]. The target composition corresponds to a fully depolymerized crystalline material consisting of Li^+ and PO_4^{3-} ions, which cannot be amorphized using melt quenching. The ionic conductivity of the Lipon thin films is 2×10^{-6} S/cm. Their electronic conductivity is 10^{-13} S/cm. This material is stable in contact with lithium, and has an electrochemical stability window of 5.5 V. It has been used by many research groups in several microbatteries [59–62] and electrochromic systems [63]. A systematic study of the influence of the sputtering parameters (power, pressure, target–substrate distance: see Figure 11.8) on Lipon properties has been performed recently [64]. It is worth noting that a pure nitride material, such as crystalline Li_3N, has a very narrow electrochemical window; however, lithium sulfide bulk glasses doped with Li_3N have an improved ionic conductivity as well as a wide electrochemical stability window [65].

The very high quenching speed afforded by sputtering has been a factor in the success in increasing the modifier amount in order to prepare some of these fully modified glasses.

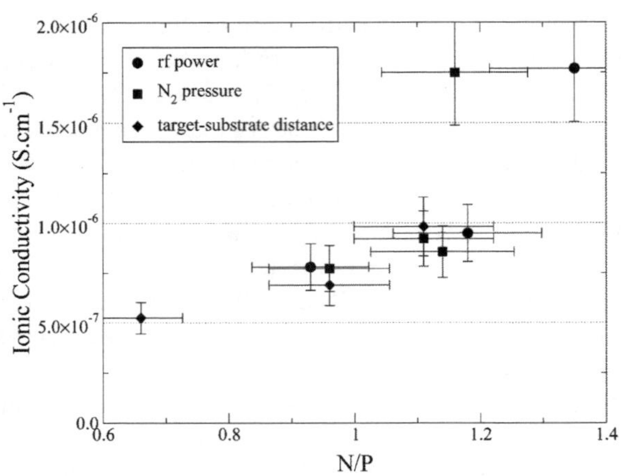

Figure 11.8 The variation of conductivity of LiPON thin films with nitrogen/phosphorus ratio, at various sputtering parameters (power, pressure, target–substrate distance) (redrawn from [64] with permission from Elsevier)

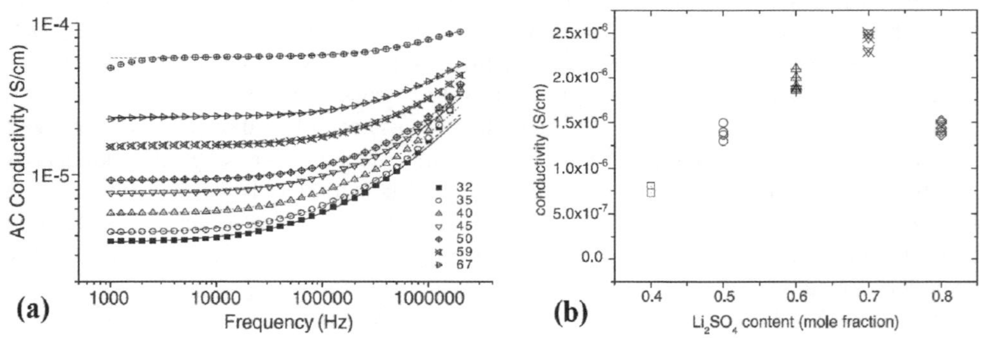

Figure 11.9 (a) Frequency dependence of AC conductivity of $0.3LiBO_2$–$0.7Li_2SO_4$ thin film. (b) The room-temperature DC conductivity of (1-x)$LiBO_2$–xLi_2SO_4 samples vs the amount of Li_2SO_4 (reproduced from [15] with permission from Elsevier)

Similarly, the vitreous domain of the B_2O_3–Li_2O–Li_2SO_4 system can be extended towards compositions richer in doping salt, resulting in a higher ionic conductivity [15]. Materials along the (1-y)[$0.5B_2O_3$ $0.5Li_2O$] yLi_2SO_4 line have been prepared (see Figure 11.2a). The $y = 0.7$ thin film is amorphous; this proves that sputtering enables extension of the vitreous domain, which is limited to $y = 0.5$ when melt quenching is used. Its room temperature ionic conductivity is 2.5×10^{-6} S/cm. Regarding conductivity measurements, this ultradoped thin film behaves exactly like a bulk glass, as evidenced in Figure 11.9a, which shows its conductivity spectra. Its electronic conductivity is 10^{-13} S/cm. This material is stable in contact with lithium, with an electrochemical stability window of 5.8 V. For a composition even richer in lithium, $y = 0.8$, the thin film is partially crystallized and the ionic conductivity decreases, as shown in Figure 11.9b. The 0.3[$0.5B_2O_3$ $0.5Li_2O$] $0.7Li_2SO_4$ electrolyte has been tested in a TiS_2//Li battery. It is worth noting that a similar approach can be applied to any vitreous system, and that a deposition under nitrogen could even improve the ionic conductivity.

A combination of some of the approaches outlined above—fully modified glass, mixed former, nitridation—has proven successful. Lisipon thin films have been prepared from a $0.25[0.8P_2O_5\ 0.2SiO_2]\ 0.75Li_2O$ target deposited under N_2 [66]. Their ionic conductivity is 1.2×10^{-5} S/cm. This electrolyte was used in a $LiCoO_2//Si_{0.7}V_{0.3}$ microbattery [67].

All the aforementioned materials are derived from bulk glasses. A new approach to improving ionic conductivity consists in taking advantage of the capacity of sputtering to prepare original materials with no bulk homologs. A lithium ion amorphous thin film can be designed by selecting a highly polarizable anion carrying a high negative charge compensated by two, three, or even more lithium ions. A first attempt using this approach has been the sputtering of a Li_2SO_4 target in nitrogen. The Lison thin films ($Li_{0.29}S_{0.28}O_{0.35}N_{0.09}$) obtained reach 2×10^{-5} S/cm. The high temperature α-Li_2SO_4 phase in the bulk form is known for its very high ionic conductivity, due to the rotational freedom of the sulphate groups [68]; the nature of the mechanisms in fully modified glasses can be questioned: is it a glass-like mechanism or a paddle-wheel mechanism as in α-Li_2SO_4? The role of nitrogen here is two-fold: it increases the polarizability and, as a second anion, prevents crystallization. The material obtained opens the door to new materials that do not mimic existing glasses. Other promising avenues of inquiry include glass–ceramic thin films similar to the glass–ceramic bulk materials already mentioned [34], as well as composite electrolytes made of a dispersion of a conducting phase in nanodispersed biphasic materials with an interface conductivity mechanism [69, 70].

11.3.4 Resulting devices

The first microbattery with a fully demonstrated efficiency was the TiS_2/P_2S_3–Li_3PO_4–LiI/Li system. These microbatteries release a capacity of $170\,\mu A\,h\,g^{-1}$ when cycled at a current density up to $100\,\mu A/cm^2$, between 2.8 V and 1.8 V [49]. However, this system has not yet left the laboratory.

It seems that two companies, at least, are about to launch microbatteries on the market. The French company HEF has recently built a pilot machine capable of producing 300 microbatteries per day (size $2\,cm^2$) [71]. The system is TiO_yS_z/borate glass/Li. Little data on its performances is available, but the amorphous positive electrode material has been investigated in detail [72]. Its possible advantages include its amorphous state (no annealing required), and its ability to operate at quite high thicknesses, which increases surface capacity.

Several systems have been developed at Oak Ridge National Laboratory using Lipon as an electrolyte. One of the most promising systems is the $LiCoO_2$/Lipon/Li microbattery [6], typical cycling curves are shown in Figure 11.10. The crystallized $LiCoO_2$ positive electrode has been extensively used in commercial lithium ion bulk batteries and it is quite well known. An alternative negative electrode is silicon tin oxynitride. The performances of this device are excellent. The capacity is about $150\,\mu A\,h\,cm^{-2}$ for a 2.4-μm-thick positive electrode under a current density of $100\,\mu A/cm^2$; it is still $50\,\mu A\,h\,cm^{-2}$ under a current density of $4\,mA/cm^2$. The microbatteries have been cycled 4000 times between 4.2 and 3 V, and even 40 000 times for a 0.05-μm-thick positive electrode.

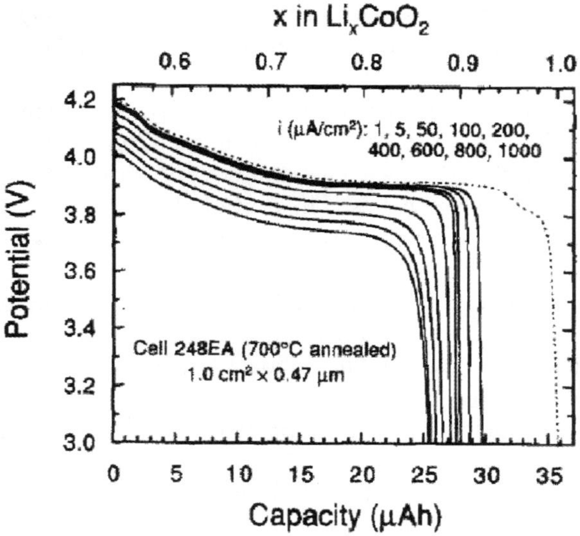

Figure 11.10 Cycling of a $LiCoO_2$/Lipon/Li microbattery (reproduced from [61] by permission of the Electrochemical Society, Inc.)

11.4 SILVER-ION-CONDUCTING CHALCOGENIDE GLASSES IN SOLID-STATE IONIC MEMORIES AND SENSORS

Silver chalcogenide glasses and thin films have many applications in optics, biology, chemistry and electronics. They have received considerable attention during recent years, and a review of their properties and applications is given in [73]. In optics, silver mobility in chalcogenide glasses has been used to fabricate *in situ* waveguides [74]. This is another interesting application of ion transport in glass. In this section we will describe two applications of ion transport related to electronics: solid-state ionic memory and sensors.

11.4.1 Solid-state ionic memory

The memory market is characterized by a continual increase of bit density, and is concerned with several requirements such as nonvolatibility, size and endurance. None of the current semiconductor technologies based on charge storage (e.g., DRAM, SRAM, Flash) can fulfill all these requirements at the same time [75]. Therefore, intense research efforts are aimed at developing a new generation of memory devices. A material-based memory effect requires a material that exists in two stable states, and is capable of switching from one state to the other under low-energy excitation. The two states should be easily readable. Chalcogenide glasses are interesting materials because they exist in states between which the energy difference is not very large. They are used in two competing new nonvolatile memories, the programmable metallization cell memory (PMCm or PMC-RAM) and the phase-change memory (PC-RAM). In the latter, the two states are the amorphous and the crystallized state of a chalcogenide glass [76]. Such amorphous–crystalline phase transformations induced by optical means are the working principle of CDs and DVDs. These transformation are induced by a small bias in a PC-RAM. The former is closely related to ion transport, and will be discussed in the following paragraphs. This technology has been developed for more than 10 years at Arizona State University, in collaboration with the electronic industry giant Infineon Technologies AG [10] and with the semiconductor fabricant Micron Technology Inc. It could arrive on the market within the next few years.

(a) Principle

The electrical and structural properties of silver chalcogenide glasses are at the core of the operational principle of PMC memory. Schematically, a PMCm element consists of three layers: a cathode made of an inert metal, a silver chalcogenide glassy electrolyte and a silver anode, as shown in Figure 11.11 [77]. If a low bias is applied to the device, the silver chalcogenide glass behaves as a standard electrolyte. In this state, it has a very low electronic conductivity, and the resistance of the device is high. This is the 'off' state. If the bias is increased above a defined threshold (a few hundred mV), electrons from the cathode reduce silver ions from the electrolyte at the electrolyte–cathode interface to form a silver deposit. Simultaneously, silver atoms from the anode are oxidized at the silver anode–electrolyte interface and silver ions are injected into the electrolyte to compensate for those reduced at the cathode. The deposit grows from the cathode to form a high-conductivity metallic pathway between the two electrodes. This process leads to an 'on' state of low resistance; it is therefore a writing process. The resistance change of the device is of several orders of magnitude, occurring within about 10 ns, even for currents in the mA range [78]. The application of a reversed bias makes it possible to drive the electrolyte back to its initial state; this is the erasing process. The 'on' and 'off' states can easily be read by the application of a smaller bias. These operations are necessary to the creation of memory, which is indeed nonvolatile.

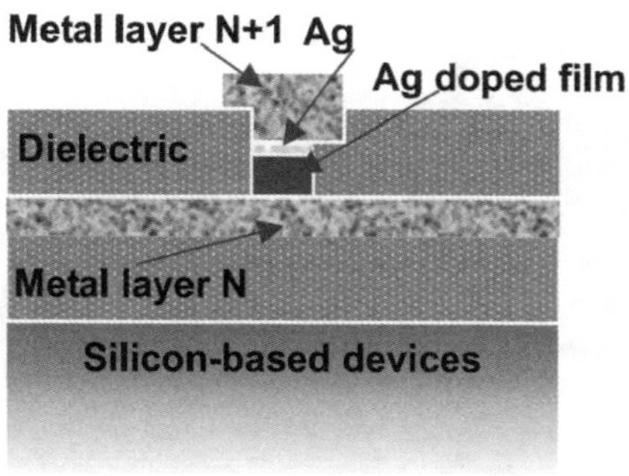

Figure 11.11 Sketch of a cross-section of a single programmable metallization cell in an active-in-via configuration (reproduced from [77] with permission from Elsevier)

(b) Materials and mechanisms

We shall now examine more closely the electrolyte preparation, the microstructure, and the mechanisms that occur during the writing process. Electrolyte deposition is a three-step photodiffusion process. First, a former glass layer is thermally evaporated onto an inert metallic electrode. A silver layer is then deposited on the top of it. For instance, a 50-nm-thick $Ge_{30}Se_{70}$ thin film is covered by a 25-nm-thick silver layer, without breaking the vacuum in the evaporation machine. Finally, a photodissolution process is carried out to introduce silver into the chalcogenide glass. This thickness combination ensures the saturation of the chalcogenide layer. A tungsten lamp is used to provide both heat and light to form the final $Ag_{33}Ge_{20}Se_{47}$ (i.e., $0.37Ge_2Se_3$ $0.63Ag_2Se$ [78]) thin film, leaving a 10-nm-thick silver layer. The thermal and photoinduced diffusion processes have been carefully analyzed using RBS, AES and XPS [79, 80]. Alternatively, silver can be diffused into the glass using a UV irradiation.

The microstructure of the $Ag_{33}Ge_{20}Se_{47}$ thin film has been studied using XRD [19, 20, 78]. It is, in fact, nanostructured: it consists of Ag_2Se crystallites dispersed in a Ge rich amorphous phase, as seen in Figure 11.12a [81]. The size of the crystallites, 7.5 nm, is determined from the width of the XRD peaks. The Ag-rich conducting pathways may be made of Ag_2Se regions in series with metallic Ag clusters containing not more than 100 atoms [78]. Indeed, the distance between the crystallites is estimated to be around 1.5 nm. Therefore, the distance that an ion has to travel in order to be reduced and incorporated into the metallic pathway during a writing process is very low.

(c) Device fabrication and electrical characterization

The nanometric structure of the electrolyte material and processes suggests that it should be possible to scale down a single cell to around 10 nm. The fabrication uses standard

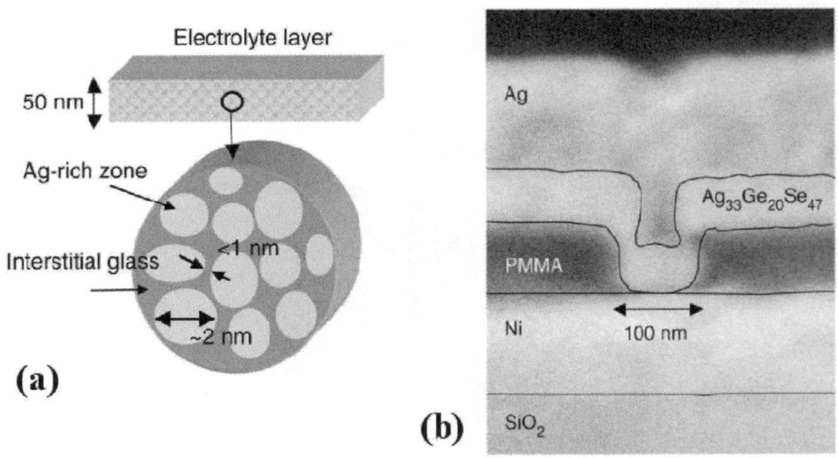

Figure 11.12 (a) Scheme of the nanostructure of the Ag–Ge–Se thin film electrolyte. In a further study, the size of the Ag_2Se crystallites was more precisely determined to be 7.5 nm [78]. (b) A cross-section of a PCM cell (reproduced from [81] with permission from Elsevier)

CMOS equipment. It starts with the deposition of an inert nickel electrode onto a SiO_2-coated Si wafer. Next, a dielectric layer is deposited, then etched to create openings that will be filled by the electrolyte, so that each hole will be a single PMCm cell. The dielectric can be either SiO_2 or a 100-nm-thick PMMA (poly-methylmethacrylate) layer deposited using spin coating. The advantage of the polymer dielectric is that holes, as small as 40 nm diameter, can be made easily using electron beam lithography. The electrolyte is then formed as described above, so that the holes are filled, and the silver electrode is deposited followed by a gold protective layer. Specially designed membrane evaporation sources are used to preserve the composition of the source bulk glass: a $Ge_{20}Se_{80}$ source yields a $Ge_{22}Se_{78}$ thin film [79]. The stack is then patterned using optical lithography as well as sputter and wet etching. A field-emission scanning electronic micrograph in Figure 11.12b shows that the walls in the 40-nm-diameter dielectric holes are virtually vertical and very well filled by the electrolyte.

The current–voltage characteristics of the test-chips are shown in Figure 11.13a. The cycle starts at a negative voltage; the low conductivity 'off' state leads to a very low current. The voltage is then increased. The nanoscaled device is written at an applied bias of 0.2 V, which drives it to the high-conductivity 'on' state. The current increases until the compliance level is reached. Then the bias is reversed and the current decreases. The device is erased at −0.5 V: the metallic pathways are destroyed, the resistance recovers its high value, and the current reverts to its initial very low value. The device resistance changes from over $10^7\,\Omega$ to $10^4\,\Omega$ as shown in Figure 11.13b. The high resistance value depends on the thickness of the electrolyte (50 nm), on the cell dimensions (40 nm diameter) and on electrolyte conductivity in the 'off' state (in the 10^{-6} S/cm range). On the contrary, the low resistance value depends only on the metallic pathway formation, i.e., on the writing current intensity and duration. The particular nanostructure of the film explains its high conductivity in the 'on' state: 10^2 S/cm.

The PMCm characteristics make it a very attractive candidate for future nonvolatile memories. Key attributes are low voltage and current (0.2 V and 10 μA), scalability (down to 10 nm), retention (10 years) and endurance (10^{10} cycles), rapid writing and erasing (less than 100 ns), and simple fabrication [10].

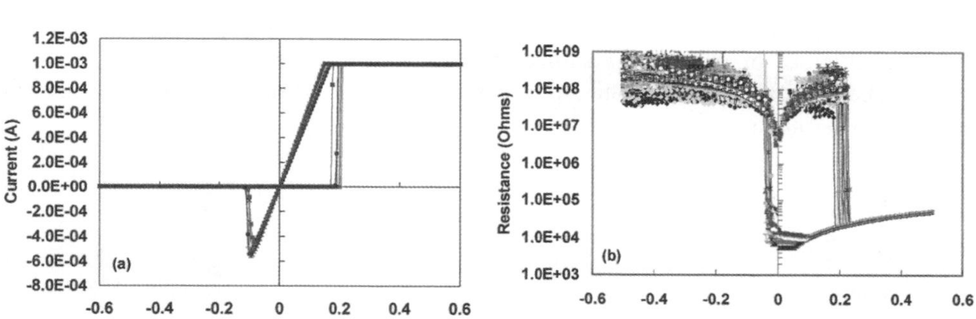

Figure 11.13 (a) Current–voltage plots for a 40 nm and a 75-nm Ag–Ge–Se chip; (b) resistance–voltage for a 75-nm Ag–Ge–Se cell (reproduced from [78] by permission of IEEE)

(d) Latest trends

The Ag–Ge–Se thin films cannot withstand temperatures above 200°C, which prevents soldering operations [12]. Other systems have been investigated to circumvent this limitation: Ag–Ge–S and Cu–Ge–S [82]. In addition, a substantially different material, Cu–WO_3, has been tested. In this case, a WO_3 layer is formed between a copper electrode and an inert electrode. The application of a small bias allows copper to be introduced into the WO_3 layer. The conductivity change is also very large for this system. However, WO_3 is not strictly an electrolyte, it is even used as an electrode in electrochromic systems or batteries. In this context, the introduction of copper ions (and electrons) into WO_3 is called intercalation. The utilization of an intercalation material in a solid-state ionic memory underlines the fact that the key is to utilize a material whose conductivity can be easily switched. This is the case for an electrolyte such as Ag–Ge–Se, but also for electrode materials such as WO_3 or $LiCoO_2$ whose conductivity depends drastically on the amount of intercalated metal ions. Besides, photodissolution itself is not a purely electrolyte process, since electron transport should be involved.

A very similar device has been designed recently [83]. A silver nanobridge can be formed and erased at the crossing point between two electrodes spaced 1 nm apart, by applying a small bias. The device, schematized in Figure 11.14, is described as a nanomechanical relay. Its fabrication is quite simple. An Ag_2S wire is first formed by sulfurization of a silver wire, and covered by a silver layer. The crossing platinum electrode is then deposited. When a bias is applied, silver atoms are oxidized and silver ions move inside the Ag_2S solid electrolyte, creating a gap between the two electrodes. The nanorelay has been opened. Further operations involve the creation and annihilation of an atomic silver bridge between the two electrodes, defining 'off' and 'on' states. Basic circuits, 'and', 'or' and 'not' gates, have been constructed using conventional nanofabrication methods. The device works at room temperature and in the air; it is fast, requires low power and is highly scalable. Moreover, conductance is quantized, which could open up new opportunities.

11.4.2 Sensors

Potentiometric chemical sensors are increasingly used in numerous applications, such as chemical processes control [84], environmental control [85, 86], food inspection [87, 88], and biomedical applications. They are electrochemical devices that are used to determine the concentrations of particular chemical species which are usually found in gases or liquids. These systems must have the highest sensitivity to, and selectivity for, the element under study as possible.

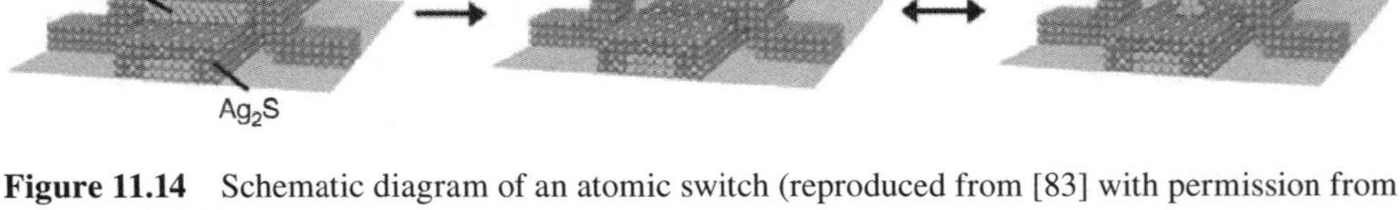

Figure 11.14 Schematic diagram of an atomic switch (reproduced from [83] with permission from Nature Publishing Group)

(a) Principle

The typical device (Figure 11.15) consists of an i-ion solid electrolyte and a reference electrode where the chemical potential (μ_i) is known and usually fixed. The other electrode is formed by the compound under study. The measurement of the electrical potential difference between the electrodes is directly related to the difference in chemical potential of mobile species at the interface of two electrolyte surfaces. The chemical potential in the reference electrode is known; the chemical potential in the compound under study depends on the i-ion concentration.

Type-I sensors are restricted to the detection of species for which a convenient electrolyte exists. Such a material can not be found for all chemical species. Therefore, in order to overcome this problem, indirect type-II sensors have been proposed [90]. These sensors can be built if there exists a simple relation between the chemical potential of the nonmobile species under study (μ_x) and the chemical potential of the mobile i-species. When the species to be detected are not involved in the electrolyte, an auxiliary solid-phase layer can be used. This corresponds to type-III sensors.

(b) Materials

There are distinct benefits to using glassy materials in chemical sensors. Due to the high chemical stability of vitreous materials, they can be applied in many cases where potentiometric sensors could not be used before. For example, they exhibit better chemical durability

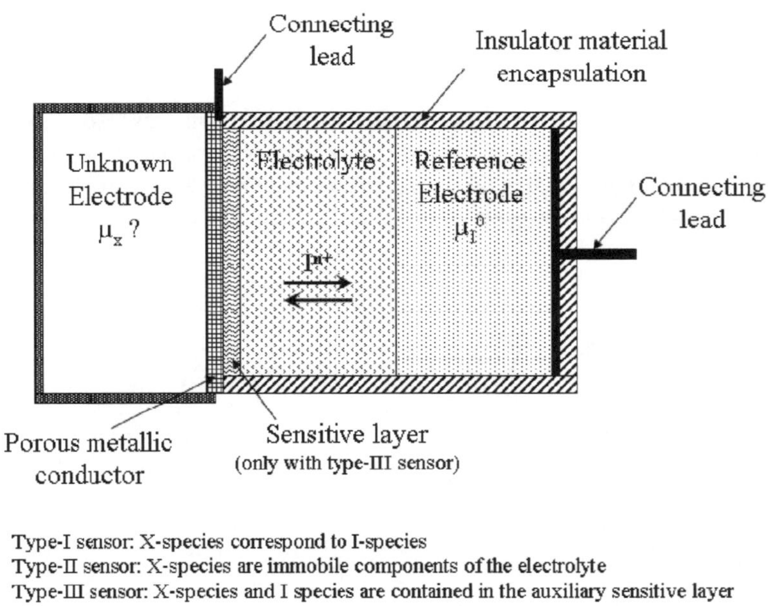

Figure 11.15 Conceptual drawing of a solid-state ionic electrochemical sensor (redrawn from [89] with permission from Elsevier)

in acidic and redox media, better selectivity and reproducibility of analytical characteristics, and a longer life than the polycrystalline chalcogenides, which are widely used as membrane materials. In particular, chalcogenide glasses were proven to be very promising ion selective membranes for the detection of heavy metals in solutions (Ag^+, Cd^{2+}, Cu^{2+}, Pb^{2+} . . .), but other glasses have also been used (aluminosilicate glasses in Na^+ sensors [91], for example). Moreover, as is typical for glassy materials, the possibility to greatly vary the composition of the glass systems can provide a wide spectrum of desirable properties, depending on the problem to be solved. Moreover, these sensors are simple in construction, and special conditions are required neither for their storage nor for surface recovering, even after continuous measurements.

(c) Mechanism

The potential generation concepts are developed according to the generally accepted theory that only the active groups localized on the working membrane surface participate in the exchange processes [92]. A model has been proposed to explain the sensitivity mechanism of sensors with chalcogenide glassy membranes by accepting the existence of a modified surface layer [93]. This layer appears on the membranes surface after the contact with the analyzed gas or liquid, and results from the interaction between the solution of the potential determining ion and the partially destroyed glassy network, which is accompanied by the creation of active exchange centers. A sketch of the modified surface layer formation is shown in Figure 11.16 [3]. The potential generation is based on the assumption of a chemical exchange of ionic species between the electrolyte and the centers on the glassy membrane surface.

(d) Devices

Recently, considerable interest has been attracted by chemical microsensors, a development which is closely related to the trend toward device miniaturization, ecology of production and natural raw materials. A general microsensor scheme is shown in Figure 11.17 [94], and several glasses used in microsensors are listed in Table 11.1 [3]. The sensitive thin film

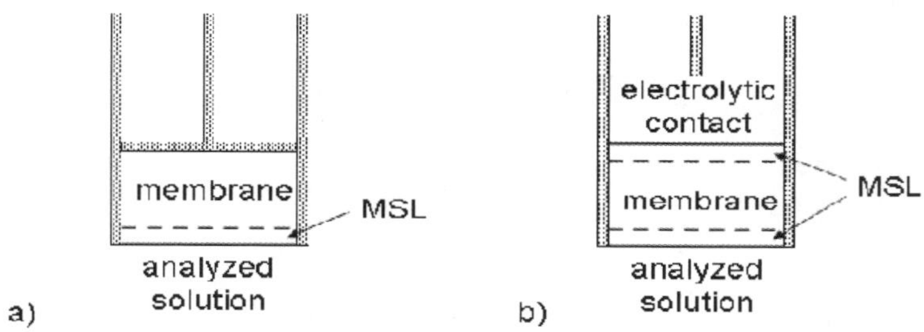

Figure 11.16 Modified surface layer (MSL) formation at: (a) solid-state ion-selective electrode; (b) ion-selective electrode with liquid inner connection (redrawn from [3] with permission from Elsevier)

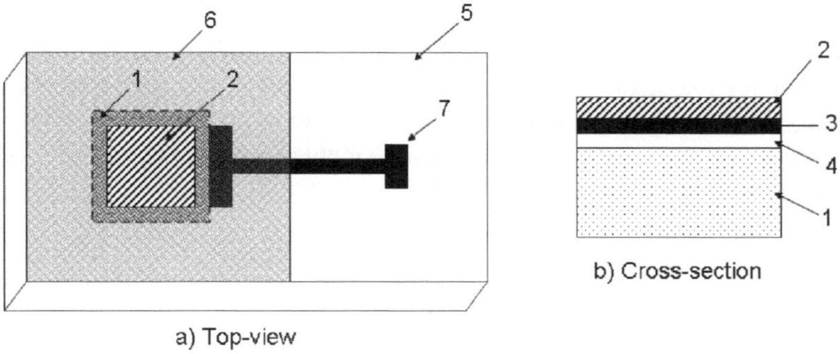

Figure 11.17 Thin-film microsensor: 1 Si wafer; 2 chalcogenide glassy film; 3 contact metal film; 4 insolating film; 5 chip carrier; 6 hermetic film; 7 metal film (redrawn from [94] with permission from Springer)

Table 11.1 Several chalcogenide glasses systems investigated as active membrane materials in microsensors [3]

Ion	Glass composition (at%)	Reference
Ag^+	Ag–As–S	[95]
$Cd2^+$	$(CdS)_x(AgI–Sb_2S_3)_{1-x}$, $x = 0–5$	[96]
	CdS–AgI–AsS	[97]
	Cu–Ag–As–Se–Te	[94; 98]
Cu^{2+}	$Cu_xAg_y(As_2S_3)_{100-(x+y)}$, $x = 15–30$, $y = 10–20$	[99]
	$Cu_xAg_y(As_2Se_3)_{100-(x+y)}$, $x = 15–30$, $y = 10–20$	[99]
$Hg2^+$	$As_2S_3–Ag_2S–AgBr$	[100]
	$5HgTe–95Ge_{0.2}Te_{0.3}Se_{0.5}$	[101]
$Pb2^+$	Pb–AgI–AsS	[102]
	Pb–AgI–AsS	[97]

is deposited by standard microelectronics techniques, e.g., thermal vacuum evaporation, cathode sputtering or laser ablation, onto an insulator substrate (Si/Si_3N_4, Si/SiO_2) covered with contact metal films (Cr, Ag, Ti/Pt). This kind of sensor has no inner reference electrodes, and an external reference electrode must be used. The most important point in microsensor construction is to ensure stable contact on the metal/functional film interface, which is held together by coating adhesion.

Among the microsensors, there exists a system called ISFETs (ion-selective field-effect transistors) which has been developed on the basis of MOSFET (metal oxide silicon field-effect transistor) [11, 103]. The basic principle is the control of a current flowing between two semiconductor electrodes. Drain and source are placed on the same element, with a third electrode, the gate, between them. The gate is insulated against drain and source by silicon dioxide and can only influence the drain-source current electrostatically. In ISFETs, the gate is the sensitive membrane. The membrane's control of the strain current is solely based on the electrostatic effect and allows an ion concentration to be determined.

Recently, multisensor systems based on an array of nonspecific chemical sensors have been studied [104]. This approach consists in combining a number of cross-selective sensors matched with a suitable data processing technique to analyze the multidimensional output of this sensor array. With these sensors it is possible to retrieve, with a certain accuracy, the concentrations of several elements in multicomponent environments [105], or to distinguish different solutions from each other [106].

11.5 CONCLUSIONS

The design of materials for applications such as microbatteries, solid-state ionic memories and sensors, has come to maturity, and devices are marketed or about to be launched. In particular, chemical compositions have been optimized in view of improving performances. In addition to the homogeneous traditional glasses, new nanostructured materials are emerging, especially chalcogenide thin films for solid-state ionic memories and also glass–ceramic electrolytes for batteries. Search for the mechanisms of ion transport in these materials continues to be challenging. Although the applications cited above belong to very different fields, operations are intimately related to electrochemistry and the materials used are closely similar. Up to now, research in these fields has progressed rather independently, but further progress in one of them will probably apply to the others since the key issues seem to converge.

REFERENCES

[1] E. Warburg, Über die Electrolyse des festen Glases, *Wiedemann Ann. Phys.*, **21**, 622–646 (1884).

[2] M. Cremer, Über die Ursache der elektromotorischen Eingenschaften der Gewebe, zugleich ein Beitrag zur Lehre von den polyphasischen Elektrolytketten, *Z. Biol. (Munich)*, **47**, 562–608 (1906).

[3] V.S. Vassilev and S.V. Boycheva, Chemical sensors with chalcogenide glassy membranes, *Talanta*, **67**, 20–27 (2005).

[4] G.-A. Nazri and G. Pistoia (eds), *Lithium batteries: science and technology*, Kluwer Academic Publishers, Boston, 2003.

[5] W. Vielstich, A. Lamm and H. Gasteiger (eds), *Handbook of Fuel Cells*, John Wiley & Sons, Ltd, Chichester, 2003.

[6] J.B. Bates, N.J. Dudney, B.J. Neudecker, and B. Wang, Thin-Film Lithium Batteries, in: *New Trands in Electrochemical Technology: Energy Storage in Electronics*, T. Osaka, and M. Datta (eds), Gordon and Breach, Singapore, 2000, 453–485.

[7] R.A. Huggins, Supercapacitors and electrochemical pulse sources, *Solid State Ionics*, **134**, 179–195 (2000).

[8] C.G. Granqvist, Electrochromic devices, *J. Eur. Ceram. Soc.*, **25**, 2907–2912 (2005).

[9] B. Price, *Emerging Memories: Technologies and Trends*, Springer, 2002.

[10] R. Symanczyk, M. Balakrishnan, C. Gopalan, T. Happ, M. Kozicki, M. Kund, T. Mikolajick, M. Mitkova, M. Park, C.U. Pinnow, J. Robertson, and K.D. Ufert, Electrical characterization of solid state ionic memory elements, *Proceedings of the non-volatile memory technology symposium*, San Diego, 17–1, 2003.

[11] M. Kleitz, J.F. Million-Brodaz, and P. Fabry, New compounds for ISFETS, *Solid State Ionics*, **22**, 295–303 (1987).

[12] T. Sakamoto, H. Sunamura, and H. Kawaura, Nanometer-scale switches using copper sulfide, *Appl. Phys. Lett.*, **82**, 3032–3034 (2003).

[13] A.C. Wright, Neutron and X-Ray amorphography, in: *Experimental techniques of glass science*, C.J. Simmons and O.H. El-Bayoumi (eds), American Ceramic Society, Westerville, 1993, 205–314.

[14] G.D. Chryssikos, E.I. Kamitsos, and A.P. Patsis, Effect of Li_2SO_4 on the structure of Li_2O–B_2O_3 glasses, *J. Non-Cryst. Solids*, **202**, 222–232 (1996).

[15] K.-H. Joo, P. Vinatier, B. Pecquenard, A. Levasseur, and H.-J. Sohn, Thin film lithium ion conducting LiBSO solid electrolyte, *Solid State Ionics*, **160**, 51–59 (2003).

[16] T. Kawaguchi and S. Maruno, Optical, electrical, and structural properties of amorphous Ag–Ge–S and Ag–Ge–Se films and comparison of photoinduced and thermally induced phenomena of both systems, *J. Appl. Phys.*, **79**, 9096–9104 (1996).

[17] M. Ribes, E. Bychkov, and A. Pradel, Ion transport in chalcogenide glasses: dynamics and structural studies, *Journal of Optoelectronics and Advanced Materials*, **3**, 665–674 (2001).

[18] A. Levasseur, M. Kbala, J.C. Brethous, and J.M. Réau, Etudes électriques et raman des verres du système B_2O_3–Li_2O–Li_2SO_4, *Solid State Common*, **32**, 839–844 (1979).

[19] P. Boolchand and W.J. Bresser, Mobile silver ions and glass formation in solid electrolytes, *Nature*, **410**, 1070–1073 (2001).

[20] P. Boolchand and W.J. Bresser, The structural origin of broken chemical order in $GeSe_2$ glass, *Philos. Mag. B*, **80**, 1757–1772 (2000).

[21] M. Mitkova, Y. Wang, and P. Boolchand, Dual chemical role of Ag as an additive in chalcogenide glasses, *Phys. Rev. Lett.*, **83**, 3848–3851 (1999).

[22] M. Ganguli, and K.J. Rao, Studies on the effect of Li_2SO_4 on the structure of lithium borate glasses, *J. Phys. Chem. B*, **103**, 920–930 (1999).

[23] M.D. Ingram, G.D. Chryssikos, and E.I. Kamitsos, Evidence from vibrational spectroscopy for cluster and tissue pseudophases in glass, *J. Non-Cryst. Solids*, **131–133**, 1089–1091 (1991).

[24] T. Kudo, Survey of types of solid electrolytes, in: *The CRC handbook of solid state electrochemistry*, P.J. Gellings and H.J.M. Bouwmeester (eds), CRC, Boca Raton, 1997, 195–221.

[25] J.P. Malugani, B. Fahys, R. Mercier, G. Robert, J.P. Duchange, S. Baudry, M. Broussely, and J.P. Gabano, De nouveaux verres conducteurs par l'ion lithium et leurs applications dans des générateurs électrochimiques., *Solid State Ionics*, **9–10**, 659–666 (1983).

[26] A. Bunde, M.D. Ingram, and P. Maass, The dynamic structure model for ion transport in glasses, *J. Non-Cryst. Solids*, **172–174**, 1222–1236 (1994).

[27] E. Bychkov and D.L. Price, Neutron diffraction studies of Ag_2S–As_2S_3 glasses in the percolation and modifier-controlled domains, *Solid State Ionics*, **136–137**, 1041–1048 (2000).

[28] P. Vinatier, M. Ménétrier, and A. Levasseur, Aggregation of doping salt in B_2S_3–Li_2S–LiI glasses, effect on the dynamical properties, *Solid State Ionics*, **116**, 35–45 (1999).

[29] P. Vinatier, M. Ménétrier, and A. Levasseur, Structure and ionic conduction in lithium thioborate glasses and crystals, *Phys. Chem. Glasses*, **44**, 135–142 (2003).

[30] N. Machida, S. Nishida, T. Shigematsu, H. Sakai, M. Tatsumisago, and T. Minami, Mechano chemical synthesis of a silver ion conductor in the system AgI–Ag_3PO_4, *Solid State Ionics*, **136–137**, 381–386 (2000).

[31] K. Funke, Transport and relaxation in ionic cristals, *Phys. Chem.*, **93**, 1197–1205 (1989).

[32] G. Robert, J.P. Malugani, and A. Saida, Fast ionic silver and lithium conduction in glasses, *Solid State Ionics*, **3–4**, 311–315 (1981).

[33] J. Kincs and S.W. Martin, Non-Arrhenius conductivity in glass: mobility and conductivity saturation effects, *Phys. Rev. Lett.*, **76**, 70–73 (1996).

[34] F. Mizuno, A. Hayashi, K. Tadanaga, and M. Tatsumisago, New, highly ion-conductive crystals precipitated from Li_2S–P_2S_5 glasses, *Adv. Mater.*, **17**, 918–921 (2005).

[35] J.W. Long, B. Dunn, D.R. Rolison, and H.S. White, Three-dimensional battery architectures, *Chem. Rev.*, **104**, 4463–4492 (2004).

[36] J.B. Bates, N.J. Dudney, G.R. Gruzalski, R.A. Zuhr, A. Choudhury, and C.F. Luck, Electrical properties of amorphous lithium electrolyte thin films, *Solid State Ionics*, **53–56**, 647–654 (1992).

[37] M.-H. Lindic, B. Pecquenard, P. Vinatier, A. Levasseur, H. Martinez, D. Gonbeau, P.-E. Petit, and G. Ouvrard, Characterization of rf sputtered TiO_xS_y thin films, *Thin Solid Films*, **484**, 113–123 (2005).

[38] A. Hohl, T. Wieder, P.A. Van Aken, T.E. Weirich, G. Denninger, M. Vidal, S. Oswald, C. Deneke, J. Mayer, and H. Fuess, An interface clusters mixture model for the structure of amorphous silicon monoxide (SiO), *J. Non-Cryst. Solids*, **320**, 255–280 (2003).

[39] J.B. Bates, N.J. Dudney, B. Neudecker, A. Ueda, and C.D. Evans, Thin-film lithium and lithium-ion batteries, *Solid State Ionics*, **135**, 33–45 (2000).

[40] J.-M. Tarascon and M. Armand, Issues and challenges facing rechargeable lithium batteries, *Nature*, **414**, 359–367 (2001).

[41] P. Soudan, T. Brousse, G. Taillades, J. Sarradin, and K. Zaghib, Ag/$Li_zMn_{1.5}Ni_{0.5}O_4$ thin film lithium ion microbattery, *Proceedings Electrochemical Society*, **20**, 633–640 (2003).

[42] B.B. Owens, Solid state electrolytes: overview of materials and applications during the last third of the twentieth century, *J. Power Sources*, **90**, 2–8 (2000).

[43] H.-K. Kim, S.-H. Cho, Y.-W. Ok, and T.-Y. Seong, All solid-state rechargeable thin-film microsupercapacitor fabricated with tungsten cosputtered ruthenium oxide electrodes, *J. Vac. Sci. Technol. B*, **21**(3), 949–952 (2003).

[44] J.H. Lim, D.J. Choi, H.-K. Kim, W.I. Cho, and Y. Soo, Thin film supercapacitors using a sputtered RuO_2 electrode, *J. Electrochem. Soc.*, **143**(3), A275–A278 (2001).

[45] B.E. Conway, V. Birss, and J. Wojtowicz, The role and utilization of pseudocapacitance for energy storage by supercapacitors, *J. Power Sources*, **66**, 1–14 (1997).

[46] C.G. Granqvist, E. Avendano, and A. Azens, Electrochromic coatings and devices: survey of some recent advances, *Thin Solid Films*, **442**, 201–211 (2003).

[47] X. Yu, J.B. Bates, G.E. Jellison, and F.X. Hart, A stable thin-film lithium electrolyte: lithium phosphorus oxynitride, *J. Electrochem. Soc.*, **144**(2), 524–532 (1997).

[48] H. Ohtsuka and J.-I. Yamaki, Electrical characteristics of Li_2O–V_2O_5–SiO_2 thin films, *Solid State Ionics*, **35**, 201–206 (1989).

[49] S.D. Jones and J.R. Akridge, Thin film rechargeable Li batteries, *Solid State Ionics*, **69**, 357–368 (1994).

[50] A. Karthikeyan, P. Vinatier, and A. Levasseur, Study of lithium glassy solid electrolyte/electrode interface by impedance analysis, *Bull. Mater. Sci.*, **23**, 179–183 (2000).

[51] R. Creus, J. Sarradin, R. Astier, and A. Pradel, The use of ionic and mixed conductive glasses in microbatteries, *Mat. Sci. Eng. B-Solid*, **3**, 109–112 (1989).

[52] S.D. Jones and J.R. Akridge, A thin film solid state microbattery, *Solid State Ionics*, **53–56**, 628–634 (1992).

[53] A. Levasseur, M. Kbala, P. Hagenmuller, G. Couturier, and Y. Danto, Elaboration and characterization of lithium conducting thin film glasses, *Solid State Ionics*, **9–10**, 1439–1444 (1983).

[54] K. Kanehori, K. Matsumoto, K. Miyauchi, and T. Kudo, Thin film solid electrolyte and its application to secondary lithium cell, *Solid State Ionics*, **9–10**, 1445–1448 (1983).

[55] A. Levasseur, R. Olazcuaga, M. Kbala, M. Zahir, and P. Hagenmuller, Synthèse et propriétés électriques de nouveaux verres souffrés de conductivité ionique élevée, *C.R. Acad. Sc.*, **293**, 563–565 (1981).

[56] M. Yamashita, H. Yamanaka, and H. Wakabayashi, Thin film preparation of the Li_2S–GeS_2–GaS_3 glass system by sputtering, *Solid State Ionics*, **89**, 299–304 (1996).

[57] R. Komiya, A. Hayashi, H. Morimoto, M. Tatsumisago, and T. Minami, Solid state lithium secondary batteries using an amorphous solid electrolyte in the system (100-x)(0.6Li_2S-0.4SiS_2)–$x Li_4SiO_4$ obtained by mechanochemical synthesis, *Solid State Ionics*, **140**, 83–87 (2001).
[58] H. Unuma, K. Komori, and S. Sakka, Electrical conductivity and chemical durability in alkali-silicate oxynitride glasses, *J. Non-Cryst. Solids*, **95–96**, 913–920 (1987).
[59] J.B. Bates, N.J. Dudney, G.R. Gruzalski, and R.A. Zuhr, Fabrication and characterization of amorphous lithium electrolyte thin films and rechargeable thin-film batteries, *J. Power Sources*, **43–44**, 103–110 (1993).
[60] Y.-S. Park, S.-H. Lee, B.-I. Lee, and S.-K. Joo, All-solid-state lithium thin-film rechargeable battery with lithium manganese oxyde, *Electrochem. Solid St.*, **2**(2), 58–59 (1999).
[61] B. Wang, J.B. Bates, B.C. Sales, and R.A. Zuhr, Characterization of thin-film rechargeable lithium batteries with lithium cobalt oxide cathodes, *J. Electrochem. Soc.*, **143**(10), 3203–3213 (1996).
[62] S.-H. Lee, P. Liu, C.E. Tracy, and D.K. Benson, All-solid-state rocking chair lithium battery on a flexible Al substrate, *Electrochem. Solid St.*, **2**(9), 425–427 (1999).
[63] A. Gerouki and R.B. Goldner, High reflectivity modulation electrochromic windows, *Mat. Res. Soc. Symp. Proc.*, **548**, 679–685 (1999).
[64] Y. Hamon, A. Douard, F. Sabary, C. Marcel, P. Vinatier, B. Pecquenard, A. Levasseur, Influence of sputtering conditions on ionic conductivity of LiPON thin flims, *Solid State Ionics*, **177**, 257–261 (2006).
[65] R. Sakamoto, M. Tatsumisago, and T. Minami, Preparation of fast lithium ion conducting glasses in the system Li_2S–SiS_2–Li_3N, *J. Phys. Chem. B*, **103**, 4029–4031 (1999).
[66] S.-J. Lee, J.-H. Bae, H.-W. Lee, H.-K. Baik, and S.-M. Lee, Electrical conductivity in Li–Si–P–O–N oxynitride thin-films, *J. Power Sources*, **123**, 61–64 (2003).
[67] S.-J. Lee, H.-K. Baik, and S.-M. Lee, An all-solid-state thin film battery using LISIPON electrolyte and Si–V negative electrode films, *Electrochem. Commun.*, **5**, 32–35 (2003).
[68] L. Nilsson, J.O. Thomas, and B.C. Tofield, The structure of the high-temperature solid electrolyte lithium sulfate at 908 K, *J. Phys. C*, **13**, 6441 (1980).
[69] K. Tadagana, K. Imai, M. Tatsumisago, and T. Minami, Preparation of AgI–Al_2O_3 composites with high ionic conductivity using Al_2O_3 aerogel and xerogel, *J. Electrochem. Soc.*, **147**, 4061–4064 (2000).
[70] C.C. Liang, Conduction characteristics of the lithium iodide–aluminum oxide solid electrolytes, *J. Electrochem. Soc.*, **120**, 1289 (1973).
[71] M. Martin and F. Faverjon, A multilayer semi-industrial vacuum deposition equipment for producing ultrathin batteries, *Thin Solid Films*, **398–399**, 572–574 (2001).
[72] M.-H. Lindic, B. Pecquenard, P. Vinatier, A. Levasseur, H. Martinez, D. Gonbeau, P.-E. Petit, and G. Ouvrard, Electrochemical mechanisms during lithium insertion into $TiO_{0.6}S_{2.8}$ thin films positive electrode in lithium microbatteries, *J. Electrochem. Soc.*, **152**, A141–A146 (2005).
[73] M. Frumar and T. Wagner, Ag doped chalcogenide glasses and their applications, *Curr. Opin. Solid State Mater. Sci.*, **7**, 117–126 (2003).
[74] J. Fick, B. Nicolas, C. Rivero, E. Elshot, R. Irwin, K.A. Richardson, M. Fischer, and R. Vallée, Thermally activated silver diffusion in chalcogenide thin films, *Thin Solid Films*, **418**, 215–221 (2002).
[75] P.S. Peercy, The drive to miniaturization, *Nature*, **406**, 1023–1026 (2000).
[76] A.L. Greer and N. Mathur, Changing face of the cameleon, *Nature*, **437**, 1246–1247 (2005).
[77] M. Mitkova and M.N. Kozicki, Silver incorporation in Ge–Se glasses used in programmable metallization cell devices, *J. Non-Cryst. Solids*, **299–302**, 1023–1027 (2002).

[78] M.N. Kozicki, M. Park, and M. Mitkova, Nanoscale memory elements based on solid-state electrolytes, *IEEE Trans. Nanotechnology*, **4**, 331–338 (2005).
[79] M. Mitkova, M.N. Kozicki, H.C. Kim, and T.L. Alford, Thermal and photodiffusion of Ag in S-rich Ge–S amorphous films, *Thin Solid Films*, **449**, 248–253 (2004).
[80] M. Mitkova, M.N. Kozicki, H.C. Kim, and T.L. Alford, Local structure resulting from photo and thermal diffusion of Ag in Ge–Se thin films, *J. Non-Cryst. Solids*, **338–340**, 552–556 (2004).
[81] M.N. Kozicki, M. Mitkova, M. Park, M. Balakrishnan, and C. Gopalan, Information storage using nanoscale electrodeposition of metal in solid electrolytes, *Superlattices Microstruct.*, **34**, 459–465 (2003).
[82] M.N. Kozicki, C. Gopalan, M. Balakrishnan, M. Park, and M. Mitkova, Non-volatile memory based on solid electrolytes, *Proceedings of the non-volatile memory technology symposium*, Orlando, 10–17, 2004.
[83] K. Terabe, T. Hasegawa, T. Nakayama, and M. Aono, Quantized conductance atomic switch, *Nature*, **433**, 47–50 (2005).
[84] A.V. Legin, E.A. Bychkov, and Y.G. Vlasov, Analytical applications of chacogenide glass chemical sensors in environnemental monitoring and process control, *Sens. Actuators B*, **24–25**, 309–311 (1995).
[85] R. De Marco and J. Shackleton, Calibration of the Hg chalcogenide glass membrane ion-selective electrode in seawater media, *Talanta*, **49**, 385–391 (1999).
[86] R. De Marco and D.J. Mackey, Calibration of a chalcogenide glass membrane ion-selective electrode for the determination of free Fe^{3+} in seawater I. Measurements in UV photooxidised seawater, *Mar. Chem.*, **68**, 283–294 (2000).
[87] L. Lvova, S.S. Kim, A. Legin, Y. Vlasov, J.S. Yang, G.S. Cha, and H. Nam, All-solid-state electronic tongue and its applications for beverage analysis, *Anal. Chim. Acta*, **468**, 303–314 (2002).
[88] A. Legin, A. Rudnitskaya, L. Lvova, Y. Vlasov, C. Di Natale, and A. D'Amico, Evaluation of Italian wine by the electronic tongue: recognition, quantitative analysis and correlation with human sensory perception, *Anal. Chim. Acta*, **484**, 33–44 (2003).
[89] O. Tillement, Solid state ionics electrochemical devices, *Solid State Ionics*, **68**, 9–33 (1994).
[90] W. Weppner, Surface modification of solid electrolytes for gas sensors, *Solid State Ionics*, **40–41**, 369–374 (1990).
[91] A. Guessous, J. Sarradin, P. Papet, A. Pradel, and M. Ribes, Electrochemical microsensor for the detection of Na^+ ions, *Sens. Actuators B*, **26–27**, 360–363 (1995).
[92] E. Pungor, Ion-selective electrodes—history and conclusions, *Fresenius J. Anal. Chem.*, **357**, 184–188 (1997).
[93] Y.G. Vlasov, New solid-state ion-selective electrodes—Sensors for chemical analysis of solutions, *Fresenius Z. Anal. Chem.*, **335**, 92–99 (1989).
[94] J. Schubert, M.J. Schöning, C. Schmidt, M. Siegert, S. Mesters, W. Zander, P. Kordos, H. Lüth, A. Legin, Y.G. Mourzina, B. Seleznev, and Y.G. Vlasov, Chalcogenide-based thin film sensors prepared by pulsed laser deposition technique, *Appl. Phys. A*, **69**, S803–S805 (1999).
[95] R. Tomova, R. Stoycheva-Topalova, and A. Buroff, Thin-film sensors based on evaporated chalcogenide glasses, *Journal of Materials Science: Material in Electronics*, **14**, 843–845 (2003).
[96] A. Guessous, J. Sarradin, P. Papet, K. Elkacemi, S. Belcadi, A. Pradel, and M. Ribes, Chemical microsensors based on chalcogenide glasses for the detection of cadmium ions in solution, *Sens. Actuators B*, **53**, 13–18 (1998).
[97] J. Schubert, M.J. Schöning, Y.G. Mourzina, A.V. Legin, Y.G. Vlasov, W. Zander, and H. Lüth, Multicomponent thin films for electrochemical sensor applications prepared by pulsed laser deposition, *Sens. Actuators B*, **76**, 327–330 (2001).

[98] M.J. Shöning, C. Schmidt, J. Schubert, W. Zander, S. Mesters, P. Kordos, H. Lüth, A. Legin, B. Seleznev, and Y.G. Vlasov, Thin film sensors on the basis of chalcogenide glass materials prepared by pulsed laser deposition technique, *Sens. Actuators B*, **68**, 254–259 (2000).

[99] R. Tomova, G. Spasov, R. Stoycheva-Topalova, and A. Buroff, Copper-doped vacuum evaporated chalcogenide layers as sensitive ion-selective membranes, *J. Non-Cryst. Solids*, **266–269**, 985–988 (2000).

[100] M. Miloshova, E. Bychkov, V. Tsegelnik, V. Strykanov, H. Klewe-Nebenius, M. Bruns, W. Hoffmann, P. Papet, J. Sarradin, A. Pradel, and M. Ribes, Tracer and surface spectroscopy studies of sensitivity mechanism of mercury ion chalcogenide glass sensors, *Sens. Actuators B*, **57**, 171–178 (1999).

[101] A. Guessous, P. Papet, and J. Sarradin, Thin film of chalcogenide glass as sensitive membranes for the detection of mercuric ions in solution, *Sens. Actuators B*, **24–25**, 296–299 (1995).

[102] Y. Mourzina, M.J. Schöning, J. Schubert, W. Zander, A.V. Legin, Y.G. Vlasov, P. Kordos, and H. Lüth, A new thin film Pb microsensor based on chalcogenide glasses, *Sens. Actuators B*, **71**, 13–18 (2000).

[103] G. Taillades, O. Valls, A. Bratov, C. Dominguez, A. Pradel, and M. Ribes, ISE and ISFET microsensors based on a sensitive chalcogenide glass for copper ion detection in solution, *Sens. Actuators B*, **59**, 123–127 (1999).

[104] C. Di Natale, F. Davide, J.A. Brunink, A. D'Amico, Y.G. Vlasov, A.V. Legin, and A.M. Rudnitskaya, Multicomponent analysis of heavy metal cations and inorganic anions in liquids by a non-selective chalcogenide glass sensor array, *Sens. Actuators B*, **34**, 539–542 (1996).

[105] A. Rudnitskaya, A. Ehlert, Y. Vlasov, and S. Büttgenbach, Multisensor system on the basis of an array of non-specific chemical sensors and artificial neural networks for determination of inorganic pollutants in a model groundwater, *Talanta*, **55**, 425–431 (2001).

[106] A. Legin, A. Rudnitskaya, Y. Vlasov, C. Di Natale, E. Mazzone, and A. D'Amico, Application of electronic tongue for qualitative and quantitative analysis of complex liquid media, *Sens. Actuators B*, **65**, 232–234 (2000).

12 DNA Conduction: the Issue of Static Disorder, Dynamic Fluctuations and Environmental Effects

Rafael Gutiérrez[1], **Danny Porath**[2] **and Gianaurelio Cuniberti**[1]

[1]*Institute for Theoretical Physics, University of Regensburg, D-93040 Regensburg, Germany*
[2]*Physical Chemistry Department, The Hebrew University, IL-91904, Jerusalem, Israel*

12.1 INTRODUCTION

The past decade has seen an increasing interest in DNA molecules as possible components of electronic devices at the molecular scale. Applications as templates, but also as wiring systems have been envisioned. In view of its exciting potential applications, it is crucial to understand the relevant charge transport mechanisms in this molecule as well as the most efficient ways to control charge migration. This requires a close collaboration of experimentalists and theoreticians belonging to different research fields: biochemistry, chemistry, material sciences and physics. Experiments indicate in general that a large variety of factors may influence the electronic transport properties of DNA, in particular, the attachment to surfaces and covalent vs. physical charge injection into the molecular π-stack. Theoretically, though full first-principle approaches are the ideal tool for a full characterization of the

Charge Transport in Disordered Solids with Applications in Electronics Edited by S. Baranovski.

structural and electronic properties of DNA, the complexity of this molecule makes these methods very time consuming. Consequently, model Hamiltonian approaches offer an efficient complementary way to study charge transport in DNA oligomers. In this chapter, we will give an overview of some experimental and theoretical work carried out in the last decade on charge transport through DNA molecules. We will especially focus on three classes of factors which have been identified as being crucial in modifying the electrical behavior of DNA: static disorder, dynamical disorder and environmental effects.

The increasing demands on the integration densities of electronic devices impose considerable limitations to conventional semiconductor-based electronics. Apart from the intrinsic technical limitations, the down-scaling of the devices to nanometer sizes leads to a change in the physical mechanisms controlling charge propagation. Indeed, in reduced dimensions the full quantum nature of the charge carriers has to be taken into account. In the last decade, new alternatives have been explored [1] which have led to the emergence of the new field of molecular electronics [2, 3]. This strong interdisciplinary research field relies basically on the idea of using single molecules or molecular groups to perform the same operations as those done by silicon-based devices [4–7]. Another conceptual idea advanced by molecular electronics is the switch from a top–bottom approach, where the devices are extracted from a single large-scale building block, to a bottom-up approach in which the whole system is composed of small basic building blocks with recognition and self-assembly properties. A variety of different candidates for molecular electronic devices are currently the subject of highly interdisciplinary investigation efforts, including small organic polymers [4, 8–11], nanotubes and fullerenes [12–15], and biomolecules [16–23].

Among biomolecules, DNA occupies an outstanding position for its crucial role as a carrier of the genetic code and as such, it has been the focus of intensive research during decades within the biochemistry and biophysics community. DNA oligomers consist of four building blocks (oligonucleotides): adenyne (A), tymine (T), cytosine (C) and guanine (G). As is well known, they have specific binding properties, i.e., only A-T and G-C pairs are possible, see Figure 12.1. Sugar and phosphate groups ensure the mechanical stability of the double helix and protect the base pairs. Since the phosphate groups are negatively charged, the topology of the duplex is conserved only if it is immersed into an aqueous solution containing counterions (Na^+, Mg^+) that neutralize the phosphate groups.

With the advent of molecular electronics, DNA has also become a potential candidate for the design of electronic circuits at the molecular scale. Mainly two properties, recognition and self-assembling, make this molecule especially attractive for applications in molecular electronics [24, 25]. The former denotes the capability of a molecule to form selective bonds with other molecules or with substrates, based on the information stored in the structural features of the interacting partners. This property builds the foundation of the DNA-replication process. Self-assembly is the capability to spontaneously organize in supramolecular aggregates under appropriate conditions. Some promising developments have been recently achieved in controlling the self-assembly of DNA, in controlling the desired lengths, sequence and accessibility to specific sites on the chain [17, 18, 26–28] as well as in coupling the molecules to metal contacts [16, 29, 30].

There is however another fundamental aspect which is a precondition for a full use of DNA in molecular electronic circuits, namely, to clarify if this molecule is able to support an electric current and how to control it [5]. The idea that the DNA stack might support charge propagation can be traced back to 1962 [31]. However, subsequent low-temperature experiments showed that radiation-induced conductivity was related to mobile charge car-

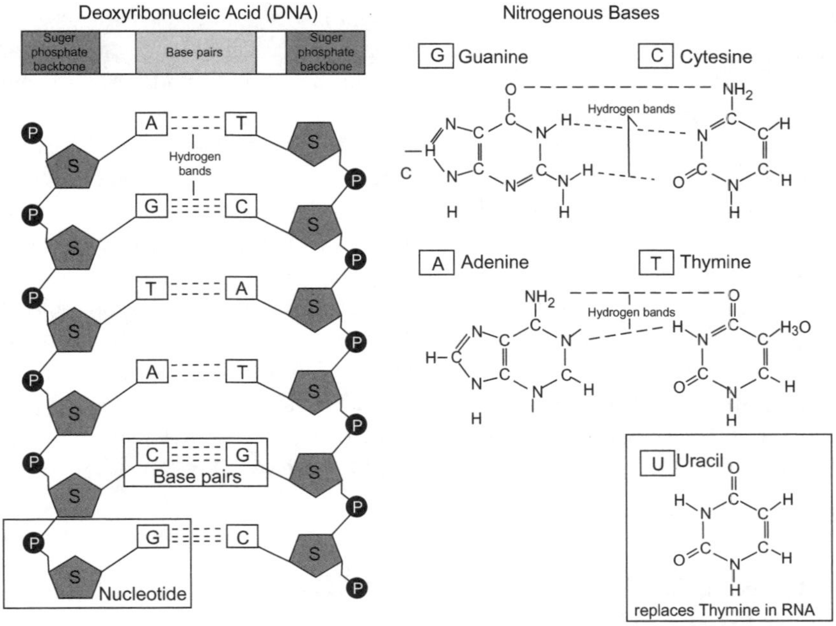

Figure 12.1 Schematic representation of a double-stranded DNA oligomer with an arbitrary base-pair sequence (reproduced from [153] with permission)

riers migrating within the frozen water layer surrounding the helix, rather than through the base-pair core [32]. Charge migration in DNA had been mainly addressed in order to clarify the mechanisms of damage repair which are essential to maintain the integrity of the molecule [33]. In the early 1990s, the striking fact of electron transfer between intercalated donor and acceptor centers in DNA oligomers in solution over unexpected long length scales [34, 35] led to a revival of interest in the conduction properties of this molecule. The almost distance-independent transfer rates suggested that DNA might act as a molecular wire [36]. Several possible mechanisms have been advanced meanwhile to explain the experimental findings: single-step superexchange [34], multistep hole hopping [37], phonon-assisted polaron hopping [38] and polaron drift [39].

In the last years, several attempts to measure the *I–V* characteristics of DNA molecules in different conformations have been carried out, see a recent review on the experimental aspects [40]. The results were however so controversial that a clear picture of the conduction mechanisms in DNA has so far not been attained. Thus, DNA was characterized as, e.g., an insulator [16, 41], a wide-bandgap semiconductor [19] or as a metallic system [42, 43]. This is a clear indication of the extreme sensitivity of transport through DNA to a variety of intrinsic and extrinsic factors.

From a theoretical point of view, the investigation of charge propagation in DNA oligomers requires the combined use of powerful first-principle methodologies [44–54] and model-based Hamiltonian approaches. [55–72] The former can yield an accurate description of the molecular electronic and atomic structure as well as of the interactions with the substrates to which the molecules are bound. The latter can address more complex physical situations where a single-particle picture, usually assumed in *ab initio* approaches, is not sustainable. The ultimate goal of the theoretical approaches is to disclose the most efficient charge transport mechanisms at the molecular scale and its dependence on the atomic and electronic structure of the individual molecules, thus assisting to find a way to control the charge transport through a molecular unit.

In this chapter, we will review experimental and theoretical work on charge transport in DNA oligomers. For reasons of space, we are not aiming at a thorough review of the research done on this molecule concerning its charge migration properties, a task that would require a book on its own. We will for instance not treat research related to charge transfer in the 'chemical community'. For the interested reader, there are several excellent reviews on this issue. [73–78] First-principle approaches will not be treated in detail, neither. Here, we also refer to other recent reviews [79–81]. In the next section, a general overview of the experimental situation concerning charge transport in DNA molecules will be given, focusing on experiments on single molecules as well as on bundles. In Section 12.3 we will then address theoretical issues related to the influence of static (Section 12.3.1) and dynamic disorder (Section 12.3.2) as well as environmental effects (Section 12.3.3) on charge transport.

12.2 CHARGE TRANSPORT EXPERIMENTS IN DNA OLIGOMERS

Conductivity experiments in DNA are based on several techniques, including imaging, SPM spectroscopy, and electrical transport measurements that reveal the electric current flux through the molecule under an external field. The results pertain to single molecules (or bundles) and can be remeasured many times. Two- and three-terminal setups are usually employed. A typical two-terminal setup consists of two metal leads to which a molecule or a bundle of molecules is attached, this is the case of, e.g., break junction experiments, or of a metallic substrate on which the molecules are deposited and a metallic tip acting as a second electrode, this being the situation of scanning tunneling (STM) or atomic force (AFM) microscopy experiments. In contrast to electron *transfer* experiments where a single charge (electrons or holes) is transferred from an excited donor to the acceptor, in *transport* experiments there is a continuum of states in the electrodes, so that by applying a potential difference the Fermi level at one of the leads may be fixed in such a way that a very large number of charge carriers can be injected and transported along the molecule. In addition, the attachment of the molecule leads to an equalization of the Fermi levels on the electrode and the molecule, molecular level bending, charge transfer and as a result a reorganization of the electronic level structure of the molecule. This reorganization, which is difficult to calculate due to the complexity of the combined system, may lead to a smaller HOMO–LUMO gap and wider band widths than for the bare molecule. Moreover, application of a voltage between the electrodes, in which the Fermi level is fixed, leads to a lowering of the inter-base barrier, to a shorter distance and faster hopping due to a minimized phase loss at each hopping stage. The combination of level and barriers reorganization together with 'faster' hopping may, e.g., explain the high currents measured in recent experiments [43, 82, 83].

The interpretation of the experiments is generally given in terms of conductivity, which is basically determined by the electronic energy levels (if the molecular structure supports the existence of localized orbitals and discrete energy levels) or band structure (if the intramolecular interactions support the formation of delocalized states described by continuous energy levels, i.e., dispersive bands). It is not specified *a priori* if the mobile charges are electrons or holes: this depends on the availability of electron states, on their filling, and on the alignment to the Fermi levels of the reservoirs.

From this we see that the effectivity of charge injection into the molecular region crucially depends on two variables: (i) the electronic structure of the molecule(s) which can support or hinder charge propagation in dependence on how the molecular wavefunctions spread out along the molecular frame; and (ii) the topology of the molecule-electrode interface, i.e., if the molecular region is connected to the electrodes through a covalent chemical bond or a physical contact [83, 84].

A series of direct electrical transport measurements through DNA molecules that commenced in 1998 was motivated by new technological achievements in the field of electron beam lithography and scanning probe microscopy, as well as by encouraging experimental data suggesting high electron-transfer rates. The latter were based on the interpretation of results of charge-transfer experiments conducted on large numbers of very short DNA molecules in solution, in particular by Barton's group at Caltech and by other colleagues [34–37, 85–93].

In perspective, it seems now that care should be taken when projecting from those experiments onto the electron transport properties of various single DNA molecules in different situations and structures, e.g., long vs short, on surfaces vs suspended, in bundles vs single, in various environmental conditions such as a dry environment, or in other exotic configurations.

Several studies have been published since 1998 describing direct electrical transport measurements conducted on single DNA molecules [16, 19, 29, 41, 43, 82, 83, 94–103]. In such measurements one has to bring (at least) two metal electrodes to a physical contact with a single molecule, apply a voltage and measure an electrical current (or vice versa). Poor conductivity, which seems to be the case for long DNA attached to surfaces, provides a small measured signal. In such cases the electrode separation should be small, preferably in the range of few to tens of nanometers, yet beyond direct tunneling distance and without any parallel conduction path. Performing good and reliable experiments on single segmented molecules is a demanding task and the interpretation of the experiments on the basis of the current data is even harder. Not only that—each segmented molecule is intrinsically different from the others in the specific details of its structure and changes dynamically from measurement to measurement when not frozen. Consequently, the details of its properties also bear some uniqueness and result in molecule-to-molecule differences. Furthermore, the properties of these molecules are sensitive to the environment and environmental conditions, e.g., humidity, buffer composition, etc. Another difficulty that arises in these measurements is that the contacts to a single molecule are very important for the transport, but difficult to perform and nearly impossible to control microscopically. For example, the electrical coupling strength between the molecule and the electrodes will determine whether a Coulomb blockade effect (weak coupling) or a mixing of energy states between the molecule and the electrodes (strong coupling) is measured. For these reasons, we find a large variety in the results of the reported experiments, most of which done by excellent scientists in leading laboratories.

The question whether DNA is an insulator, a semiconductor or a metal is often raised. This terminology originates from the field of solid-state physics where it refers to the electronic structure of semi-infinite periodic lattices. It has been successfully applied to describe the electrical behavior of one-dimensional wires such as carbon nanotubes, where a coherent band structure is formed. However, it is questionable whether or not this notion describes well, with a similar meaning, the orbital energetics and the electronic transport through one-dimensional soft polymers, which are formed of a large number of sequential segments. In these polymers the number of junctions and phase-coherent 'islands' is large and may determine the electronic structure and the transport mechanisms along the wire.

DNA in particular is sometimes said to be an insulator or a semiconductor. In the bulk, the difference between a wide-bandgap semiconductor and an insulator is mainly quantitative with regard to the resistivity. For DNA and other one-dimensional polymers we may instead introduce the following distinction. If we apply a voltage (even high) across a wide-bandgap polymer and successfully induce charge transport through it without changing the polymer structure and its properties in an irreversible way, then it would be a wide-bandgap semiconductor. However, if the structure is permanently damaged or changed upon this voltage application then it is an insulator. This distinction is important with regard to the relevant experiments, where very high fields are present, and to the methods to check whether or not the conduction properties of the molecule are reproducible.

12.2.1 Single-molecule transport experiments

The first direct electrical transport measurement on a single, 16-μm-long λ-DNA, was published in 1998 by Braun *et al.* [16] In this fascinating experiment the λ-DNA was stretched on a mica surface and connected to two metal electrodes, 12 μm apart. This was accomplished using the double-strand recognition between a short single-strand in the end of the long λ-DNA and a complementary single-strand that was connected to the metal electrode on each side of the molecule (see Figure 12.2). Electrical transport measurements through the single molecule that was placed on the surface yielded no observable current up to 10 V.

Later on in 1999 Fink and Schonenberger [94] reported nearly ohmic behavior in λ-DNA molecules with a resistance in the MΩ range. The molecules were a few hundred nanometers long and were stretched across ~2-μm-wide holes in a metal-covered transmission electron microscope (TEM) grid, as shown in Figure 12.3. This fantastic technical accom-

Figure 12.3 (a) The LEEPS microscope used to investigate the conductivity of DNA. The atomic size electron point source is placed close to a sample holder with holes spanned by DNA molecules. Due to the sharpness of the source and its closeness to the sample, a small voltage U_e (20–300 V) is sufficient to create a spherical low-energy electron wave. The projection image created by the low-energy electrons is observed at a distant detector. Between the sample holder and the detector, a manipulation-tip is incorporated. This tip is placed at an electrical potential U_m with respect to the grounded sample holder and is used to mechanically and electrically manipulate the DNA ropes that are stretched over the holes in the sample holder. (b) A projection image of λ-DNA ropes spanning a 2-μm-diameter hole. The kinetic energy of the imaging electrons is 70 eV. (c) SEM image, showing the sample support with its 2-μm-diameter holes. (d) SEM image of the end of a tungsten manipulation-tip used to contact the DNA ropes. Scale bar 200 nm. (e) The metal tip is attached to the λ-DNA molecule. (f) *I–V* curves taken for a 600-nm-long DNA rope. In the range of ±20 mV, the curves are linear; above this voltage, large fluctuations are apparent. A resistance of about 2.5 MΩ was derived from the linear dependence at low voltage (reproduced from [94] by permission, copyright 1999 Nature Macmillan Publishers Ltd)

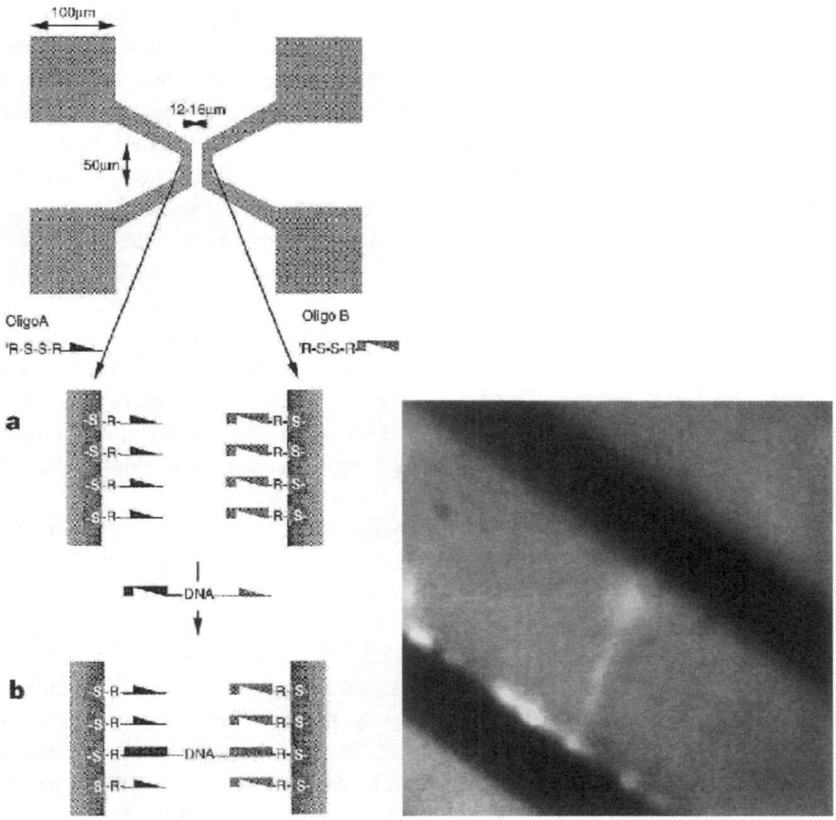

Figure 12.2 (a–c) 16-μm-long λ-DNA was stretched between two metal electrodes using short hangover single strands complementary to single strands that were pre-attached to the metal electrodes; (d) a fluorescent image of the DNA molecule, connecting the metal electrodes (reproduced from [16] with permission, Copyright 1998 Nature Macmillan Publishers Ltd)

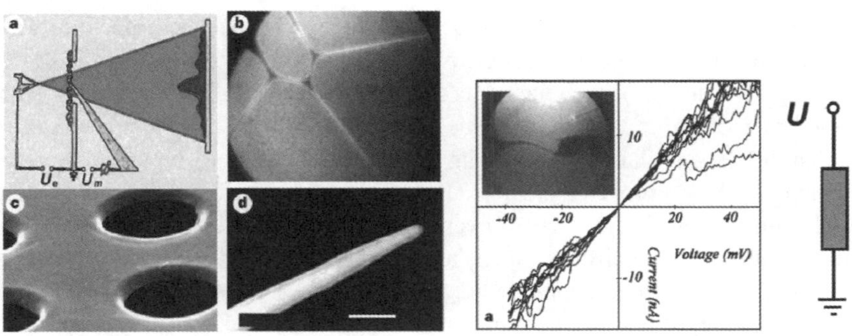

plishment was achieved in a high-vacuum chamber where a holographic image was created with a low-energy electron point source (LEEPS) claimed not to radiatively damage the DNA. Note, however, that the bright parts of the DNA in the images may suggest scattering of the beam electrons from the molecule, which may indicate the presence of scattering points along the DNA that could affect the charge transport along the molecule. The actual measurement was performed between a sharp tungsten tip, which was connected to the stretched molecule in the middle of one of the grid holes, and the metal covering the TEM grid. The tungsten tip was aligned using the holographic image. A nearly ohmic behavior was observed in the current–voltage (I–V) curves, sustained up to 40 mV, and then disappeared. The resistance division between two DNA branches appeared consistent with the ohmic behavior. This result seemed very promising. However, while conduction over long distances was observed later in bundles, it was not repeated in further measurements of single long (~nm) DNA molecules with one exception of a superconducting behavior that is discussed later [96]. The resolution of the LEEPS in this measurement did not enable to determine whether it was a single molecule or a bundle that was suspended between the metal tip and the metal grid.

In a further experiment published in 2000 by Porath *et al.* [19], electrical transport was measured through ~10-nm-long (30 base-pairs) homogeneous poly(dG)-poly(dC) molecules that were electrostatically trapped [104, 105] between two Pt electrodes (see Figures 12.4 and 12.5). The measurements were performed at temperatures ranging from room temperature and down to 4 K. Current was observed beyond a threshold voltage of 0.5–1 V indicating that the molecules transported charge carriers. At room temperature in ambient atmosphere, the general shape of the current–voltage curves was preserved for tens of samples, but the details of the curves varied from curve to curve. The possibility of ionic conduction was ruled out by measurements that were performed in vacuum and at low temperature, where no ionic conduction is possible. High reproducibility of the I–V curves was obtained at low temperature for tens of measurements on individual samples, followed by a sudden switching to a different curve shape (see insert of Figure 12.5) that was again reproducible (e.g., peak position and height in the dI/dV curves, Figure 12.5). This variation of the curves in different samples can originate from the individual structural conformation of each single molecule, or from the different formation of the specific contact. The variation of the curves measured on the same sample may be also due to switching of the exact overlap of the wavefunctions that are localized on the bases. A rather comprehensive set of control experiments helped to verify the results and ensure their validity. The existence of the DNA between the electrodes was verified by incubating the DNA devices with DNase I, an enzyme that specifically cuts DNA (and not any other organic or inorganic material). Following incubation of the sample with the enzyme the electrical signal was suppressed, indicating that the molecule through which the current was measured before is indeed DNA. The procedure was cross-checked by repeating this control experiment in the absence of Mg ions in the enzyme solution so that the action of the enzyme could not be activated. In this case the signal was not affected by incubation with the enzyme. This procedure ensured that it was indeed the enzyme that did the cut, thus confirming again that it was the DNA between the electrodes. This experiment clearly proves that short DNA molecules with homogeneous structure are capable of transporting charge carriers over a length of at least 10 nm.

In another attempt to resolve the puzzle around the DNA conduction properties, de Pablo *et al.* [95] applied a different technique to measure single λ-DNA molecules on the surface in ambient. They deposited a large number of DNA molecules on mica, covered some of them partly with gold and, using a metal covered AFM tip as a second mobile electrode,

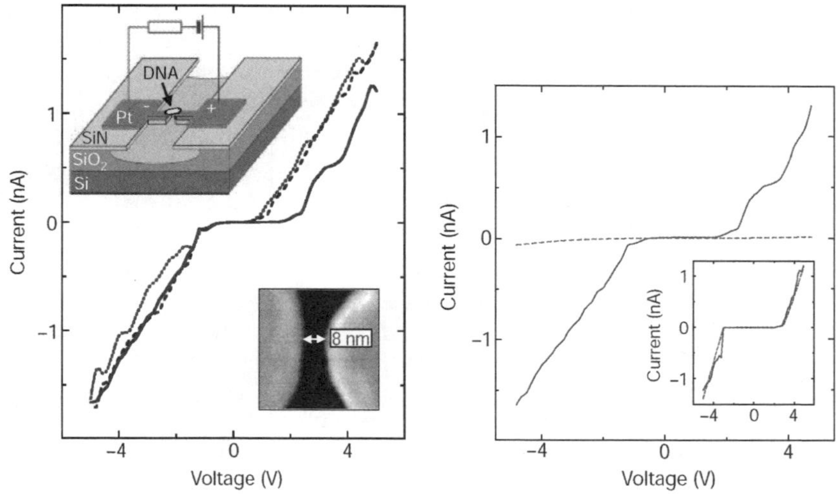

Figure 12.4 (a) Current–voltage curves measured at room temperature on a 10.4-nm-long DNA molecule (30 base-pairs, double-stranded poly(dG)-poly(dC)) trapped between two metal nanoelectrodes that are 8 nm apart. Subsequent *I–V* curves (different curves) show similar behavior, but with a variation of the width of the gap. The upper inset shows a schematic of the sample layout. Using electron-beam lithography, a local 30 nm narrow segment in a slit in the SiN layer is created. Underetching the SiO_2 layer leads to two opposite freestanding SiN 'fingers' that become the metallic nanoelectrodes after sputtering Pt through a Si mask. The lower inset is a SEM image of the two metal electrodes (light area) and the 8-nm gap between them (dark area). Deposition of a DNA molecule between the electrodes was achieved with electrostatic trapping. A 1-μl droplet of dilute DNA solution is positioned on top of the sample. Subsequently, a voltage of up to 5 V is applied between the electrodes. The electrostatic field polarizes a nearby molecule, which is then attracted to the gap between the electrodes due to the field gradient. When a DNA molecule is trapped and current starts to flow through it, a large part of the voltage drops across a large (2 GΩ) series resistor, which reduces the field between the electrodes and prevents other molecules from being trapped. Trapping of DNA molecules using this method is almost always successful. (b) Current–voltage curves that demonstrate that transport is indeed measured on DNA trapped between the electrodes. The solid curve is measured after trapping a DNA molecule as in (a). The dashed curve is measured after incubation of the same sample for 1 h in a solution with 10 mg/ml DNase I enzyme. The clear suppression of the current indicates that the double-stranded DNA was cut by the enzyme. This experiment was carried out for four different samples. The inset shows two curves measured in a complementary experiment where the above experiment was repeated, but in the absence of the Mg ions that activate the enzyme and in the presence of 10 mM EDTA (ethylenediamine tetraacetic acid) that complexes any residual Mg ions. In this case, the shape of the curve did not change. This observation verifies that the DNA was indeed cut by the enzyme in the original control experiment (reproduced from [19] with permission, Copyright 2000 Nature Macmillan Publishers Ltd)

measured the conductance along individual molecules. The minimum distance between the macroscopic gold electrode and the AFM tip was 70 nm. No current was observed in these experiments, suggesting that charge transport through DNA molecules longer than 70 nm which are attached to surfaces is blocked. Furthermore, they covered ~1000 parallel molecules on both ends with metal electrodes (~2 μm apart) and again no current was observed.

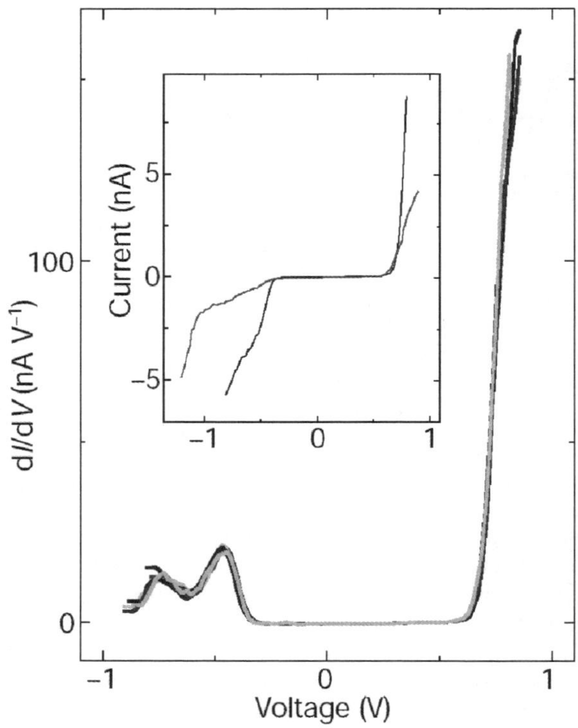

Figure 12.5 Differential conductance dI/dV versus applied voltage V at 100 K. The differential conductance manifests a clear peak structure. Good reproducibility can be seen from the six nearly overlapping curves. Peak structures were observed in four samples measured at low temperatures although details were different from sample to sample. Subsequent sets of I–V measurements can show a sudden change, possibly due to conformational changes of the DNA. The inset shows an example of two typical I–V curves that were measured before and after such an abrupt change. Switching between stable and reproducible shapes can occur upon an abrupt switch of the voltage or by high current (reproduced from [19] with permission, Copyright 2000 Nature Macmillan Publishers Ltd)

Additional experiments were performed in 2001 by Storm *et al.* [41], in which DNA molecules (longer than 40 nm) with various lengths and sequence compositions were stretched on different surfaces between planar electrodes in various configurations. No current was measured in these experiments, suggesting again that charge transport through DNA molecules, longer than 40 nm and adsorbed on a hard surface, is blocked.

Yet another negative result published in 2002 was obtained in a similar experiment by Zhang *et al.* [29] who stretched many single DNA molecules in parallel between metal electrodes, covalently bonded to them with thiols and measured no current upon voltage application. Both results were consistent with the de Pablo *et al.* experiment [95].

In parallel, Kasumov *et al.* [96] reported ohmic behavior of the resistance of λ-DNA molecules deposited on a mica surface and stretched between rhenium–carbon electrodes. This behavior was measured at temperatures ranging from room temperature and down to 1 K. Below 1 K a particularly unexpected result was observed: proximity-induced superconductivity. The resistance was measured directly with a lock-in technique and no current–voltage curves were presented. This surprising proximity-induced superconductivity is in contrast to all the other data published so far, and with theory. No similar result was reported later by this or any other group.

Beautiful and quite detailed measurements with different results on shorter molecules were reported in a series of papers by Watanabe *et al.* [97, 99], Shigematsu *et al.* [98] and Shimotani *et al.* [100], using a rather sophisticated technique. A short, single DNA molecule was laid on the surface and contacted with a triple-probe AFM consisting of three conducting CNTs (see Figure 12.6a, b). Two of them, 20 nm apart, were attached to the DNA (see Figure 12.6c). A third carbon nanotube was attached to the AFM tip. In one case

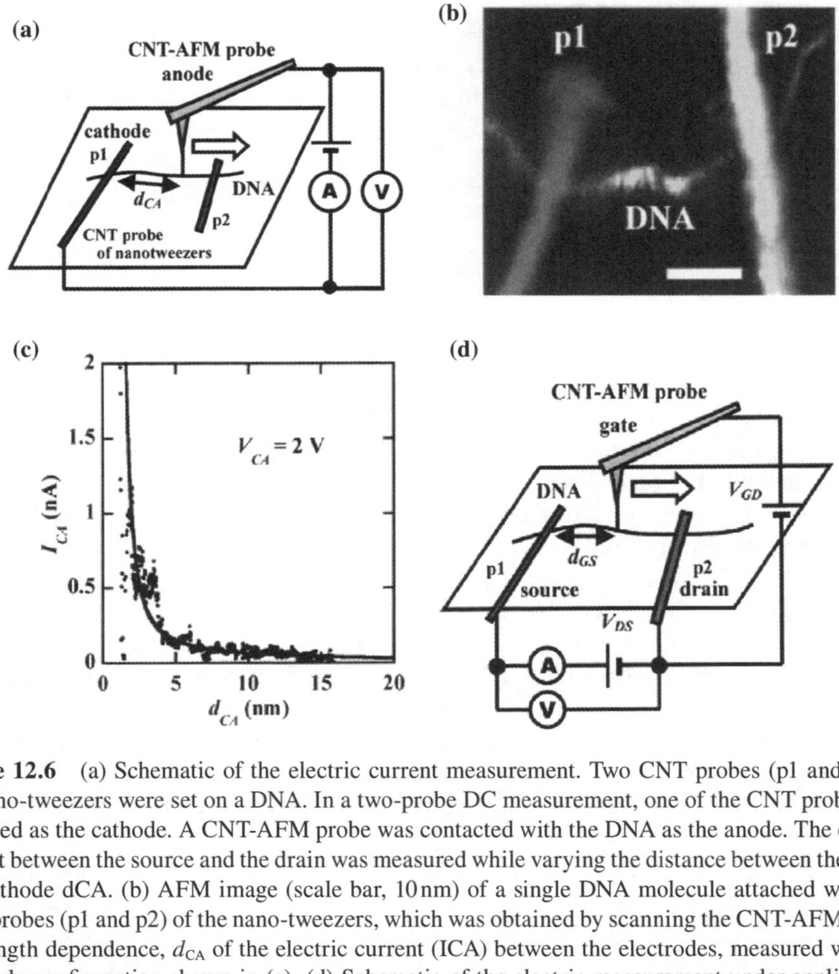

Figure 12.6 (a) Schematic of the electric current measurement. Two CNT probes (p1 and p2) of the nano-tweezers were set on a DNA. In a two-probe DC measurement, one of the CNT probes (p1) was used as the cathode. A CNT-AFM probe was contacted with the DNA as the anode. The electric current between the source and the drain was measured while varying the distance between the anode and cathode dCA. (b) AFM image (scale bar, 10 nm) of a single DNA molecule attached with two CNT probes (p1 and p2) of the nano-tweezers, which was obtained by scanning the CNT-AFM probe. (c) Length dependence, d_{CA} of the electric current (ICA) between the electrodes, measured with the electrode configuration shown in (a). (d) Schematic of the electric measurement under applied gate bias. Two nano-tweezer probes (p1 and p2) and the CNT-AFM probe were used as source, drain, and gate electrodes, respectively. The electric current between the source and drain was measured with varying the distance between source and gate dGS, (reproduced from [98] with permission, Copyright 2003 by the American Institute of Physics)

voltage was applied between the nanotube on one side of the molecule and the tip-nanotube that contacted the DNA molecule at a certain distance from the side electrode, so that the dependence of the current on the DNA length was measured under a bias voltage of 2 V between the two electrodes. The current dropped from 2 nA at ~2 nm to less than 0.1 nA in the length range 6–20 nm. In a second experiment [98], current was measured between the side nanotubes (20 nm apart) under a bias voltage of 2 V while moving the tip-nanotube that served this time as a gate along the DNA molecule. A clear variation of the current due to the effect of the gate electrode, reproducible forwards and backwards, was observed. The current–voltage curves in this experiment are measured through carbon nanotubes. Their conductivity is indeed much higher than that of the DNA molecule and therefore likely to have only a small effect on the *I–Vs*. However, this and the contacts of the nanotubes to the AFM tip and metal electrodes might still have an effect on the specific measured results.

Watanabe *et al.* [99] further demonstrated the operation of a single DNA molecule transistor in which three CNT terminals were connected with the DNA molecule, as outlined in Figure 12.7. At room temperature, they observed quantum steps in the current measured while scanning the source–drain voltage, when the distance between the source and drain was less than 10 nm.

The conclusion of poor conductivity in long single molecules on surfaces is further supported by indirect electrostatic force microscope (EFM) measurements, reported by Bockrath *et al.* [106] and Gomez-Navarro *et al.* [101]. In these measurements no attraction was found between a voltage-biased metal tip and the λ-DNA molecules lying on the surface. This indicates that the electric field at the tip failed to induce long-range polarization in the molecules on the surface, which would in turn indicate charge mobility along the molecule, as was found for carbon nanotubes.

The absence of conductance over the 40 nm length is not too surprising if we recall that DNA is a soft segmented molecule and is therefore likely to have distortions and defects when subjected to the surface force field. This is also manifested in AFM imaging where the measured height of the molecule is different from its nominal height [41, 101, 106], partly due to the effect of the pushing tip and partly due to the effect of the surface force field. This force field may be the culprit for blocking current transport through DNA molecules when attached to surfaces (though other effects may play a role as well). In later experiments to evaluate in depth the conduction properties of DNA, various investigators tried to address specific issues that may have a crucial impact on the measured results. Such issues include the surface force field and its effects on the DNA structure, the influence of humidity (aqueous environment [43]) and temperature on the deposited DNA structure and on possible conduction mechanisms as well as the type of contact between the DNA molecules and the electrodes (physical vs chemical bonding) and its contribution to the overall measurement.

Kasumov *et al.* confirmed that the height of DNA deposited on mica is about half the height that is expected based on DNA crystal structure (about 1 nm vs about 2.4 nm, respectively), using TEM to image a sample prepared by a shadow evaporation technique [107]. This finding supports numerous reports by various groups who measured the height of DNA deposited directly on solid substrate (i.e., mica or gold) by AFM. Kasumov and his coworkers tried to improve the DNA conductivity by minimizing the surface interaction with the DNA. This was achieved by depositing an intermediate thin, discontinuous layer between the DNA strands and the underlying mica. The layer was formed by glow discharge of

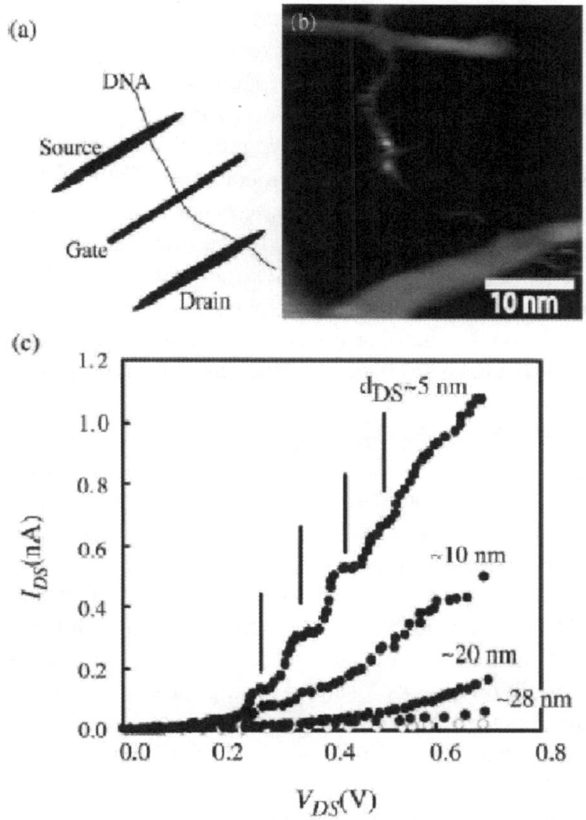

Figure 12.7 (a) A schematic of the DNA transistor. A single DNA molecule deposited on SiO_2 (10 nm)/*n*-Si(100)/Au (20 nm, gate terminal) was connected with source and drain terminals using TAFM. (b) AFM image of a single molecule DNA transistor. The source–drain distance, d_{DS}, is ~20 nm. (c) The I_{DS}–V_{DS} curves of the DNA transistor for several d_{DS} = ~5, ~10, ~20, ~28 nm at gate voltage, V_G = 2 V. The white circles indicates the leak current plots at d_{DS} = ~5 nm (reproduced from [99] with permission from Elsevier)

pentylamine vapor. The obtained film constitutes mainly of ionized NH_3^+ on which the negatively charged phosphate groups of the DNA backbone get attached. They report that the height of DNA molecules deposited on this film is 2.4–0.5 nm, close to the value expected from X-ray structure of B-form DNA. Over the organic elevation layer, Kasumov *et al.* deposited native DNA to obtain between one and five molecules that lay between two platinum contacts. They concentrated in their study on charge transport at the very low temperature range between 0.1 and 1 K (ionic conduction can be neglected below 250 K). They report scaling of the conductance obtained under low bias (10^{-5}–4×10^{-4} V) that is similar to the scaling obtained with CNTs, and conclude from that the DNA behaves like a 1D conductor.

Yang *et al.* [108] tried to evaluated the effects of drying DNA helix on its structure and on the stacking of base pairs, as required for utilization of DNA as molecular wire in electronic devices. For that purpose, they studied the X-ray diffraction of aligned films of DNA complexes within a surfactant, under different conditions: dry, wet and at low temperature. They find that dehydration of prehydrated samples causes the base pairs to go from planar to edge stacking and vice versa. The base-pair spacing within the DNA helix are 0.41 nm in dry sample in air at 50% relative humidity, while it is 0.34 nm in aqueous environment at room temperature. At low temperatures, below −16°C, the wet sample becomes dehydrated by crystallization and phase separation of the weakly bound water. This leads to rotation of the bases from planar to edge stacking, which corresponds to poor π-electron overlap. The findings indicate that the structure of the DNA helix when deposited on dry surfaces may be very different from that found by crystallization of DNA in solution, and may be another reason for the poor conductivity found in a nonaqueous environment. The issue of type and transmission quality of the contact point between the DNA strands and the metal electrodes was also recognized as a major factor in performing conductivity measurements in single DNA molecules. Several groups tried to address this issue through chemical bonding of functionalized DNA molecules to the metal surface [41, 43, 82, 83, 102, 103]. This is usually done by attaching a thiol or an S–S end group to the DNA strands, utilizing the well-known tendency of sulfur to form a covalent bond with gold surfaces and with gold nanoparticles (GNP) [84].

Xu *et al.* [43] used a very elegant method for measuring electrical transport through double-strand (ds) DNA in aqueous solution, where the native form of the DNA is preserved. The measurement approach enables to accumulate larger statistics than most previous experiments. In addition the DNA sequences are varied by comparing measurements where AT base-pairs replace the GC base-pairs. The authors studied short sequences of DNA (8, 10, 12 and 14 base-pairs) that are modified at their 3′ end with C_3H_6SH. The experimental approach addresses the two issues emphasized above: the contact is formed through a chemical bond between the electrode and the DNA molecule, and the DNA molecules are suspended in a buffer solution that maintains their native structure. This approach presents a powerful investigation tool for short molecules. It allows thiolated dsDNA to adsorb to a gold electrode in a buffer solution that supports annealing of the two strands and gives preference to the B-form structure of the double helix over other possible structures. In the same buffer solution, a gold STM tip, which is covered with an insulating layer over most of the tip surface except for its end, was brought into contact. Once contact is formed with a molecule, a junction the tip is pulled backwards and the current is monitored (see Figure 12.8). Distinct steps were observed in the current when the tip is pulled away, which are interpreted as consecutive breaking of junctions (e.g., the number of DNA strands that connect both electrodes is gradually reduced until none remain and the current drops). Xu *et al.* provide statistics of over 500 individual measurements. They observed some dispersion in the values of the peaks, which is attributed to variations in the microscopic details of the molecules. A control experiment is carried out in buffer solution with no DNA, and does not show any current steps. To measure a single molecule, they halt the tip retraction at the position of the last peak (assuming that only one dsDNA connects both electrodes at that position) and measure I–V curves through the molecule. The curves obtained through three different single molecules show a rather smooth ohmic profile, coinciding with the average values of conductivity obtained from the pulling experiments. We note that although STM was used here, no imaging characterization of single molecules or the structure and organization of the molecules on the surface is reported. No similar experiment with single-strand (ss) DNA that can only

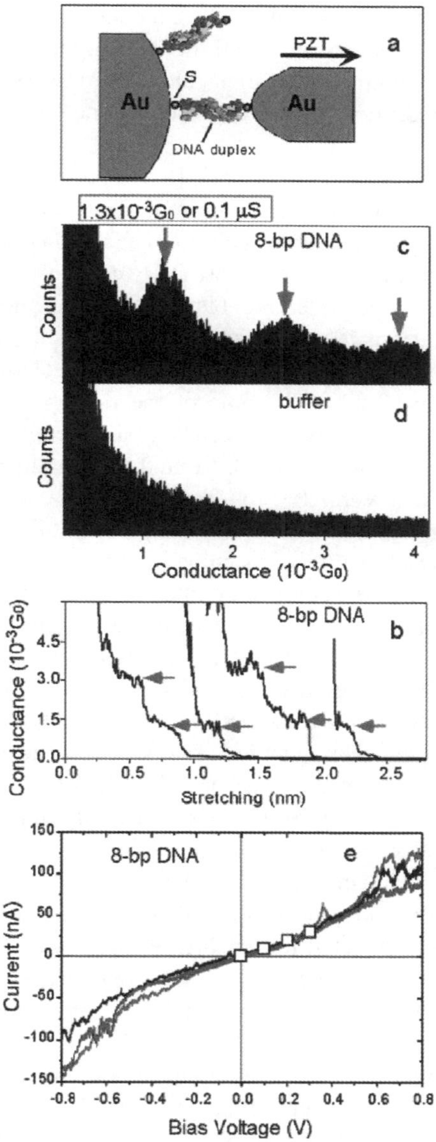

Figure 12.8 (a) Schematic illustration of a single DNA conductance measurement (bp formation of molecular junctions shown as discrete steps in the conductance (eight-bp DNA duplex). (c) Conductance histogram constructed form more than 500 individual measurements revealing well-defined peaks near integer multiples of a fundamental value, $1.3 \times 10^{-3} G_0$ (0.1 μS), which is identified as the conductance of a single eight-bp DNA duplex. (d) Conductance histogram in buffer solution revealing a smooth background. (e) Current–voltage characteristic curves of a single eight-bp DNA. Lines with different colors are obtained by recording current vs bias voltage for three different DNA junctions. The open squares are from the peak positions of the conductance histograms obtained at different bias voltages (reproduced from [43] with permission, Copyright 2003 American Chemical Society)

connect to one side is reported as a control. Such a control could validate the transport of charge through the double-stranded helix, and shed light on the role of the chemical contact between the electrode and the measured molecule. Nevertheless, we find this report to be one of the most elegant and informative experiments done on electrical transport in DNA so far.

Recently, Nogues *et al.* [102] and Cohen *et al.* [82, 83] published a different experimental approach which is also based on measuring current through dsDNA molecules connected between a metal substrate and a metal tip. This approach allows measurement of conductivity in a systematic manner through many short DNA molecules and obtaining a large statistics. They formed well-characterized monolayers of 5′ end thiol-modified 26-bases-long ssDNA. The complementary oligomer, which is also modified with a thiol at its 5′ end, is adsorbed onto a GNP of 10 nm in diameter. Hybridization of the two strands affords an insulating ssDNA monolayer in which some of the ssDNA were hybridized with their complementary strands to form dsDNA that can be easily identified by the GNPs connected to them, and where a direct contact of the GNP and the metal substrate is prevented. The formed monolayer is scanned with AFM to locate the GNPs indicating the dsDNA. A conductive tip is used to form contact to the GNP, and through this contact the *I–V* curves are measured through the dsDNA while approaching the tip to the GNP in contact mode. The control over the tip motion and the possible pressing of the GNP towards the surface during the electrical measurement is limited in this way, but during the approach, current of a few nA can be measured through the dsDNA. The experimental setup and the measured *I–V*s are presented schematically in Figure 12.9. It is estimated that up to 10 dsDNA molecules can connect simultaneously between the GNP and the underlying gold surface although it is likely that the number of connecting molecules is smaller. The curves that were measured through the DNA show current of the order of 220 nA at 2.0 V. This is significantly larger than the current values in previous measurements (by different experimental setups) on single DNA molecules, with exception of the experiments by Xu *et al.* [43] discussed above. The large currents might be attributed, e.g., to the fact that the bonding to the molecule is covalent and not just physical adsorption, that DNA is not attached to the surface along its length and is therefore not subjected to the surface forces or, eventually, to some kind of vibrationally assisted mechanism. These results seem to indicate that the electrical transport mechanism is different and faster than those that account for the donor–bridge–acceptor systems in solution chemistry experiments.

From the direct electrical transport measurements on single DNA molecules reported so far one can draw some conclusions. First, it is possible to transport charge carriers through single DNA molecules. This was observed however, only for short molecules in the range of up to 20 nm in the experiments of Porath *et al.* [19], Watanabe *et al.* [97, 99], Shigematsu *et al.* [98], Xu *et al.* [43], and Cohen *et al.* [82, 83]. The first three experiments demonstrated currents of order 1 nA upon application of a voltage of ~1 V. The reported results in the last two show however much larger currents of the order of 150 and 220 nA at 0.8 and 1 V, respectively. The experimental results by Fink *et al.* [94] and Kasumov *et al.* [96] showed high currents and lower resistivities over longer molecules, but they were never reconfirmed for individual molecules. In all the other experiments, by de Pablo *et al.* [95], Storm *et al.* [41] and Zhang *et al.* [29] that were conducted for molecule lengths longer than 40 nm on single DNA oligomers attached to surfaces, no current was measured.

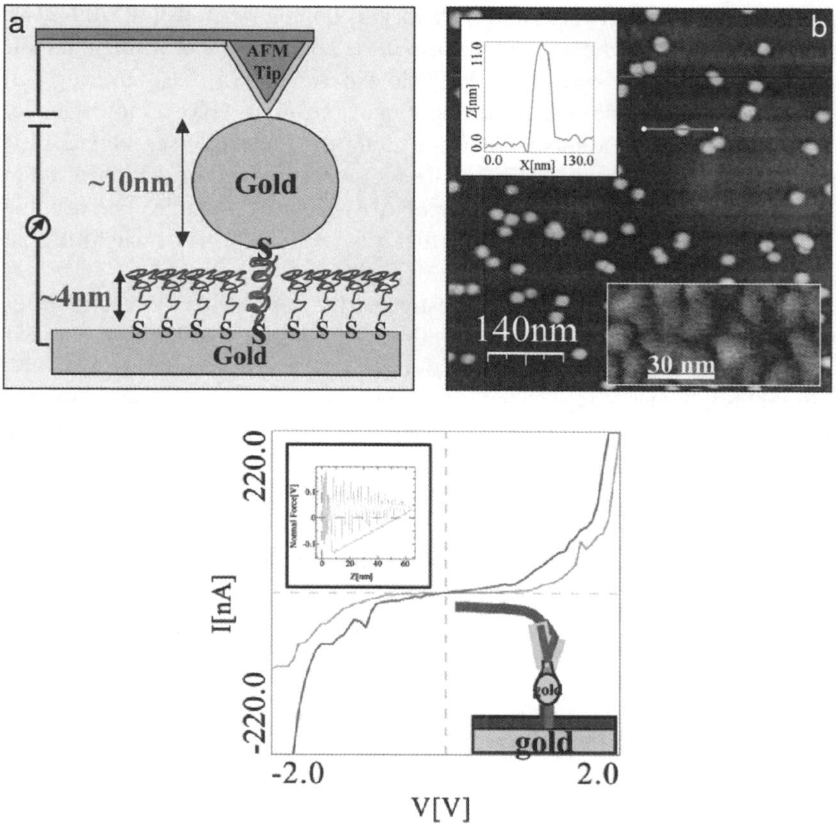

Figure 12.9 (See also colour plates.) Upper panel (left): Schematic presentation of the gold particle-connected dsDNA that is embedded in an ssDNA monolayer and the measurement configuration; (right) AFM topography image showing a top view of the GNPs connected to dsDNA molecules on the background of the ssDNA monolayer. Lower panel: Collection of current–voltage curves measured through various dsDNA molecules connected on opposite ends to a gold substrate and a GNP using a metallized AFM tip. This measurement was performed on a metal particle without pressing on it, as seen in the inset. The current measured is larger than 220 nA at 2 V. The hysteresis in the backward curves in the insets is caused by adhesion and indicates that good contact was established with the surface or the particle underneath. The green curves are the 'forward' and were measured first; the red curves are the 'backward' and were measured after the green ones (reproduced from [82] with permission, Copyright 2005 National Academy of Sciences)

12.2.2 Transport experiments on bundles and networks

A few measurements of direct electrical transport were also performed on single bundles. Other measurements were done on networks formed of either double-stranded DNA [109] or alternative polynucleotides [110]. All the reported measurements showed current flowing through the bundles. We will show a few examples here. The most productive group in this category is the group of T. Kawai from Osaka who published an extended series of experiments on different networks and with various doping methods [42, 109–114]. In one of their early experiments the authors measured the conductivity of a single bundle [109]. This was done in a similar way to the de Pablo experiment [95], i.e., covering part of the bundle by a metal (electrode) and attaching a metal-covered AFM tip to the molecules at a various distances from the metal electrode along the molecule (see Figure 12.10). The conductivity of a poly(dG)–poly(dC) bundle was measured as a function of length (50–250 nm) and was compared with that of a poly(dA)–poly(dT) bundle. The results showed a very clear length-dependent conductivity that was about an order of magnitude larger for the poly(dG)–poly(dC) bundle.

More recently, [114] Kawai's group investigated the dependence of electrical resistivity of poly(dG)–poly(dC) thin films on humidity. This study was carried out using AC-impedance and DC time-dependence measurements at various humidity levels (40–80% relative humidity). In these experiments the authors report strong dependence of the conductivity on the relative humidity, in agreement with the findings of Jo *et al.* [115]. It was also concluded that ionic conduction through the water layers dominates the total resistivity. In another study [113] Kawai's group studied the electrical properties of the contact between a gold-coated AFM tip and networks of 50-bases-long oligomers of poly(dA-dT)–poly(dA-dT) and poly(dG-dC)–poly(dG-dC) that form on a highly oriented pyrolytic graphite (HOPG) surface. In this study, the resistance of DNA in air was found to be inversely pro-

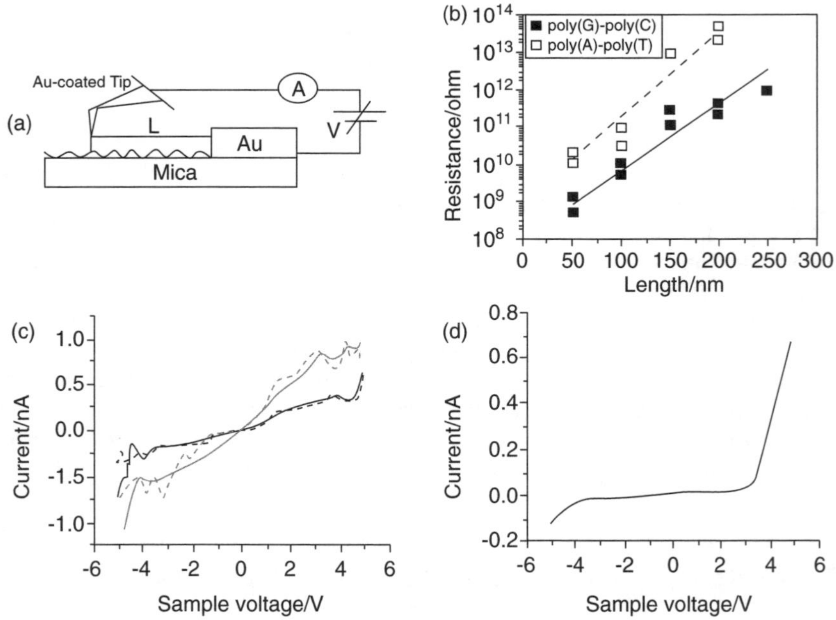

Figure 12.10 (a) Schematic illustration of the measurement with a conducting-probe AFM. (b) Relationship between resistance and DNA length for poly(G)–poly(C) (dark marks) and for poly(A)–poly(T) (empty marks). The exponential fitting plots of the data are also shown. (c) Typical *I–V* curves of poly(dG)–poly(dC), the linear Ohmic behaviors on $L = 100\,\text{nm}$ at the repeat measurement of five samples. (d) Rectifying curves of poly(dG)–poly(dC) at $L = 100$ nm (reproduced from [109] with permission, Copyright 2000 by the American Institute of Physics)

portional to the contact area at tip forces between 0 and 50 nN. When higher force was applied a direct contact with the HOPG was established. The resistance of the DNA networks at 40 nN was found to be 1–5 MΩ. The authors conclude that in order to utilize DNA in molecular devices, it is necessary to modify the DNA so as to enable improved conductivity at the interface to the metal electrode with a lower potential barrier.

One of the interesting measurements among the 'bundle experiments' was done by Rakitin *et al.* [116]. They compared the conductivity of a λ-DNA bundle to that of an M-DNA bundle [117–119] (DNA that contains an additional metal ion in each base-pair, developed by the group of J. Lee [118]). The actual measurement was performed over a physical gap between two metal electrodes in vacuum. Metallic behavior was observed for the M-DNA bundle over 15 μm, while for the λ-DNA bundle a gap of ~0.5 V in the *I–V* curve was observed followed by a rise of the current.

Another measurement that follows the line of the Porath *et al.* [19] experiment was performed by Yoo *et al.* [42]. In this experiment, long poly(dG)–poly(dC) and poly(dA)–poly(dT) molecules were electrostatically trapped between two planar metal electrodes that were 20 nm apart (see Figure 12.11) on a SiO_2 surface, such that a bundle, 10 nm wide, was formed. A planar gate electrode added another dimension to this measurement. The current–voltage curves showed a clear current flow through the bundle and both temperature and gate dependencies. The resistivity for the poly(dG)–poly(dC) was calculated to be 0.025 Ωcm. Very high currents are measured in an experiment reported by Jo *et al.* [115], who attempted to measure conductivity of λ-DNA versus the conductivity of poly(dG)–poly(dC) and poly(dA)–poly(dT) in vacuum and under controlled humidity conditions. Electrostatic trapping was used to attach the molecules to gold electrodes separated by a gap of 150 nm. The authors present *I–V* curves obtained immediately after dropping the solution containing the DNA samples, after drying the sample and then putting it in a controlled humidity chamber and in vacuum. S-shaped curves (1 mA at 1 V) are reported for all types of DNA at relatively high humidity, which the authors relate to ionic charge transport by the H^+ ions. Under vacuum the conductivity of poly(dG)–poly(dC) for a given bias

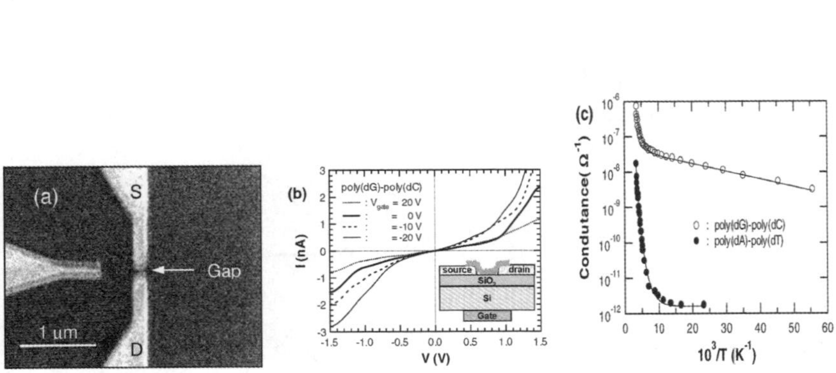

Figure 12.11 (a) SEM image of an Au/Ti nanoelectrode with a 20 nm spacing. Three electrodes are shown, S and D stand for source and drain. (b) *I–V* curves measured at room temperature for various values of the gate voltage (V_{gate}) for poly(dG)–poly(dC). The inset of (b) is the schematic diagram of electrode arrangement for gate dependent transport experiments. (c) Conductance versus inverse temperature for poly(dA)–poly(dT) and poly(dG)–poly(dC), where the conductance at $V = 0$ was numerically calculated from the *I–V* curve (reproduced from [42] with permission, Copyright 2001 by the American Physical Society)

(300 nA at 1 V) is found to be three and four orders of magnitudes larger than that of λ-DNA and poly(dA)–poly(dT), respectively, with a rectifying behavior. The authors relate all the changes in the curves to the changes in the humidity conditions. No information is provided regarding the number of molecules that are attached between the electrodes or independent evidence that DNA only is connecting the electrodes. Since the molecules are much larger than the gap between the electrodes, it is likely that bundles are measured in this experiment.

Hwang and co-workers [120] investigated conductivity of thiol-modified poly(dG)–poly(dC) of 60 base-pairs. They adsorbed the dsDNA onto gold nanoparticles of 20 nm in diameter. These DNA coated nanoparticles were then dropped onto a 50-nm gap between two gold electrodes, and the *I–V* of the system was measured. Hwang *et al.* report currents of about 100 nA for 1 V bias for this system. The curves are generally gapless with some nonlinearity. The number of parallel molecules bridging the GNPs and the metal electrodes is, however, not clear, but quite large. Therefore, this experiment may reflect a network behavior.

Lei *et al.* [121] reported high resistivity of λ-DNA network as measured by EFM using a method similar to that of Gomez-Navarro [101]. An experiment on a DNA-based network embedded in a cast film was done by Okahata *et al.* as early as 1998 [122]. In this pioneering experiment the DNA molecules were embedded (with side groups) in a polymer matrix that was stretched between electrodes (see Figure 12.12). It was found that the conductivity parallel to the stretching direction (along the DNA) was 4.5 orders of magnitude larger than the perpendicular conductivity.

Measurements on a different type of DNA-based material were reported by Rinaldi *et al.* [123, 124]. In this experiment a few layers of deoxyguanosine ribbons were deposited in the gap between two planar metal electrodes, 100 nm apart. The current voltage curves

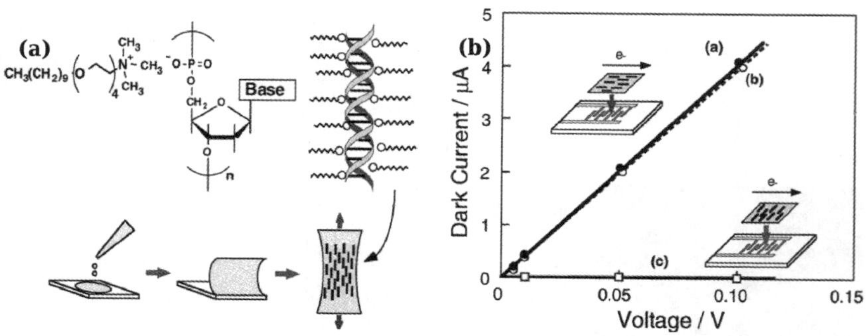

Figure 12.12 (a) Schematic illustration of a flexible, aligned DNA film prepared from casting organic-soluble DNA–lipid complexes with subsequent uniaxial stretching. (b). Experimental geometries and measured dark currents for aligned DNA films (20 × 10 mm, thickness 30 ± 5 μm) on comb-type electrodes at 25 C. In the dark-current plot, the three curves (a,b,c) represent different experimental settings and environments: (a) DNA strands in the film placed perpendicular to the two electrodes (scheme in the upper inset) and measured in ambient; (b) the same film as in (a) measured in a vacuum at 0.1 mm Hg; (c) DNA strands in the film placed parallel to the two electrodes, both in vacuum and in ambient (reproduced from [122] with permission, Copyright 1998 by the American Chemical Society)

showed a gap, followed by rise of the current-beyond a threshold of a few volts. The curves depended strongly on the concentration of the deoxyguanosine in the solution.

12.3 THEORETICAL ASPECTS OF DNA CONDUCTION

We proceed now to address, from a theoretical point of view, some intrinsic and extrinsic factors that have been identified as being essential in determining the electrical transport properties of DNA molecules.

12.3.1 Static disorder

The specificity in the base-pair sequence is essential for DNA to fulfill its function as a carrier of the genetic code. However, this same fact can be detrimental for charge transport. The apparent random way in which the DNA sequence is composed, strongly suggests that a charge propagating along the double helix may basically feel a random potential leading to backscattering. In one-dimensional systems with uncorrelated disorder all electronic states are completely localized (Anderson localization). However, correlated disorder [125] may lead to delocalized states within some special energy windows in the thermodynamic limit.

Two important issues to be addressed when investigating the role of disorder in DNA are, in our view, the following: (i) is the base-pair sequence in DNA completely random (Anderson-like) or do there exist (long- or short-ranged) correlations?; (ii) a measure for the degree of localization of the electronic wave function is given by the localization length ξ [126]. How do the localization lengths compare with the actual lengths L of the DNA segments studied in transport experiments? Assuming that $\xi >> L$, the molecule may appear as effectively conducting, despite the presence of disorder, though in the thermodynamic limit all states remain localized. We review in this section some theoretical studies addressing these problems.

The simplest way to mimic a DNA wire is by assuming that after charge injection, the electron (hole) will basically propagate along the π-stack of one of the strands (the inter-strand coupling being much smaller), so that one-dimensional tight-binding chains can be a good starting point to minimally describe a DNA wire. The generic Hamiltonian is given then by:

$$H = \sum_{\ell} E_{\ell} c_{\ell}^{\dagger} c_{\ell} - \sum_{\ell,j} t_{\ell,j} \left(c_{\ell}^{\dagger} c_{j} + \text{h.c.} \right)$$

where both on-site energies E_ℓ and hopping integrals $t_{\ell,j}$ are assumed to be in general site dependents and 'h.c.' is the Hermition conjugate.

Roche [65] investigated such a model for poly(GC) and λ-phage DNA, with on-site disorder arising from the differences in the ionization potentials of the base pairs, and bond disorder $t_{n,n+1} \sim \cos \theta_{n,n+1}$ related to random twisting fluctuations of nearest-neighbor bases along the strand. While for poly(GC) the effect of disorder does not appear to be very dramatic, the situation changes when considering λ-phage. In this case the transmission peaks are considerably reduced in intensity and in number with increasing chain length at zero temperature, since only few electronic states are not backscattered by the random potential profile of the chain. To some degree it is surprising that averaged Lyapunov exponents increase with increasing temperature, suggesting that despite thermal fluctuations many states are still contributing to the charge transport.

In an early paper Roche *et al.* [65] used rescaling coefficients, which give a measure for the existence of long-range correlations in disordered systems. The calculations show that, e.g., DNA built from Fibonacci sequences has a very small Hurst exponent (indicating strong correlations). Uncorrelated random sequences show a strong fragmentation and suppression of the transmission with increasing length, while in correlated sequences several states appear to be rather robust against the increasing rate of backscattering. Hence, it may be expected that correlated disorder will be more favorable for long-distance carrier transport in DNA wires.

Zhu *et al.* [69] formulated an effective tight-binding model poly(GC) together with on-site Coulomb interactions. On-site and off-diagonal disorder, related to fluctuations of the local electrostatic potential [127] and to twisting motion of the base pairs at finite temperatures, respectively, were also included. The main effect of the Coulomb interaction was to first reduce the bandgap, so that the system goes over to a metallic state, but finally the gap reappears as a Coulomb blockade gap.

Caetano and Schulz [128] investigated a double-strand model with uncorrelated disorder along the individual strands, but taking into account the binding specificity of the four bases when considering the complementary strand (A–T and G–C). The analysis of the computed participation ratios $P(E)$ suggest that inter-strand correlations may give rise to bands of delocalized states, with $P(E)$ not appreciably decaying with increasing length.

A very detailed study of the localization properties of electronic states in two minimal models of different DNA oligomers (poly(GC), λ-DNA, telomeric DNA) was presented by D. Klotsa *et al.* [72]: a fishbone model [60, 63, 64] and a ladder model. The authors addressed environmentally induced disorder. Hence, they assume that only the backbone sites were affected by disorder, while the nucleotide core was well screened. Nevertheless, as shown by a decimation procedure [72], disorder in the backbone sites can induce local fluctuations of the on-site energies on the base pairs. Uniform disorder (where the on-site energies of the backbones vary continuously over an interval $[-W, W]$, W being the disorder strength) is shown to reduce the localization length, as expected. For binary disorder (on-site energies take only two possible values $\pm W/2$), as may arise by the binding of counterions to the backbone sites, the situation is similar up to some critical disorder strength W_c. However, further increase of W leads to an unexpected behavior: the localization length on the electronic side bands is suppressed, but a new band around the midgap with *increasing* localization length shows up. Thus, disorder-induced delocalization of the electronic states is observed in some energy window, a quite striking result.

12.3.2 Dynamical disorder

While in the previous section we addressed problems related to time-independent disorder effects in DNA molecules, we will discuss here the influence of dynamical disorder on the conduction properties of DNA oligomers. The origin of this time-dependent effects lies basically in the 'softness' of the DNA double helix which allow for the existence of a large number of low-energy vibrational excitations that can couple to a tunneling charge. A recent detailed analysis of electron–phonon coupling in B-DNA, based on semiempirical quantum chemistry [129] has shown that there exist conformational modes with large electron–phonon matrix elements, which may be thus capable of influencing charge transport via a modification of the electronic hopping integrals and eventually lead to polaron formation.

The considerably small decay rates found in electron-transfer experiments [34] and mentioned in Section 12.1 have led to the proposal that, besides uni-step superexchange mechanisms, phonon-assisted hole hopping might also be of importance [76]. The hole can occupy a specific molecular orbital, localized on a given molecular site; it can also, however, extend over several molecular sites and build a (small) polaron. The polarons is the result of two competing tendencies: charge delocalization with a consequent gain in kinetic energy, and charge localization with gain in elastic energy. Conwell and Rakhmanova investigated this issue using the Su–Heeger–Schrieffer (SSH) model [39], which is known to entail a rich nonlinear physics and that has extensively been applied to study polaron formation in conducting polymers. The SSH model deals classically with the lattice degrees of freedom while treating the electrons quantum mechanically. The calculations showed that a polaron may be built and be robust within a wide range of model parameters. The influence of random base sequences was apparently not strong enough to destroy it. Thus, polaron drifting may constitute a possible transport mechanism in DNA oligomers.

While the majority of the vibrational modes can be well described within a harmonic approximation, there are special modes, like inter-strand H-bond stretching, whose anharmonic character can not be neglected. The nonlinear H-bond fluctuations can lead to a local breaking of the double strand and have thus been investigated in relation to the DNA denaturation problem [130]. Komineas *et al.* [131] studied a model with strong anharmonic potentials and local coupling of the lattice to the charge density. The strong nonlinearity of the problem led to a *dynamical* opening of bubbles with different sizes that may eventually trap the polaron and thus considerably affect this charge transport channel.

Zhang and co-workers [62, 132] studied a simple model that describes the coupling of torsional excitations (twistons) in DNA to propagating charges and showed that this interaction leads to polaron formation. Twistons modify the inter-base electronic coupling, though this effect is apparently less strong than, e.g., in the Holstein model [133], because of the strong nonlinearity of the twiston restoring forces as well as of the twiston–electron coupling. For low-frequency twisting modes and in the nonadiabatic limit where the 'spring constant' is much bigger than the electronic coupling, the inter-base coupling is maximally perturbed and an algebraic band reduction is found, considerably weaker than the exponential dependence known from the Holstein model. Thus, it may be expected that the polaron will have a higher mobility along the chain.

These results suggest that the nonlinear dynamics of the DNA vibrational excitations can be a key element in understanding charge transport over long length scales. Indeed, the role of solitons and breathers in the denaturation of DNA [130, 134, 135] as well as in the transmission of 'chemical' information between remote DNA segments [57] have been addressed in the earlier literature.

The observation of two very different time scales (5 and 75 ps) in the decay rates of electron transfer processes in DNA, as measured by femtosecond spectroscopy [136], led Bruinsma *et al.* [137] to investigate the coupling of the electronic system to collective modes of the DNA cage. For this, they considered a tight-binding model of electrons interacting with two modes: a twisting mode which mainly couples to the inter-base π-orbital matrix elements, and a linear displacement coupling to the on-site energies of the radical and acting as a local gating of the latter. In the strong-coupling, high-temperature limit, charge transport is basically an incoherent process, where phonon-assisted electron (hole) motion between nearest-neighbor sites takes place. The authors provide a picture where there are two reaction coordinates related to the above-mentioned linear and angular modes. The strong thermal fluctuations associated with the twisting motion are shown to introduce two time scales for electron transfer that can be roughly related to optimal (short) and nonoptimal (long) relative orientation of neighboring base pairs.

In a series of papers, Hennig *et al.* [55, 66, 138] proposed a model Hamiltonian where only the relative transverse vibrations of bases belonging to the same pair are included. The authors suggested that poly(GC) should be more effective in supporting polaron-mediated charge transport than poly(AT), since for the latter the electron–lattice coupling was found to be about one order of magnitude smaller. The authors remarked that no appreciable coupling to twisting distortions was found by their quantum chemical calculations; however, this issue requires further investigation in view of the previously presented results [62, 132, 137], which clearly show that twisting motion can indeed have a non-negligible influence.

Asai [139] proposed a small polaron model to describe the experimental findings of Yoo *et al.* [42] concerning the temperature-dependence of the electric current and of the linear conductance. Basically, he assumed that in poly(GC) completely incoherent polaron hopping dominates while in poly(AT) quasi-coherent hopping, i.e., with total phonon number conservation, is more important. As a result, the temperature dependence of the above quantities in both molecules is considerably different.

Matulewski *et al.* [140, 141] have investigated the effect of dynamic disorder on the charge transfer between donor and acceptor centers in a model DNA molecule. They calculated the charge transfer rates between two G-bases separated by an AT-bridge. It was found that the transition rate decays exponentially with the number of AT-pairs N as long as $N < 4$, while for longer bridges with dynamic disorder, the rate is basically length independent. These calculations nicely illustrate how a nontrivial effect can be obtained from a simple quantum mechanical model by directly solving the Schrödinger equation. Similar results have been obtained using more involved techniques such as reduced density matrices [142].

12.3.3 Environmental effects

Complementary to the research of the previous section which mainly addressed individual vibrational modes of the DNA cage, other studies have focused on the influence of environmental effects [47, 48, 59, 81, 143–147].

As shown in [47], the existence of rather different time scales of the environment may have a strong impact on a charge propagating along the DNA molecule. First-principle simulations were performed, including four base-pairs of B-DNA in the sequence GAGG, together with counterions and a hydration shell. It turns out that holes can be gated by the temperature-dependent dynamics of the environment. In other words, dynamical fluctuations of the counterions can create configurations that support or hinder the propagation of the hole along the double strand. These results have been partly confirmed by recent *ab initio* simulations [147]. The authors have additionally pointed out at a different, proton-mediated mechanism for hole localization, which may be quite effective in Poly(GC) DNA.

The first-principle studies [48, 81, 146] have yielded further insight into the role played by water and counterions in modifying the low-energy electronic structure of DNA. Despite the differences in the DNA conformations (Z- [48] vs B-DNA [81, 146]) as well as in computational approaches (different basis sets and approximations for the exchange-correlation potentials), they nevertheless indicate that the environment can introduce midgap states. These electronic states do not really form extended bands; they are thus similar to shallow impurity levels in conventional bulk semiconductors. They could however support thermal activated hopping leading to an increase of the conductivity.

Basko and Conwell [143] used a semiclassical model to describe the interaction of an injected hole in DNA which is placed in a polar solvent. Their basic conclusions pointed out that the main contribution was given by the interaction with water molecules and not with counterions; further, polaron formation was not hindered by the charge–solvent coupling, the interaction rather increased the binding energy (self-localization) of the polaron by around 0.5 eV, which is much larger than relevant temperature scales. Li and Yan [59] as well as Zhang *et al.* [144] investigated the role of dephasing reservoirs. Zhang *et al.* showed that a change in the length scaling of the conductance can be induced by the dephasing reservoirs as a result of incoherent phonon-mediated transport, a result known from electron transfer theories [142, 148].

As an example of the influence of a dissipative environment on charge transport, we will address in some detail a model we have recently proposed to describe the low-energy transport properties of a DNA nanowire in interaction with a complex environment, which is not further specified [63, 64]. We considered poly(GC) oligomers as our reference point because of: (i) its periodic structure which should make optimal the interbase coupling; and (ii) the availability of very nice single-molecule experiments performed by Xu *et al.* [43] in aqueous environment (see also Section 2.1). The experimental results, revealing power-law dependence of the conductance on the wire length, are at variance with the *exponential* dependence predicted by *ab initio* calculations [53] and call for possible explanations.

In our model, we exclusively focused on the low-energy transport, when the charge injection energies are small compared with the molecular bandgap of the isolated molecule which is of the order of ~2–3 eV. We used an effective ladder model where only one of the legs is available for propagation of a tunneling charge, see Figure 12.13. The size of the

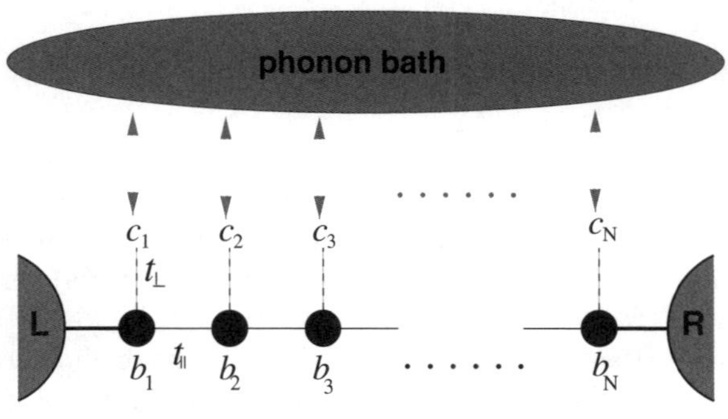

Figure 12.13 Schematic representation of the tight-binding model used to describe the low-energy transport properties of a short DNA wire in presence of a dissipative environment. Notice that only one of the legs of this 'broken' ladder is connected to the electrodes. Charges injected into the chain propagate only along the contacted leg and not along the disrupted one (reproduced from [64] with permission, Copyright 2005 by the American Physical Society)

bandgap can be tuned by the strength of the transversal coupling between the two local levels at each site, $t_{\perp}$. The environment is described by a collection of harmonic oscillators. The Hamiltonian reads:

$$
\begin{aligned}
\mathcal{H} &= \varepsilon_b \sum_j b_j^\dagger b_j + \varepsilon \sum_j c_j^\dagger c_j - t_\pi \sum_j [c_j^\dagger c_{j+1} + \text{h.c.}] - t_\perp \sum_j [b_j^\dagger c_j + \text{h.c.}] \\
&+ \sum_\alpha \Omega_\alpha B_\alpha^\dagger B_\alpha + \sum_{\alpha,j} \lambda_\alpha b_j^\dagger b_j (B_\alpha + B_\alpha^\dagger) \\
&+ \sum_{\mathbf{k}\in \mathrm{L,R},\sigma} \varepsilon_{\mathbf{k}\sigma} d_{\mathbf{k}\sigma}^\dagger d_{\mathbf{k}\sigma} + \sum_{\mathbf{k}\in \mathrm{L},\sigma} (V_{\mathbf{k},1} d_{\mathbf{k}\sigma}^\dagger c_1 + \text{h.c.}) + \sum_{\mathbf{k}\in \mathrm{R},\sigma} (V_{\mathbf{k},N} d_{\mathbf{k}\sigma}^\dagger c_N + \text{h.c.}) \\
&= \mathcal{H}_{\text{el}} + \mathcal{H}_{\text{B}} + \mathcal{H}_{\text{leads}}.
\end{aligned}
$$

In the above equation, $\mathcal{H}_{\text{el}} = \mathcal{H}_{\text{c}} + \mathcal{H}_{\text{b}}$ is the electronic Hamiltonian for the central leg $\mathcal{H}_{\text{c}}$ (that coupled to the electrodes) and the side 'broken' leg, $\mathcal{H}_{\text{b}}$. $\mathcal{H}_{\text{B}}$ contains both the Hamiltonian of the bath and the mutual interaction of the bath with the electronic degrees of freedom (second row). Finally, $\mathcal{H}_{\text{leads}}$ describes the leads Hamiltonians as well as the tunneling Hamiltonian describing the propagation of a charge from the leads onto the wire and vice versa. In absence of coupling to the bath, the eigenstates of $\mathcal{H}_{\text{el}}$ yield two manifolds containing N states each and separated by a bandgap. The bath is described by introducing its spectral density as given by [149]: $J(\omega) = J_0\left(\frac{\omega}{\omega_{\text{c}}}\right)\exp^{-\omega/\omega_{\text{c}}}$, where ω_{c} is a high-frequency cutoff and we assume ohmic dissipation, $J(\omega) \sim \omega$. By performing a unitary transformation, the linear coupling to the bath can be eliminated. However, the transversal coupling terms will be renormalized by exponential bosonic operators [63, 64]. Using equation of motion techniques for the Green functions, it can be shown to lowest order in $t_{\perp}$, that the Green function of the wire satisfies the following Dyson equation:

$$
\mathbf{G}^{-1}(E) = E\mathbf{1} - \mathcal{H}_{\text{c}} - \Sigma_{\text{L}}(E) - \Sigma_{\text{R}}(E) - t_\perp^2 \mathbf{P}(E).
$$

Notice that the influence of the electrodes is captured by the complex self-energy functions $\Sigma_{\text{L/R}}(E)$. The function $P(E)$ which acts as an additional self-energy is an entangled electron–phonon Green function:

$$
\begin{aligned}
P_{j\ell}(t) &= -i\Theta(t)\langle [b_j(t)\chi(t), b_\ell^\dagger(0)\chi^\dagger(0)]_+ \rangle \\
\chi &= \exp\left[\sum_\alpha (\lambda_\alpha/\Omega_\alpha)(B_\alpha - B_\alpha^\dagger)\right].
\end{aligned}
$$

We have mainly focus on the strong-coupling limit (SCL) to the bath degrees of freedom, which is defined by the condition $J_0/\omega_{\text{c}} > 1$. We refer to [63, 64] for further technical details.

There are two main effects arising from the strong charge–bath interaction [63, 64]: (i) bath-induced electronic states *within* the wire's bandgap emerge; and (ii) these states are strongly damped by the dissipative action of the bath. This latter effect has the consequence that the bath-induced states do not appear as resonances in the density of states or, correspondingly in the transmission spectrum, see Figure 12.14. Notwithstanding, a temperature-dependent incoherent background survives and offers additional states to charges tunneling at low energies. We have called this low-density-of-states energy region a pseudo-gap to

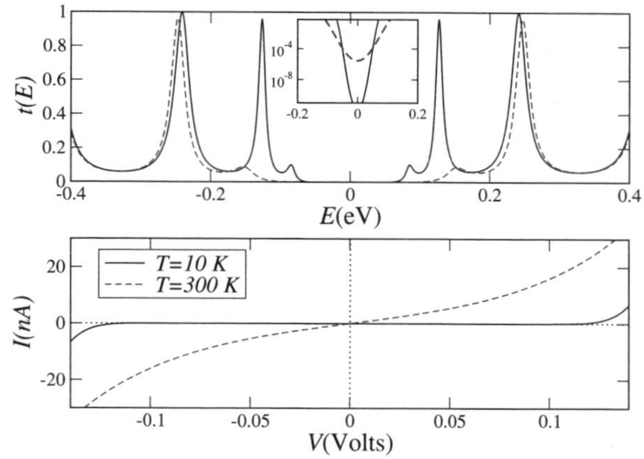

Figure 12.14 Upper panel: the function $t(E)$ for two different temperatures; the inset is a logarithmic plot around $E = 0$ showing the strong temperature dependence of the pseudo-gap. Lower panel: I–V characteristics. Parameters: $N = 20$, $J_0/\omega_c = 20$, $t_{\|} = 0.6\,eV$, $t_{\perp}/t_{\|} = 0.2$, $\Gamma_{L/R}/t_{\|} = 0.16$ (reproduced from [63] with permission, Copyright 2005 by the American Physical Society)

stress that its weight is a function of temperature. The existence of the pseudo-gap results in a finite current around zero bias at high temperatures and, more important, in a very weak exponential dependence $g \sim e^{-\gamma N}$ or even a power-law scaling $g \sim N^{-\alpha}$, $\alpha \sim 1$–2 of the linear conductance with the number of sites in the wire, reflecting the strong contribution of incoherent processes, see Figure 12.15, middle and lower panels.

12.4 CONCLUSIONS

The continuous refinement of the experimental techniques, especially the better control of the molecule–metal contact topology, as well as the close interdisciplinary collaboration of chemists, material scientists and physicists has led to more reliable assertions concerning the electrical transport properties of individual DNA molecules. In particular, the very recent experiments of Xu *et al.* [43] and Cohen *et al.* [82, 83] have demonstrated that, contrary to the early accepted view, DNA may support surprisingly large currents. Nevertheless, a lot still has to be done to clarify which are the relevant charge transport channels in this molecule by probing in a controlled manner the influence of the base-pair sequence, different linker groups to the metal electrodes, environmental effects and thermal fluctuations, among others. On the theory side, considerable progress has been achieved along different paths, either by focusing on a more accurate description of the electronic structure via *ab initio* methods, or by addressing, within model Hamiltonian approaches, individual factors which are thought to be relevant for charge transport. Another exciting research direction, which we have not discussed in detail are artificial DNA molecules, such as G4 [150, 151] and M-DNA [116, 118, 119], which open new exciting possibilities for engineering DNA into molecular nanodevices. In M-DNA, for instance, interesting new effects such as spin ordering [152] or even many-body effects such as the Kondo effect might be expected.

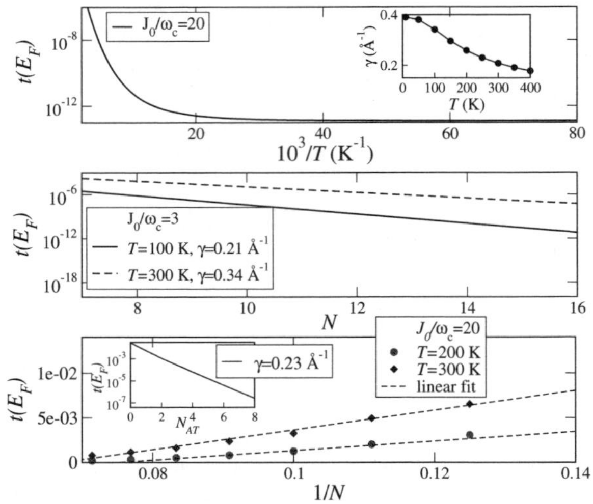

Figure 12.15 Upper panel: Arrhenius plot for *t(EF)*. Parameters: $N = 20$, $t_{\|} = 0.6\,\text{eV}$, $t_{\perp}/t_{\|} = 0.2$, $\Gamma_{L/R}/t_{\|} = 0.16$. Inset: temperature dependence of the exponential decay length γ (see text). Middle and lower panels: length dependence of $t(E_F)$ at different temperatures for two different strengths of the electron–bath coupling J_0/ω_c. The electronic coupling parameters are the same as in the upper panel. The inset in the lower panel shows the effect of introducing tunnel barriers to simulate the insertion of N_{AT} A–T base pairs in an otherwise homogeneous chain of length N (T) 300 K). The algebraic length dependence goes over onto an exponential one. For this, the on-site energies along a finite segment of the wire were shifted by $\varepsilon = -1.5\,\text{eV}$. The number of unperturbed sites N_{GC} in a sequence such as N_{GC}–N_{AT}–N_{GC} was kept constant ($N_{GC} = 4$) while varying $N_{AT} = 1, \ldots, 8$ (reproduced from [63] with permission, Copyright 2005 by the American Physical Society)

ACKNOWLEDGEMENT

This work has been supported by the Volkswagen Foundation (grant Nr. I/78-340), by the Deutsch Forschungsgemeinschaft (grant Nr. CU 4415-1) and by the EU Commission under contracts IST-2001-38951 and IST-029192-2.

REFERENCES

[1] S. Luryi, J. Xu, and A. Zaslavsky, *Future Trends in Microelectronics: The Road Ahead*, John Wiley & Sons, Inc., New York, USA, 1999.

[2] C. Joachim and M.A. Ratner, *Proc. Natl. Acad. Sci. USA*, **102**, 8801 (2005).

[3] *Introducing Molecular Electronics*, *Lecture Notes in Physics*, Vol. 680, G. Cuniberti, G. Fagas, and K. Richter, (eds) Springer, Berlin, 2005.

[4] A. Aviram and M.A. Ratner, *Chem. Phys. Lett.*, **29**, 277 (1974).

[5] C. Dekker and M. Ratner, Physics *World*, **14**, 29 (2001).

[6] *Molecular Electronics Science and Technology, Annals of the New York Academy of Sciences*, Vol. 852, A. Aviram and M.A. Ratner (eds), New York Academy of Sciences, New York, 1998.

[7] *Molecular Electronics II, Annals of the New York Academy of Sciences*, Vol. 960, A. Aviram, M.A. Ratner, and V. Mujica (eds), New York Academy of Sciences, New York, 2002.

[8] R.M. Metzger, *Acc. Chem. Res.*, **9**, 2027 (1999).
[9] C.P. Collier, E.W. Wong, M. Bolohradsky, F.M. Raymo, J.F. Stoddart, P.J. Kuekes, R.S. Williams, and J.R. *Heat Science*, **285**, 391 (1999).
[10] M.A. Reed, C. Zhou, C.J. Muller, T.P. Burgin, and J.M. Tour, *Science*, **278**, 252 (1997).
[11] J. Chen and M.A. Reed, *Chem. Phys.*, **281**, 127 (2002).
[12] D. Porath and O. Millo, *J. Appl. Phys.*, **81**, 2241 (1997).
[13] S.G. Lemay, J.W. Janssen, M. van den Hout, M. Mooji, M.J. Bronikowski, P.A. Willis, R.E. Smalley, L.P. Kouwenhoven, and C. Dekker, *Nature*, **412**, 617 (2001).
[14] W. Liang, M.P. Shores, M. Bockrath, J.R. Long, and H. Park, *Nature*, **417**, 725 (2002).
[15] M. Thorwart, M. Grifoni, G. Cuniberti, H.W.C. Postma, and C. Dekker, *Phys. Rev. Lett.*, **89**, 196402 (2002).
[16] E. Braun, Y. Eichen, U. Sivan, and G. Ben-Yoseph, *Nature*, **391**, 775 (1998).
[17] K. Keren, R.S. Berman, E. Buchstab, U. Sivan, and E. Braun, *Science*, **302**, 1380 (2003).
[18] K. Keren, M. Krueger, R. Gilad, G. Ben-Yoseph, U. Sivan, and E. Braun, *Science*, **297**, 72 (2002).
[19] D. Porath, A. Bezryadin, S.D. Vries, and C. Dekker, *Nature*, **403**, 635 (2000).
[20] R. Rinaldi, A. Biasco, G. Maruccio, R. Cingolani, D. Alliata, L. Andolfi, P. Facci, F.D. Rienzo, R. Di Felice, and E. Molinari, *Adv. Mater.*, **14**, 1453 (2002).
[21] R. Rinaldi, E. Branca, R. Cingolani, R. Di Felice, A. Calzolari, E. Molinari, S. MAsiera, G. Spada, G. Gottarelli, and A. Garbesi, *Ann. New York Acad. of Sci.*, **960**, 184 (2002).
[22] Y. Benenson, T. Paz-Elizur, R. Adar, E. Keinan, Z. Livneh, and E. Shapiro, *Nature*, **414**, 430 (2001).
[23] Y. Benenson, R. Adar, T. Paz-Elizur, Z. Livneh, and E. Shapiro, *Proc. Natl. Acad. Sci. USA*, **100**, 2191 (1997).
[24] E. Di Mauro and C.P. Hollenberg, *Adv. Mat.*, **5**, 384 (1993).
[25] C.M. Niemeyer, *Angew. Chem. Int. Edn.*, **36**, 585 (1997).
[26] Y. Zhang and N.C. Seeman, *J. Am. Chem. Soc.*, **116**, 1661 (1994).
[27] N.C. Seeman, *Nano letters*, **1**, 22 (2001).
[28] M. Hazani, F. Hennrich, M. Kappes, R. Naaman, D. Peled, V. Sidorov, and D. Shvarts, *Chem. Phys. Lett.*, **391**, 389 (2004).
[29] Y. Zhang, R.H.A.J.K.E.C. Cox, and N.P. Ong, *Phys. Rev. Lett.*, **89**, 198102 (2002).
[30] M. Mertig, R. Kirsch, W. Pompe, and H. Engelhardt, *Eur. Phys. J. D*, **9**, 45 (1999).
[31] D.D. Eley and D.I. Spivey, *Trans. Faraday Soc.*, **58**, 411 (1962).
[32] J.M. Warman, M.P. de Haas MP, and A. Rupprecht, *Chem. Phys. Lett.*, **249**, 319 (1996).
[33] E.C. Friedberg, *Nature*, **421**, 436 (2003).
[34] C.J. Murphy, M.R. Arkin, Y. Jenkins, N.D. Ghatlia, S.H. Bossmann, N.J. Turro, and J.K. Barton, *Science*, **262**, 1025 (1993).
[35] D.B. Hall, R.E. Holmlin, and J.K. Barton, *Nature*, **382**, 731 (1996).
[36] S.O. Kelley, N.M. Jackson, M.G. Hill, and J.K. Barton, *Angew. Chem. Int. Edn.*, **38**, 941 (1999).
[37] M. Bixon, B. Giese, S. Wessely, T. Langenbacher, M.E. Michel-Beyerle, and J. Jortner, *Proc. Natl. Acad. Sci. USA*, **96**, 11713 (1999).
[38] G.B. Schuster, *Acc. Chem. Res.*, **33**, 253 (2000).
[39] E.M. Conwell and S.V. Rakhmanova, *Proc. Natl. Acad. Sci. USA*, **97**, 4556 (2000).
[40] D. Porath, N. Lapidot, and J. Gomez-Herrero, in *Introducing Molecular Electronics, Lecture Notes in Physics*, Vol. 680, G. Cuniberti, G. Fagas, and K. Richter (eds), Springer, Berlin, 2005, p. 411 (cited in [3]).
[41] A.J. Storm, J.V. Noort, S.D. Vries, and C. Dekker, *Appl. Phys. Lett.*, **79**, 3881 (2001).
[42] K.-H. Yoo, D.H. Ha, J.-O. Lee, J.W. Park, J. Kim, J.J. Kim, H.-Y. Lee, T. Kawai, and H.Y. Choi, *Phys. Rev. Lett.*, **87**, (2001).
[43] B. Xu, P. Zhang, X. Li, and N. Tao, *Nano Letters*, **4**, 1105 (2004).

[44] R. Di Felice, A. Calzolari, and E. Molinari, *Phys. Rev. B*, **65**, 045104 (2002).
[45] A. Calzolari, R. Di Felice, E. Molinari, and A. Garbesi, *Appl. Phys. Lett.*, **80**, 3331 (2002).
[46] R. Di Felice, A. Calzolari, and H. Zhang, *Nanotechnology*, **15**, 1256 (2004).
[47] R.N. Barnett, C.L. Cleveland, A. Joy, U. Landman, and G.B. Schuster, *Science*, **294**, 567 (2001).
[48] F.L. Gervasio, P. Carolini, and M. Parrinello, *Phys. Rev. Lett.*, **89**, 108102 (2002).
[49] E. Artacho, M. Machado, D. Sanchez-Portal, P. Ordejon, and J.M. Soler, *Mol. Phys.*, **101**, 1587 (2003).
[50] S.S. Alexandre, E. Artacho, J.M. Soler, and H. Chacham, *Phys. Rev. Lett.*, **91**, 108105 (2003).
[51] J.P. Lewis, P. Ordejon, and O.F. Sankey, *Phys. Rev. B*, **55**, 6880 (1997).
[52] E.B. Starikov, *Phil. Mag. Lett.*, **83**, 699 (2003).
[53] H. Wang, J.P. Lewis, and O. Sankey, *Phys. Rev. Lett.*, **93**, 016401 (2004).
[54] H. Mehrez and M.P. Anantram, *Phys. Rev. B*, **71**, 115405 (2005).
[55] F. Palmero, J.F.R. Archilla, D. Hennig, and F.R. Romero, *New J. Phys.*, **6**, 13 (2004).
[56] S. Komineas, G. Kalosakas, and A.R. Bishop, *Physical Review E*, **65**, 061905 (2002).
[57] Z. Hermon, S. Caspi, and E. Ben-Jacob, *Europhys. Lett.*, **43**, 482 (1998).
[58] J. Yi, Phys. *Rev. B*, **68**, 193103 (2003).
[59] X.Q. Li and Y. Yan, *Appl. Phys. Lett.*, **79**, 2190 (2001).
[60] G. Cuniberti, L. Craco, D. Porath, and C. Dekker, *Phys. Rev. B*, **65**, 241314 (2002).
[61] P. Ao, S. Grundberg, and J. Rammer, *Phys. Rev. B*, **53**, 10042 (1996).
[62] W. Zhang and S.E. Ulloa, *Microelec. Journal*, **35**, 23 (2004).
[63] R. Gutierrez, S. Mandal, and G. Cuniberti, *Nano Letters*, **5**, 1093 (2005).
[64] R. Gutierrez, S. Mandal, and G. Cuniberti, *Phys. Rev. B*, **71**, 235116 (2005).
[65] S. Roche, *Phys. Rev. Lett.*, **91**, 108101 (2003).
[66] H. Yamada, E.B. Starikov, D. Hennig, and J.F Archilla, *Eur. Phys. J. E.*, **17**, 149 (2005).
[67] W. Zhang and S.E. Ulloa, *Phys. Rev. B*, **69**, 153203 (2004).
[68] M. Unge and S. Stafstrom, *Nano letters*, **3**, 1417 (2003).
[69] Y. Zhu, C.C. Kaun, and H. Guo, *Phys. Rev. B*, **69**, 245112 (2004).
[70] V.M. Apalkov and T. Chakraborty, *Phys. Rev. B*, **71**, 033102 (2005).
[71] V.M. Apalkov and T. Chakraborty, cond-mat/0508475 (2005).
[72] D. Klotsa, R.A. Roemer, and M.S. Turner, q-bio. GN/0504004 (2005).
[73] G.B. Schuster, *Long-Range Charge Transfer in DNA I and II, Topics in Current Chemistry*, Vol. 237, (ed.), Springer, Berlin, 2004.
[74] A. Nitzan, *Ann. Rev. Phys. Chem.*, **52**, 681 (2001).
[75] A. Nitzan and M. Ratner, *Science*, **300**, 1384 (2003).
[76] J. Jortner, M. Bixon, T. Langenbacher, and M.E. Michel-Beyerle, *Proc. Natl. Acad. Sci. USA*, **95**, 12759 (1998).
[77] J. Jortner and M. Bixon, *Chem. Phys.*, **281**, 393 (2002).
[78] Y.A. Berlin, A.L. Burin, L.D.A. Siebbeles, and M.A. Ratner, *J. Phys. Chem. A*, **105**, 5666 (2001).
[79] D. Porath, G. Cuniberti, and R. Di Felice, in. *Long-Range Charge Transfer in DNA I and II*, Vol. 237 of *Topics in Current Chemistry*, G.B. Schuster (ed.), Springer, Berlin, 2004, p. 183, (cited in [73]).
[80] M. Di Ventra and M. Zwolak, *Encyclopedia of Nanoscience and Nanotechnology* (H.S. Nalwa (ed.), 2004), Vol. X.
[81] R.G. Endres, D.L. Cox, and R.R.P. Singh, *Rev. Mod. Phys.*, **76**, 195 (2004).
[82] H. Cohen, C. Nogues, R. Naaman, and D. Porath, *Proc. Natl. Acad. Sci. USA*, **102**, 11589 (2005).
[83] H. Cohen, C. Nogues, D. Ullien, S. Daube, R. Naaman, and D. Porath, *Faraday Discuss.*, **131**, 367 (2006).

[84] A. Salomon, D. Cahen, S. Lindsay, J. Tomfohr, V.B. Engelkes, and C.D. Frisbie, *Adv. Mater.*, **15**, 1881 (2003).
[85] N.J. Turro and J.K. Barton, *J. Biol. Inorg. Chem.*, **3**, 201 (1998).
[86] S.O. Kelley and J.K. Barton, *Science*, **283**, 375 (1999).
[87] P.F. Barbara and E.J.C. Olson, *Adv. Chem. Phys.*, **107**, 647 (1999).
[88] E. Meggers, M.E. Michel-Beyerle, and B. Giese, *J. Am. Chem. Soc.*, **120**, 12950 (1998).
[89] B. Giese, J. Amaudrut, A.K. Köhler, and M.S.S. Wessely, *Nature*, **412**, 318 (2001).
[90] W.B. Davis, I. Naydenova, R. Haselbeger, A. Ogrodnik, B. Giese, and M.E. Michel-Beyerle, *Angew. Chem. Int. Edn.*, **39**, 3649 (2000).
[91] B. Giese, *Curr. Opin. Chem. Biol.*, **6**, 612 (2002).
[92] M.A. O'Neill and J.K. Barton, *Proc. Natl. Acad. Sci. USA*, **99**, 16543 (2002).
[93] P.T. Henderson, G.H.D. Jones, Y. Kan, and G.B. Schuster, *Proc. Natl. Acad. Sci. USA*, **96**, 8353 (1999).
[94] H.-W. Fink and C. Schonenberger, *Nature*, **398**, 407 (1999).
[95] P.J. de Pablo, F. Moreno-Herrero, J. Colchero, J.G. Herrero, P. Herrero, A.M. Baro, P. Ordejn, J.M. Soler, and E. Artacho, *Phys. Rev. Lett.*, **85**, 4992 (2000).
[96] A.Y. Kasumov, M. Kociak, S. Gueron, B. Reulet, V.T. Volkov, D.V. Klinov, and H. Bouchiat, *Science*, **291**, 280 (2001).
[97] H. Watanabe, C. Manabe, T. Shigematsu, K. Shimotani, and M. Shimizu, *Appl. Phys. Lett.*, **79**, 2462 (2001).
[98] T. Shigematsu, K. Shimotani, C. Manabe, H. Watanabe, and M. Shimitzu, *J. Chem. Phys.*, **118**, 4245 (2003).
[99] H. Watanabe, K. Shimotani, T. Shigematsu, and C. Manabe, *Thin Solid Films*, **438**, 462 (2003).
[100] K. Shimotani, T. Shigematsu, H. Watanabe, and M. Shimitzu, *J. Chem. Phys.*, **118**, 8016 (2003).
[101] C. Gomez-Navarro, F. Moreno-Herrero, P.J. de Pablo, J. Colchero, J. Gomez-Herrero, and A.M. Baro, *Proc. Natl. Acad. Sci. USA*, **99**, 8484 (2002).
[102] C. Nogues, S.R. Cohen, S.S. Daube, and R. Naaman, *Phys. Chem. Chem. Phys.*, **6**, 4459 (2004).
[103] H. van Zalinge, D.J. Schiffrin, A.D. Bates, W. Haiss, J. Ulstrup, and R.J. Nichols, *Phys. Chem. Chem. Phys.*, **7**, 94 (2006).
[104] A. Bezryadin and C. Dekker, *J. Vac. Sci. Technol.*, **15**, 793 (1997).
[105] A. Bezryadin, C. Dekker, and G. Schmid, *Appl. Phys. Lett.*, **7**, 1273 (1997).
[106] M. Bockrath, N. Markovic, A. Shepard, M. Tinkham, L. Gurevich, L.P. Kouvenhoven, M.W. Wu, and L.L. Sohn, *Nano letters*, **2**, 187 (2002).
[107] A.Y. Kasumov, D.V. Klinov,, P.-E. Roche, S. Gueron, and H. Bouchiat, *Appl. Phys. Lett.*, **84**, 1007 (2004).
[108] C. Yang, D. Moses, and A.J. Heeger, *Adv. Mater.*, **15**, 1364 (2003).
[109] L. Cai, H. Tabata, and T. Kawai, *Appl. Phys. Lett.*, **77**, 3105 (2000).
[110] H.Y. Lee, H. Tanaka, Y. Otsuka, K.-H. Yoo, J.-O. Lee, and T. Kawai, *Appl. Phys. Lett.*, **80**, 1670 (2002).
[111] H. Tabata, L.T. Cai, J.-H. Gu, S. Tanaka, Y. Otsuka, Y. Sacho, M. Taniguchi, and T. Kawai, *Synthetic Metals*, **133**, 469 (2003).
[112] M. Taniguchi, H.Y. Lee, H. Tanaka, and T. Kawai, *Jpn. J. Appl. Phys.*, **42**, L215 (2003).
[113] S. Tanaka, L.-T. Cai, H. Tabata, and T. Kawai, *Jpn. J. Appl. Phys.*, **42**, 2818 (2003).
[114] M. Taniguchi, Y. Otsuka, H. Tabata, and T. Kawai, *Jpn. J. Appl. Phys.*, **42**, 6629 (2003).
[115] Y.-S. Jo, Y. Lee, and Y. Roh, *Mater. Sci. Eng.*, **C23**, 841 (2003).
[116] A. Rakitin, P. Aich, C. Papadopoulos, Y. Kobzar, A.S. Vedeneev, J.S. Lee, and J.M. Xu, *Phys. Rev. Lett.*, **86**, 3670 (2001).

[117] P. Aich, S.L. Labiuk, L.W. Tari, L.J.T. Delbaere, W.J. Roesler, K.J. Falk, R.P. Steer, and J.S. Lee, *J. Mol. Biol.*, **294**, 477 (1999).
[118] S.D. Wettig, D.O. Wood, and J.S. Lee, *J. Inorg. Biochem.*, **94**, 94 (2003).
[119] C.-Z. Li, Y.-T. Long, H.-B. Kraatz, and J.S. Lee, *J. Phys. Chem. B*, **107**, 2291 (2003).
[120] J.S. Hwang, S.W. Hwang, and D. Ahn, *Superlattices and Microstructures*, **34**, 433 (2003).
[121] C.H. Lei, A. Das, M. Elliott, and J.E. Mcdonald, *Appl. Phys. Lett.*, **83**, 482 (2003).
[122] Y. Okahata, T. Kobayashi, and K.T.M. Shimomura, *J. Am. Chem. Soc.*, **120**, 6165 (1998).
[123] R. Rinaldi, E. Branca, R.C.R.S. Masiero, G.P. Spada, and G. Gottarelli, *Appl. Phys. Lett.*, **78**, 3541 (2001).
[124] G. Maruccio, P. Visconti, V. Arima, S. D'Amico, A. Biasco, E. D'Amone, R. Cingolani, R. Rinaldi, S. Masiero, T. Giorgi, and G. Gottarelli, *Nano letters*, **3**, 479 (2003).
[125] P. Carpena, P. Bernaola-Galvan, P.C. Ivanov, and H.E. Stanley, *Nature*, **418**, 955 (2002).
[126] P. Phillips, *Advanced Solid State Physics*, Boulder, CO, Westview Press (2003).
[127] C. Adessi, S. Walch, and M.P. Anantram, *Phys. Rev. B*, **67**, 081405 (R) (2003).
[128] R.A. Caetano and P.A. Schulz, cond-mat/0409245 (2004).
[129] E.B. Starikov, *Phys. Rev. Lett.*, **95**, 126601 (2005).
[130] M. Peyrard and A.R. Bishop, *Phys. Rev. Lett.*, **62**, 2755 (1989).
[131] S. Komineas, G. Kalosakas, and A.R. Bishop, *Phys. Rev. E*, **65**, 061905 (2002).
[132] W. Zhang, A.O. Govorov, and S.E. Ulloa, *Phys. Rev. B*, **66**, 060303(R) (2002).
[133] T. Holstein, *Ann. Phys. N.Y.*, **8**, 325 (1959).
[134] J.-X. Xiao, J.-T. Lin, and G.-X. Zhang, *J. Phys. A: Math. Gen.*, **20**, 2425 (1987).
[135] L.V. Yakushevich, A.V. Savin, and L.I. Manevitch, *Phys. Rev. E*, **66**, 016614 (2002).
[136] C. Wan, T. Fiebig, S.O. Kelley, C.R. Treadway, and J.K. Barton, *Proc. Natl. Acad. Sci.*, **96**, 6014 (1999).
[137] R. Bruinsma, G. Gruener, M.R. D'Orsogna, and J. Rudnick, *Phys. Rev. Lett.*, **85**, 4393 (2000).
[138] D. Hennig, *Phys. Rev. E*, **64**, 041908 (2001).
[139] Y. Asai, *J. Phys. Chem. B*, **107**, 4647 (2003).
[140] J. Matulewski, S.D. Baranovski, and P. Thomas, *Phys. Stat. Sol.*, (b), **241**, R46 (2004).
[141] J. Matulewski, S.D. Baranovski, and P. Thomas, *Phys. Chem. Chem. Phys.*, **7**, 1514 (2005).
[142] D. Segal and A. Nitzan, *Chem. Phys.*, **281**, 235 (2002).
[143] D.M. Basko and E.M. Conwell, *Phys. Rev. Lett.*, **88**, 098102 (2002).
[144] H.Y. Zhang, X.-Q. Li, P. Han, X.Y. Yu, and Y.-J. Yan, *J. Chem. Phys.*, **117**, 4578 (2002).
[145] J.-F. Feng and S.-J. Xiong, *Phys. Rev. E*, **66**, 021908 (2002).
[146] A. Huebsch, R.G. Endres, D.L. Cox, and R.R.P. Singh, *Phys. Rev. Lett.*, **94**, 178102 (2005).
[147] F.L. Gervasio, A. Laio, M. Parrinello, and M. Boero, *Phys. Rev. Lett.*, **94**, 158103 (2005).
[148] D. Segal, A. Nitzan, W.B. Davies, M.R. Wasielewski, and M.A. Ratner, *J. Phys. Chem. B*, **104**, 3817 (2000).
[149] U. Weiss, *Quantum Dissipative Systems*, Vol. 10 of *Series in Modern Condensed Matter Physics*, World Scientific, 1999.
[150] K. Phillips, Z. Dauter, A.I.H. Morchie, D.M.J. Lilley, and B. Luisi, *J. Mol. Biol.*, **273**, 171 (1997).
[151] K. Tanaka, Y. Yamada, and M. Shionoya, *J. Am. Chem. Soc.*, **124**, 8802 (2002).
[152] A.C.H. Zhang and R. Di Felice, *J. Phys. Chem. B*, **109**, 15345 (2005).
[153] http://www.accessexcellence.org/RC/VL/GG/

Index

Note: Plain type locators refer to text treatments; page references in **bold** or *italic* type indicate references to tables and figures, respectively.